# 2022 2nd International Conference on Fluid and Chemical Engineering

Journal of Physics: Conference Series Volume 2329

Online
16-18 June 2022

**Editors:**

**Omid Mahian**
**Sébastien Déon**

ISBN: 978-1-7138-6048-8
ISSN: 1742-6588

**Printed from e-media with permission by:**

Curran Associates, Inc.
57 Morehouse Lane
Red Hook, NY 12571

proceedings.com

**Some format issues inherent in the e-media version may also appear in this print version.**



Institute of Physics
Dirac House, Temple Back
Bristol BS1 6BE UK

Phone: 44 1 17 929 7481
Fax: 44 1 17 920 0979

techtracking@iop.org

**Additional copies of this publication are available from:**

Curran Associates, Inc.
57 Morehouse Lane
Red Hook, NY 12571 USA
Phone: 845-758-0400
Fax: 845-758-2633
Email: curran@proceedings.com
Web: www.proceedings.com

# TABLE OF CONTENTS

# Preface

Due to the continuing effect of Covid-19 epidemic, 2022 2nd International Conference on Fluid and Chemical Engineering (ICFCE 2022) has been successfully held on 17th June 2022 as an online conference. Organized by Hubei Zhongke Institute of Geology and Environment Technology, ICFCE 2022 intends to invite worldwide famous scientists, experts, scholars, and researchers for academic presentations.

ICFCE 2022 proceeding is a collection of outstanding submissions in the field of biofluid mechanics, chemically reacting fluids and combustion, computational fluid dynamics (CFD), experimental fluid mechanics, flow through porous media, fluid–solid interactions (FSI), geophysical fluid dynamics, granular/suspension flows, heat and mass transfer, hydrodynamics, physical, theoretical and computational chemistry, chemical engineering fundamentals, chemical reaction engineering, chemical engineering equipment design and process design, thermodynamics, catalysis & reaction engineering, particulate systems, crystallization, etc. 49 articles were selected from 118 submissions after peer-review, and authors participated in ICFCE 2022 with oral presentations and posters, promoting the communication among researchers and experts from institutes and universities, including Xi'an Research Institute, National Engineering Research Center of Dredging Technology and Equipment, State Key Laboratory of Aerodynamics, China Aerodynamics Research and Development Center, Xi'an University of Architecture and Technology, Jiangsu University of Science and Technology, Beijing Institute of Technology, etc.

Cordial gratitude is delivered to Technical Program Committee of ICFCE 2022 for the professional work on guarantying the high standard of conference as well as the proceedings. We believe the proceedings will enhance the exchange of ideas and state-of-art knowledge on Fluid and Chemical Engineering.

Professor Omid Mahian
Xian Jiaotong University, China

Associate Professor Sébastien Déon
University of Bourgogne-Franche-Comté, France

**Technical Committee**

**Chair**

Professor Mohamed Kheireddine Aroua, Sunway University, Malaysia

**Editor**

Professor Omid Mahian, Xian Jiaotong University, China.

Associate Professor Sébastien Déon, University of Bourgogne-Franche-Comté, France.

**International Committee Members**

Professor Mohamed Gamal El-Din, University of Alberta, Canada

Professor T V V L N Rao, Madanapalle Institute of Technology & Science, India

Professor Mohsen Sharifpur, University of Pretoria, South Africa

Professor Shengli Pu, University of Shanghai for Science and Technology, China

Professor Ilmutdin M. Abdulagatov, Dagestan State University, Russia

Professor Dr. Mohammad Mansur Rahman, Sultan Qaboos University, Sultanate of Oman

Professor Jianzhong Lin, Zhejiang University, China

Professor Alberto Brucato, University of Palermo, Italy

Professor Olivier SIMONIN, Institut de Mécanique des Fluides de Toulouse (IMFT), France

Professor P. Selvakumar, PSN College of Engineering and Technology (Autonomous), India

Professor Mohammad Mehdi Rashidi, University of Electronic Science and Technology of China, Chengdu, China

Associate Professor Mario Versaci, “Mediterranea” University, Italy

Associate Professor Ehsan Roohi, Xi'an Jiaotong University, China

Associate Professor Zhipeng Qin, Guangxi University, China

Associate Professor Dr. CHARMY S. KOTHARI, Institute of Pharmacy, Nirma University, India

Associate Professor Ramesh Kasi, University of Malaya, Malaysia

Associate Professor Renu Gupta, Dr B R Ambedkar National Institute of Technology Jalandhar

Associate Professor Dr. Pralay Kumar Karmakar, Tezpur University, India

First class Research Director, Dr. Lingai LUO, Centre National de la Recherche Scientifique (CNRS), Laboratoire de Thermique et Energie de Nantes, France

Senior Lecturer (Assistant Prof.) Noor Fadiya Mohd Noor, University of Malaya, Malaysia

Senior Lecturer (Assistant Professor) Mohd Amiruddin Abd Rahman, Universiti Putra Malaysia, Malaysia

Associate Professor Zafar Said, University of Sharjah, Sharjah, UAE

Assistant Professor Dhananjay Yadav, University of Nizwa, Sultanate of Oman

Assistant Professor Dr. Ehnber Naheed Maraj, HITEC University, Pakistan

Assistant Professor Dr. Rashid Mehmood, HITEC University Taxila, Pakistan

Assistant Professor Muhammed Abdullah Hassan Ahmed, Cairo University, Egypt

Assistant Professor Ehtsham Azhar, PMAS-Arid Agriculture University, Pakistan

Assistant Professor Md Readul Mahmud, Independent University, Bangladesh (IUB), Bangladesh

Assistant Professor Muhammad Anser Bashir, Mirpur University of Engineering & Technology, Pakistan

Journal of Physics: Conference Series **2329** (2022) 011001 doi:10.1088/1742-6596/2329/1/011001

# Peer Review Statement

All papers published in this volume have been reviewed through processes administered by the Editors. Reviews were conducted by expert referees to the professional and scientific standards expected of a proceedings journal published by IOP Publishing Publishing.

- **Type of peer review:** Double Anonymous
- **Conference submission management system:** https://zentao.meetconf.com.cn/?conf_id=337#/login?redirect=%2Fcontribute
- **Number of submissions received:** 118
- **Number of submissions sent for review:** 102
- **Number of submissions accepted:** 49
- **Acceptance Rate (Submissions Accepted / Submissions Received × 100):** 41.5
- **Average number of reviews per paper:** 2
- **Total number of reviewers involved:** 75
- **Contact person for queries:**
  **Name:** Terasa Lei
  **Email:** academic@icfce-conf.org
  **Affiliation:** Hubei Zhongke Institute of Geology and Environment Technology

# Numerical Simulation of Acoustic Scattering by a Rotating Cylinder Based on the Enhanced Optimized Scheme

**Siqi Yuan[1*], Ruixuan Ma[1,2], Conghai Wu[1] and Shuhai Zhang[1]**

[1] State Key Laboratory of Aerodynamics, China Aerodynamics Research and Development Center, MianYang 621000, China;
[2] Key Laboratory of Aerodynamic Noise Control, China Aerodynamics Research and Development Center, MianYang 621000, China.
Email: ysq047@mail.ustc.edu.cn

**Abstract.** In this paper, we develop a new physical model which involves a Taylor vortex around a rotating cylinder. The scattering of sound waves from this new model is investigated by solving the linearized Euler equations under the polar coordinate system in the time domain. We use the enhanced optimized scheme for numerical simulations. Numerical results show that scattering causes a change in the spatial distribution of sound energy as the sound wave passes through a rotating cylinder. The directionality of the scattered field varies significantly with the length ratio of the acoustic wavelength to the radius of the rotating cylinder and the vortex intensity. The influences of the ratio and the strength on the directivities and intensities of scattered fields are analyzed. It is shown that the intensity of the scattering cross-section increases with the strength of the vortex. The pressure of the scattered sound fields has a phase shift induced by the couple scattering from the cylinder and the vortex.
**Keywords.** Numerical simulation, Rotating cylinder, Vortex, Acoustic scattering.

## 1. Introduction

In the propagation process, sound waves will change their original propagation direction and scatter when they pass through regions with different densities or different sound speeds. The difference between the sound waves after passing through this region and those without through this region is called a scattered wave [1]. The sound wave will scatter after passing through the cylinder as a rigid solid [2]. Vortex also has a scattering effect on sound waves. Vortex is the basic component unit of a complex flow field, and the study of vortex acoustic scattering is the basis of the analysis of difficult flow field results. At present, most of the research only focuses on sound scattering of rigid solid or vortex. In recent years, many scholars have studied the problem of vortex acoustic scattering using numerical simulation. Cande [3] studied the distortion characteristics of the sound field after a sound wave passes through the Rankine vortex. Colonius [4] studied the acoustic scattering characteristics of compact and noncompact vortices. Researchers mostly carry out a series of studies on the relationship between the radius of the vortex core and the wavelength of the incident acoustic wave. Wang Yimin [5] carried out a numerical simulation of spatial scale characteristics of vortex acoustic scattering. Ma Ruixuan [6, 7] carried out a study of vortex acoustic scattering based on high-order precision and high-resolution numerical methods.

The rotation of the cylinder drives the flow field to rotate and generates vortices, which also scatters sound waves. Therefore, in this paper, the numerical calculation method is adopted to simulate acoustic scattering under the action of vortex-solid coupling, and the scattering sound field is studied to analyze the scattering situation under the ratio of different vortex strength, incident wave, and cylinder radius length scale. We use the linearized Euler equation in polar coordinates. The EO2 scheme is used for spatial discretization [8-10]. We use fourth-order Runge-Kutta scheme for numerical simulations [11]. The distortion characteristics of the scattering sound field are analyzed by numerical simulation of the ratio of incident wavelength to cylinder radius and vortex strength.

## 2. Methodology

### 2.1. Model

This paper uses the cylinder-vortex coupling structure as the research model. Taylor vortex is the exact solution of the unsteady Euler equation and can be used as the ideal background flow in the calculation process. Therefore, the mean-entropy Taylor vortex model is adopted to simulate the flow field outside the cylinder. We drive distributions of the Taylor vortex in cylindrical coordinates $(r,\theta)$, where $r$ and $\theta$ are the radial and azimuthal coordinates. The vortex center is the coordinate centre, and its expression is

$$\begin{aligned} v_r &= 0 \\ v_\theta &= M_v \frac{r}{R} e^{\left[\frac{1}{2}\left(1-\frac{r^2}{R^2}\right)\right]} \end{aligned} \tag{1}$$

where $v_r$ is the radial velocity. $v_\theta$ is the tangential velocity. R is cylindrical radius. $M_v$ is the strength of the vortex.

$$M_v = U_v / c_\infty \tag{2}$$

where $U_v$ is the maximal velocity. $\rho(r)$ is density and $p(r)$ is pressure.

$$\begin{aligned} \rho(r) &= \left[1 - \frac{\gamma-1}{2} M_v^2 \exp\left(1-r^2\right)\right]^{1/\gamma-1} \\ p(r) &= \frac{1}{\gamma}\left[1 - \frac{\gamma-1}{2} M_v^2 \exp\left(1-r^2\right)\right]^{\gamma/\gamma-1} \end{aligned} \tag{3}$$

The flow field of the Taylor vortex is shown in figure 1. The density, tangential velocity and pressure distribution of the flow field at $M_v$=0.5 are shown.

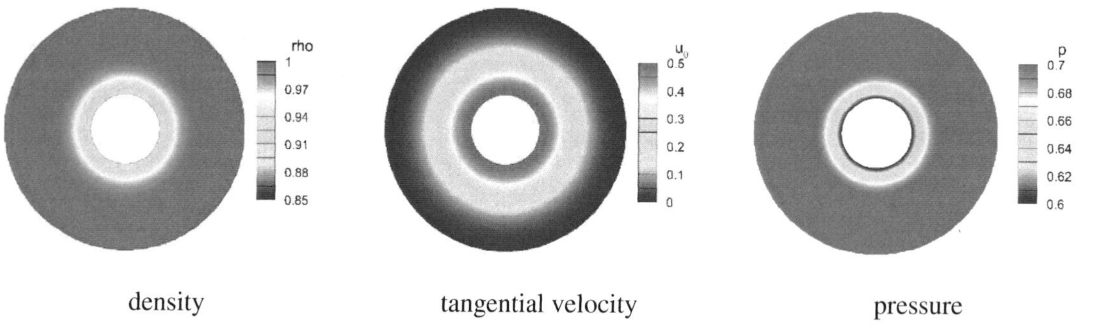


**Figure 1.** The flow field of the Taylor vortex.

### *2.2. Enhanced Optimized Scheme*

In this paper, the enhanced optimized scheme is used for spatial discretization [9]. It is a nonlinear combination of the 6-th order-optimized r scheme and 4-th order optimized scheme. It can be expressed as

$$\tilde{f}_0' = (1-\lambda)\tilde{f}'^{S} + \lambda \tilde{f}'^{O} \tag{4}$$

where $\tilde{f}'^{S}$ is the order-optimized one, $\tilde{f}'^{O}$ is a order optimized scheme.

## 3. Results and Discussions

### *3.1. The Scattering Cross-Section*

We introduce $\Sigma(r)$ that is the scattering cross-section to describe the power of the scattered sound field.

$$\Sigma(r) = \frac{1}{\pi R}\int_0^{2\pi} \frac{p_{rms}^2(r,\theta)}{p_i^2(r,\theta)} r\mathrm{d}\theta \tag{5}$$

where $p_i$ is the scattered sound pressure and $p_{rms}$ is the RMS of the pressure of the scattered wave. Figure 2 shows the scattering cross-section when the wavelength is 0.25, 0.5, 1, 2, 4, and 8. The coordinate system adopts a logarithmic coordinate system. As can be seen from the figure, the overall trend is to increase the incident wavelength within the range of calculated parameters. When $\lambda$ =0.25, the scattering cross-section intensity reaches the maximum, and after being stable at the order of 1, it does not increase significantly. The scattering wave intensity is equivalent to the incident wave intensity. When $\lambda$ =1, there is a slight decrease at $M_v$ =0.5, and when $\lambda$ >4, the increasing trend of the scattering cross-section conforms to the exponential growth rate.

Figure 3 shows the relationship between scattering cross-section and incident wavelength $\lambda$ when vortex strength is 0.015625, 0.03125, 0.0625, 0.125, 0.25, and 0.5 in the logarithmic coordinate system. It is clear that for all vortex strength within the range of calculated parameters, the scattering cross-section decreases with the increase of the incident wave, and the decreasing velocity increases with the rise of the incident wave.

When $M_v$ =0, 0.0156245, 0.03125, the increasing rate of scattering cross-section intensity increases with the increase of $\lambda$. When $M_v$ =0.125, 0.625, the scattering cross-section intensity and the wavelength of the incident sound wave present a "fourth power law". When $M_v$ =0.5, there is a fluctuation at $\lambda$ =1, and the overall trend conforms to the law of exponential attenuation.

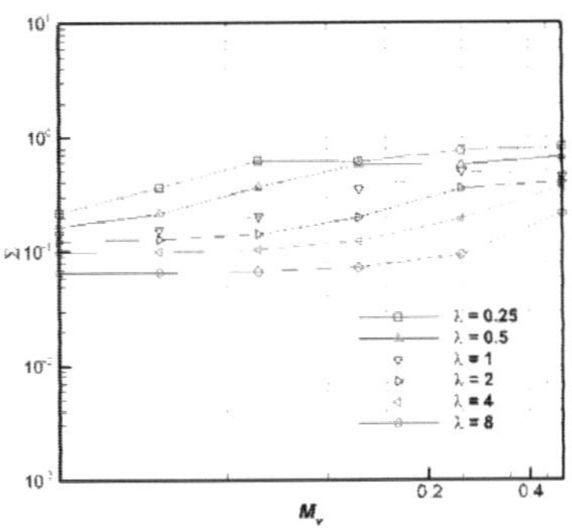


**Figure 2.** Scattering cross-section with $M_v$.

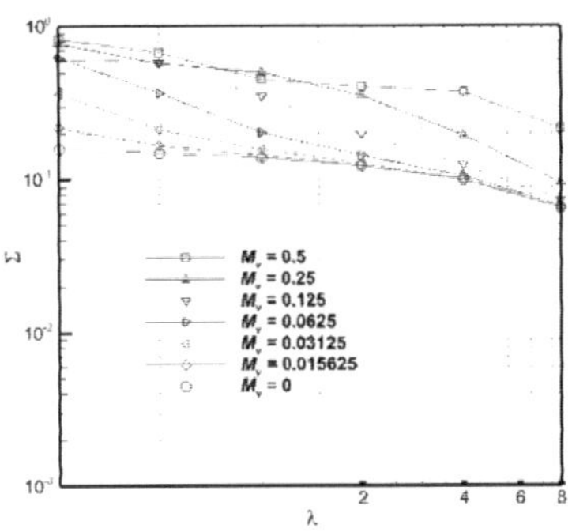


**Figure 3.** Scattering cross-section with $\lambda$.

### *3.2. RMS Pressure of the Total Pressure*

This paper adopts the control variable method without changing the vortex strength. The influence of incident wave wavelength on the sound field is studied by changing incident wave wavelength $\lambda$. We calculate the RMS of total sound pressure over a long period to find a more general rule. The figure 4 shows the RMS of the total pressure under different incident wave intensities when $M_v$=0.5 and 0.0015625. It can be seen that with the increase of incident acoustic waves, the total sound field sound pressure gradually decreases. When $M_v$=0.5, the entire sound field presents a rotary distribution. When $M_v$=0.015625, the total sound pressure at various wavelengths is symmetric.

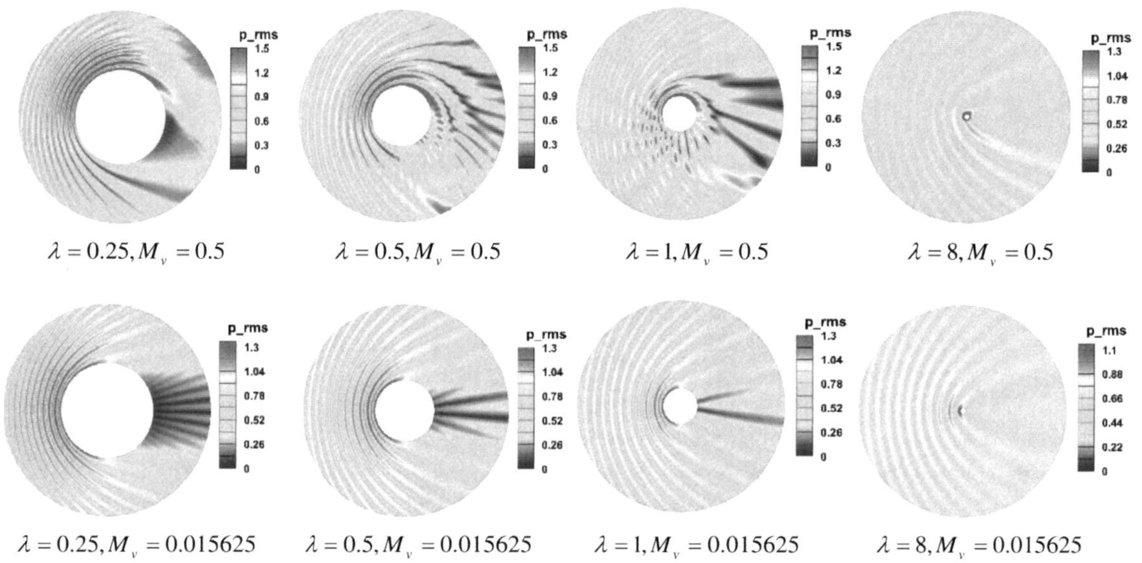


**Figure 4.** RMS of the total pressure.

### *3.3. RMS Pressure of the Scattered Pressure*

To analyze the distortion characteristics of the scattered sound field more clearly, we extracted the scattered sound field from the total sound field. Figure 5 is RMS pressure of the scattered pressure. When $M_v$=0.015625, it can be seen that with the decrease of the incident wave, the scattered sound pressure gradually moves from the rear to the front of the cylinder.

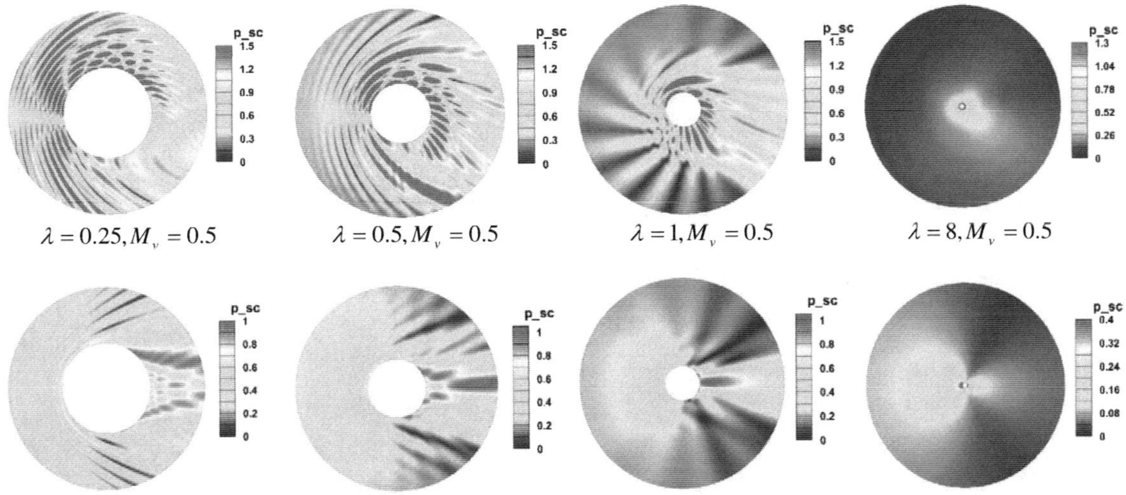


$\lambda = 0.25, M_v = 0.5$ $\lambda = 0.5, M_v = 0.5$ $\lambda = 1, M_v = 0.5$ $\lambda = 8, M_v = 0.5$

$\lambda = 0.25, M_v = 0.015625$ $\lambda = 0.5, M_v = 0.015625$ $\lambda = 1, M_v = 0.015625$ $\lambda = 8, M_v = 0.015625$

**Figure 5.** RMS pressure of the scattered fields.

### *3.4. The Instantaneous Total Pressure*

To better observe the scattering characteristics of sound waves, we make an instantaneous scattering sound field. Figure 6 shows the instantaneous total pressure. It can be see that the phase and amplitude of the sound wave are distorted after passing through the rotating cylinder.

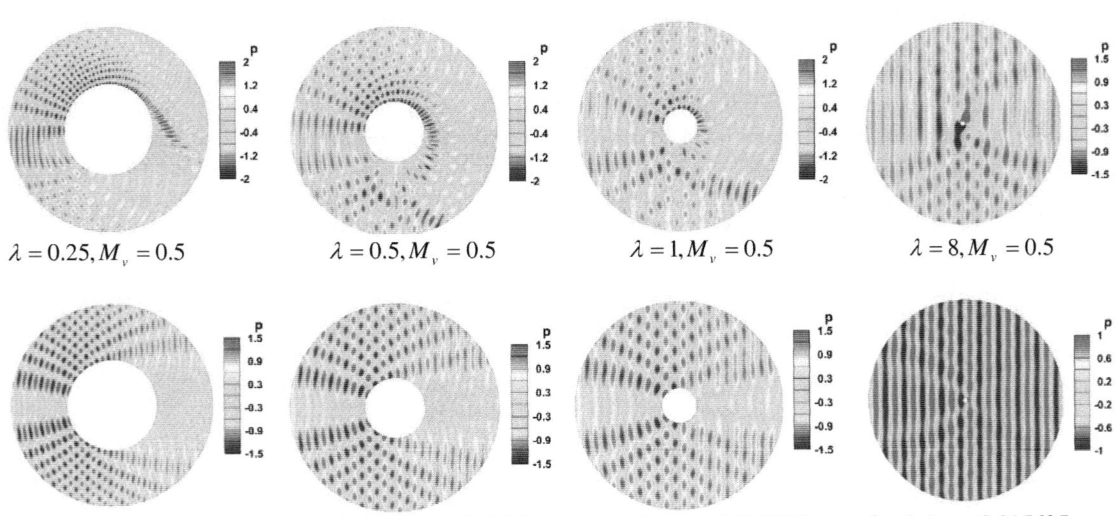


$\lambda = 0.25, M_v = 0.5$ $\lambda = 0.5, M_v = 0.5$ $\lambda = 1, M_v = 0.5$ $\lambda = 8, M_v = 0.5$

$\lambda = 0.25, M_v = 0.015625$ $\lambda = 0.5, M_v = 0.015625$ $\lambda = 1, M_v = 0.015625$ $\lambda = 8, M_v = 0.015625$

**Figure 6.** The instantaneous total pressure.

### *3.5. The Instantaneous Scattered Pressure*

Figure 7 shows the instantaneous scattered pressure. With the increase of the incident sound wave, the scattered sound field spreads around the cylinder as a typical cylindrical wave. The results are consistent with cylindrical acoustic scattering.

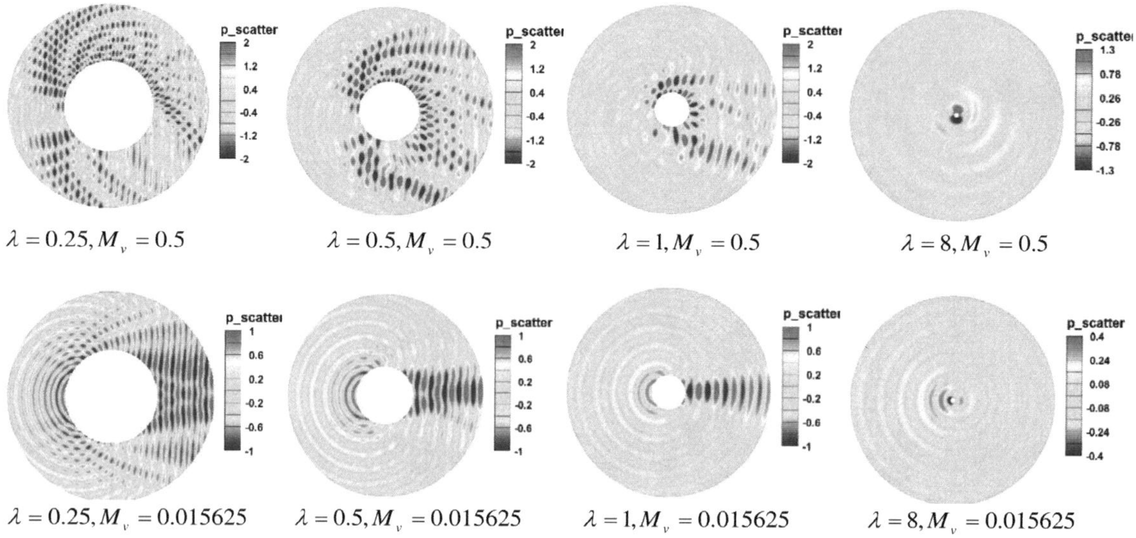


**Figure 7.** The instantaneous scattered pressure.

## 4. Conclusion

This paper analyzes the effects of incident wave wavelength $\lambda$ and vortex strength $M_v$ in the scattering sound field. The results show that the scattering cross-section intensity decreases with λ and increases with $M_v$. $\lambda < 0.5$, the scattering cross-section intensity does not increase significantly and stabilizes at the order of 1. When $\lambda > 4$, the scattering cross-section grows logarithmically with the rise of $\lambda$. $M_v < 0.25$, the increased rate of wavelength, and the intensity of the scattering cross-section conform to the characteristics of linear growth. It can be seen that the directivity distribution of scattered sound fields becomes complicated with the increase of incident wavelength. When the incident wavelength is unchanged, the intensity of the vortex increases so does the scattered sound pressure. Due to the influence of the vortex strength, the scattered sound pressure appears to have a phase difference. When the vortex strength is constant, the scattered sound pressure decreases with the increase of incident wavelength.

## Acknowledgments

The authors gratefully acknowledge the support for this research provided by National Natural Science Foundation of China (Grant Nos. 11732016 and 12172374), Sichuan Science and Technology Program (Grant No. 2018JZ0076), CARDC Fundamental and Frontier Technology Research Fund (Grant No. PJD20200185), and National Numerical Windtunnel project.

## References

[1] Tam C K W 2006 Recent advances in computational aeroacoustics *Fluid Dynamics Research* **38** 591–615.

[2] Yuan S Q, Ma R X and Wu C H 2021 Numerical simulation of acoustic scattering by a cylinder based on the enhanced optimized scheme 2nd International Conference on Mechanical Engineering and Materials *Journal of Physics: Conference Series* **2101** 012030

[3] Candel S M 1979 Numerical-solution of wave scattering problems in the parabolic appoximation *Journal of Fluid Mechanics* **90** 465-507.

[4] Colonius T, Lele S K and Moin P 1994 The scattering of sound-waves by a vortex-numerical simulations and analytical solutions *Journal of Fluid Mechanic* **260** 271-298.

[5] Wang Y M, Ma R X and Wu C H 2021 Numerical study on spatial scale characteristics of sound scattering by a static isentropic vortex *Acta Physica Sinica* **70**(19) 105-117
[6] Ma R X, Wang Y M, Zhang S H, Wu C H and Wang X N 2021 Numerical investigation of scale effect on acoustic scattering by vortex *Acta Physica Sinica* **70**(10) 104301.
[7] Ma R, Zhang S, Luo Y, et al. 2022 Numerical study on multiple acoustic scattering by a vortex array *Journal of Sound and Vibration* **527** 1-8.
[8] Wu C H, Wang Y M and Zhang S H 2019 A hybrid optimized difference scheme *Physics of Gases* **4**(6) 22-28.
[9] Wu C H, Ma R X, Wang Y M, et al. 2020 Application of enhanced optimized difference schemes in scoustic scattering *Aerodynamica Sinica* **38**(6) 1120-1128.
[10] Wu C H 2012 *Research on High Order Accurate and High Resolution Difference Methods of Fluid Dynamics* (Nanjing: Nanjing University of aeronautics and astronautics).
[11] Jiang G S and Shu C W 1996 Efficient implementation of weighted ENO schemes *Journal of Computational Physics* **126**(1) 202-22.

# Research on the Influence of Sliding Piece Gap on Dredging Performance and Installation Technology Improvement

**Mingming Liu[1*], Jingzhao Hu[1], Haifei Zhuang[1] and Chunlei Yang[2]**

[1] National Engineering Research Center of Dredging Technology and Equipment, Shanghai, China,
[2] CHEC Dredging Co., Ltd, Shanghai, China.
Email: lmm611love@126.com

**Abstract.** In order to ensure the gap between the sliding piece of trailing suction hopper dredger and the hull suction to meet the operation requirements, taking a 4500 $m^3$ trailing suction hopper dredger and a 12888 $m^3$ trailing suction hopper dredger as examples, the solid-liquid two-phase flow method was used to simulate and analyze the influence of different gap between sliding piece and the hull suction on the dredging performance. On this basis, the sliding piece installation procedure was improved and successfully applied to a 6500 $m^3$ trailing suction hopper dredger. The results indicate that the gap has a great influence on the production. The larger the gap, the lower the production and the stronger the non-uniformity of flow field distribution; When the gap is less than 4 mm, the production reduction rate is slow, and the dredging production decreases by about 2% when the gap is 2 mm,. When the gap is more than 4 mm, the dredging production decreases linearly with the increase of the gap. The application effect of the improved sliding piece installation technology is better, and the maximum gap of the sliding piece of the 6500 $m^3$ trailing suction hopper dredger is only 0.45 mm, which meets the needs of dredging engineering.
**Keywords.** Sliding piece, trailing suction hopper dredger, solid-liquid two-phase flow, numerical simulation, installation technology.

## 1. Introduction

The trailing suction hopper dredger requires the dredger to operate stably for a long time in the dredging process, so as to realize the whole dredging and transportation process efficiently. As an integral part of the delivery system, the matching state between the sliding piece and the hull suction directly affects the dredging efficiency. During the upward and downward movement of the sliding piece, the friction between the sliding piece and the hull intensifies the uneven wear of the wear ring of the sliding piece and the retaining ring of the hull suction, resulting in a gap between the sliding piece and the hull suction, Thus, the surrounding liquid enters the transmission pipeline through the gap, which changes the dredger construction parameters and directly affects the dredging efficiency. The key to decide whether to replace or adjust the wearing parts is to determine the influence of the gap between the sliding piece and the hull side suction on the dredging efficiency.

Because there is no theoretical calculation method for the pipe resistance when there is a gap between the sliding piece and the hull suction, the adjustment gap usually depends on experience, so it is necessary to analyze and study it. At present, the numerical simulation technology of solid-liquid two-phase flow has been widely used in the dredging industry [1-6]. Chen [7] used the mixture multiphase flow model to analyze the solid particles distribution under different volume concentrations, and obtained the law that the centrifugal pump slurry head decreases with the increase of particle concentration; Liu [8] used Euler method to analyze the solid-liquid two-phase flow field in centrifugal pump, obtained the particle concentration distribution under different particle size and volume concentration, and studied the changes of external characteristics. It was found that the pump slurry head decreased with the increase of particle diameter and concentration; Xiong [9] used Euler model to simulate solid-liquid two-phase flow in dredger slurry pipeline, and compared with the calculated value and test data of Durand model, it is found that the calculated value of pressure drop of CFD model is closer to the test value than Durand model. Deng [10] respectively used Euler method and Lagrange method to simulate the pipeline conveying high concentration mud, and compared with the test. It is concluded that the Euler model is closer to the actual working condition when conveying high concentration fine particle, and the Lagrange model is more suitable for conveying coarse particle. Therefore, the project team plans to use the numerical simulation method of solid-liquid two-phase flow to quantitatively study the internal matching relationship of dredger pump, pipeline and sliding piece, so as to obtain the variation of dredger construction parameters with the gap between sliding piece and hull side suction.

In the current sliding piece installation process, due to the insufficient understanding of the importance of the fit between the sliding piece and the hull suction, the sliding piece is required to be lowered to the hull suction during the installation process, and there are no specific requirements for the gap between the sliding piece and the hull suction, which leads to the poor fit between the sliding piece and the hull suction after installation, which affects the vacuum degree of the dredge pump and the dredging efficiency of the dredger. Based on the above calculation results, combined with the working principle of the sliding piece and the technical requirements of sliding piece installation, the project team improved the existing sliding piece installation process, which improved the sliding piece installation accuracy.

## 2. Numerical Scheme

### *2.1. Computation Model*

The configuration of excavation and transportation systems for different types of trailing suction hopper dredger is basically similar. Taking a 4500 $m^3$ trailing suction hopper dredger as an example, the excavation and transportation system is composed of drag head, drag arm pipe, sliding piece, suction pipe, dredge pump, hopper inlet pipe and other components, as shown in figure 1. Considering that the calculation of the overall transportation system is complex and time-consuming, in order to improve the calculation efficiency, the model is simplified. Only the position where the gap is generated is analyzed, and an extension section of five times the pipe diameter is added before and after the gap to increase the calculation stability. The simplified model is shown in figure 2. It can be seen from the figure that the inlet 1 is behind the drag arm pipe and the sliding piece, and the inlet 2 is where the surrounding liquid enters due to the existence of gap.

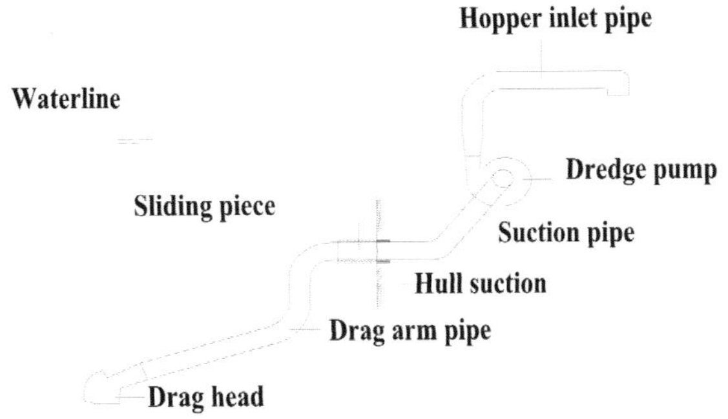


**Figure 1.** Excavation and transportation systems.

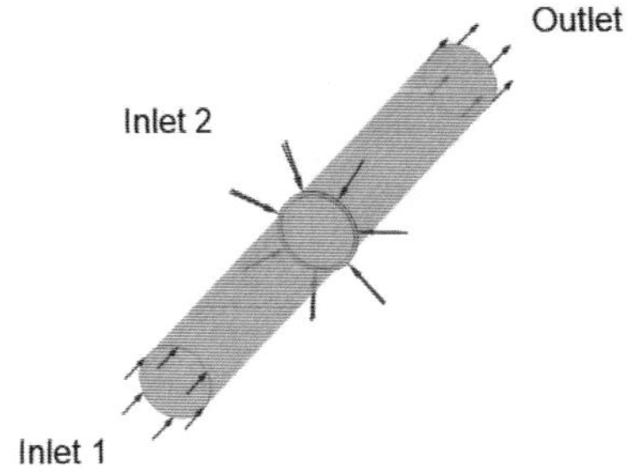


**Figure 2.** Computation model.

### *2.2. Numerical Method*

In order to improve the convergence of calculation and the accuracy of calculation results, the hexahedral structured grid is adopted for the computation model, the Euler solid-liquid two-phase flow is numerically analyzed by ANSYS CFX software, the turbulence of liquid phase and solid phase is solved by SST model and discrete phase zero equation model respectively, and the drag between solid and liquid is calculated by Wen Yu model, The influence of virtual mass force and lift force is ignored in the calculation. The wall surface is set as non-slip wall surface, and the wall surface roughness is 0.5 mm. The convergence accuracy is set to 10-5. The boundary conditions are set as total pressure inlet and static pressure outlet. After the model is simplified, the parameters such as dredge pump, pipeline, dredging depth, discharge height and particle in the excavation and transportation system need to be input into the boundary conditions in the form of fitting formula. As shown in figure 2, the total pressure condition of inlet 1 includes dredging depth, average waterline, drag arm pipe loss, etc., the total pressure condition of inlet 2 is related to the dredging depth and average waterline, and the static pressure condition of outlet includes dredge pump head, discharge height, loss of suction pipe, loss of hopper inlet pipe, etc. Among the above parameters, the dredging depth, average waterline, discharge height and the dredge pump head are known, and the rest include the slurry head of the dredge pump, the slurry head of the pipeline, the local loss of elbow and gate valve, etc., which are calculated according to the engineering experience and relevant specifications.

### *2.3. Computation Conditions*

The transported slurry is fine sand, with natural soil density of 1.95 t/m$^3$ and soil particle density of 2.7 t/m$^3$. The density of liquid phase (seawater) is 1.025 t/m$^3$. For a 4500 m$^3$ trailing suction hopper dredger, the dredging production and suction vacuum of dredge pump under different dredging depth, different gap and different soil density are calculated. In addition, the numerical simulation is carried out for a 12888 m$^3$ trailing suction hopper dredger with a specific dredging depth, different gap and different soil density. The numerical simulation conditions are shown in table 1. It is assumed that the dredging pump speed and inlet concentration remain unchanged in the calculation.

**Table 1.** Numerical simulation conditions.

| Dredger | Pump speed (rpm) | Dredging depth (m) | Gap (mm) | Slurry density (t/m$^3$) |
|---|---|---|---|---|
| 4500 m$^3$ | 214 | 15, 20, 30 | 0, 2, 4, 10 | 1.025, 1.1, 1.2, 1.3, 1.4 |
| 12888 m$^3$ | 181 | 35 | 0, 2, 4, 10 | 1.025, 1.1, 1.2, 1.3, 1.4 |

## 3. Numerical Results Analysis

### *3.1. Influence of Gap on Construction Parameters*

Figure 3 shows the influence of gap change on inlet 1 flow, outlet flow and total pressure when the dredging depth is 15 m and the slurry density is 1.2 t/m$^3$. It can be seen from the figure that the inlet 1 flow decreases and the outlet flow increases with the increase of gap; As the inlet 1 concentration remains unchanged, the decrease of inlet flow makes the drag arm pipe resistance smaller, resulting in the increase of inlet 1 total pressure and the increase of outlet flow makes the outlet concentration decrease; The outlet total pressure also increases with the increase of gap, but the difference between the inlet 1 total pressure and the outlet total pressure changes relatively little.

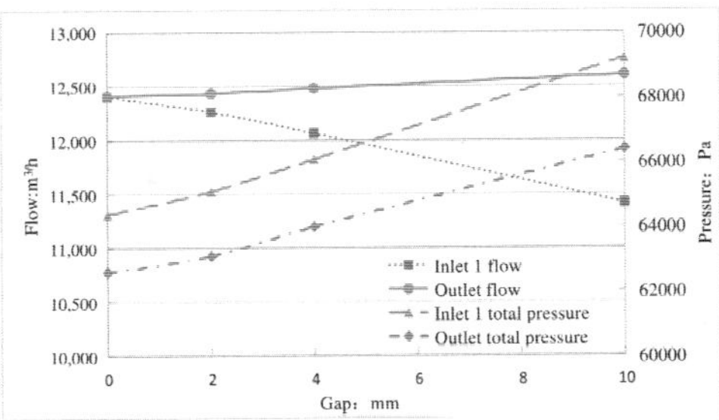


**Figure 3.** Variation of flow and total pressure with gap.

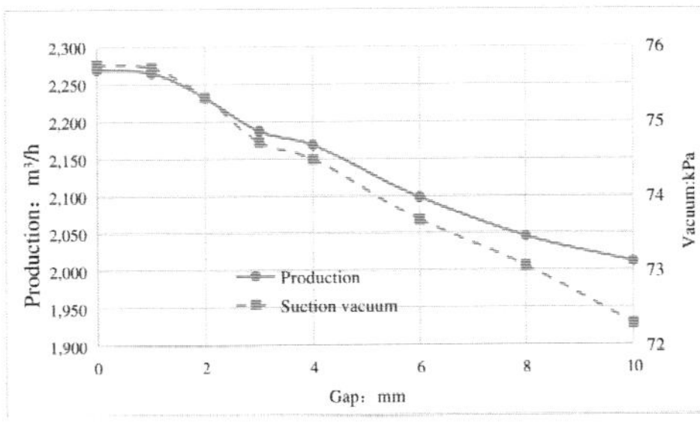


**Figure 4.** Variation of production and suction vacuum with gap.

Figure 4 shows the changes of production and suction vacuum under different gaps under the same working conditions (dredging depth 20 m and dredge pump speed 214 rpm). The absolute values of production and suction vacuum decrease with the increase of the gap. When the gap is less than 4 mm, the production and suction vacuum change slowly with the increase of the gap. When the gap is greater than 4 mm, the production and suction vacuum decrease linearly with the increase of the gap. When the gap is 4 mm, the production and suction vacuum are reduced by 4.5% and 1.7% respectively compared with that without gap. And the production and suction vacuum are reduced by 1.7% and 0.6% respectively when the gap is 2 mm.

Figure 5 shows the influence of different densities and gaps on production and suction vacuum when the dredging depth is 15 m. It can be seen that the production gradually increases with the increase of density; Under the same concentration, the larger the gap, the smaller the production; When the gap is 2 mm, 4 mm and 10 mm, The average production (Average suction vacuum (or production) = (suction vacuum (or production) at 1.1 t/m$^3$ + suction vacuum (or production) at 1.2 t/m$^3$ + suction vacuum (or production) at 1.3 t/m$^3$ + suction vacuum (or production) at 1.4 t/m$^3$ (or production)) / 4) decreased by 1.7%, 4.4% and 11.2% respectively. The suction vacuum increased with the increase of density; The larger the gap, the smaller the suction vacuum under the same concentration,. When the gap was 2 mm, 4 mm and 10 mm, the average suction vacuum of different densities decreased by 0.7%, 1.8% and 4.8% respectively.

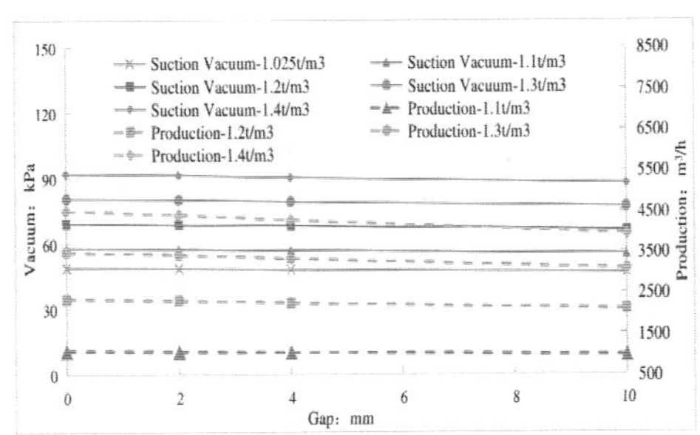


**Figure 5.** Variation of production and suction vacuum with gap and density.

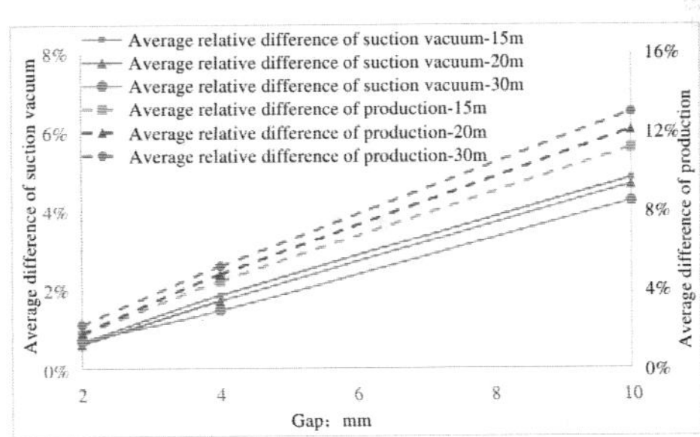


**Figure 6.** Variation of production and suction vacuum with gap and depth.

Figure 6 shows influence of different dredging depths and gaps on the average relative difference of suction vacuum and average output. The relative difference refers to the relative difference between the calculated value of a gap and the calculated value when the gap is zero. Assuming that the dredging depth is 15 m and the density is 1.1 t/m$^3$, the production without gap and with gap of 2 mm are 1000 m$^3$/h and 900 m$^3$/h respectively, Then the relative difference of production when gap is 2 mm is | 900-1000 | / 1000 * 100%=10%. The calculation method of relative difference of vacuum is the same. The average relative difference of suction vacuum and average relative difference of production refer to the average value of suction vacuum relative difference and production relative difference of different densities at the same gap and dredging depth. It can be seen from the figure that the average relative difference of suction vacuum and production have the trend of decreasing and increasing respectively with the increase of dredging depth at the same gap; Generally speaking, the influence of dredging depth on the average relative difference of suction vacuum is relatively small at the same gap. When the gap is 2 mm, the average relative difference of suction vacuum and production at different dredging depths are almost the same. When the gap is 4 mm, the maximum difference of average suction vacuum and yield relative difference at different dredging depths is 0.4% and 0.7% respectively; When the common dredging depth is 20 m, the average production of 2 mm, 4 mm and 10 mm is reduced by 1.79%, 4.76% and 12.12% respectively.

Under the same soil quality, the specific dredging depth of 12888 $m^3$ trailing suction hopper dredger is 35 m compared with that of 4500 $m^3$ dredging depth of 20 m. Figure 7 shows the comparison of the average relative difference of suction vacuum and average relative difference of production at different gaps. It can be seen that the average relative difference of suction vacuum of 12888 $m^3$ dredger is lower than that of 4500 $m^3$, and with the increase of gap, the larger the difference is, the difference is 0.3%, 0.8% and 2% at 2 mm, 4 mm and 10 mm respectively; The average relative difference of the production of 12888 $m^3$ dredger is higher than that of 4500 $m^3$. With the increase of the gap, the difference is greater. When the gap is 2 mm, the difference of the two dredgers are almost equal, and when the gap is 4 mm and 10 mm, the difference is 1.25% and 2.6% respectively. To sum up, when the gap is less than 2 mm, the reduction ratio of dredger dredging production to that without gap is less than 2%, and when the gap is less than 4 mm, the reduction ratio of dredging production to that without gap is less than 6%.

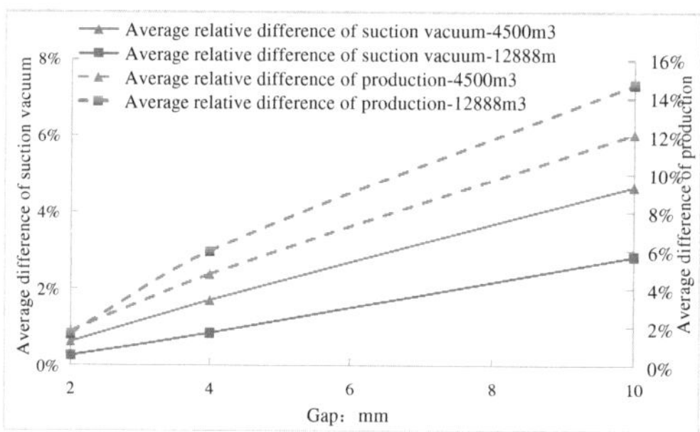


**Figure 7.** Variation of production and suction vacuum with gap and dredger.

### 3.2. *Influence of Gap on Flow Field*

Figure 8 shows the static pressure distribution of different gaps when the dredging depth is 20 m and the soil density is 1.2 t/$m^3$, and the flow direction is from left to right. It can be seen that the surrounding liquid enters the pipe through the gap, resulting in pressure stratification in the pipe. There are local high-pressure and local low-pressure areas on the left and right sides of the pipe wall near the gap inlet; The pressure in the main flow area in the pipe decreases gradually near the gap from left to right; With the increase of the gap, the uneven pressure distribution in the pipe is gradually strengthened, and the influence area is also gradually enlarged.

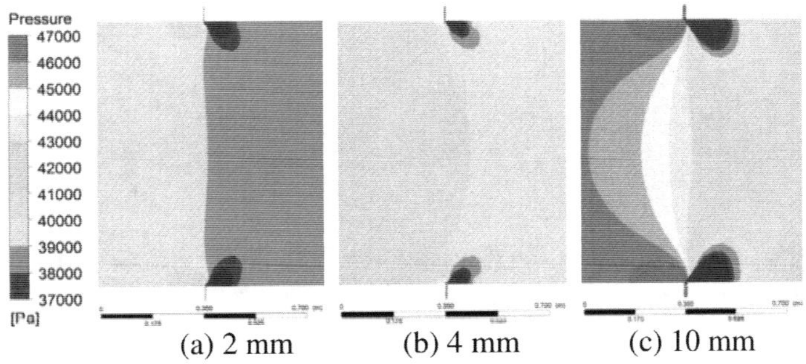


**Figure 8.** Static pressure distribution at different gaps.

Figure 9 shows the particle concentration distribution of different gaps. It can be seen from the figure that the concentration of the main flow area in the pipe with different gaps are basically the same. There is a low concentration area on the inner wall of the pipe near the right side of the gap inlet, and with the increase of the gap, the low concentration area becomes larger, resulting in the gradual decrease of the cross-sectional average concentration in the area on the right side of the gap.

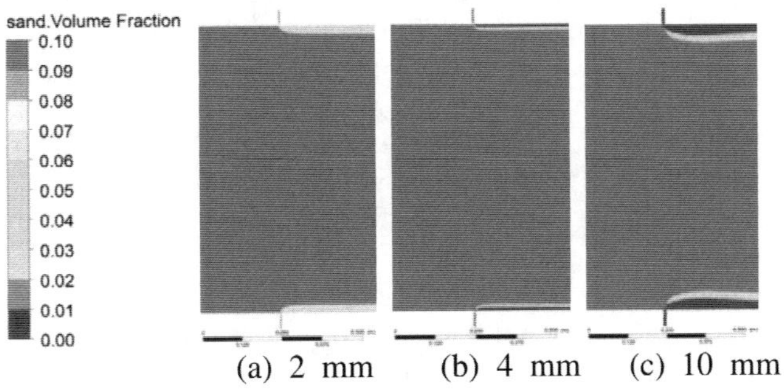


**Figure 9.** Concentration distribution at different gaps.

Figure 10 and figure 11 respectively show the velocity distribution of liquid phase and solid phase at different gaps. It can be seen from the way that the gap has almost the same influence on liquid phase and solid phase; The non-uniformity of velocity distribution in the pipe increases with the increase of gap; There is a high-velocity area at the gap position. With the increase of the gap, the high-velocity area gradually becomes larger.

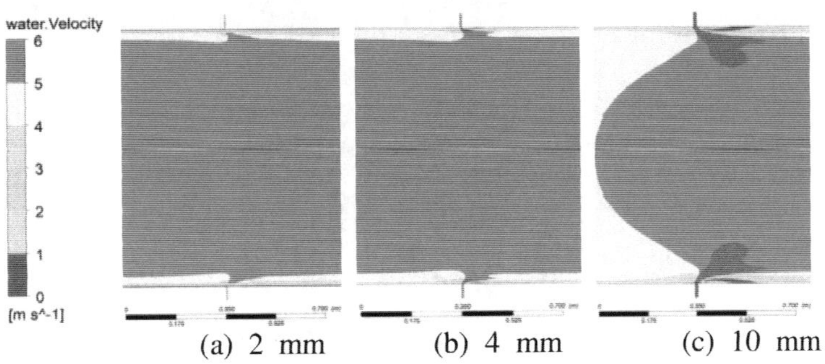


**Figure 10.** Liquid velocity distribution at different gaps.

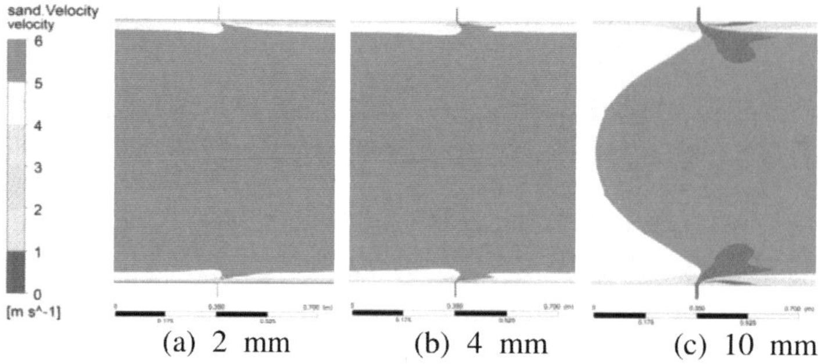


**Figure 11.** Solid velocity distribution at different gaps.

## 4. Improvement on Installation Technology of Sliding Piece

In view of the downward and stern offset trend of the sliding piece, allowance is reserved for the relative position between the sliding piece suction and the hull suction. It is required that after positioning, the center of the sliding piece suction is pre offset 10mm to the bow direction and height direction relative to the hull suction center. During the construction process, the coincidence between the sliding piece suction and the hull suction will be increased gradually. In addition, the control requirements for the fitting gap are improved. Based on the view of the numerical calculation results in the previous section, it is specified that the average gap shall not be greater than 2mm for the first installation of sliding piece, and when the average gap is greater than 4mm, the sliding piece needs to be repaired and maintained.

The sliding piece wedge block fits evenly with the corresponding surface of the hull steel wedge, the larger the contact area, the more uniform the transfer of physical force between the sliding piece and the dredger. The positioning of the sliding piece shall be measured in two times: The first time, and the wedge block shall be fixed on the sliding piece body by spot welding. After the sliding piece is hoisted in place, the vertical deviation shall be measured with a steel ruler; After adjusting the wedge position to the design requirements according to the vertical deviation measured in the first lifting, start the second lifting of the sliding piece, apply blue oil on the contact surface of the hull steel wedge, then lower the sliding piece to the crane cable without stress, recheck the vertical deviation and record it and check the fitting gap between the sliding piece suction and the hull suction port with a feeler gauge. Lift out the sliding piece and check the blue oil area of the contact surface of the sliding piece wedge. If the blue oil contact is uniform and the blue oil contact area exceeds 70%, the requirements are met, as shown in figure 12.

**Figure 12**. Sliding piece wedge block blue oil brush.

The improved installation process of sliding piece was successfully applied to a 6500 $m^3$ trailing suction hopper dredger. After installation, the sliding piece suction fits well with the hull suction, and the maximum gap is 0.45 mm, which meets the installation requirements. The contact between the sliding piece wedge and the hull steel wedge is uniform. During the wharf test and offshore test, the dredge pump suction vacuum degree reached 0.8 bar, and the maximum slurry density reached 1.45 t/m$^3$.

## 5. Conclusion

- The larger the gap, the lower the dredging production, and the smaller the dredge pump suction vacuum; And when the gap is less than 4 mm, the production and suction vacuum change relatively slowly with the increase of the gap; When the clearance is greater than 4 mm, the production and suction vacuum will decrease linearly with the increase of gap.
- The average relative difference of suction vacuum of 12888m$^3$ trailing suction hopper dredger is lower than that of 4500m$^3$, and the average relative difference of production is higher than that of 4500m$^3$; Compared with no gap, when the gap is less than 2mm, the average dredging production decreases by about 2%, and when the gap is 4mm, the average dredging production decreases by about 6%.
- With the increase of the gap, the uneven distribution of pressure and velocity in the pipe increases gradually, and the low concentration area on the inner wall of the pipe increases gradually; The effect of gap on liquid phase and solid phase is the same. There is a high-velocity region at the gap position, the gap increases, and the high-velocity region increases.
- The sliding piece installation process was improved and successfully applied to the newly built trailing suction hopper dredger. The maximum gap after installation was only 0.45 mm. During the test, the dredge pump suction vacuum degree reached 0.8 bar and the maximum slurry density was 1.45 t/m$^3$, which met the actual needs of the dredging project.

## Reference

[1] Yu D, Yu L, Shi Q Z, Zheng J L, Sui Y, Zhao Z J and Mao H Y 2020 Overflow effect test and numerical analysis of environmental protection valve of trailing suction hopper dredger *Marine Technology* **6** 6-11.
[2] Zhang Q Y, Shi H D, Gao W and Li J F 2017 Numerical study on critical velocity of coral reef fragment hydraulic transport in dredging engineering *Port & Waterway Engineering* **6** 34-38.
[3] Liu M M, Guo T, Hu J Z and Xia C 2021 Application of DPM method in wear resistance selection of dredge pump impeller *Pump Technology* **4**(06) 10-15.
[4] Zuo J L, Yang H, Wei B Q, Hou J M and Zhang K 2020 Numerical simulation of gas-liquid two-phase flow in gas lift system *Acta Physica Sinica* **69**(06) 187-196.
[5] Yu D and Yu L 2018 Effect analysis of green valve of TSHD based on CFD *Ship Engineering* **40**(07) 29-33-45.
[6] Wang Y J and Jin X 2011 Kinematic simulation and calculation of two-phase flow in hydraulic lifting pipelines *Mining & Processing Equipment* **39**(12) 50-54.
[7] Chen C C, Yang C M and Xiong M T 2006 Numcrical analysis on two-phase flow through impeller of low-specific-speed centrifugal pump *Journal of Drainage and Irrigation Machinery Engineering* **6** 1-3.
[8] Liu J R, Xu Y G, Wang D M and Su Q Q 2010 Numerical simulation of solid-liquid two-phase turbulent flow in impeller channel and pump characteristics analysis *Transactions of the Chinese Society for Agricultural Machinery* **41**(03) 86-90.
[9] Xiong T G, Zhang M D, Wei Wi and An B 2018 CFD simulation of characteristics of pipeline slurry transportation *Yellow River* **40**(04) 19-23.
[10] Deng Y, Liu J W, Cao F and Luo C J 2015 Numerical simulation of high concentration slurry pipeline transport *Journal of Wuhan University of Technology (Transportation Science & Engineering)* **39**(06) 1130-1134.

# Study on Interaction Analysis Method between Fog Drops and Tree Trunk Surface for Precision Pesticide Application by Plant Protection UAV

**Yalan Jia*, Wei Xiong, Bo Luo, Yue Wang, Shujie Liu and Ming Cheng**

Research Center of Unmanned Aerial Vehicle, Army Aviation Institute, Beijing, China.
Email: jiayalany@163.com

**Abstract.** In order to improve pesticide application efficiency and reduce environmental pollution, precise pesticide application has become the core concept of forestry pest control. In this paper, based on the precise target research requirements of the pesticide liquid of plant protection UAV, the VOF model of droplet impact was established. The two-phase flow model was used to simulate the three processes (contact, spread and contraction) of droplet impact on bark, and analyse the state of droplet impact on bark surface at different speeds and heights. Through the study, the behaviour state of fog droplets in different conditions can be obtained, which can maximize the use of liquid medicine, reduce environmental pollution, and provide a new research method for fog droplets deposition on tree trunk surface.
**Keywords**. Drone, Precision applying pesticide, Droplet behaviour, VOF model.

## 1. Introduction

In recent years, UAV technology has developed rapidly, and its application scope in the civil field has also been expanding, especially in the application of plant protection UAV. Compared with traditional plant protection operation methods, plant protection UAV has many advantages such as efficiency, convenience, accuracy and environmental protection. Its key technologies are mainly reflected in two aspects: the stability and precise control of UAV platform, and the accuracy and effectiveness of liquid medicine on target. It is of great significance to study the interaction between fog drops and tree trunk surface, to study the deposition, sputtering and rebound of fog drops after they impact on tree trunk surface, and to analyze the interaction between tree trunk epidermis and fog drops, which is of great significance to improve the accuracy and effectiveness of drug liquid on target. Willow tree three-dimensional model is presented in this paper, on the basis of modeling based on Fluent software, through the simulation experiment to simulate droplets impact behavior process on the surface of the trunk, the analysis of the effect of different conditions on the impact process, as the droplets deposited on the surface of the trunk provides a new research method, for the nozzle design and solution precision of target to provide important reference [1, 2].

## 2. Droplet Impact Modelling

Fluent provides a wide variety of turbulence models, including single-equation (Spalart-Allmaras) model, RNG $k-\varepsilon$ model, two-way equation model (Standard $k-\omega$ model, including Standard $k-\omega$ model, Realizable $k-\varepsilon$ model) and other models. Among them, RNG $k-\varepsilon$ model has the advantage of high precision and is suitable for this study [1, 3].

### *2.1. Flow Field Division and Gas-Liquid Two Phase Parameter Determination*

The simulation model is established according to the established three-dimensional model of tree trunk. However, because the trunk model is too large and the contact area when a single fog drop impacts is limited, in order to reduce the model size and mesh calculation amount in CFD calculation, a piece of bark from the trunk is randomly selected for simulation study.

Assuming that the air is an incompressible, isothermal and constant Newtonian fluid, the momentum, turbulent kinetic energy and vortex dissipation rate equations of $k-\varepsilon$ turbulence model are controlled by using three-dimensional time and volume averaging Navier-Stokes in the Cartesian coordinate system [4].

The cut bark is shown in figure 1 with a size of 30 mm×10 mm×2 mm.

In this paper, the behavior of fog droplets impacting tree trunk surface is studied. The droplets fall free because of gravity. In the three-dimensional space, assuming that the initial state of fog droplets is spherical, the process of fog droplets hitting bark surface is observed and its change rule is studied with the given initial velocity. The established flow field can not only conform to the actual situation but also cannot affect the motion state of fog droplets. Based on theoretical research and previous research results, the flow field (figure 2) is established in this paper for simulation. The flow field was set according to the shape of the cut bark. The bottom of the flow field was just wrapped around the bark, and the height of the flow field was h=6mm. This method can reduce the size of the flow field, reduce the excess flow field at the junction between the flow field and the bark, and improve the mesh quality and calculation accuracy.

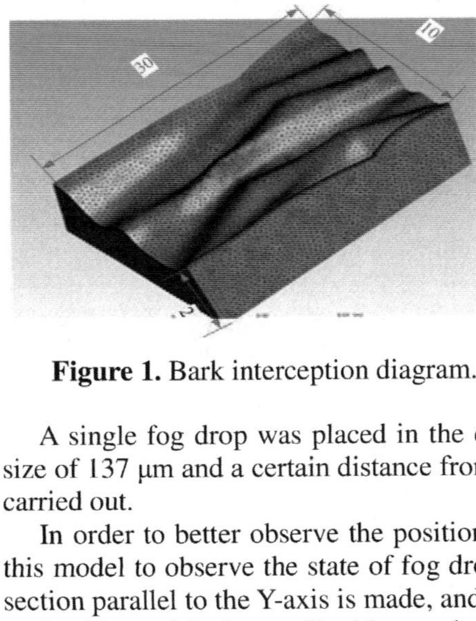


**Figure 1.** Bark interception diagram.

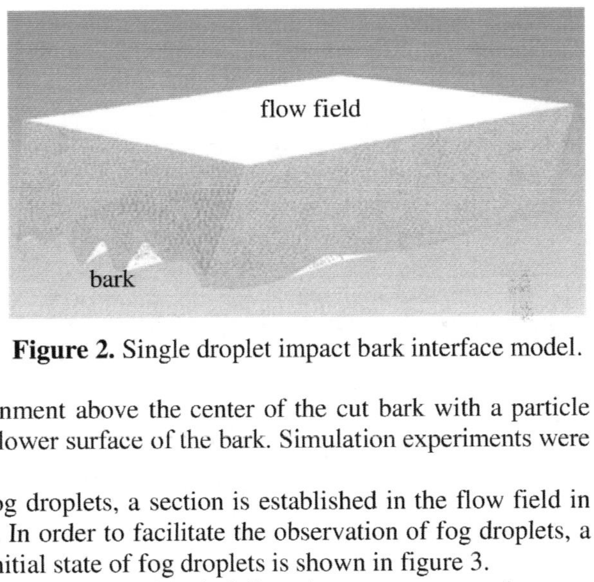


**Figure 2.** Single droplet impact bark interface model.

A single fog drop was placed in the environment above the center of the cut bark with a particle size of 137 μm and a certain distance from the lower surface of the bark. Simulation experiments were carried out.

In order to better observe the position of fog droplets, a section is established in the flow field in this model to observe the state of fog droplets. In order to facilitate the observation of fog droplets, a section parallel to the Y-axis is made, and the initial state of fog droplets is shown in figure 3.

In this model, the gas-liquid two phases are constant and both follow the momentum and energy equation. This study mainly discusses the rebound characteristics of liquid phase when fog drops impact bark surface, and does not involve the energy problem, but only considers the momentum equation. It is assumed that the experiment is carried out under constant temperature, the temperature value is certain, the two phases of gas and liquid are incompressible Newtonian fluids, the phase surface of gas-liquid contact surface does not transfer, the temperature is constant during the experiment, and there is no heat and mass transfer between the two phases.

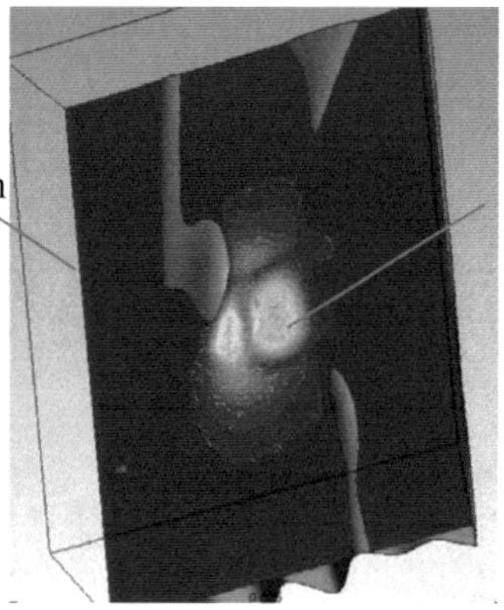


**Figure 3.** Flow field droplet location map.

The gas-liquid two-phase parameter Settings required in the simulation experiment are shown in table 1.

**Table 1.** Main parameters of gas-liquid two phase.

| Parameter | Gas phase (air) | Liquid phase (fog droplets) |
|---|---|---|
| Temperature (°C) | 25 | 19.7 |
| Density (kg/m$^3$) | 1.225 | 998.2 |
| Viscosity (kg/m$^{-s}$) | 1.7894e-05 | |
| Intensity of turbulence (%) | 10 | 10 |

*2.2. Computational Grid and Boundary Condition Setting*

The simulation experiment mainly is to simulate droplets collision bark on the surface of the form, the internal process and distribution of liquid phase, because bark surface shape is irregular and not easy to differentiate, so using unstructured grid and structured grid combined with partitioning (hexahedral and tetrahedral) model is helpful to control the size of the grid and node density and closer to the actual model. At the junction between the flow field and the bark surface, structured mesh is used to encrypt the whole area, which reduces the difficulty of fitting the boundary of the area, improves the mesh quality, and is closer to the actual model to improve the accuracy of the simulation experiment. The model combines unstructured grid with structured grid, which can reduce the number of grids and speed up the computation. The total number of grids is 32,575.

The boundary conditions of multiphase flow are different from those of single-phase flow. Each single-phase flow must be set separately. The other conditions are mixture shared by all phases.

In this paper, the VOF model is used, and conditions must be set for the region of each class to specify boundaries:

(1) For exhaust fan, inlet fan, outlet vent, pressure outlet, velocity inlet boundaries do not have to specify conditions for the main phase. For the second phase of each, a volume fraction must be set, either a constant, a profile distribution, or a user-defined function, and the rest is specified for the mixed phase [5].

(2) For Axis, fan, Outflow, periodic, porous jump, Radiator, solid, symmetry, wall zone boundary, all conditions were specified for mixed phase and none was specified for single phase [6].

(3) For fluid region, mass source term (mass sources) is specified for single phase, and all other source terms except mass source term are specified for mixed phase. If the fluid region is not porous, all other conditions are specified for the mixed phase. If fluid Zones are Porous, select Porous Zones for the mixed phase under the Fluid dialog. Porosity inputs are also specified for the mixed phase. And Resistance Coefficients and Direction Vectors are specified for each phase, respectively. All other conditions are specified for the mixed phase. For the Wall region, if Wall adhesion is considered, the contact Angle should be set for the mixed phase [7].

(4) When the open channel model is selected, wave generation parameters can be set in the inlet boundary, and Multiphase can be set at the interface of the inlet boundary. There are two types of Wave generation models, Wave Theory and Wave Spectrum.

In the gas-liquid two-phase model studied in this paper, due to the large density difference between liquid and gas, the surface of fog droplets was regarded as a free surface, and the inlet and outlet pressures were 101325Mpa, as shown in figure 4.

When the droplets hit the bark surface, their motion state was affected by the bump and roughness of the wall. Assuming that the fog droplet impacting wall is a non-slip wall, it can be defined as [7]:

$$\frac{u_p u^*}{\tau_w / p} = \frac{1}{\tau} \ln\left( E \frac{\rho u^* y_p}{\mu} \right) - \Delta B \tag{1}$$

$$u^* = C_\mu^{1/4} k^{1/2} \tag{2}$$

According to dong Xiang's research results and the roughness of bark in this paper, the roughness of wall surface is set as the completely rough area ( $K_S^+ > 90$ ) [1, 8, 9]:

$$\Delta B = \frac{1}{\kappa} \ln\left(1 + C_{k_s} K_s^+\right) \tag{3}$$

$$K_s^+ = \rho K_s u^* / \mu \tag{4}$$

where $C_{k_s}$ is the roughness constant, which is related to the material of the impacted wall.

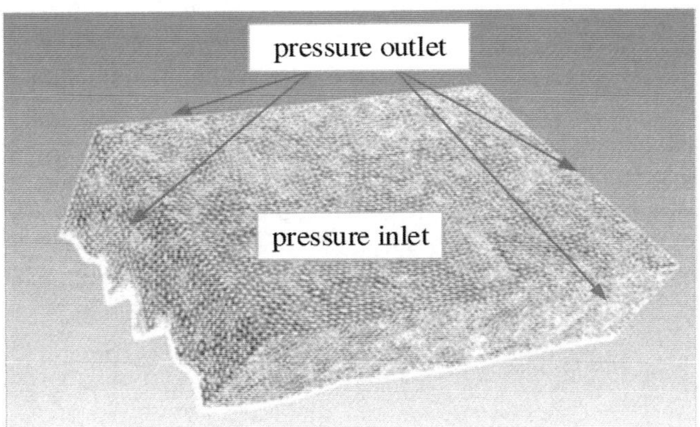


**Figure 4.** Boundary condition setting.

*2.3. Calculation Method*

In order to improve the solution accuracy and convergence of the VOF model, the following points should be paid attention to when choosing the solution parameters:

(1) VOF model pressure interpolation using PRESTO! Plan;

(2) Implicit volume fraction parameter equation is used in this paper. In order to improve the clarity of interphase interface, CICSAM scheme was used to discretize the volume fraction equation, and QUICK scheme was used to capture the diffusion surface.

(3) When setting the solver, the relaxation factor should be considered and the velocity pressure coupling scheme should be selected. Since transient flow calculation is adopted in this paper, PISO scheme is adopted (table 2). When PISO is used, increasing the value of relaxation factor will not weaken the stability of the solution. For the model of single droplet impacting bark surface,

unstructured mesh and structured mesh were used to divide the model, and structured mesh was used at the boundary to improve the mesh quality. In order to achieve the expected convergence speed (i.e. fewer iterations per time step) and relative stability, the pressure under relaxation factor was set as 0.7 (table 3).

**Table 2.** Solution mode parameter setting.

| Parameter | Pressure | Scheme | Momentum | Type | Time |
|---|---|---|---|---|---|
| Solution | PRISTO! | PISO | QUICK | Pressure-Based | Transient |

**Table 3.** Solve control parameter settings.

| Parameter | Pressure | Body Forces | Density | Momentum | Turbulent Kinetic Energy | Turbulent Viscosity |
|---|---|---|---|---|---|---|
| Solution Controls | 0.3 | 1 | 1 | 0.7 | 0.8 | 1 |

$$U_{db} = \sqrt{\frac{8\delta_l}{C_d \rho_g D}} \tag{5}$$

Equation 5 is the critical condition that fog droplets do not break when they impact on the surface, Where $\frac{\delta_l}{D}$ is the ratio of surface tension coefficient to fog droplet diameter; $C_d$ is a constant depending on the rupture condition, generally 0.09; $\rho_g$ is the gas density, 1.29 kg/m$^3$; B is the surface tension, the standard value of pure water at 19.7°C is $7.2\times10^{-2}$ N/m;

According to equation 5, the critical velocity of fog droplet is 4.3m/s. Therefore, the height of fog drops is set as 5mm, and the initial speed is 2.1 m/s, so that the fog drops will not break when acting on the surface of the tree trunk. Under this parameter, the interaction of fog droplet impinging on tree trunk surface was observed and analyzed.

(4) The fog droplets are subjected to the negative gravity acceleration of z-axis: -9.81 m/s$^2$. Given droplet velocity value: 2.1 m/s, the direction is negative of the Z axis.

(5) The number of iterations was10-5s, and the number of running steps was set to 100,000 according to the comprehensive factors such as the stability required in the research process, the simplicity of the calculation process, time saving and efficiency improvement, etc. Since the simulated droplet impact is in liquid phase, that is, transient state, the transient mode is set to observe the simulation results.

## 3. Simulation Calculation and Analysis

The deposition process of fog droplets is usually divided into three stages: interface contact stage, spreading stage and contraction stage. The state of fog droplets hitting tree trunk surface will be described from these three stages respectively. In this experiment, fog droplets were simulated to impact the bark at a height of 2.1 m/s and 5 mm, and the process before falling was ignored. The moment when fog droplets touched the bark surface was 0ms, and observation began. The three phases will be discussed separately below.

### *3.1. Interface Contact Stage*

Figure 5 shows the liquid phase distribution at the contact stage when a single fog droplet impacts the bark surface. At this stage, the fog droplets have touched the bark surface and remain roughly in their original shape (spherical), but their shape is slightly changed due to the influence of air pressure and gravity. Figure 6 shows the pressure distribution inside the droplet and the flow field. When the droplet impacts, the droplet pressure increases, and the pressure in the flow field also increases due to the influence of the airflow. The pressure inside the droplets is evenly distributed. Given the initial speed and height of fog droplets, when they contact the surface of the tree trunk, the velocity increases by about 2.11 m/s. The fog droplets are affected by the flow field, and the velocity of the surrounding airflow also increases with the increase of time, as shown in figure 7.

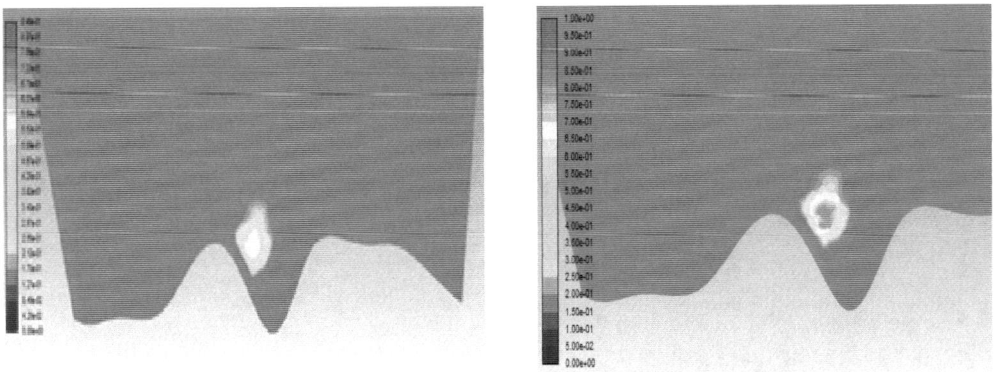

(a) 0 s (b) 0.25 s

**Figure 5.** Liquid phase distribution in the contact phase.

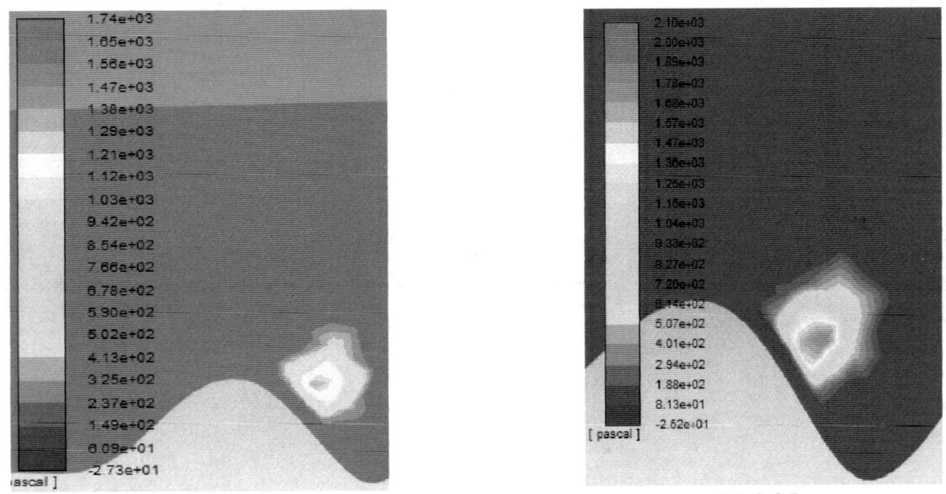


(a) 0 s (b) 0.25 s

**Figure 6.** Contact stage pressure distribution map.

Journal of Physics: Conference Series **2329** (2022) 012003 doi:10.1088/1742-6596/2329/1/012003

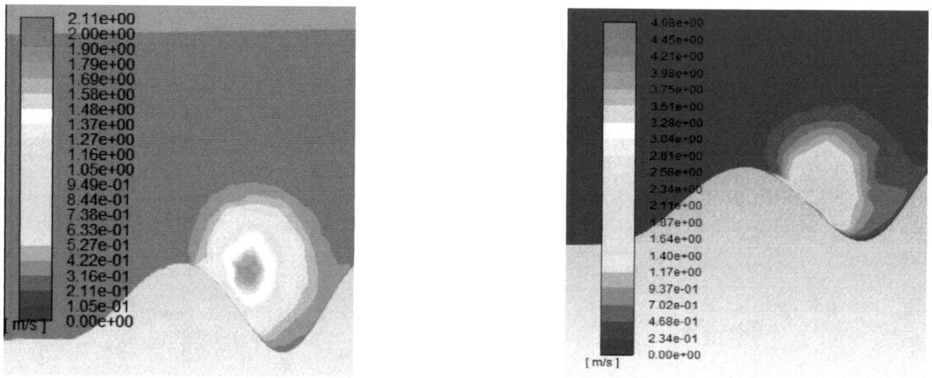


(a) 0 s (b) 0.25 s

**Figure 7.** Velocity distribution of the contact phase.

### *3.2. Spreading Stage*

Figure 8 depicts the spreading stage of droplets striking the tree trunk surface. Since the surface of the trunk is a complex and variable structure, it is composed of many curved surfaces. Under the action of gravity, fog droplets will move down the surface, forming a spread in the process. Due to the surface tension, viscous resistance, gravity and complex characteristics of the surface of the tree trunk, fog droplets change their original spherical shape and begin to move in the direction of gravity (below in the figure), resulting in asymmetric shapes of fog droplets. In the concave surface at the bottom of the trunk, the spreading process is completed. When the fog drops fall completely, with the disappearance of inertia force, there is a certain supporting force at the bottom to slow down the influence of gravity, and the fog drops reach the maximum spread area.

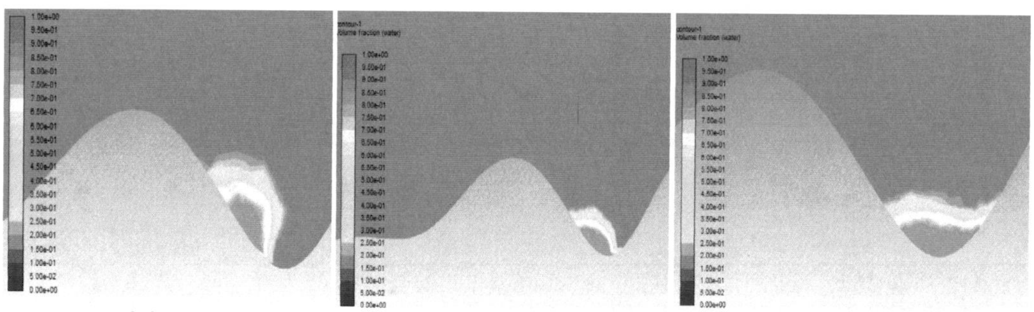

(a) 0.05 s (b) 0.25 s (c) 1.05 s

**Figure 8.** Liquid phase distribution in the spreading stage.

Figure 9 is the state diagram of pressure distribution in the process of spreading. At the beginning of the spread, the droplet shape began to change. Initially, the overall pressure distribution was uniform, and the pressure in contact with the tree surface was relatively high. As can be seen from the above liquid phase diagram, due to the special shape of the impact surface, air is mixed in the droplets and the contact surface during the impact of the droplets, and the pressure is different during the movement, and the pressure changes significantly when the droplets are squeezed inside. As the spreading process continues, the total pressure in the flow field increases, but due to the decrease of inertia force, the pressure inside the droplets is significantly less than that at the beginning of spreading stage. When spread to the largest area, the internal pressure of fog drops is closest to the pressure of flow field.

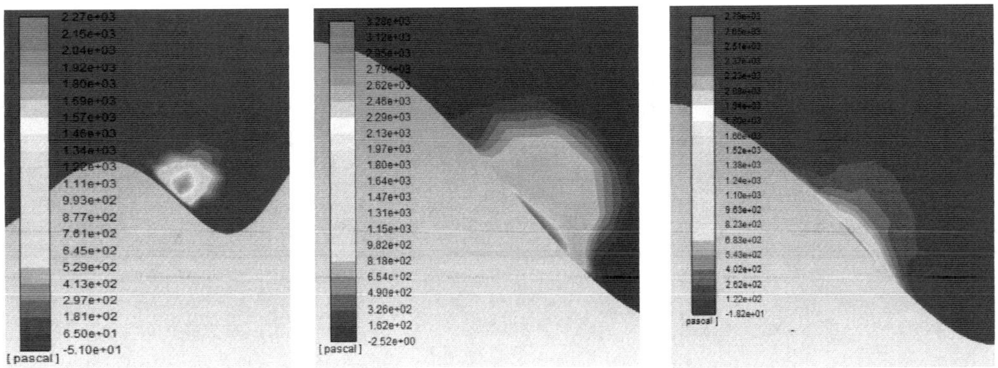

(a) 0.05 s (b) 0.25 s (c) 1.05 s

**Figure 9.** Pressure distribution map during the spreading stage.

Figure 10 shows the velocity cloud. As the spread begins, the droplets move to the right and their speed decreases. In the spreading process of fog droplets, due to the conservation of energy, the kinetic energy and potential energy of fog droplets are reduced and converted into surface energy, and viscous dissipation is caused along with the viscous force of fog droplets (the viscous dissipation energy comes from the surface free energy). When the droplet spreading speed decreases to the lowest value (approximately 0), the droplet reaches the maximum spreading.

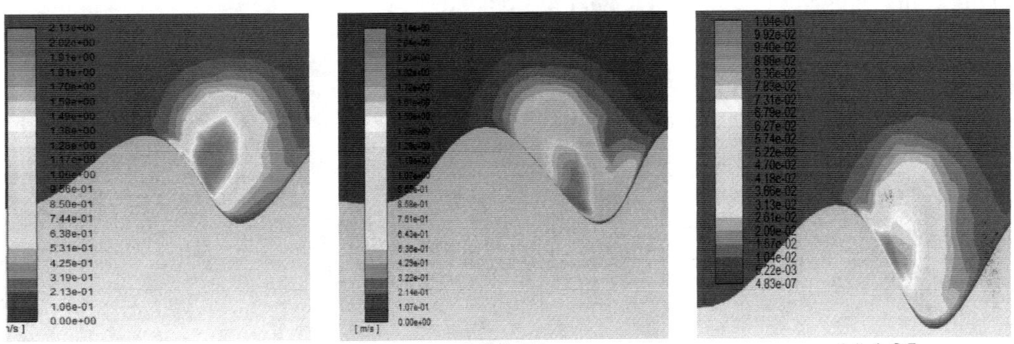

(a) 0.05 s (b) 0.25 s (c) 1.05 s

**Figure 10.** Speed distribution of the spreading stage.

### *3.3. Contraction Phase*

Figure 11 shows the cloud image of liquid phase when fog droplets shrink. When the droplet spreading speed decreases to the minimum value, the droplet reaches the maximum spreading area and begins to shrink under the action of surface free energy. As the surface tension of the solid and liquid phases is greater than that of the gas and liquid phases, the droplets contract to the center. Due to the influence of gravity and inertia force, the left side of the fog droplet shrinks faster than the right side, resulting in an asymmetric shape. Then the two forces reach the equilibrium state, the contraction rate on both sides is basically the same, and the fog drops return to the original ball shape. At this point, the contraction ends.

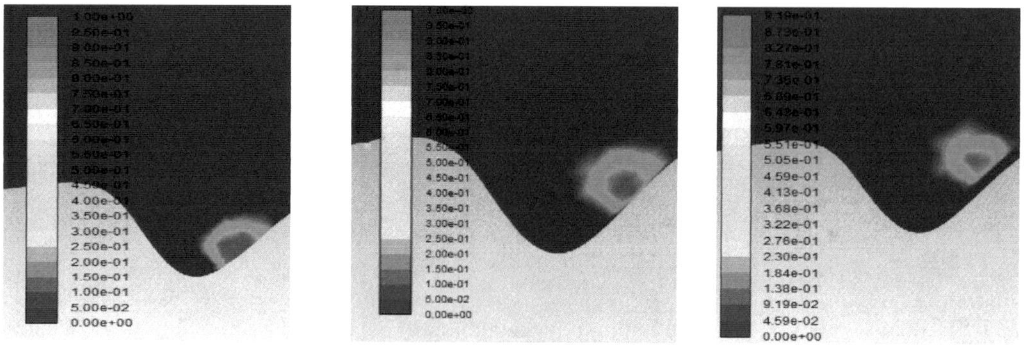

(a) 0.8 s (b) 1.25 s (c) 1.35 s

**Figure 11.** Liquid phase cloud diagram in the contraction phase.

Figure 12 is the cloud diagram of pressure distribution at the contraction stage. At the beginning of the contraction, the pressure inside the droplets was close to the pressure in the flow field, and the droplets tended to converge from the center. When the droplet shrinks by 1.25 s, the total pressure of the droplet increases. Because the droplet center is subjected to the double pressure of atmospheric pressure and contact surface, the inner central pressure is higher than the pressure on the upper and lower sides. When the shrinkage reaches the minimum, the internal pressure and surface tension of the fog drops reach the maximum, so that the internal kinetic energy of the fog drops reaches the maximum value and saves energy for the next movement.

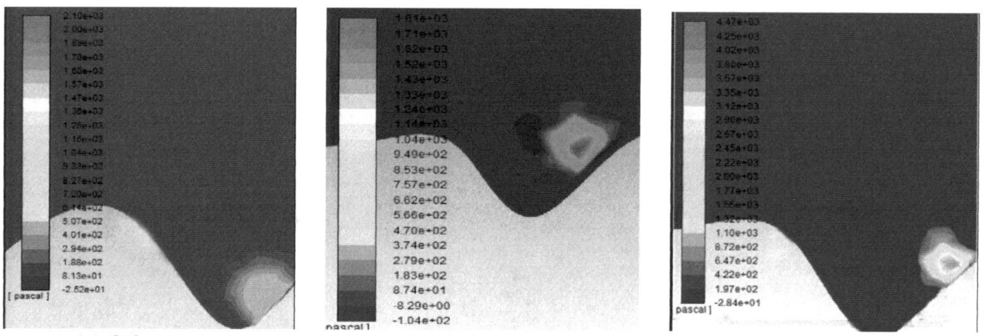

(a) 0.8 s (b) 1.25 s (c) 1.35 s

**Figure 12.** Pressure map of the contraction phase.

It can be seen from the velocity cloud diagram at the contraction stage (figure 13) that the fog droplet velocity is the smallest at the beginning of contraction. After some time, the shrinkage rate on the left side is significantly higher than that on the right side, due to surface tension and inertial forces. As the contraction progresses, the rate gradually increases, and the droplets gradually converge from both sides to the center. At 1.35s, the droplets became spherical and their velocity reached the maximum.

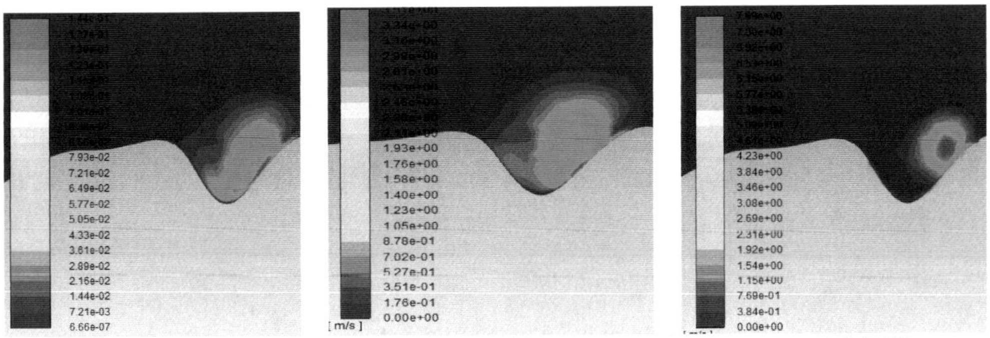


(a) 0.8 s (b) 1.25 s (c) 1.35 s

**Figure 13.** Shrinking stage speed cloud.

## 4. Conclusion

Based on the THREE-DIMENSIONAL model of willow trunk and Fluent modeling software, this paper established the VOF model of a single fog drop impacting on the trunk surface, and studied and analyzed the interaction of fog drop impacting on the trunk. According to the simulation experiment, make the fog drops hit the tree trunk surface at different heights and speeds, observe the liquid phase, pressure and velocity cloud of the three stages after the fog drops hit, and finally get the basic parameter range of the fog drops, the diameter of the fog drops is 137μm, the distance from the contact surface 5mm, the initial speed of 2.1m/s impact, ensure that the fog drops do not rebound. It can be deposited on the surface of tree trunk to maximize the utilization of liquid medicine.

## Reference

[1] Endalew A M, Debaer C and Rutten N 2010 Modelling pesticide flow and deposition from air-assisted orchard spraying in orchards: A new integrated CFD approach *Agricultural and Forest Meteorology* **150** 1-19.

[2] Friedli M, Kirchgessner N and Grieder C 2016 Terrestrial 3D laser scanning to track the increase in canopy height of both monocot and dicot crop species under field conditions *Plant Methods* **12** 9-10.

[3] Endalew A M, Hertog M and Delele M A 2009 CFD modelling and wind tunnel validation of airflow through plant canopies using 3D canopy architecture *International Journal of Heat and Fluid Flow* **30** 356-368.

[4] Byers B A, Savage V M and Enquist B J 2006 A general model for metabolic scaling in self-similar asymmetric networks *PLOS Computational Biology* **13** 177-179.

[5] Melese A, Hertog M and Delele M A 2007 3D measurement and representation of pear canopy for modelling air-assisted orchard spraying *Communications in Agricultural & Applied Biological Sciences* **72** 245-247.

[6] Liang T, Gang Y, Zheng P 2008 Two-phase Flow Coupling Simulation Based on VOF Model and Moving Grid Technology *Journal of Wuhan University of Technology* **30** 525-528.

[7] Hackenberg J, Morhart C, Sheppard J and Spiecker H 2014 Highly accurate tree models derived from terrestrial laser scan data a method description *Forests* **5** 1069-1105.

[8] Shlyakhter I, Rozenoer M and Dorsey J 2001 Reconstructing 3D tree models from instrumented photographs *Computer Graphics and Applications* **4** 53- 61

[9] Pesci A, Casula G and Boschi E 2011 Laser scanning the Garisenda and Asinelli towers in Bologna detailed deformation patterns of two ancient leaning buildings *Journal of Cultural Heritage* **12** 117-127.

# Modification and Optimization of Cylindrical Blade of Dredge Pump Based on CFD

**Haifei Zhuang*, Mingming Liu and Tao Guo**

CCCC National Engineering Research Center of Dredging Technology and Equipment CO., TD, room 2902, 1088 Yangshupu Road, Shanghai, China
Email: zhuanghaifei@cccc-drc.com

**Abstract.** The 3500 $m^3/h$ series cutter suction dredger pump adopts cylindrical blade impeller with low efficiency and serious wear of impeller channel; The twisted blade is optimized by conformal transformation method, and the blade wrap angle is increased by 75°; The efficiency of dredge pump with twisted blades is increased by 9% compared with cylindrical blades when the flow is 8000-12000 $m^3/h$; The relative velocity nephogram shows that the twisted blade eliminates the vortex in the impeller channel and improves the flow field; Through the erosion simulation analysis of solid-liquid two-phase flow, through the surface equalization treatment of the erosion wear of the impeller wall, it is found that the average wear rate of the twisted blades is 19% lower than that of the cylindrical blades, indicating that the wear resistance of the designed twisted blade impeller has been improved and the service life of the dredge pump has been increased.

**Keywords.** Dredge pump, twisted blade, CFD, worn.

## 1. Introduction

A large number of 3500 $m^3/h$ series cutter suction dredgers were built in China from 2006 to 2010, which greatly met the domestic dredging market demand at that time. This series of cutter suction dredgers applied the dredging technology of the 1990s. The impeller of the dredge pump adopts cylindrical blades, and the maximum test water efficiency is only 74%. With the development of dredging technology, the newly-built dredgers have adopted high-efficiency dredge pumps with twisted blades, with an efficiency of 86%, and the dredging efficiency of dredgers has been increased by more than 10%; The high-efficiency dredge pump has also been independently developed in China, which has been successfully applied to the customized transformation of several cutter suction and trailing suction dredgers, and achieved good results [1-3]; However, due to the time limit for a dredging, transformation cost and other factors, there are still a large number of dredgers that fail to carry out the transformation of the dredge pump, the relative efficiency in the dredging is low, and the impeller of the dredge pump is seriously worn. The dredge pump impeller is a vulnerable part. Replacing the newly developed dredge pump impeller in the form of spare parts will not affect the normal dredging of the dredger, but also improve the dredging efficiency of the dredger at a lower cost. How to optimize the impeller blades of the dredge pump, improve the efficiency of the dredge pump and reduce the impeller wear without changing the overall dimension of the impeller is of great significance to the energy conservation and efficiency increase of this series of cutter suction dredgers.

Many domestic universities and scientific research institutes have carried out numerical simulation and experimental research on the design of centrifugal pump blades: Zhuang Haifei et al. [4] has developed a special impeller with short conveying distance for the pump in the cutter suction dredger, which improves the dredging efficiency under the condition of short conveying distance; Li Longhua et al. [5] found that cavitation and wear of pump occurred at the same time, and the solid content had little effect on the cavitation performance, but with the increase of particle size, the pump head and efficiency decreased rapidly, and the anti cavitation performance decreased rapidly; Cao Weidong et al. [6] uses CFX simulation analysis, with the increase of blade outlet angle, the head increases gradually, and vortices appear in the impeller channel near the blade pressure surface; Zhao Wanyong et al. [7] numerically simulated and analyzed the influence of solid particles on blade inlet wear. In a certain range, reducing the inlet placement angle of flow line in the middle of the impeller can improve the wear resistance of the impeller; Based on numerical simulation and external characteristic test, Wang Yanyan et al. [8] improved the design of impeller by optimizing blade wrap angle and outlet placement angle. After improvement, the head was increased by 2.7 m and the efficiency was increased by 5.8%; Li Jing et al. [9] studies the influence of blade profile on the hydraulic performance and impeller wear characteristics of slurry pump. The blade with small wrap angle leads to the decline of pump hydraulic performance, but the anti-wear performance of impeller is improved. Zhuang Haifei et al. [10] modified the blade inlet of impeller to improve the wear resistance of dredge pump under coarse sand condition.

The blade of dredge pump is modified into twisted blade by changing the wrap angle, inlet and outlet placement angle of blade, without changing the overall dimension of the impeller, optimize the parameters of the impeller and improve the efficiency and the wear properties of the dredge pump.

## 2. Twist Modification of Cylindrical Blade

The common flow range of the dredge pump of 3500 $m^3$/h series cutter suction dredger is 8000-12000 $m^3$/h, the rated speed of the dredge pump is 257 rpm, the head before modification is 72-76 m, the main dimensional parameters of the dredge pump are: inlet diameter 900 mm, outlet diameter 850 mm, impeller outer diameter 2500 mm and impeller outlet width 435 mm; The modification and optimization of the impeller blade of the dredge pump is carried out, and the conformal transformation method is used to design the three-dimensional twisted blade, which is more in line with the fluid law [11]; The boundary dimension of impeller remains unchanged before and after modification, and the number of blades remains unchanged, all of which are 5 blades; Compared with that before the modification as shown in figure 1 and figure 2, the wrap angle of the blade working surface increases by 75°, and the warp angle of the intersection between the blade working surface and the front cover plate increases by 70°.

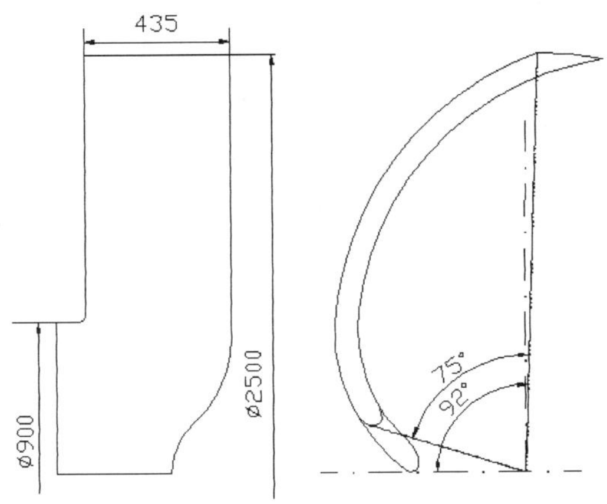


**Figure 1.** Axial surface of impeller channel and cylindrical blade before modification (unit: mm).

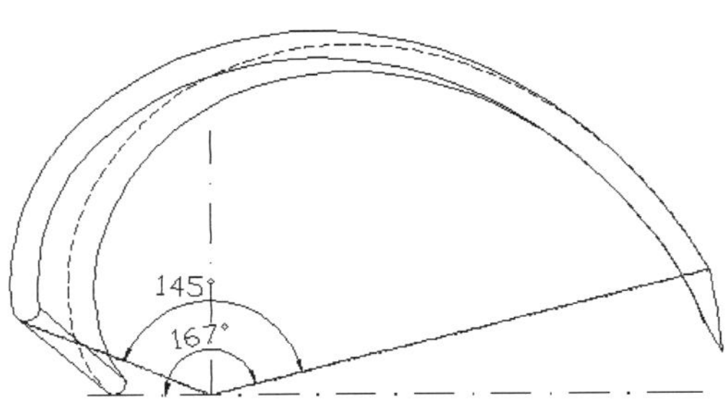


**Figure 2.** Twisted blade after modification.

Due to the spatial overlap of the inlet part of the twisted blade, the passing capacity of the impeller channel is reduced. The maximum passing ball diameter of the cylindrical blade impeller before the modification is 330 mm, and that of the twisted blade impeller after the modification is 300 mm. Although the ball diameter of the modified impeller is relatively reduced by 9%, the impeller passing capacity still meets the dredging requirements of the dredger.

## 3. Clean Water Performance Prediction

The water performance of the dredge pump with cylindrical blades before modification and twisted blades after modification is numerically simulated and analyzed, and the external characteristics and relative velocity field are predicted and analyzed with ANSYS 17.2 software. The boundary conditions of numerically simulation are set as inlet speed and outlet free flow. Calculate three flow points of 8000 $m^3/h$, 10000 $m^3/h$ and 12000 $m^3/h$ respectively to obtain the comparison of external characteristic parameters of head and efficiency, as shown in table 1. After the blade is modified into twisted blade, the head is increased by 3.4-4.0 m and the efficiency is increased by 8.7-9.2 percent points; After the blade is modified into twisted blade, the simulated water efficiency reaches 82%.

**Table 1**. Comparison of performance data before and after blade modification.

| | Cylindrical blade | | Twisted blade | | Contrast | |
|---|---|---|---|---|---|---|
| Flow<br>$m^3/h$ | Head<br>m | Efficiency<br>% | Head<br>m | Efficiency<br>% | Head difference<br>m | Efficiency difference<br>% |
| 8000 | 75.7 | 65.3 | 79.1 | 74.0 | 3.4 | 8.7 |
| 10000 | 74.4 | 70.3 | 77.7 | 79.5 | 3.3 | 9.2 |
| 12000 | 72.2 | 73.1 | 76.2 | 82.2 | 4.0 | 9.1 |

Figure 3 shows the relative velocity field of the impeller channel: there are obvious vortices in the channel of the cylindrical blade impeller before modification, and the vortex area is large, which almost fills the channel between the two adjacent blades of the impeller. There are no vortices in the channel of the twisted blade impeller after modification, and the relative velocities are all along the direction of the blades, indicating that the modified twisted blade channel is more in line with the fluid law, It is beneficial to improve the hydraulic efficiency of dredge pump and reduce the wear of impeller channel wall.

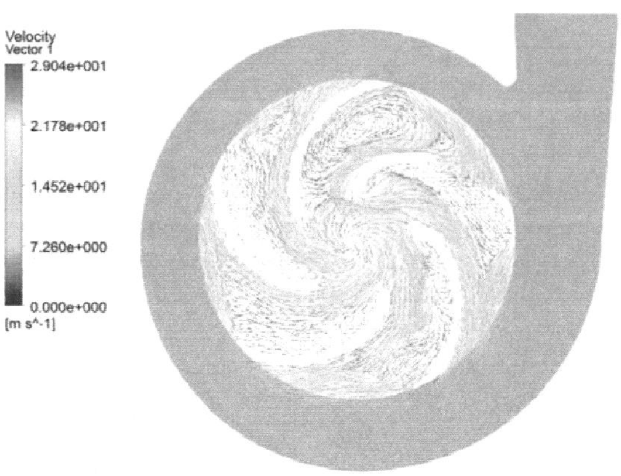


**Figure 3. (a)** Cloud diagram of relative velocity of cylindrical blade impeller channel.

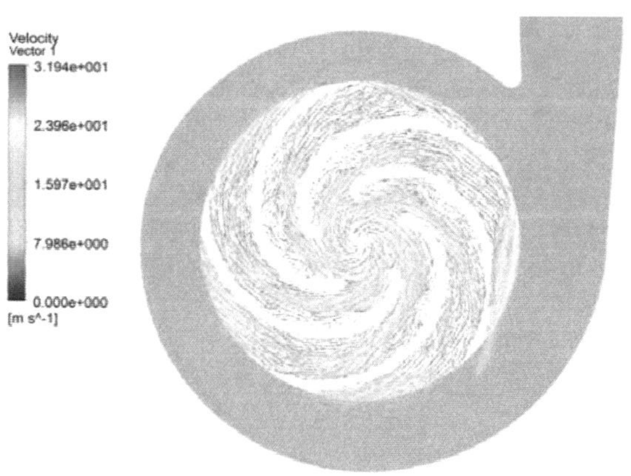


**Figure 3. (b)** Cloud diagram of relative velocity of twisted blade impeller channel.

## 4. CFD Simulation Analysis and Comparison

### *4.1. Numerical Calculation Method and Verification*

The fluid model Euler-Euler and the turbulence model SST K-ω is used to solve the two-phase flow in the pump transiently. Taking the impeller rotation of 1.5° as a time step, a total of 5 cycles are calculated, and the data of the last calculation cycle is used as the basis for analysis. The boundary conditions of the dredge pump are velocity inlet and free outlet. It is assumed that the inlet velocity has no tangential and radial components and the outlet flow is fully developed; The wall surface is non slip wall surface and treated by standard wall function; The second-order upwind scheme is discretized in space to solve the turbulence equation; Based on the two-phase flow calculation results, the discrete phase model (DPM) model is introduced to track the particle trajectory in a steady state and calculate the wall erosion rate. During the construction of the mud pump, the particle concentration is generally about 20%. Due to the non-uniformity of flow, the particle concentration in the local flow field is higher. Therefore, the Huilin-Gidaspow model with smooth transition from Wen-Yu model with low particle concentration to Ergun model with high particle concentration is adopted for the drag calculation of two phases from low concentration to high concentration. In order to improve the calculation efficiency, the influence of interphase forces such as virtual mass force and lift force is ignored in the calculation.

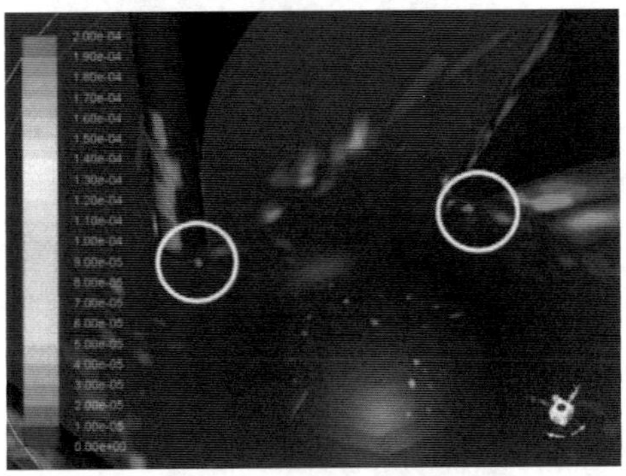

**Figure 4. (a)** Simulates the wear position of the blade inlet near the low pressure surface of the rear cover plate.

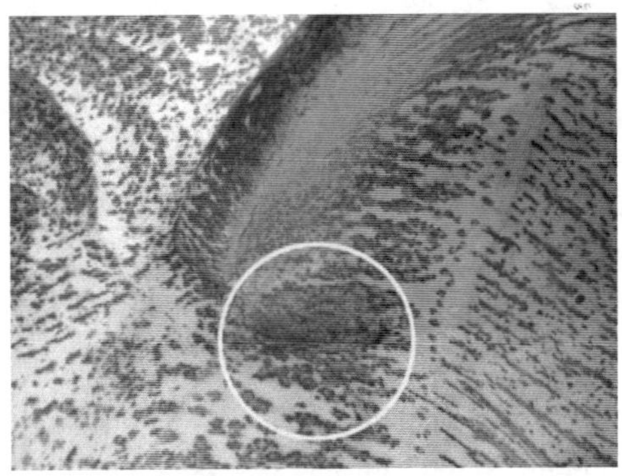

**Figure 4. (b)** Wear position of the actual blade inlet close to the low pressure surface of the rear cover plate.

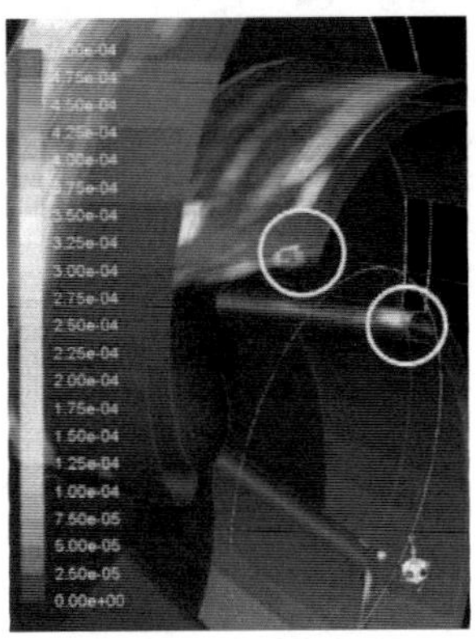

**Figure 5. (a)** Simulate the damaged position of the impeller inlet near the blade root of the front cover plate.

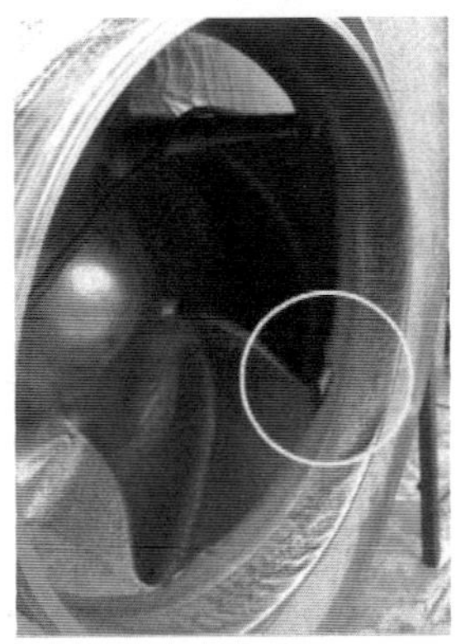

**Figure 5. (b)** Damage position of actual impeller inlet close to blade root of front cover plate.

Figure 4 and figure 5 show the comparison between the actual wear of the cylindrical blade impeller of the dredge pump and the simulated erosion wear of the impeller. It can be seen from figures that through the comparison between the CFD predicted wear and the actual pictures, the prediction of most positions is accurate, which shows that the erosion simulation calculation method is feasible. Among them, the cylindrical impeller wear mainly occurs at the root of the blade inlet and near the suction surface of the blade inlet.

### *4.2. Erosion Wear Simulation of Twisted Blade of Mud Pump*

Based on the same calculation conditions, the comparative analysis of erosion wear of impeller channel is carried out. The flow of dredge pump is 12000 $m^3/h$, the rotating speed is 257 rpm, the particle size is 0.3 mm, the particle concentration is 15%, and the particle density is 2700 $kg/m^3$.

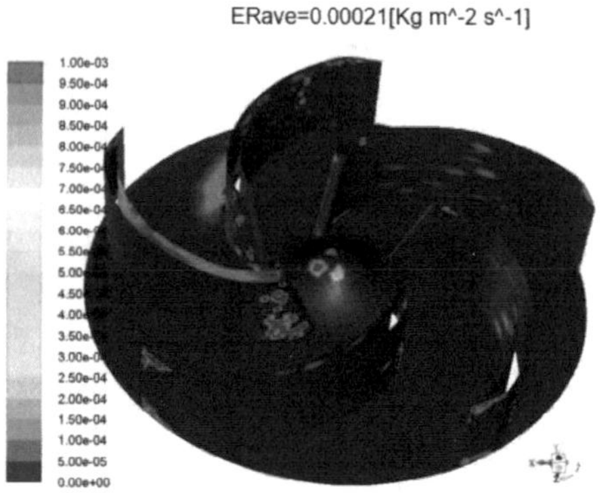


**Figure 6. (a)** Erosion wear rate - Cylindrical Blade.

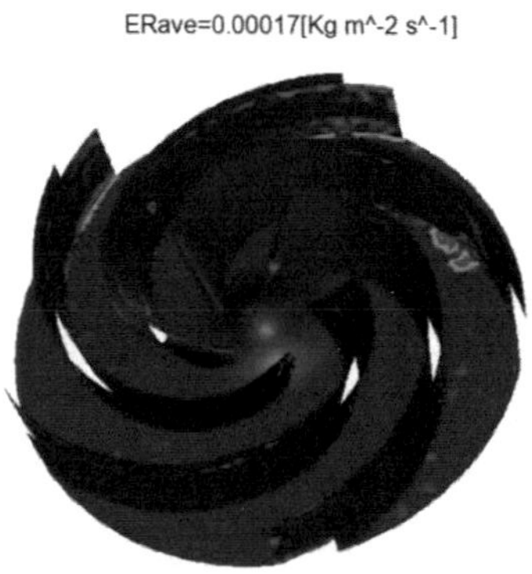


**Figure 6. (b)** Erosion wear rate - twisted blade.

Figure 6 shows the comparison of erosion wear rates of cylindrical blade and twisted blade. The simulation results show that compared with cylindrical blade, the blade suction port, blade wall and wear phenomenon of twisted blade impeller are significantly reduced, the wear of impeller wall is more uniform, and the wear is mainly concentrated at the intersection of suction surface near blade outlet and rear cover plate; Through the surface equalization treatment of the erosion wear on the impeller wall, it is found that the average wear rate of the twisted impeller wall is 19% lower than that of the cylindrical blade impeller, indicating that the wear resistance of the designed twisted blade impeller has been improved and the service life of the mud pump has been improved.

## 5. Conclusion

- The cylindrical blade is modified and optimized into the twisted blade of the dredge pump impeller. The blade wrap angle is increased by 75° and the carrying capacity is reduced by 9% after modification;
- When the flow is 8000-12000 $m^3/h$, the efficiency of the dredge pump with twisted blade is 9 percent points higher than that of cylindrical blade, and the maximum water efficiency reaches 82%; The relative velocity shows that the twisted blade eliminates the vortex in the impeller channel and improves the flow field;
- Erosion simulation shows that the average wall wear rate of twisted blade is 19% lower than that of cylindrical blade, and the wear resistance of impeller is improved.

## References

[1] Zhuang H F, Wang W K, Zheng L Z, Yang C and Hu J Z 2015 Optimization design of dredge pump based on CFD and model experiment *China Rural Water and Hydropower* (10) 130-132

[2] Zhuang H F, Yang C and Hu J Z 2017 Research and application of dredge pump for 4500m$^3$ trailing suction hopper dredger *China Harbour Engineering* (1) 63-66

[3] Liu M M, Zhuang H F, Zheng L Z and Wang W K 2018 Development and application of high efficient and high wear-resisting dredge pump for 3500m$^3$/h cutter suction dredger *Science & Technology of Ports* (2) 15-22

[4] Zhuang H F, Liu M M and Jin B X 2020 Research and application of centrifugal pump impeller modification for cutter-suction dredger *Fluid Machinery* **48**(4) 42-45

[5] Li L H, Miao Y and Gong Y F 2006 Approach to factors influencing the cavitation in centrifugal slurry pumps *Pump Technology* (6) 8-11

[6] Cao W D, Zhang Y N and Yao L J 2017 Influence of outlet blade angel on performance of centrifugal pump and correction of slip factor *Journal of Drainage and Irrigation Machinery Engineering* **35**(9) 755-760

[7] Zhao W Y, Guo Q, Song Q B and Hu X N 2018 Effects of blade entrance geometrical parameter on centrifugal pump wear rule *Journal of Gansu Sciences* **30**(4) 126-131

[8] Wang Y Y, Zhao W G, Han X D and Zheng Y J 2019 Improved centrifugal pump impeller design in terms of blade wrap and exit angles *Journal of Drainage and Irrigation Machinery Engineering* **37**(7) 574-579.

[9] Li J, Zhang R H, Guo R and Li R N 2020 Influence of blade camber profile on hydraulic performance of slurry pump and impeller wear characteristic *Journal of Drainage and Irrigation Machinery Engineering* **38**(1) 21-27.

[10] Zhuang H F, Liu M M and Wu Y D 2021 Optimization and application of D850 dredge pump blades under medium coarse sand condition of cutter suction dredger *2021 International Conference on Fluid and Chemical Engineering (ICFCE 2021)* **2097**(2021) 012005

[11] Guan X F 2011 *Modern Pumps Theory and Design* (Beijing: China Aerospace Press) chapter 11 p 387–388

# Numerical Investigation on the Effect of the Diversion Layer on the Flow Field of the α-Type Cyclone Separator

**J L Fan[1], W H Li[1], P Zhang[1] and G H Chen[2*]**

[1] School of Electromechanical Engineering, Qingdao University of Science & Technology, Qingdao 266042, Shandong, China
[2] School of Chemical Engineering, Qingdao University of Science & Technology, Qingdao 266042, Shandong, China
Email: guanghui@qust.edu.cn

**Abstract.** The flow field stability and separation performance of the cyclone separator decrease significantly as its size increases. An inner guider was added in α-type cyclone separator, and in this paper, a numerical investigation on the gas-solid two-phase flow in the α-type cyclone was conducted. The Reynolds stress model and the DPM model were used to simulate the gas flow and the motion of particles, respectively. The influence of the guider on the velocity field, pressure field, turbulent kinetic energy field, and particle separation efficiency of the cyclone separator was analyzed. The results showed that after adding an inner guider, the tangential velocity increases, the turbulent kinetic energy decreases, the downward area of the axial airflow increases, and the number of airflow-carrying particles downward increases.
**Keywords.** Cyclone separator, inner guider, numerical study.

## 1. Introduction

As a kind of dry gas-solid separation equipment, the cyclone separator is widely used in petroleum, chemical, coal, metallurgy, machinery, textile, and other industries due to its simple structure, reasonable price, and convenient maintenance. It is mainly used to recover valuable substances, obtain clean gas, and purify gas to protect the environment [1]. The flow field in the cyclone separator is very complicated, and there are a lot of vortices. These vortices will affect the separation efficiency and pressure loss to varying degrees [2]. As the size of the cyclone separator increases, its separation efficiency will decrease significantly, which is commonly referred to as the size effect [3].

Some researchers have studied the size effect of the cyclone separator. Halil et al. [4] believed that the correlation analysis results of small-diameter cyclone separators were not suitable for large-diameter's one by comparing the collection efficiencies of two Stairmand cyclones with 254mm and 700 mm diameters. Faulkner et al. [5] also compared the collection efficiency of cyclones with cylindrical diameters of 6, 12, 24, and 36 inches and found that the collection efficiency decreased nonlinearly as the diameter increased. Ali Sakin et al. [6] conducted a numerical analysis on a new countercurrent tangential inlet cyclone with different separation space diameters under different flow conditions. They found that the maximum axial velocity and tangential velocity, and the change of turbulence intensity were inversely proportional to the transformation of the separation space diameter. L.S. Brar et al. [7] compared the pressure drop and collection efficiency of the cyclone separator under three different parameters that are 0.8, 1, and 1.2 times the standard diameter. Authors found that increasing the diameter can improve collection efficiency but at the expense of increased pressure drop.

Safikhani et al. [8] used numerical and experimental simulations to explore the flow field information under the three standard separator forms (1D3D, 2D2D, and 1D2D) of cyclones with two separation space diameters of 200 mm and 900 mm. The results show that the cyclone separator's pressure drop, speed, and turbulence intensity are almost independent of the diameter of the cyclone (the error is less than 11%).

Based on the volute cyclone separator, Jin et al. [9, 10] studied the effect of the expansion of the separation space diameter on the separation efficiency under geometrically similar conditions. They proposed and optimized the calculation expression of the separation efficiency. Jin et al. found through numerical simulation that only by increasing the diameter of the cylinder, the pressure drop of the long cone cyclone was almost unchanged, and the separation efficiency was improved [11]. Yuan et al. [12] studied the effect of diameter on the pressure drop of the Stairmand cyclone separator. Studies have shown that at the same inlet gas velocity, the larger the diameter of the cyclone separator, the more significant the pressure drop. Yuan et al. conducted experiments on Stairmand cyclones with different cylinder diameters. The results showed that after the cyclone separator was enlarged according to the similar geometric structure when the inlet gas velocity is the same, the increase of the diameter of the cylinder led to the decrease of the particle separation efficiency and the rise in the pressure drop. When the volume of the processed gas was the same, the increase in diameter led to reduced particle separation efficiency and pressure drop [13]. Based on the fundamental equation of dust particle movement, Shao [14] conducted a theoretical study on the amplification effect of the cyclone separator and proposed a dimensionless number and a dimensionless time to express the separation effect. Song et al. [15] investigated the gas flow field characteristics of the geometrically similar cyclone when the diameter size changed. They found that the flow field would enter the self-mode region when the diameter was large enough, and the flow field and flow state would remain similar.

Usually, small-diameter cyclones are combined in parallel for particle separation to avoid the amplification effect of cyclones. Studies have shown that when two or more cyclones are arranged side by side, uneven gas-solid flow distribution will occur, reducing their separation efficiency [16]. Liu et al. [17] compared the gas flow field of two and four separators in parallel. Authors found that the flow rate and pressure drop of each unit have slight deviations. The four cyclones are arranged in a symmetrical manner to make the flow field more stable. Liao et al. [18] conducted simulation and experimental research on the parallel arrangement of six cyclones. They found that the pressure drop and particle flow rate of the separator at different locations are different. The particle flow rate and separation efficiency of the two side separators are higher than those of the middle separator. Zhou et al. [19] studied the uneven distribution of gas-solid two-phase flow in six cold model cyclones in parallel. They found that the solid volume fraction, flux, and circulation rate of the intermediate cyclone were the lowest on the same side and believed that the uneven distribution of the apparent solid flow rate and volume fraction at the outlet of the riser may be the initial cause of the instability of the flow between the cyclones. Chen et al. [20] designed three separator layout schemes: the parallel design of the same cyclone separator, the cyclone separator with different internal gas rotation directions, and the cyclone separator with different core tube lengths, and performed simulations and experiments. The results show that the influence of the structure size and rotation direction on the working efficiency of the parallel separator decreases in decreasing importance. Zhong [21] studied the effect of the number of microcyclone separators with a diameter of 30mm on the overall separation performance. The results show that when the inlet velocity of each microcyclone separator is the same, as the number of microcyclone separators increases, the separation efficiency of particles smaller than 3 μm decreases. The gradual decrease in the distance between the feed legs leads to intensified air turbulence near the dust outlet, which makes the stability of the flow field worse, increases the probability of "particle backmixing", and makes the separation effect worse.

In summary, the cyclone separator has obvious size effect. The improved method of connecting multiple small cyclones in parallel also has the problem of uneven intake air distribution. Therefore, based on the principle of shortening the radial movement distance of particles [22], we added a guider in the α-type cyclone separator [23] to improve its separation efficiency. Numerical simulation analyzes its influence on the flow field inside the separator and the influence of particle movement.

## 2. Simulation Set-up

### *2.1. Geometry Model*

We have designed a α-type cyclone separator with a diameter of 320 mm, referred as Type-A. An inner guider is installed to improve its separation efficiency, referred as Type-B. The upper part of the inner guider is welded with the spiral roof, and the bottom is parallel to the bottom end of the vortex finder. figure 1 shows the structure comparison diagram of the two models. Table 1 listed the main structure dimensions.

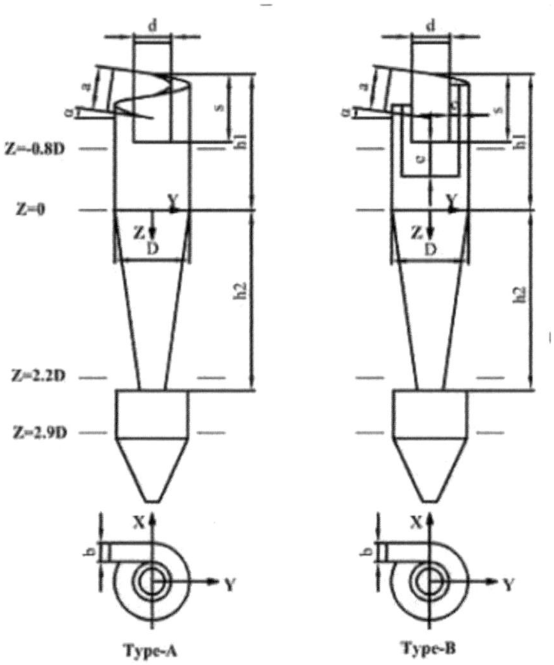


**Figure 1.** Schematic diagram of cyclone separator structure.

**Table 1.** Geometric dimensions of cyclone separator.

| D/mm | Q/m³·h$^{-1}$ | a/mm | b/mm | d/D | h1/D | h2/D | c/mm | e/mm |
|---|---|---|---|---|---|---|---|---|
| 320 | 970 | 160 | 80 | 0.5 | 1.5 | 2 | 37 | 142 |

### *2.2. Governing Equations*

(1) Continuity equation:

$$\frac{\partial \rho}{\partial t}+\frac{\partial(\rho u_i)}{\partial x_i}=0 \tag{1}$$

Where $\rho$ is the fluid density; $u$ is fluid velocity.

(2) Momentum equation:

$$\rho\left(\frac{\partial u}{\partial t}+u\frac{\partial u}{\partial x}+v\frac{\partial u}{\partial y}+w\frac{\partial u}{\partial z}\right)=\rho f_x-\frac{\partial p}{\partial x}+\mu\left(\frac{\partial^2 u}{\partial x^2}+\frac{\partial^2 u}{\partial y^2}+\frac{\partial^2 u}{\partial z^2}\right) \tag{2}$$

$$\rho\left(\frac{\partial v}{\partial t}+u\frac{\partial v}{\partial x}+v\frac{\partial v}{\partial y}+w\frac{\partial v}{\partial z}\right)=\rho f_y-\frac{\partial p}{\partial y}+\mu\left(\frac{\partial^2 v}{\partial x^2}+\frac{\partial^2 v}{\partial y^2}+\frac{\partial^2 v}{\partial z^2}\right) \quad (3)$$

$$\rho\left(\frac{\partial w}{\partial t}+u\frac{\partial w}{\partial x}+v\frac{\partial w}{\partial y}+w\frac{\partial w}{\partial z}\right)=\rho f_z-\frac{\partial p}{\partial z}+\mu\left(\frac{\partial^2 w}{\partial x^2}+\frac{\partial^2 w}{\partial y^2}+\frac{\partial^2 w}{\partial z^2}\right) \quad (4)$$

Where $\rho$ is the fluid density; $u$ , $v$ , $w$ are the velocity components of the fluid at (x, y, z) at time t; $p$ is the pressure; $f$ is the external force per unit volume of the fluid; the constant $\mu$ is the dynamic viscosity.

(3) The turbulence model

The commonly used turbulence models are the standard $k$ - $\varepsilon$ model, RNG $k$ - $\varepsilon$ model, Reynolds stress model (RSM), and large eddy model (LES) in the literature. The Reynolds stress model (RSM) does not adopt the homogeneity assumption of vortex viscosity but directly solves the Reynolds stress term in the N-S equation and simultaneously solves the dissipation rate equation. Compared with the two-eqations turbulence models, the Reynolds Stress Model (RSM) provides better prediction in the region dominated by convex curvature such as the cyclone separator. Furthermore, it needs less computational resource than the Large Eddy Model (LES) [24]. The turbulent kinetic energy transport equation and dissipation rate transport equation of the RSM model are:

$$\frac{\partial(\rho k)}{\partial t}+\frac{\partial(\rho k\mu_i)}{\partial x_i}=\frac{\partial}{\partial x_j}\left[\left(\mu+\frac{\mu_t}{\sigma_k}\right)\frac{\partial k}{\partial x_j}\right]+\frac{1}{2}P_{ij}-\rho\varepsilon \quad (5)$$

$$\frac{\partial(\rho\varepsilon)}{\partial t}+\frac{\partial(\rho\varepsilon u_i)}{\partial x_i}=\frac{\partial}{\partial x_j}\left[\left(\mu+\frac{\mu_t}{\sigma_\varepsilon}\right)\frac{\partial\varepsilon}{\partial x_j}\right]+C_{1\varepsilon}\frac{1}{2}P_{ij}\frac{\varepsilon}{k}-C_{2\varepsilon}\rho\frac{\varepsilon^2}{k} \quad (6)$$

$P_{ij}$ is the stress generation term, and its expression is:

$$P_{ij}=-\rho\left(\overline{u_i'u_k'}\frac{\partial u_j}{\partial x_k}+\overline{u_j'u_k'}\frac{\partial u_i}{\partial x_k}\right) \quad (7)$$

$\sigma_k=0.82$, $\sigma_\varepsilon=1.0$, $C_{1\varepsilon}=1.44$, $C_{2\varepsilon}=1.92$

(4) Particle motion governing equations

Particle motion equation is expressed as:

$$m_p\frac{dV_P}{dt}=F_D+F_G+F_B+F_A+F_C+F_M+F_S \quad (8)$$

where $F_D$ represents drag force, $F_G$ represents gravity, $F_B$ represents buoyancy, $F_A$ represents additional mass force, $F_C$ represents the force generated by the collision between particles and particles and the wall surface, $F_M$ represents Magnus force, which is the force generated by the particle due to the rotational motion, $F_S$ represents Saffman force, which is generated by the particle due to the speed gradient.

Due to the small particle size and low concentration of the particles in the cyclone, only drag and gravity forces are considered. Other forces are much smaller than drag and gravity in order of magnitude and can be ignored.

Journal of Physics: Conference Series **2329** (2022) 012005 doi:10.1088/1742-6596/2329/1/012005

### *2.3. Numerical Method*

The discrete format was the QUICK format, and the pressure interpolation format was PRESTO! Format. The SIMPLE algorithm was utilized to solve the pressure-velocity coupling in the momentum equations. Air was used as the gas phase. A uniform inlet velocity condition was used at the inlet, and the inlet gas velocity was 20 m/s. The outflow condition was used at the exhaust. The standard wall function was applied at the wall surfaces, and the roughness was 0.5. For the particle phase, the escape boundary condition was used at the exhaust. The trap boundary condition and reflect condition were applied at the ash hopper wall and the other walls, respectively. The particle collision recovery coefficient was set to 0.5.

### *2.4. Grid Independence Verification*

Figure 2 shows this paper conducts numerical simulations on the α-type cyclone separator with a diameter of 320mm with grid numbers of 0.91million, 1.52 million and 1.94 million, respectively. Considering the grid quality and calculation time, this paper uses a grid with a grid number of 0.91 million to perform numerical simulation calculations for the α-type cyclone separator with a diameter of 320 mm.

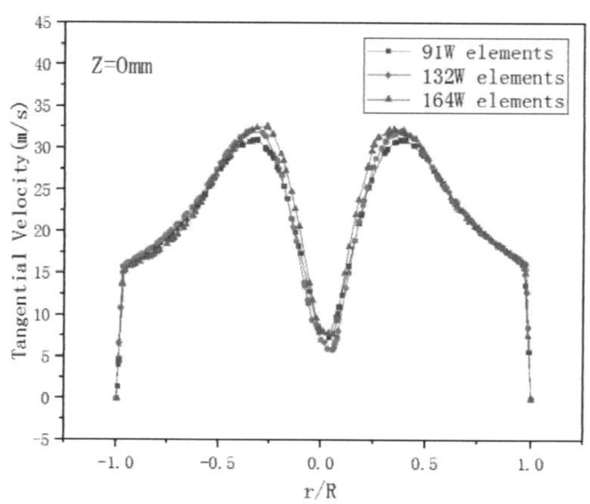


**Figure 2.** Tangential velocity profiles of α-type cyclone separator with different mesh numbers (Z=0mm, D=320mm).

## 3. Results

### *3.1. Velocity Field*

*3.1.1. Tangential Velocity.* Figure 3 shows the cloud diagram of the tangential velocity distribution of the Type-A and Type-B in the Y=0 section (longitudinal cross-section). as shown, the inner guider has little effect on the precession of the vortex core. However, the influence range of the secondary vortex at the lower end of the vortex finder is reduced. The secondary vortex scale means the length of the secondary vortex in the radial direction of the cyclone separator. The secondary vortex scale of Type-A is about 0.25, and the secondary vortex scale of Type-B is about 0.13, but the position does not change significantly.

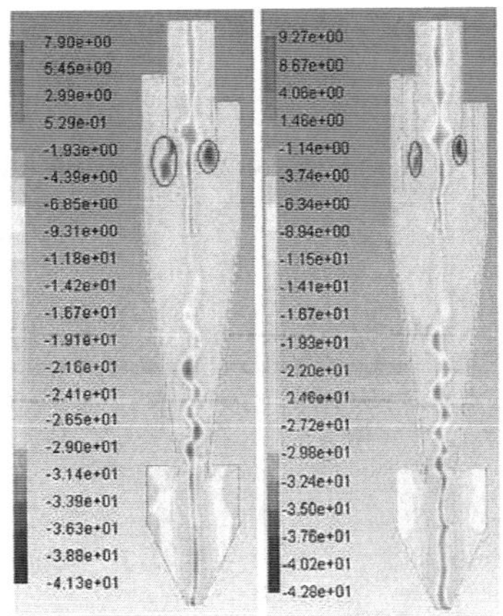


(a)Type-A (b)Type-B

**Figure 3.** The tangential velocity distribution of the two cyclone separators at section Y=0.

Figure 4 shows the contour plots of the tangential velocity at the Y=0 section (longitudinal section) of the A and B cylinders. It can be seen from figure 4(a) that the velocity contours near the Type-A's end of the vortex finder are denser, the velocity gradient is larger. The maximum tangential velocity appears at the lower end of the vortex finder, and the secondary vortex has an extensive range of influence, which may causes the separation efficiency to drop. It can be seen from figure 4(b) that a diversion channel is formed inside Type-B. The velocity contours in the diversion channel are relatively sparse and evenly distributed, and the fluid flow is relatively regular.

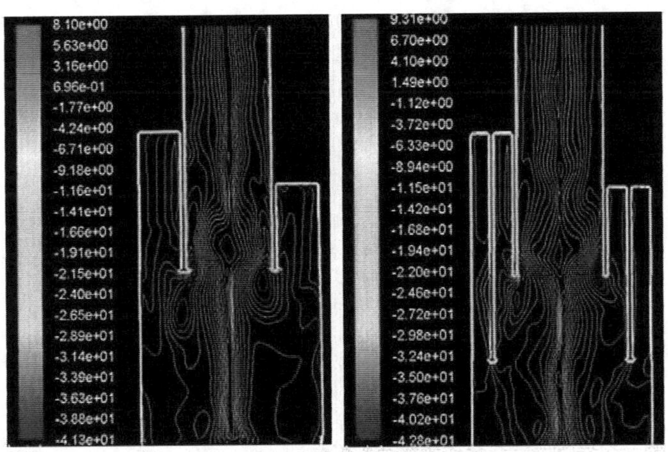


(a)Type-A (b)Type-B

**Figure 4.** Velocity contour maps of the barrel parts of two cyclone separators at section Y=0.

Figure 5 compares the velocity vectors on the Z=-0.8D section in Type-A and Type-B. The velocity vector distribution in Type-A is relatively irregular. The closer to the central axis, the more chaotic the fluid flow. This phenomenon is caused by the mutual interference of swirling flows in the separation zone. As shown in figure 5(b), the velocity vector distribution of Type-B is more regular and orderly than type A. The inner deflector divides the flow field into multiple flow channels between the inner deflector and the cylindrical section wall and between the inner deflector and the riser. This reduces the non-interference between the flows in the guide channels, which is conducive to the improvement of particle separation efficiency.

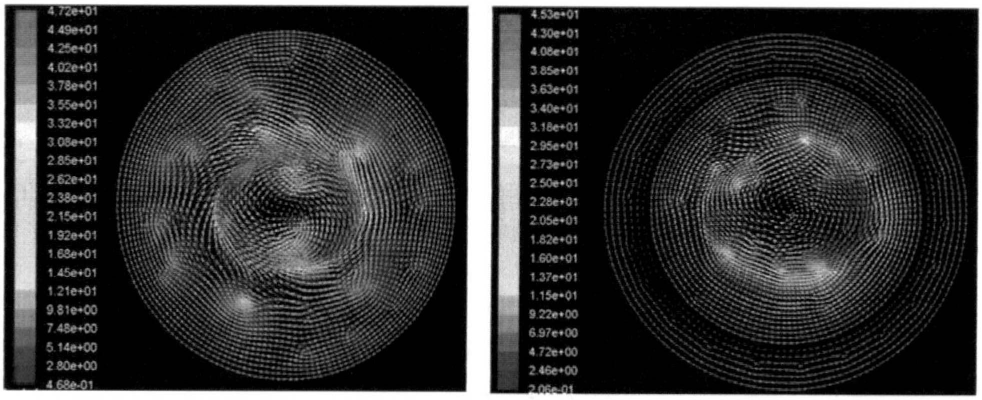


(a)Type-A (b)Type-B

**Figure 5.** The velocity vector diagram of the cyclone separator at section Z=-0.8D.

Figure 6(a) compares the tangential velocity distribution along the diameter at Z=-0.8D in Type-A and Type-B. It can be seen from the figure that the tangential velocity at the boundary of the A-type inner and outer swirling flow is 33.81 m/s, and the relative position of the radial maximum tangential velocity r/R is 0.30; the B-type outer swirling flow tangential velocity is higher than that of A-Type, the maximum tangential speed is increased to 34.96 m/s. The increase in tangential velocity increases the centrifugal force obtained by the particles, making them easier to be thrown towards the wall. In the diversion channel, the tangential velocity distribution also presents similar characteristics to the Rankine combined vortex. As the radius decreases, the tangential velocity first increases, then decreases, and finally becomes zero.

Figure 6(b) further compares the tangential velocity distribution along the diameter at Z=-2.2D in Type-A and Type-B. It can be seen from the figure that the additional inner deflector has little effect, and the vortex core tail swing phenomenon still exists. Nevertheless, the inner guide makes the tangential velocity change range smaller. The maximum tangential velocity is reduced from 23.56 m/s to 21.58 m/s, which helps to reduce the entrapment of separated particles, thereby improving the separation efficiency.

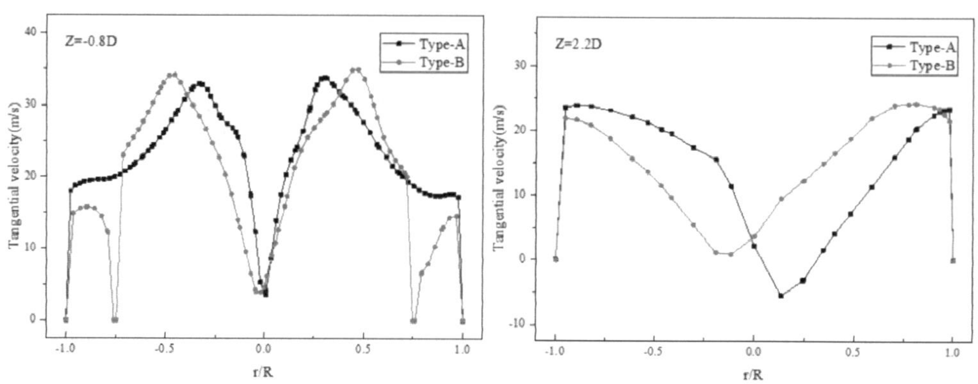


(a) (b)

**Figure 6.** The tangential velocity distribution of the two cyclone separators at section Z=-0.8D and Z=2.2D.

*3.1.2. Axial Velocity.* Figure 7 (a) compares the axial velocity distribution along the diameter axis at Z=-0.8D between Type-A and Type-B. In Type-A, there is an upward axial velocity near the wall, which will cause the backmixing of particles, thereby reducing separation efficiency. The inner guide avoids the above phenomenon in Type B and also makes the axial velocity distribution more symmetrical.

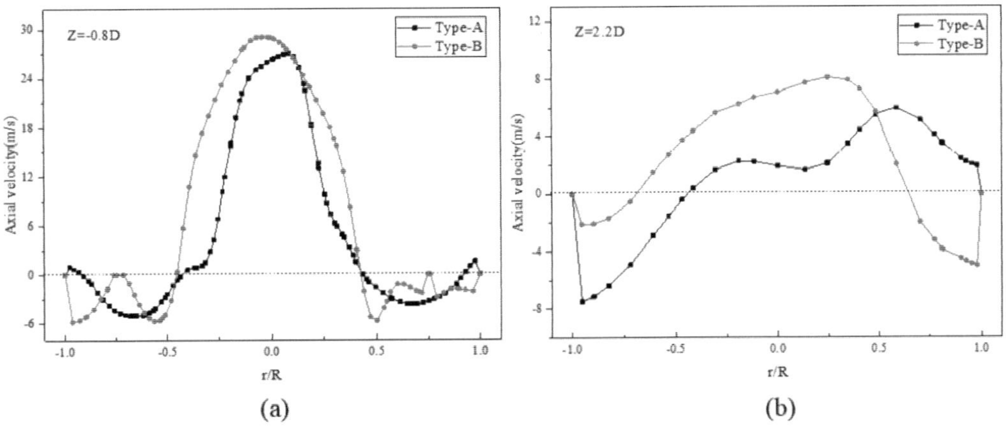


**Figure 7.** The axial velocity distribution of the cyclone separator at section Z=-0.8D and Z=2.2D.

Figure 7(b) further compares the axial velocity distribution at Z=-2.2D along the diameter axis between Type-A and Type-B. The inner deflector makes the fluid close to the side wall flow downward along the axis, reducing the possibility of particles on the wall being re-carried upward.

*3.2. Pressure Field*

Figure 8 plots the cloud diagram of the total pressure distribution of Type-A and Type-B at the Y=0 section (longitudinal section). The symmetry of the total pressure distribution cloud image is good.The maximum total pressure reduces from 1.93 kPa to 1.69 kPa, and the maximum total pressure area transfers from the cyclone separator's wall to the inner guider's inner wall.

Figure 9 shows the total pressure distribution of Type-A and Type-B at the Z=-0.8D section (near the bottom end of the vortex finder). After adding the inner guider, the overall total pressure value of the cyclone separator decreases, and the total pressure value in the diversion channel decreases most obviously. The maximum total pressure on the wall decreased from 1,740.80 kPa before adding the inner guider to 1,431.78 kPa.

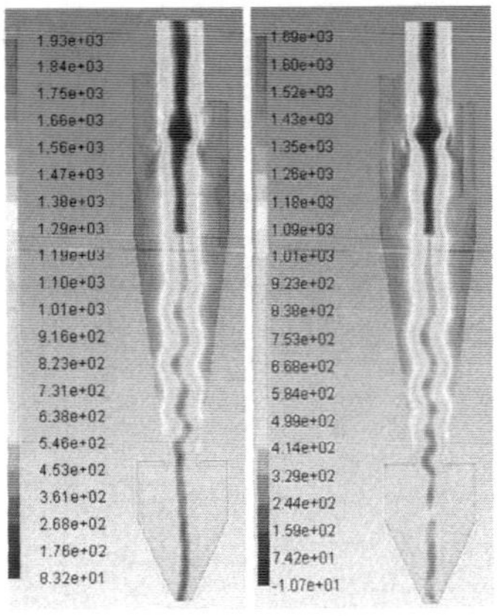


**Figure 8.** The total pressure distribution cloud image of the two cyclone separators at section Y=0.

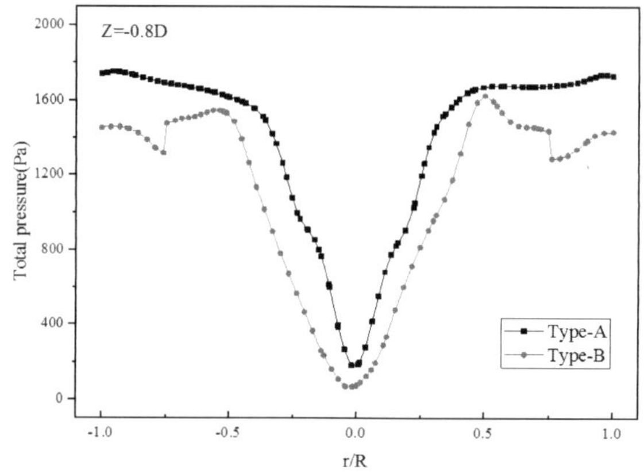


**Figure 9.** The total pressure distribution of the two cyclone separators at section Z=-0.8D.

Figure 10 shows the total pressure distribution of Type-A and Type-B at the Z=2.2D section (near the dust outlet). The maximum total pressure of the Type-B's wall decreased from 1,235.84kPa before adding the inner guider to 1107.07 kPa. But in the radial direction, Type-A and Type-B did not show a clear pressure difference.

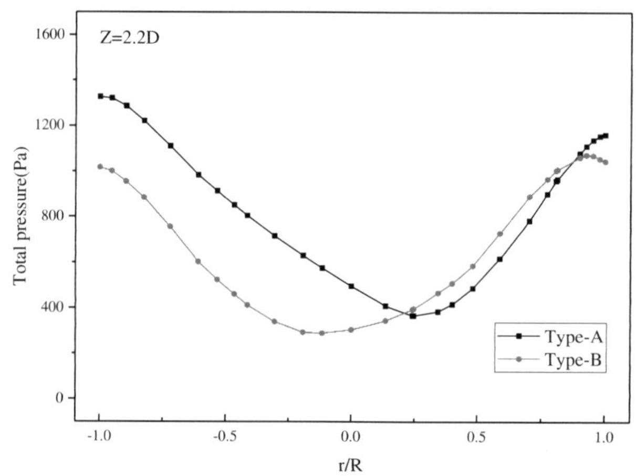


**Figure 10.** The total pressure distribution of the two cyclone separators at section Z=2.2D.

*3.3. Turbulent Kinetic Energy*

Figure 11 compares the turbulent kinetic energy distribution of Type-A and Type-B on the Y=0 section (longitudinal section). The inner guide reduces the maximum turbulent kinetic energy from 60.1 $m^2/s^2$ to 40.1 $m^2/s^2$. Lower turbulence can make the flow field more stable and reduce the turbulent diffusion of particles.

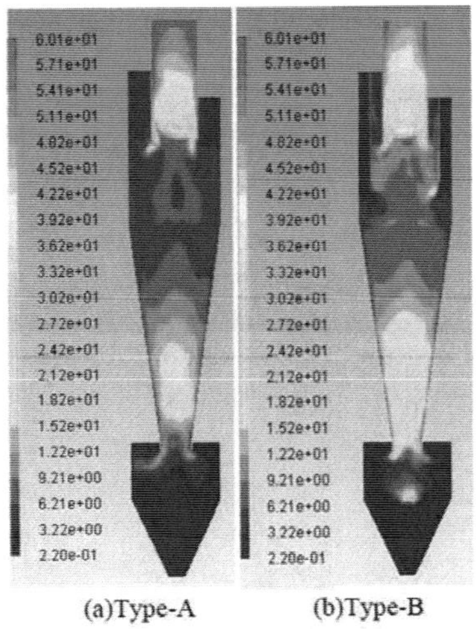


**Figure 11.** The turbulence kinetic energy distribution cloud image of the two cyclone separators at section Y=0.

Figure 12 shows the turbulent energy distribution of Type-A and Type-B at the Z=-0.8D section (near the bottom end of the vortex finder). The turbulent energy of the Type-B diversion channel is significantly reduced, and the maximum turbulence energy is reduced from 20.3 $m^2/s^2$ before the addition of the inner guider to 12.5 $m^2/s^2$.

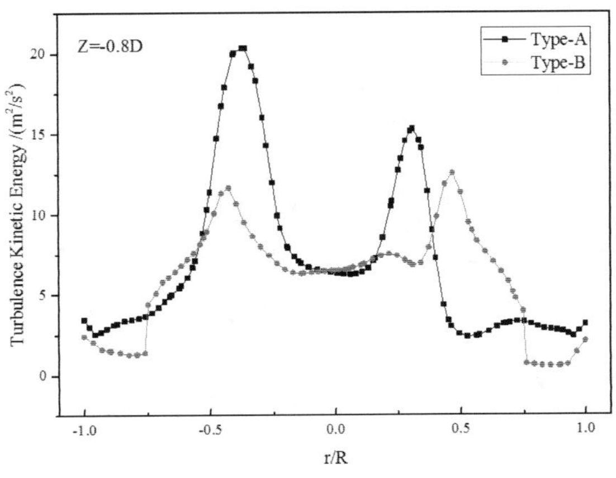


**Figure 12.** The turbulence kinetic energy distribution of the two cyclone separators at section Z=-0.8D.

### *3.4. The Separation Performance*

Figure 13 shows the trajectory of particles with a diameter of 1 μm in Type-A. It can be seen that no matter the particles enter at the upper, middle and lower part of the air inlet, they may be collected or separated by the cyclone separator.

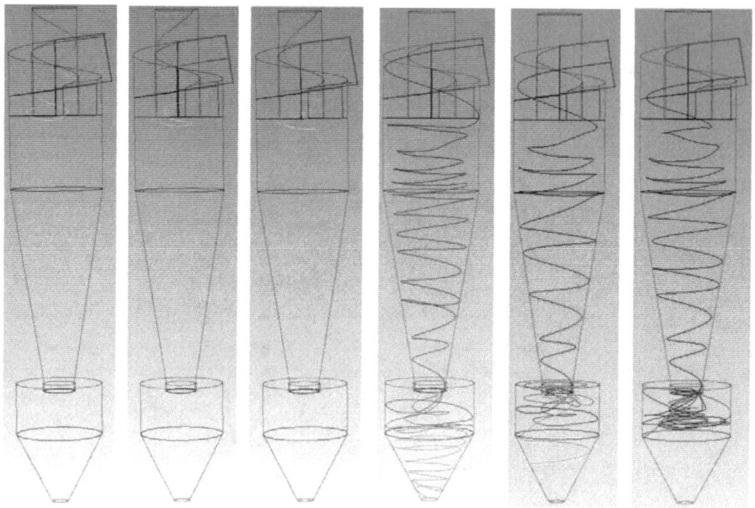

(a)Enter from the top of the inlet (b)Enter from the middle of the inlet (c)Enter from the bottom of the inlet (d)Enter from the top of the inlet (e)Enter from the middle of the inlet (f)Enter from the bottom of the inlet

**Figure 13.** The trajectories of particles with a particle size of 1μm in Type-A.

Figure 14 shows the trajectory of particles with a particle size of 1μm in Type-B. The particles entering between the inner deflector and the riser are basically affected by the short-circuit flow and escape from the vortex detector, as shown in figure 14(a), (b) and (c). Although the particles entering the space between the inner deflector and the separator wall can easily escape, some particles can still be collected successfully, as shown in figure 14 (d), (e) and (f).

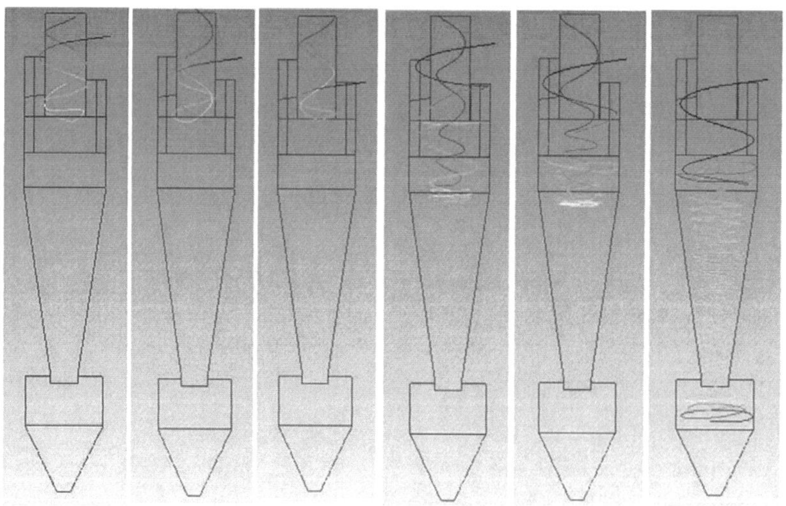

(a)Enter from the top of the inlet (b)Enter from the middle of the inlet (c)Enter from the bottom of the inlet (d)Enter from the top of the inlet (e)Enter from the middle of the inlet (f)Enter from the bottom of the inlet

**Figure 14.** Movement trajectories of particles with a particle size of 1μm in Type-B.

Figure 15 shows the classification efficiency curves of Type-A and Type-B. Figure 15 shows the classification efficiency of the cyclone separator before and after the addition of the inner guider. For particles below 5 μm, the separation efficiency of Type-B is lower than Type-A. This is because the centrifugal force of small particles is small, and the particles in the internal flow channel are more likely to be involved in the internal swirling flow, resulting in a reduction in separation efficiency. When the particle size is larger than 5 μm, the separation efficiency of Type-B starts to be higher than that of Type-A. At 10 μm, Type-B can basically separate all particles, while Type-A's separation efficiency is only 96.73%. The inner guider is added to the α-type cyclone separator to improve the separation efficiency of particles above 5 μm.

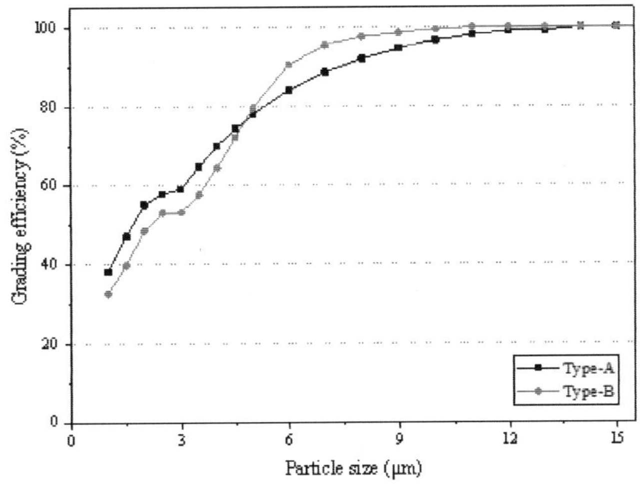


**Figure 15.** Grading efficiency curves of cyclone separators.

## 4. Conclusions

(1) The inner deflector increases the maximum tangential velocity, which is beneficial to improve the separation efficiency. The arrangement of the inner flow deflector reduces the radial distance from the particle to the wall to 1/2 of the original, making it easier for the particles to collide with the inner wall of the inner deflector.

(2) The inner guider makes the fluid close to the side wall flow downward along the axis, reducing the possibility of particles on the wall being re-carried upward, and makes the axial velocity distribution more symmetrical.

(3) After installing the inner deflector, the overall pressure of the cyclone separator is reduced, the maximum total pressure is reduced to 1.69kPa, and the pressure gradient is also reduced. The maximum turbulent energy of the cyclone separator is reduced from 60.1 $m^2/s^2$ to 40.1 $m^2/s^2$, the flow field is more stable, and the impact on particle separation is reduced. Nevertheless, the internal guide and ash hopper have higher turbulent kinetic energy. Adding a flow inner guide to the α-type cyclone improves the separation efficiency of particles above 5μm.

## References

[1] Wu C J and Ma Z F 2011 Numerical simulation of flow field and separation efficiency of cyclone separator with oblique tangential double inlets *Petroleum Processing and Petrochemicals* **42** (12) 50-55

[2] Zhu H T, Chen G H, Wang W W and Li J L 2017 Numerical Simulation and Analysis of Secondary Vortex in Different Cyclone Separators *Journal of Chemical Engineering of Chinese Universities* **31** (5) 1062-1071

[3] Chen J H 2008 *Study on Key Technologies of the Gas-Solid Flow in Large-Scale CFB Boilers* (Chongqing: Chongqing University) 15-115

[4] Erol H I, Turgut O and Unal R 2019 Experimental and numerical study of Stairmand cyclone separators: a comparison of the results of small-scale and large-scale cyclones *Heat and Mass Transfer* **55** (8) 2341-2354

[5] Faulkner W B, Buser M D, Whitelock D P and Shaw B W 2007 Effects of cyclone diameter on performance of1D3D cyclones: Collection efficiency *Transactions of the ASABE* **50** (3) 1053-1059

[6] Sakin A, Karagoz I and Avci A 2019 Effects of separation space diameter on the performance of a novel reverse flow cyclone *Separation Science and Technology* **54** (15) 2450–2460

[7] Brar L S and Sharma R P 2015 Effect of varying diameter on the performance of industrial scale gas cyclone dust separators *Materialstoday: PROCEEDINGS* **2** (4) 3230-3237

[8] Safikhani H, Akhavan-Behabadi M A, Shams M and Rahimyan M H 2010 Numerical simulation of flow field in three types of standard cyclone separators *Advanced Powder Technology* **21** 435-442

[9] Jin Y H and Shi M X 1990 Research on similar amplification test of cyclone separator *Journal of the University of Petroleum, China* **14** (5) 46-55

[10] Jin Y H and Shi M X 1990 Application of similarity theory in cyclone separator *Journal of the University of Petroleum, China* **14** (2) 33-39

[11] Jin W Lee, Hoe J Yang and Dong Y Lee 2006 Effect of the cylinder shape of a long-coned cyclone on the stable flow-field establishment *Powder Technology* **165** (1) 30-38

[12] Yuan H X and Li L L 2010 Numerical simulation of material properties and structural parameters' effects on pressure drop of the cyclone separator *Mining & Processing Equipment* **38** (5) 91-94

[13] Yuan Y, Sun G G, Zhou F Q and Sun Z P 2017 Effects of the cylinder diameter on cyclone performance *Acta Petrolei Sinica (Petroleum Processing Section)* **33** (4) 738-745

[14] Shao M W 1996 Analysis of the amplification effect of cyclone separator *Building Energy & Environment* (4) 32-33

[15] Song J F, Yang G F, Chen J Y and Wei Y D 2010 Similarity analysis of modeling of gas phase flow field in cyclone separator(ii)size parameters *CIESC Journal* **61** (9) 2274-2279

[16] Grace J R 2008 Mal-distribution of flow through parallel cyclones in circulating fluidized beds *Proceedings of the 9th International Conference on Circulating Fluidized Beds（Hamburg）* vol 1 (Germany:TuTech Innovation GmbH)p 969-974

[17] Liu F, Chen J Y, Zhang A Q and Gao R 2015 Analysis of cyclone stability of parallel cyclone separator *The Chinese Journal of Process Engineering* **15** (6) 923-928

[18] Liao L,Na Y J, Lv Q G, Sun Y K and Wang H G 2011 Experimental study on a circulating fluidized bed with six cyclone separators *Proceedings of the CSEE* **31** (11) 11-16

[19] Zhou X L, Cheng L M, Wang Q H and Luo Z Y 2012 Non-uniform distribution of gas–solid flow through six parallel cyclones in a CFB system: An experimental study *Particuology* **10** (2) 170-175

[20] Chen J Y, Gao R, Liu X L and Li Z F 2016 Measurement of separation performance and numerical analysis on flow field of different cyclone separators in parallel *CIESC Journal* **67** (8) 3287-3296

[21] Zhong X, Du H Y, Zhu S J, Liu R Z and Chen X L 2019 Effects of numbers of parallel micro-cyclone separators on separation performance *Journal of Chemical Engineering of Chinese Universities* **33** (03) 548-556

[22] Chen W, Feng J M, Han J Q and Peng X Y 2019 Effects of separation layer on the performance of multi-layer cyclone separators *Journal of Chemical Engineering of Chinese Universities* **33** (6) 1378-1385

[23] Fan J L, Wang C, Feng S K and Chen G H 2020 Numerical simulation on enlargement effects of α type cyclone separator *Mining & Processing Equipment* **48** (6) 49-53

[24] Wang H G and Liu S 2003 Application and comparison of different turbulence models in the three-dimensional numercial simulation of cyclone separators *Journal of Engineering for Thermal Energy and Power* **18** (4) 337-342

# CFD Simulation of Bubbling Process in Nitric Acid Solution in a Bubble Column

**Ruishu Tang[1], Chang Cui[1], Dagang Li[1], Dongxiang Zhang[1,*], Xiyan Xu[1,*] and Jinying Li[1,2]**

[1] School of Chemistry and Chemical Engineering, Beijing Institute of Technology, Beijing 102488, China
[2] China National Nuclear Corporation, Beijing 100822, China
Emails: boris@bit.edu.cn (D. Zhang), xiyanxu@bit.edu.cn (X. Xu).

**Abstract.** Bubble column is a commonly used multiphase reactor. Owing to the complicated nature of its flow structure, computational fluid dynamics has frequently been used as a useful instrument for studying the fluid mechanics in a bubble column. However, most of the previously studied systems are air-water system, while air-nitric acid solution system is rarely reported. In the present study, the differences in bubble size distribution of pure water and 1 mol/L nitric acid in the bubble column were observed through experiments. The bubble size was found to be smaller than that in water when the liquid phase was nitric acid. According to the actual situation, the bubbling process of nitric acid solution was successfully simulated by adjusting the coalescence factor of the bubble coalescence model (deviation less than 10%). Meanwhile, it was observed by simulation that the axial time-averaged liquid velocity distribution in nitric acid solution was flatter than that in water under the same gas flow rate.
**Keywords.** CFD, bubble column, nitric acid.

## 1. Introduction

Bubble column reactors are extensively employed in many industrial applications [1] because of its simple geometry, low cost of operation and maintenance, and large interfacial area between gas and liquid phase for heat and mass transfer [2]. Complex phenomena such as bubble deformation, bubble coalescence and breakup will occur during bubble column operation [3], which significantly affect bubble size distribution and gas holdup. Bubble size distribution and gas holdup are the key parameters of a bubble column as they determine the interface area for heat and mass transfer between the gas and liquid phases, depending on gas/liquid physical properties, gas distributor configuration and operating conditions, etc. Although many studies have been reported on bubble column reactors, many flow field parameters are difficult to obtain experimentally owing to the complicated nature of the flow, which makes it hard to investigate the fluid mechanics and mass transfer in bubble columns [4].

Computational Fluid Dynamics (CFD) can be applied to obtain the hydrodynamic by means of the solution of the control equations for fluid flow, which provides much convenience in understanding the hydrodynamics and mass transfer in bubble columns. In recent years, owing to the remarkable development of CFD technology for simulating multiphase flow, many scholars have successfully simulated the hydromechanics inside the bubble column accurately by means of CFD technology [3-5]. Nevertheless, most previously reported systems were air-water or air-organic solution systems, where the influence of the physical properties of the gas and liquid on bubble size distribution and hydrodynamics were studied [5-7], while air-nitric acid systems were rarely reported. Studying the bubbling process of the air-nitric acid system is of great significance for the reprocessing of spent fuel, because the dissolving liquid of spent fuel is a highly acidic (mainly nitric acid) and radioactive waste liquid, and its post-treatment process involves gas-liquid multiphase reaction process. In addition to changing the density, viscosity and surface tension of the solution, the presence of nitric acid also reduces the rate of bubble coalescence due to Marangoni effect [8] and electrical repulsion [9].

This study aims to develop a CFD simulation model for the bubbling process of the nitric acid solution and to investigate the impact of nitric acid on the flow pattern and bubble size distribution in the bubble column. Through bubbling experiments of water and 1 mol/L nitric acid solution in the bubble column, the pattern of flow and distribution of bubble size under different gas flow rates were investigated and the gas holdup during bubbling process was measured to validate the CFD model. In the CFD simulation, PBM model was employed to simulate the distribution of bubble size, bubble breakup and coalescence, and a correction factor was added to the bubble coalescence model to take into account the inhibitory impact of nitric acid on bubble coalescence.

## 2. Experimental Sections

### *2.1. Bubbling Experiment Setup*

This experiment is conducted in a bubble column, which is 19.6cm long, 3.0cm wide and 30.0cm high, with a gas distributor which is 10.0cm long, 1.11cm wide and 1.2cm high installed in the bottom centre of the bubble column. The initial liquid level in the bubble column is 24.0cm. The gas flow rate is regulated via a rotameter and is set to 3 L/min, 5 L/min and 8 L/min. The liquid phase is deionized water and 1 mol/L nitric acid solution. The temperature is 20°C.

### *2.2. Measurement of Gas Holdup*

In the experiment, the average gas holdup within the bubble column is determined by the liquid level difference, which is achieved through the measurement of the difference between the static liquid level and the bubbling level:

$$\alpha_g = \frac{H-h}{H} \times 100\% \tag{1}$$

Among them, $\alpha_g$ is the average gas holdup of the whole bubble column, H(cm) and h(cm) represent the bubbling liquid level and the static liquid level, respectively.

## 3. Mathematical Calculation Equation

In the present study, the Eulerian method was employed to simulate the flow of gas and liquid within the bubble column.

### *3.1. Continuity Equation and Momentum Conservation Equation*

In the Eulerian model, the continuity and momentum equation are as follows.

$$\frac{\partial(\alpha_i \rho_i)}{\partial t} + \nabla\cdot(\alpha_i \rho_i \vec{u}_i) = 0 \tag{2}$$

$$\frac{\partial(\alpha_i \rho_i \vec{u}_i)}{\partial t} + \nabla\cdot(\alpha_i \rho_i \vec{u}_i \vec{u}_i) = -\alpha_i \nabla p + \nabla\cdot[\mu_{eff,i}\alpha_i(\nabla\vec{u}_i + (\nabla\vec{u}_i)^{\mathrm{T}})] + \alpha_i \rho_i \vec{g} + F_i \tag{3}$$

where $\alpha$, $\rho$, $\vec{u}$, and $t$ represent phase holdup, density, velocity vector, and time, respectively; The subscript $i$ represents the phase(gas or liquid); $p$ represents pressure; $\mu_{eff}$ represents fluid effective viscosity; $F_i$ represents the interphase momentum exchange term, including drag force, lift force, turbulent dispersion force, and wall lubrication force. In the present study, the drag force is calculated using the model proposed by Ishii-Zuber [10], lift force is calculated using the model proposed by

Tomiyama [11], turbulent dispersion force is calculated using the model proposed by Burn [12], and wall lubrication force is calculated using the model proposed by Frank [13].

$$\mu_{eff} = \mu_L + \mu_t + \mu_{BIT} \tag{4}$$

where $\mu_L$ is dynamic viscosity, $\mu_t$ is turbulent viscosity determined by RNG k-ε turbulence model and $\mu_{BIT}$ is bubble-induced turbulence determined by Sato's model [14].

### *3.2. PBM Model*

The equation for the PBM model is as follows:

$$\frac{\partial[n(V,t)]}{\partial t} + \nabla \cdot [\vec{u}n(v,t)] = 0.5\int_0^V c(V-V',V)n(V-V',t)n(V',t)dV' - \int_0^\infty c(V,V')n(V,t)n(V',t)dV' + \int_{\Omega_V} b(V')\beta(V|V')n(V',t)dV' - b(V)n(V,t) \tag{5}$$

where $n$ represents the bubble number density; $c$ and $b$ represent the coalescence kernel function and the breakup kernel function, respectively; $\beta$ represents the distribution function of daughter bubbles when the parent bubbles break; the first and second terms to the right of the equals sign in equation (5) indicate the generation and disappearance of bubbles caused by coalescence, respectively. And the third and fourth terms indicate the generation and disappearance of bubbles caused by breakup, respectively.

In this study, Prince's coalescence model [15], Luo & Svendsen coalescence model [16] coupled with 4 coalescence factors, Luo's breakup model [17], are adopted to simulation the bubble coalescence and breakup, in order to find out which model is suitable for air-nitric acid system.

### *3.3. Numerical Details.*

In numerical simulations, the hexahedral mesh is adopted. Figure 1 displays the geometric model and mesh setup of the bubble column. The degass outlet is set for the top of the column and velocity inlet is set for the inlet, with no-slip boundary for the walls. The gas bin equation is calculated using first order windward format and the volume fraction equation is calculated using QUICK format. The other equations are calculated using second order upwind format. Unsteady state numerical simulations with a time step of 0.001 seconds were carried out.

(a)

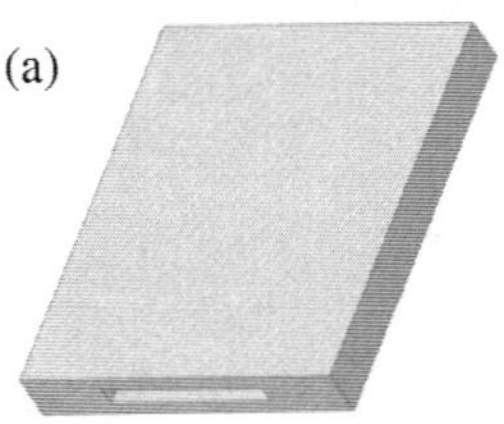

(b)

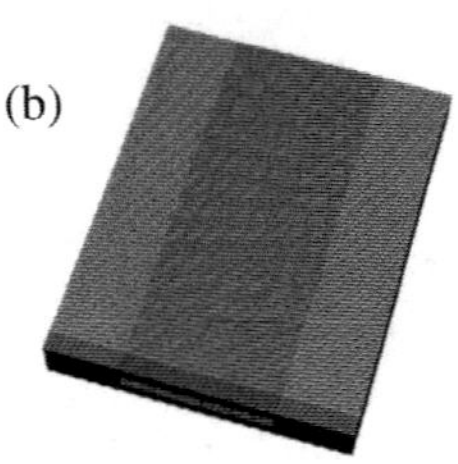

**Figure 1.** Geometric model (a) and mesh setup (b) of the bubble column.

## 4. Results and Discussion

### *4.1. Model Validation*

In the present study, CFD simulations of the bubbling process of nitric acid solution were performed, and PBM model was employed to simulate the bubble size distribution and bubble coalescence and breakup in the bubble column, so as to find an appropriate model for the bubbling process of nitric acid solution. Syed [18] pointed out that Luo's coalescence model overestimates the frequency of collisions between bubbles, so a coalescence factor is required to reduce the rate of bubble coalescence determined by Luo's coalescence model. Yang et al. [19] proposed a coalescence factor expression related to superficial gas velocity based on EMMS theory, and the results show that the coalescence factor combined with Luo's coalescence and breakup model (PBM-(EMMS)Luo-Luo) can successfully simulate the bubbling process of pure water. The results are shown in figure 2(a), and the deviation does not exceed 5%.

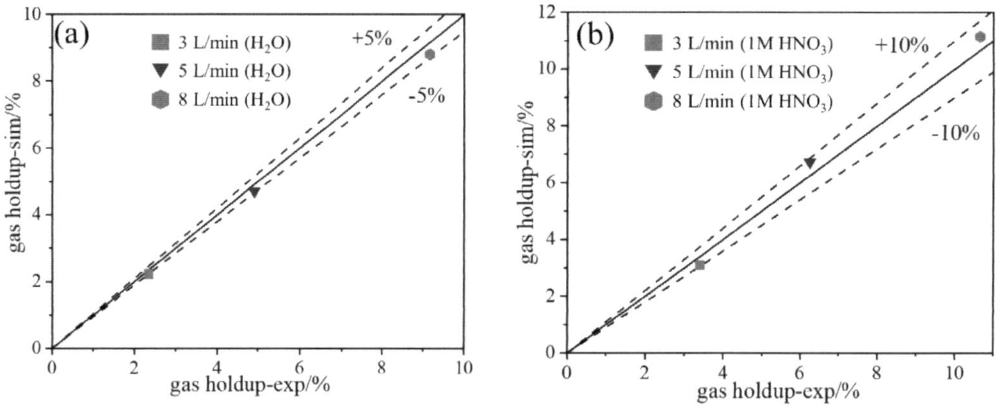


**Figure 2.** Comparison of simulated gas holdup and experimental gas holdup during bubbling in water (a) and 1 mol/L $HNO_3$ (b).

However, the PBM-(EMMS) Luo-Luo model does not apply to the air-nitric acid system. The results calculated using various bubble coalescence and breakup models are displayed in figure 3.

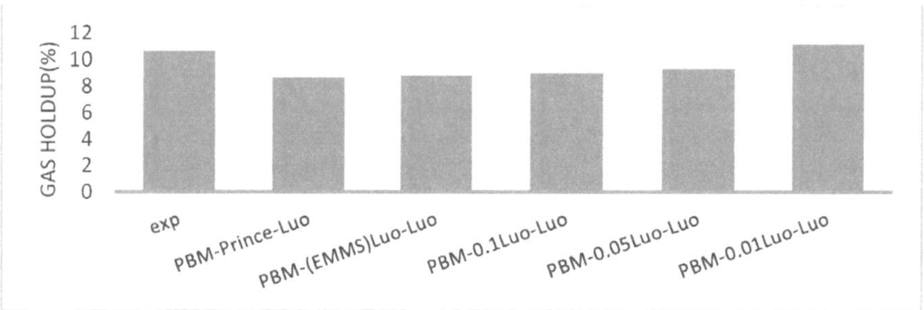


**Figure 3.** Experimental measurement and simulation value of gas holdup in 1 mol/L $HNO_3$ at 8L/min.

It can be observed from figure 3 that using PBM-Prince-Luo or PBM-(EMMS)Luo-Luo model can not accurately simulate the bubbling process of nitric acid solution, which may be caused by overestimation of the rate of bubble coalescence in the nitric acid solution. In addition, as can be noted in figure 3, even though the coalescence rate calculated using Luo's coalescence model is reduced by a factor of 0.1 and 0.05, the coalescence rate between bubbles is still overestimated. When the rate of coalescence obtained using Luo's coalescence model is reduced by a factor of 0.01, the calculations are relatively well. Therefore, PBM-0.01Luo-Luo model can be employed to simulation the bubbling process of 1 mol/L $HNO_3$ solution. The results of the comparison between the experimental and simulated values of gas holdup under other gas flow rates are presented in figure 2(b). As can be noted in figure 2(b), the deviation between the results calculated using the PBM-0.01Luo-Luo model and the results measured from experiments is about 10%.

*4.2. Flow Structures Analysis*

Figure 4 displays some images of the experimental recordings and the corresponding time-averaged gas holdup calculated from CFD simulations.

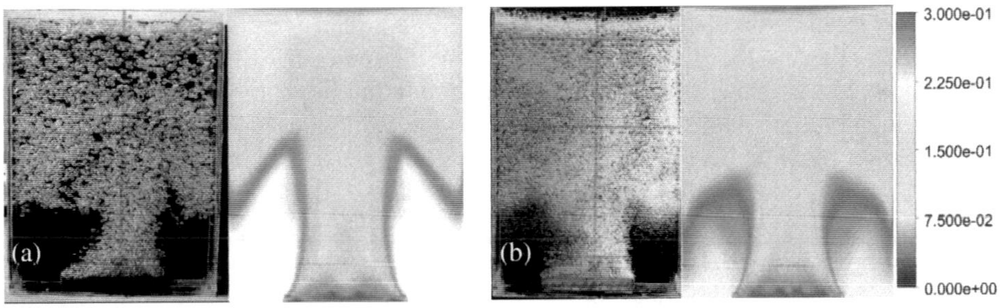


**Figure 4.** Images taken from the experiments and the corresponding time-averaged gas holdup from simulations (a): air-water at 8L/min; (b): air-1 mol/L $HNO_3$ at 8L/min.

It can be observed from figure 4 that after the gas enters the liquid phase through the gas distributor, it moves upward under the influence of buoyancy, and the liquid follows the upward movement of the gas owing to the drag force from gas. When the gas reaches the top of the bubble column, a part of the gas leaves the system, and the liquid changes its motion direction after following the gas to the top, and flows back from both sides of the gas rising path, forming vortical structures. A part of the gas flows back to the system under the influence of the liquid motion, and then moves upwards again under the influence of the vortexes and buoyancy. At the same time, as can be noticed in figure 5 that the larger the gas flow rate, the lower the liquid backflow velocity near the wall. This is attributed to the fact that with higher gas flow rates, the liquid rises at a faster speed and the kinetic energy of the liquid increases, resulting in carrying more bubbles back into the system and reducing the liquid backflow velocity.

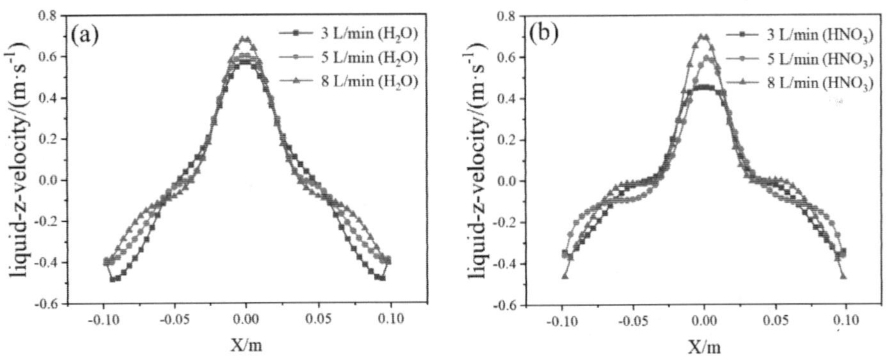


**Figure 5.** The axial time-averaged liquid velocity distribution resulting from simulation. (a): air in water at z=0.15m; (b): air in 1 mol/L $HNO_3$ at z=0.15m.

It can also be seen from the pictures taken in the experiment (figure 4) that in 1mol/L nitric acid solution, the bubble size is much smaller than that in water, which is caused by Marangoni effect [8] and electrical repulsive forces [9]. Since the bubbles in the nitric acid solution are smaller, the bubbles in the nitric acid solution are more easily affected by the liquid flow, so at the same gas flow rate, the bubbles in the nitric acid solution are more widely distributed than those in water, which maybe a favorable factor for mass transfer. Owing to the smaller bubble size in the nitric acid solution, the hydromechanics during bubbling of the nitric acid solution is somewhat different from that in water. As can be noted from figure 6, under the same gas flow rate, the axial time-averaged liquid velocity in the water is greater than that in the nitric acid solution at the central region of the bubble column, and the axial reflux velocity in the water is also greater than that in the nitric acid solution near the wall. This is because the coalescence rate of bubbles in the water is faster, so there are more large bubbles, resulting that the liquid velocity at the central region of the bubble column is faster, the kinetic energy of liquid is larger, leading to a larger liquid reflux speed. At the same time, because the motion state of large bubbles is not easily affected by the liquid flow, the number of backflow bubbles is small, the backflow liquid is less influenced by the bubbles, so the liquid backflow speed is faster. Therefore, the time-average liquid velocity distribution in the process of bubbling in water is steeper than that in the bubbling process of nitric acid solution.

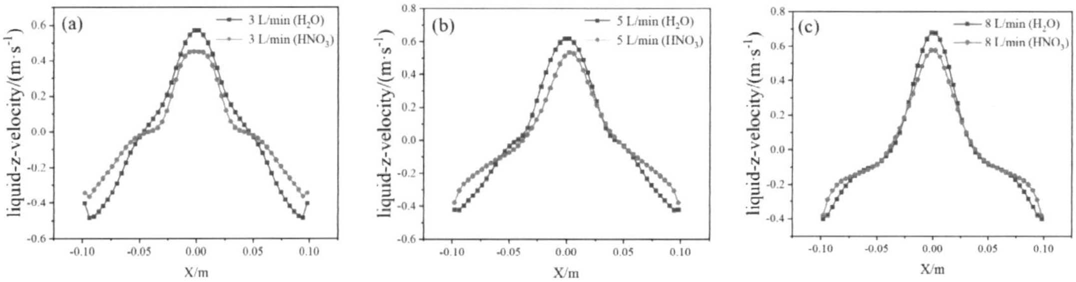


**Figure 6.** The axial time-averaged liquid velocity under different gas flow rates ((a) 3L/min, (b) 5 L/min, (c) 8 L/min) at z=0.15m resulting from simulation.

## 5. Summary

In this study, the difference between the bubbling process of 1mol/L nitric acid solution and pure water was observed through experiments, and the bubbling process was successfully simulated. It was found through experiments that due to Marangoni effect and electrical repulsive forces, the size of bubbles in nitric acid solution was much smaller than that in water, and the bubble distribution in nitric acid solution was wider than that in water during bubbling. In addition, according to the size distribution of bubbles in nitric acid solution, this study successfully simulated the bubbling process of nitric acid solution by adjusting the coalescence factor. It was found that PBM-0.01Luo-Luo model was appropriate for the bubbling process of nitric acid solution, while the PBM-(EMMS)Luo-Luo model was applicable to the water bubbling process. The simulation results revealed that the axial time-averaged liquid velocity distribution in the nitric acid solution was flatter than that in water under the same gas flow rate because of the smaller bubble size in the nitric acid solution.

## References

[1] Basha O M, Sehabiague L, Abdel-Wahab A and Morsi B 2015 Fischer–Tropsch Synthesis in Slurry Bubble Column Reactors: Experimental Investigations and Modeling–A Review *International Journal of Chemical Reactor Engineering* **13**(3) 201-88

[2] Anastasiou A D, Passos A D and Mouza A A 2013 Bubble columns with fine pore sparger and non-Newtonian liquid phase: Prediction of gas holdup *Chem. Eng. Sci.* **98** 331-8

[3] Chen J and Brooks C S 2021 Experiments and CFD simulation of mass transfer and hydrodynamics in a cylindrical bubble column *Chem. Eng. Sci.* **234** 116435

[4] Liu J, Zhou P, Liu L, Chen S, Song Y and Yan H 2021 CFD modeling of reactive absorption of $CO_2$ in aqueous NaOH in a rectangular bubble column: Comparison of mass transfer and enhancement factor model *Chem. Eng. Sci.* **230** 116218

[5] Zhang H, Sayyar A, Wang Y and Wang T 2020 Generality of the CFD-PBM coupled model for bubble column simulation *Chem. Eng. Sci.* **219** 115514

[6] Zhang H, Guo K, Wang Y, Sayyar A and Wang T 2020 Numerical simulations of the effect of liquid viscosity on gas-liquid mass transfer of a bubble column with a CFD-PBM coupled model *Int. J. Heat Mass Tran.* **161** 120229

[7] Sarhan A R, Naser J and Brooks G 2018 CFD modeling of bubble column: Influence of physico-chemical properties of the gas/liquid phases properties on bubble formation *Sep. Purif. Technol.* **201** 130-8

[8] Weissenborn P K and Pugh R J 1996 Surface Tension of Aqueous Solutions of Electrolytes: Relationship with Ion Hydration, Oxygen Solubility, and Bubble Coalescence *J. Colloid Interface Sci.* **184**(2) 550-63

[9] Marrucci G and Nicodemo L 1967 Coalescence of gas bubbles in aqueous solutions of inorganic electrolytes *Chem. Eng. Sci.* **22**(9) 1257-65

[10] Ishii M and Zuber N 1979 Drag coefficient and relative velocity in bubbly, droplet or particulate flows *AICHE J.* **25**(5) 843-55

[11] Tomiyama A, Tamai H, Zun I and Hosokawa S 2002 Transverse migration of single bubbles in simple shear flows *Chem. Eng. Sci.* **57**(11) 1849-58

[12] Burns A D, Frank T, Hamill I and Shi J-M 2004 The Favre averaged drag model for turbulent dispersion in Eulerian multi-phase flows. *5th International Conference on Multiphase Flow, ICMF* vol 4 (Japan: ICMF) pp 1-17

[13] Frank T, Zwart P J, Krepper E, Prasser H M and Lucas D 2008 Validation of CFD models for mono- and polydisperse air–water two-phase flows in pipes *Nucl. Eng. Des.* **238**(3) 647-59

[14] Sato Y and Sekoguchi K 1975 Liquid velocity distribution in two-phase bubble flow *Int. Multiphase Flow* **2**(1) 79-95

[15] Prince M J and Blanch H W 1990 Bubble coalescence and break-up in air-sparged bubble columns *AICHE J.* **36**(10) 1485-99

[16] Luo H 1993 *Coalescence, Breakup and Liquid Circulation in Bubble Column Reactors* (Norway: Norwegian Institute of Technology)

[17] Luo H and Svendsen H F 1996 Theoretical model for drop and bubble breakup in turbulent dispersions *AICHE J.* **42**(5) 1225-33

[18] Syed A H, Boulet M, Melchiori T and Lavoie J 2017 CFD Simulations of an Air-Water Bubble Column: Effect of Luo Coalescence Parameter and Breakup Kernels *Front Chem.* **5** 68

[19] Wang J and Yang N 2017 CFD-PBM simulation with EMMS correctors for bubble column reactors *CIESC Journal* **68**(7) 2667-77

# A Method for Obtaining Solution of the Spatial Optical Soliton Based on the Adomian Decomposition Method

**Yungao Cai**

College of Engineering Technology, Baoshan University, Baoshan Yunnan, 678000, China.
Email: ynmzcyg@163.com

**Abstract.** For obtaining the solution of the spatial optical solution in the nonlinear Schrödinger equation, the aomian decomposition method was used for solving the complex equation. The specific process for solving had four steps. Firstly, the nonlinear equation was appropriately decomposed into several parts and the solution also was decomposed into infinite parts. Secondly, the special aomian polynomial that equaled to the nonlinear term was produced. Thirdly, the high order components were obtained by the low order components with the inverse operator. Finally, the sum of the components that was the approximation solution was obtained. The sum of the real part (the imaginary part) of the front five items by aomian decomposition method was approximation to that of exact solution respectively. The results show that the aomian decomposition method has the characteristics of the fast convergence and the high accuracy. The aomian decomposition method was an important method for solving the spatial solution
**Keywords.** Information Optics; Spatial solition; Nonlinear Schrödinger equation; Adomian decomposition.

## 1. Introduction

Since the solitons [1] were observed in 1834 by Russell in the narrow channel, solitons had become one of the most attractive research objects in nonlinear science, soliton effect got a lot of attention in elementary particle, materials science, statistical mechanics, fluid mechanics, plasma physics, molecular biology, and nonlinear optics. The study of optical solitons attracts much attention until the 1970s due to the development of optical fiber communication.

There are time soliton and space optical soliton, spatial optical soliton is a state in which the diffraction effect and nonlinear effect are balanced when the beam propagates in the medium. Spatial optical solitons are widely used in optical switch, photon information processing, all-optical network and so on. There is great significance for the study of spatial optical solitons.

In this paper, the nonlinear Schrödinger equation was solved by a simple inverse operator method, the exact approximate solution or even exact solution of the spatial optical solitons was obtained, and then the properties of the spatial photon are studied.

## 2. Adomian Decomposition Method

Compound field $u = u(x, t) \in C$ as time and space evolve, the general form of a nonlinear Schrödinger equation [2],

$$\mathrm{i}\frac{\partial}{\partial t}u+\frac{\partial^2}{\partial x^2}u+m|u|^2u=0 \tag{1}$$

$$u(x,0)=g(x) \tag{2}$$

where $x \in R, t > 0$, it is bright soliton when $m > 0$, and it is dark soliton when $m < 0$, the nonlinear schrödinger equation is completely integrable based on the backscattering method [3-4]. There is one-dimensional solution of the nonlinear schrödinger equation,

$$u(x,t)=\sqrt{\frac{2a}{m}}\exp\{\mathrm{i}(\frac{c}{2}x-(\frac{c^2}{4}-a)t)\}\,\mathrm{sec\,h}[\sqrt{a}(x-ct)] \tag{3}$$

The inverse operation of equation (1) could be written as follow based on the adomian decomposition [5],

$$L_t^{-1}(\cdot)=\int_0^t(\cdot)\mathrm{d}t,\ t>0 \tag{4}$$

Therefore, equation (1) could be transformed into,

$$L_t u=\mathrm{i}\frac{\partial^2}{\partial x^2}u+\mathrm{i}mG(u),\ x\in R,\ t>0 \tag{5}$$

Here $L_t = \partial_t$, the cubic nonlinear term is,

$$G(u)=|u|^2u \tag{6}$$

Acorrding equation (2) and equation (6), the solution of equation (5) is,

$$u(x,t)=g(x)+\mathrm{i}L_t^{-1}\partial_x^2u+\mathrm{i}mL_t^{-1}G(u),\ x\in R,\ t>0 \tag{7}$$

$$u(x,t)=\sum_{n=0}^{\infty}u_n(x,t) \tag{8}$$

$$G(u)=\sum_{n=0}^{\infty}A_n(u) \tag{9}$$

$$A_n=\frac{1}{n!}\frac{\mathrm{d}^n}{\mathrm{d}\lambda^n}[G(\sum_{p=0}^{\infty}\lambda_p u_p)]_{\lambda=0},\ n=0,1,2,\ldots \tag{10}$$

$u_n(x,t)$ is periodic function, $A_n$ is adomian polynomial, Substitute equation (8) and equation (9) into equation (7) , it could be obtainded,

$$\sum_{n=0}^{\infty}u_n(x,t)=g(x)+\mathrm{i}\sum_{n=0}^{\infty}L_t^{-1}\partial_x^2u(x,t)+\mathrm{i}m\sum_{n=0}^{\infty}L_t^{-1}A_n,\ t>0 \tag{11}$$

Assuming that,

$$u_0(x,t)=g(x),\ t>0 \tag{12}$$

$$u_{n+1}(x,t)=\mathrm{i}L_t^{-1}\partial_x^2u_n(x,t)+\mathrm{i}mL_t^{-1}A_n,\ n=0,1,2,\ldots \tag{13}$$

Then,

$$A_0 = \frac{1}{0!} G(u_0) \tag{14a}$$

$$\begin{aligned} A_1 &= \frac{1}{1!}\frac{\mathrm{d}}{\mathrm{d}\lambda}[G(u_0 + \lambda u_1 + \lambda^2 u_2 + \lambda^3 u_3 + \cdots)]_{\lambda=0} \\ &= [(u_1 + 2\lambda u_2 + 3\lambda^2 u_3 + \cdots)G'(u_0 + \lambda u_1 + \lambda^2 u_2 + \lambda^3 u_3 + \cdots)]_{\lambda=0} \\ &= u_1 G'(u_0) \end{aligned} \tag{14b}$$

$$\begin{aligned} A_2 &= \frac{1}{2!}\frac{\mathrm{d}^2}{\mathrm{d}\lambda^2}[G(u_0 + \lambda u_1 + \lambda^2 u_2 + \lambda^3 u_3 + \cdots)]_{\lambda=0} \\ &= \frac{1}{2}\frac{\mathrm{d}}{\mathrm{d}\lambda}[(u_1 + 2\lambda u_2 + 3\lambda^2 u_3 + \cdots)G'(u_0 + \lambda u_1 + \lambda^2 u_2 + \lambda^3 u_3 + \cdots)]_{\lambda=0} \\ &= \frac{1}{2}[(2u_2 + 6\lambda u_3 + \cdots)G'(u_0 + \lambda u_1 + \lambda^2 u_2 + \lambda^3 u_3 + \cdots) \\ &\quad + (u_1 + 2\lambda u_2 + 3\lambda^2 u_3 + \cdots)(u_1 + 2\lambda u_2 + 3\lambda^2 u_3 + \cdots)G''(u_0 + \lambda u_1 + \lambda^2 u_2 + \lambda^3 u_3 + \cdots)]_{\lambda=0} \\ &= u_2 G'(u_0) + \frac{1}{2}u_1^2 G''(u_0) \end{aligned} \tag{14c}$$

$$\begin{aligned} A_3 &= \frac{1}{3!}\frac{\mathrm{d}^3}{\mathrm{d}\lambda^3}[G(u_0 + \lambda u_1 + \lambda^2 u_2 + \lambda^3 u_3 + \cdots)]_{\lambda=0} \\ &= \frac{1}{6}\frac{\mathrm{d}}{\mathrm{d}\lambda}[(2u_2 + 6\lambda u_3 + \cdots)G'(u_0 + \lambda u_1 + \lambda^2 u_2 + \lambda^3 u_3 + \cdots) \\ &\quad + (u_1 + 2\lambda u_2 + 3\lambda^2 u_3 + \cdots)(u_1 + 2\lambda u_2 + 3\lambda^2 u_3 + \cdots)G''(u_0 + \lambda u_1 + \lambda^2 u_2 + \lambda^3 u_3 + \cdots)]_{\lambda=0} \\ &= u_3 G'(u_0) + u_1 u_2 G''(u_0) + \frac{1}{6}u_1^3 G'''(u_0) \end{aligned} \tag{14d}$$

$$\begin{aligned} A_4 &= \frac{1}{4!}\frac{\mathrm{d}^4}{\mathrm{d}\lambda^4}[G(u_0 + \lambda u_1 + \lambda^2 u_2 + \lambda^3 u_3 + \cdots)]_{\lambda=0} \\ &= \frac{1}{24}\frac{\mathrm{d}}{\mathrm{d}\lambda}[(6u_3 + 24\lambda u_4 + \cdots)G'(u_0 + \lambda u_1 + \lambda^2 u_2 + \lambda^3 u_3 + \cdots) \\ &\quad + 3(2u_2 + 6\lambda u_3 + \cdots)(u_1 + 2\lambda u_2 + 3\lambda^2 u_3 + \cdots)G''(u_0 + \lambda u_1 + \lambda^2 u_2 + \lambda^3 u_3 + \cdots) \\ &\quad + (u_1 + 2\lambda u_2 + 3\lambda^2 u_3 + \cdots)^3 G'''(u_0 + \lambda u_1 + \lambda^2 u_2 + \lambda^3 u_3 + \cdots)]_{\lambda=0} \\ &= u_4 G'(u_0) + u_3 u_1 G''(u_0) + \frac{1}{2}u_2^2 G''(u_0) + \frac{1}{2}u_1^2 u_2 G'''(u_0) + \frac{1}{24}u_1^4 G''''(u_0) \end{aligned} \tag{14e}$$

Assuming that, $G(u) = u^2\bar{u},\ u_1(\bar{u}_0)' = \bar{u}_1,\ u_2(\bar{u}_0)' = \bar{u}_2,\ u_3(\bar{u}_0)' = \bar{u}_3, \ldots$ (15)

It can be calculated directly,

$$A_0(x,t) = G(u_0) = u_0^{\,2}\bar{u}_0 \tag{16a}$$

$$A_1(x,t) = u_1 G'(u_0) = u_1[2u_0\bar{u}_0 + u_0^{\,2}(\bar{u}_0)'] = 2u_0u_1\bar{u}_0 + u_0^2u_1(\bar{u}_0)' = 2u_0u_1\bar{u}_0 + u_0^2\bar{u}_1 \tag{16b}$$

$$\begin{aligned} A_2(x,t) &= u_2G'(u_0) + \frac{1}{2!}u_1^{\,2}G''(u_0) = u_2[2u_0\bar{u}_0 + u_0^{\,2}(\bar{u}_0)'] + \frac{1}{2}u_1^{\,2}[2\bar{u}_0 + 4u_0(\bar{u}_0)'] \\ &= 2u_0u_2\bar{u}_0 + u_0^{\,2}\bar{u}_2 + u_1^{\,2}\bar{u}_0 + 2u_1^{\,2}u_0(\bar{u}_1)' = 2u_0u_2\bar{u}_0 + u_1^2\bar{u}_0 + 2u_0u_1\bar{u}_1 + u_0^2\bar{u}_2 \end{aligned} \tag{16c}$$

$$\begin{aligned} A_3(x,t) &= u_3G'(u_0) + u_1u_2G''(u_0) + \frac{1}{3!}u_1^3G'''(u_0) \\ &= u_3[2u_0\bar{u}_0 + u_0^2(\bar{u}_0)'] + u_1u_2[2\bar{u}_0 + 4u_0(\bar{u}_0)'] + \frac{1}{6}u_1^3[6(\bar{u}_0)'] \\ &= 2u_0u_3\bar{u}_0 + 2u_1u_2\bar{u}_0 + 2u_0u_2\bar{u}_1 + u_1^2\bar{u}_1 + 2u_0u_1\bar{u}_2 + u_0^2\bar{u}_3 \end{aligned} \tag{16d}$$

$$\begin{aligned} A_4(x,t) &= u_4G'(u_0) + u_3u_1G''(u_0) + u_2^2G''(u_0) + \frac{1}{2}u_1^2u_2G'''(u_0) + \frac{1}{24}u_1^4G''''(u_0) \\ &= u_4[2u_0\bar{u}_0 + u_0^2(\bar{u}_0)'] + u_3u_1[2\bar{u}_0 + 4u_0(\bar{u}_0)'] + \frac{1}{2}u_2^2[2\bar{u}_0 + 4u_0(\bar{u}_0)'] + \frac{1}{2}u_1^2u_2[6(\bar{u}_0)'] \\ &= 2u_0\bar{u}_0u_4 + u_0^2\bar{u}_4 + 2\bar{u}_0u_3u_1 + 4u_0u_3\bar{u}_1 + u_2^2\bar{u}_0 + 2u_2\bar{u}_2u_0 + 3u_1^2\bar{u}_2 \end{aligned} \tag{16e}$$

$$\begin{aligned} A_5(x,t) &= u_5G'(u_0) + u_4u_1G''(u_0) + u_3u_2G''(u_0) + \frac{1}{2}u_3u_1^2G'''(u_0) + \frac{1}{2}u_2^2u_1G'''(u_0) \\ &+ \frac{1}{6}u_1^3u_2G''''(u_0) + \frac{1}{120}u_1^5G'''''(u_0) \\ &= u_5[2u_0\bar{u}_0 + u_0^2(\bar{u}_0)'] + u_4u_1[2\bar{u}_0 + 4u_0(\bar{u}_0)'] + u_3u_2[2\bar{u}_0 + 4u_0(\bar{u}_0)'] \\ &+ (\frac{1}{2}u_3u_1^2 + \frac{1}{2}u_2^2u_1)[2(\bar{u}_0)'+4(\bar{u}_0)'] \\ &= 2u_5u_0\bar{u}_0 + u_0^2\bar{u}_5 + 2u_4u_1\bar{u}_0 + 4u_1u_0\bar{u}_4 + 2u_3u_2\bar{u}_0 + 4u_0u_2\bar{u}_3 + u_1^2\bar{u}_3 + 2u_1^2\bar{u}_3 + u_2^2\bar{u}_1 + 2u_2^2\bar{u}_1 \end{aligned} \tag{16f}$$

If it kept doing this, $A_n$ $(n > 5)$ is not hard be got.

## 3. Solution of the Spatial Optical Soliton by Adomian Decomposition Method

It should be pointed out here that, all the terms $u_n(x,t)$ can be obtained by equations (12) and (13) in principle. Since Adomian decomposition method has fast convergence, it only needs to sum its limited terms for approximate solution. The one-dimensional soliton solution of equation (1) is expressed as equation (3) .

$$u(x,t) = \sqrt{\frac{2a}{m}}\exp\{i(\frac{c}{2}x - (\frac{c^2}{4} - a)t)\}\frac{1}{\cosh[\sqrt{a}(x-ct)]} \tag{17}$$

According to euations (12) (13) and (16), the components of the exact solution can be obtained, and then the higher-order solution components can be deduced from the lower-order solution components

$$u_0(x,t) = \sqrt{\frac{2a}{m}}e^{i\frac{c}{2}x}\frac{1}{\cosh(\sqrt{a}x)} \tag{18a}$$

$$\begin{aligned} u_1(x,t) &= iL_t^{-1}\frac{\partial^2}{\partial x^2}u_0(x,t) + imL_t^{-1}A_0 \\ &= \sqrt{\frac{2a}{m}}e^{i\frac{c}{2}x}it(i\frac{c}{2})^2\frac{1}{\cosh(\sqrt{a}x)} + 2\sqrt{\frac{2a}{m}}e^{i\frac{c}{2}x}it(i\frac{c}{2})(-\frac{\sqrt{a}\sinh(\sqrt{a}x)}{\cosh^2(\sqrt{a}x)}) \\ &+ \sqrt{\frac{2a}{m}}e^{i\frac{c}{2}x}it(-\frac{a}{\cosh(\sqrt{a}x)} + \frac{2a\sinh^2(\sqrt{a}x)}{\cosh^3(\sqrt{a}x)}) + \frac{it2a}{\cosh^3(\sqrt{a}x)}\sqrt{\frac{2a}{m}}e^{i\frac{c}{2}x} \end{aligned} \tag{18b}$$

$$u_2(x,t) = \mathrm{i}L_t^{-1}\frac{\partial^2}{\partial x^2}u_1(x,t) + \mathrm{i}mL_t^{-1}A_1$$
$$= \frac{t^2}{32}\sqrt{\frac{2a}{m}}\mathrm{e}^{\mathrm{i}\frac{c}{2}x}\Big[\frac{\mathrm{i}c^4}{\cosh(\sqrt{a}x)} - \frac{8\sqrt{a}c^3\sinh(\sqrt{a}x)}{\cosh^2(\sqrt{a}x)} - \frac{24iac^2}{\cosh(\sqrt{a}x)} + \frac{40iac^2}{\cosh^3(\sqrt{a}x)} + \frac{32a^{\frac{3}{2}}c\sinh(\sqrt{a}x)}{\cosh^2(\sqrt{a}x)}$$
$$- \frac{96a^{\frac{3}{2}}c\sinh(\sqrt{a}x)}{\cosh^4(\sqrt{a}x)} + \frac{16ia^2}{\cosh(\sqrt{a}x)} - \frac{32ia^2}{\cosh^3(\sqrt{a}x)}\Big]$$
$$- \frac{amt^2}{4}\sqrt{\frac{2a}{m}}\mathrm{e}^{\mathrm{i}\frac{c}{2}x}\Big[\frac{-c^2}{\cosh^3(\sqrt{a}x)} - \frac{12i\sqrt{a}c\sinh(\sqrt{a}x)}{\cosh^4(\sqrt{a}x)} + \frac{4a}{m}\frac{1}{\cosh^3(\sqrt{a}x)}\Big] \tag{18c}$$

$$u_3(x,t) = \mathrm{i}L_t^{-1}\frac{\partial^2}{\partial x^2}u_2(x,t) + \mathrm{i}mL_t^{-1}A_2$$
$$= \frac{1}{3\cdot 2^9}\sqrt{\frac{2a}{m}}\mathrm{e}^{\mathrm{i}\frac{c}{2}x}t^3\Big[-\mathrm{i}\frac{4(64a^3 - 240a^2c^2 + 60ac^4 - c^6)}{\cosh(\sqrt{a}x)} - 3\mathrm{i}\frac{(512a^2c^2 - 128ac^4)}{\cosh^3(\sqrt{a}x)}$$
$$- 16\sqrt{a}c(48a^2 - 40ac^2 + 3c^4)\frac{\sinh(\sqrt{a}x)}{\cosh^2(\sqrt{a}x)} - 1536a\sqrt{a}c^3\frac{\sinh(\sqrt{a}x)}{\cosh^4(\sqrt{a}x)}\Big] \tag{18d}$$

$$u_4(x,t) = \mathrm{i}L_t^{-1}\frac{\partial^2}{\partial x^2}u_3(x,t) + \mathrm{i}mL_t^{-1}A_3 = -\frac{1}{6144}\sqrt{2a}\mathrm{e}^{\mathrm{i}\frac{c}{2}x}t^4[u41 + \mathrm{i}u42]\frac{1}{\cosh^8(\sqrt{a}x)}$$

$$u_{41} = 577536a^4\cosh(\sqrt{a}x) - 6144a^2c^4\cosh^3(\sqrt{a}x) + (6656a^2c^4 - 3072a^3c^2 - 192ac^6)\cosh^5(\sqrt{a}x) - (256a^4 - 1792a^3c^2 + 1120a^2c^4 - 112ac^6 + c^8)\cosh^7(\sqrt{a}x) \tag{18e}$$

$$u_{42} = (6144a^{\frac{5}{2}}c^3 - 1536a^{\frac{3}{2}}c^5)\cosh^4(\sqrt{a}x)\sinh(\sqrt{a}x) + (1024a^{\frac{7}{2}}c - 1792a^{\frac{5}{2}}c^3 + 448a^{\frac{3}{2}}c^5 - 16\sqrt{a}c^7)\cosh^6(\sqrt{a}x)\sinh(\sqrt{a}x) \tag{18f}$$

When computed $n = 4$, that is, the value obtained is the highly approximate or exact solution of the soliton solution of $\sum_{n=0}^{4} u_n(x,t)$ as shown in figure 1 to figure 4 as follow.

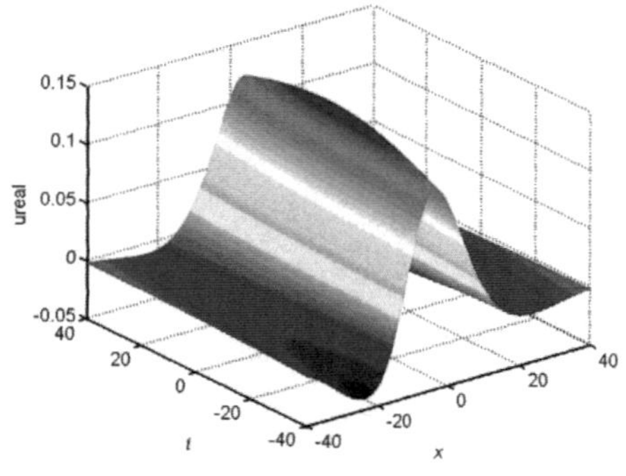


**Figure 1.** Real part of the exact solution.

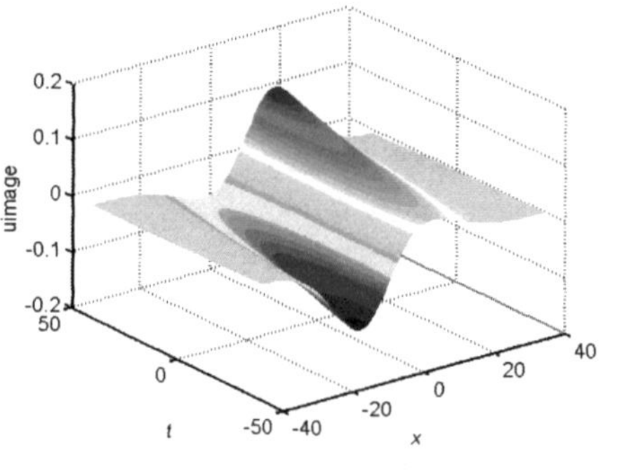


**Figure 2.** Imaginary part of the exact solution.

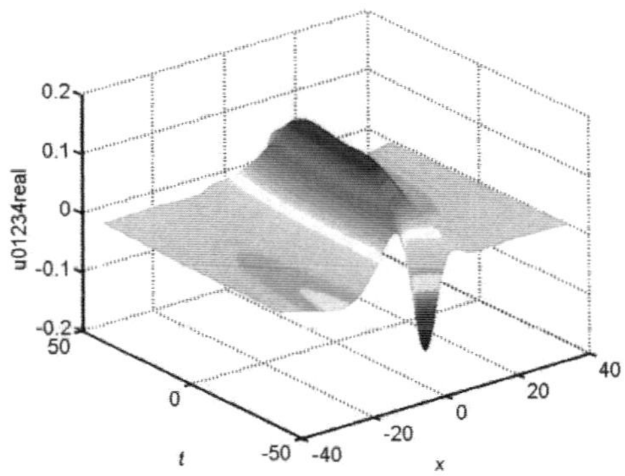


**Figure 3.** Real part of the sum of the front five items by Adomian decomposition method.

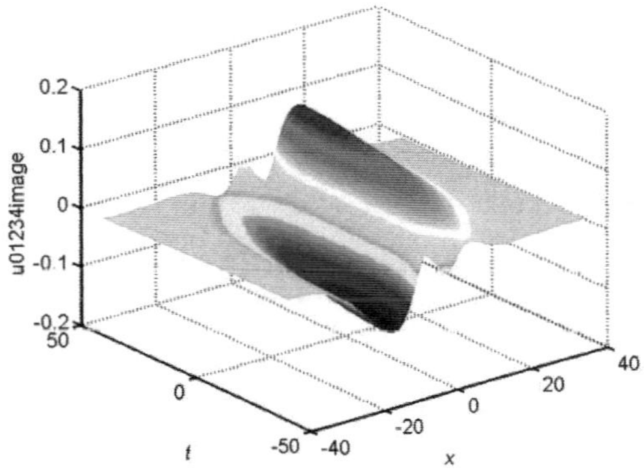


**Figure 4.** Imaginary part of the sum of the front five items by Adomian decomposition method.

## 4. Summary

Generally, it can be clearly seen from figure 1 to figure 4. that the adomian decomposition method can be used to accurately solve the approximate solution of spatial bright soliton solution and the convergence speed is fast. As long as the simplest initial form of a solution is found and its adomian polynomial is determined, its higher-order solution can be accurately calculated by using equation (12) and equation (13). Then, the finite sum of solution components can be used to successfully obtain approximate or exact solutions. This solution method is simpler and more convenient than backscattering method and fourth-order Runge-Kutta method, etc. It has become an important method for solving spatial light solitons, and has been applied in many aspects.

**Acknowledgments**
This work is supported by the Scientific Research Projects of Yunnan Education Department (Grant Number 2022J0907), Academician Workstation of Wen Bangchun of Yunnan Province (Grant Number 202205AF150032).

## References

[1] Song N, Xue H and Xue Y K. 2020 *Journal of Chongqing University of Technology (Natural Science)* **34**(4) 165-169.
[2] Argyris J, Haase M 1987 *Appl. Mech. Engrg.* **61** 71–80.
[3] Wazwaz A M 1999 *Appl. Math. Comput.* **102** 77–86.
[4] Wang J L, Zhong C X, Li R, et al. 2021 *Laser Journal* **42**(11) 36–40.
[5] Jin X J, Xu S L, Chen J X, et al. 2019 *Chinses Journal of Quantum Electronics* **36**(1) 82-86.

# Modelling and Simulation of Transverse Homogeneity of Slotted Coating Process

**Jianhong Wang[1], Yongchao Zhong[1], Yubiao Li[1] and Xiaodong Wang[2*]**

[1]Hebei Lucky Chemical Engineering Design Co., LTD, Baoding 071022, Hebei, China.
[2]Northwestern Polytechnical University, Xi'an 710129, Shaanxi, China.
Email: xiaodongwang@nwpu.edu.cn

**Abstract.** Slot coating is a kind of precise coating process, which is widely used in the production of membrane materials. The thickness deviation of coating along the transverse direction is the key index to measure the coating quality. The homogeneity of the transverse distribution of the coating velocity at the exit of the die is the most direct factor leading to the transverse thickness deviation of the coating. To predict the transverse homogeneity of slot coating, a mathematical model for the internal flow field in the die was established, where the incompressible Navier-Stokes equations were used to describe the fluid flow and generalized Newtonian fluid model was used to describe the rheological behaviour of non-Newtonian coating. An efficient explicit splitting algorithm based on finite element method is proposed to solve the model. To verify the effectiveness of the model and the algorithm, the internal flow field of a trapezoidal double cavity die with equal section was simulated. The transverse homogeneity obtained was close to the experimental measured value, which verified the validity of the model. Furthermore, the influence of measuring slot width on the transverse homogeneity of outlet velocity is studied, and the simulation results are consistent with the theoretical expectations.
**Keywords.** Non-Newtonian fluids; computational fluid dynamics; numerical simulation; slot coating; internal flow field; finite element method; transverse homogeneity.

## 1. Introduction

Slotted coating, as a popular coating method, has broad applications in manufacturing of high performance products. However, because the mechanisms of slotted coating are still poorly understood, the research and development costs are very high, and the coating quality and accuracy are very difficult to control. Therefore, it is urgent to carry out mechanical modelling and simulation research on the coating process, to provide theoretical support for the design of slotted die and the precise control of the coating process [1-3].

With the development of computer technology, numerical simulation has played an important role in the innovation of slotted coating process. The die structure can be optimized and improved according to the simulation results. The early research was limited by the computer hardware, and it was difficult to realize the 3D CFD simulation of the flow field in the die. At that time, the Navier-Stokes equation was greatly simplified according to the geometric symmetry of the die and the flow characteristics of the coating liquid inr the die. These simplified models were used to predict the flow field in the die, and many important conclusions and rules were summarized which are of guiding significance to the production practice [4-6].

The simplified models mentioned above did not consider the structure and shape of the die cavity, so it cannot be used for the quantitative analysis of the whole coating die but only applicable to the slot part. For the given coating requirements and coating liquid parameters, the simplified models cannot answer the question how to design the die structure. Due to the limitations of simplified models, many researchers begin to focus on the global integrated modelling and analysis of the flow field in the whole die. Lin et al. [7] simulated the 2D flow field of a slotted die and found that the coating window predicted by theory is larger than that observed by experiment, and both trends are qualitatively consistent. Based on Fluent software, Lee et al. [8, 9] studied the flow characteristics of Newtonian and non-Newtonian coatings in the process of slotted coating, providing a basis for the optimization design of slotted coating moulds. A fully coupled model proposed by Siqueira et al. [10, 11] was applied to the slotted coating flow. The slotted coating flow of slender particle suspension was analysed and the particle concentration and average orientation in the coating were studied. Campana et al. [12] used the finite element method to study the slotted coating flow of non-colloidal suspension by solving the mass and momentum conservation equations coupled with the particle transport equation. Reboucas et al. [13] used the stabilized finite element method to solve the conservation equation and the elliptic mesh method to calculate the position of the free surface. In a typical slotted coating flow, they analysed the influence of particle migration and rheology of the coating on the final particle distribution of the coating and its technological limitations. Kim et al. [14] established a 3D computational fluid dynamics model for analyzing the dynamics of the outlet ink and the wet thickness of the coating film of the slot-coater.

To sum up, at present, the numerical simulation research on slotted coating is still mainly 2D, and the full 3D simulation research is relatively few. For slotted coating, the thickness deviation of the coating along the transverse direction is a key index to measure the coating quality [15]. The homogeneity of the transverse distribution of the coating velocity at the exit of the die is the most direct factor leading to the transverse thickness deviation of the coating. Modelling and simulation the homogeneity of the transverse velocity distribution of the coating at the exit of the die is helpful to improve the transverse homogeneity of the coating.

Due to the lack of symmetry of the flow field in the die, 2D simulation cannot predict the transverse homogeneity of the coating velocity. However, at present, the numerical simulation research on slot coating is still mainly focused on 2D cases. In view of this, 3D modelling and simulation of the flow field in the slotted coating die are carried out in this paper. The purpose is to present a mathematical model and numerical algorithm for predicting the transverse homogeneity of slot coating. To verify the effectiveness of the model and the algorithm, the internal flow field of a trapezoidal double cavity die with equal section was simulated and analysed.

## 2. Mathematical Model of Flow Field in the Die

The real physical process is very complicated. In the range of current cognitive level and computing ability, to realize the simulation of the internal flow field of the slotted coating process, some reasonable physical assumptions need to be made first. According to the physical process of slotted coating and the physical properties of the coating liquid, we put forward the following five hypotheses:

- During the process of slotted coating, the injection flow rate of coating liquid at the die inlet is usually small, and the pressure inside the die is approximately the atmospheric pressure, so the coating liquid can be assumed to be incompressible.
- During the process of slotted coating, the flow is relatively slow, the viscous heat generated by energy dissipation can be ignored, and the ambient temperature is constant, so the coating process is assumed to be an isothermal process.
- The coating liquid is usually dilute polymer solutions with weak interaction between macromolecules and similar mechanical response behaviour in all directions. Therefore, the coating liquid is assumed to be isotropic fluid.
- It is assumed that the physical properties of the coating liquid are stable, and the rheological behaviour remains unchanged throughout the coating process.

- Assuming that the coating liquid has been evenly mixed when injecting to the die, and no phase separation and chemical reaction occurs in the coating process, so that it can be regarded as a single-phase fluid.

### *2.1. Governing Equations of the Flow Field*

Based on the above assumptions, the internal flow field of slotted coating can be described by the 3D isothermal incompressible Navier-Stokes equations. The specific form of the governing equation is as follows:

$$\nabla \cdot \mathbf{u} = 0 , \tag{1}$$

$$\rho\left(\frac{\partial \mathbf{u}}{\partial t} + \mathbf{u} \cdot \nabla \mathbf{u}\right) = -\nabla p + \nabla \cdot \mathbf{T} . \tag{2}$$

The two equations represent the conservation of mass and momentum, respectively. Where $\mathbf{u}$ is the velocity vector ( $\mathrm{m/s}$ ), $p$ is the dynamic pressure ( $\mathrm{Pa}$ ), $\mathbf{T}$ is the extra stress tensor, and $\rho$ is the liquid density ( $\mathrm{kg/m^3}$ ). For the 3D case

$$\mathbf{x} = (x,\ y,\ z)^{\mathrm{T}},\ \mathbf{u} = (u,\ v,\ w)^{\mathrm{T}},\ \mathbf{T} = \begin{pmatrix} T_{xx} & T_{xy} & T_{xz} \\ T_{xy} & T_{yy} & T_{yz} \\ T_{xz} & T_{yz} & T_{zz} \end{pmatrix},$$

where $u$, $v$, $w$ are the velocity component at the $x$, $y$, $z$ directions, respectively.

### *2.2. Boundary Conditions of the Flow Field*

The boundary conditions of the internal flow field can be divided into three parts: inlet, outlet and solid wall. For the inlet boundary, we assume the entrance is a circle of radius $r$ and the inlet flow rate is $\dot{\mathbf{q}}=\dot{q}\mathbf{n}_{in}$, where $\mathbf{n}_{in}$ is the unit normal vector of the inlet section along the flow direction. Then the inlet velocity can be set as

$$\mathbf{u} = \frac{\dot{\mathbf{q}}}{\pi r^2} = \frac{\dot{q}}{\pi r^2}\mathbf{n}_{in} . \tag{3}$$

For the outlet boundary, the boundary conditions are difficult to specify precisely. It is always assumed that the flow is fully developed at the outlet and that the flow velocity does not change along the normal direction of the outlet section, that is

$$\frac{\partial \mathbf{u}}{\partial \mathbf{n}} = 0 . \tag{4}$$

The outlet pressure is directly taken as the external environment pressure, i.e

$$p = p_e , \tag{5}$$

where $p_e$ is the ambient pressure, usually taken as one atmosphere. But in the negative pressure environment, the outlet pressure should set the value based on the actual situation. For the solid wall, because the fluid will adhere to the solid surface during the flow process, non-slip boundary condition $\mathbf{u} = 0$ is usually adopted at the solid wall, i.e.,

$$u = v = w = 0 . \tag{6}$$

In addition to the outlet, the boundary conditions for the pressure are all set as

$$\frac{\partial p}{\partial \mathbf{n}} = 0 . \tag{7}$$

### *2.3. Constitutive Equation of Coating Liquid*

In practical applications, coating liquids are of various types and their rheological properties are widely distributed. Some are typical Newtonian fluids, others exhibit generalized Newtonian fluid properties, and even some are viscoelastic fluids. The coating liquid considered in this paper is mainly generalized Newtonian fluid. For generalized Newtonian fluids, there is nonlinear dependence between stress tensor and deformation rate tensor, and the viscosity is related to flow state. If we define the shear rate

$$\dot{\gamma}=\sqrt{\frac{1}{2}\dot{\boldsymbol{\gamma}}:\dot{\boldsymbol{\gamma}}}, \tag{8}$$

then the viscosity $\eta$ is a function of $\dot{\gamma}$, namely $\eta=\eta(\dot{\gamma})$. Here, $\dot{\boldsymbol{\gamma}}=\nabla\mathbf{u}+\nabla\mathbf{u}^T$ is the deformation rate tensor. At this point, the equation for calculating the extra stress of the coating liquid becomes to

$$\mathbf{T}=\eta(\dot{\gamma})\left(\nabla\mathbf{u}+\nabla\mathbf{u}^T\right). \tag{9}$$

So far, there are dozens of viscosity models for generalized Newtonian fluids, among which the Ostwald model and the Carreau model are the most widely used and applicable for coating liquids. The Ostwald model gives

$$\eta(\dot{\gamma})=K\dot{\gamma}^{\hat{n}-1}, \tag{10}$$

where $K$ is power law coefficient and $\hat{n}$ is power law index. The above law applies to the medium shear rate range (greater than $10\mathrm{s}^{-1}$), and in the lower shear rate region the following three-parameter model is used

$$\eta(\dot{\gamma})=\frac{\eta_0}{1+\left(\dfrac{k}{\eta_0}\right)\dot{\gamma}^{1-\hat{n}}}. \tag{11}$$

The Carreau model gives

$$\eta(\dot{\gamma})=\eta_\infty+(\eta_0-\eta_\infty)\left[1+(\lambda\dot{\gamma})^2\right]^{\frac{\hat{n}-1}{2}}, \tag{12}$$

where $\eta_\infty$ is infinite shear viscosity, $\eta_0$ is infinitesimal shear viscosity (zero shear viscosity), $\lambda$ is time constant, and $\hat{n}$ is power law index.

## 3. Numerical Method

In order to improve the computational efficiency, we first discretize the governing equation explicitly, and then use operator splitting to decouple the velocity and pressure. The solving process includes the following three steps:

- Step 1: intermediate velocity calculation

$$\frac{\mathbf{u}^*-\mathbf{u}^n}{\Delta t}=-\mathbf{u}^n\cdot\nabla\mathbf{u}^n+\frac{1}{\rho}\nabla\cdot\left[\eta(\dot{\gamma}^n)\left(\nabla\mathbf{u}^n+\left(\nabla\mathbf{u}^n\right)^{\mathrm{T}}\right)\right], \tag{13}$$

- Step 2: pressure calculation

$$\nabla^2 p^{n+1}=\frac{\rho}{\Delta t}\nabla\cdot\mathbf{u}^*, \tag{14}$$

- Step 3: velocity correction

$$\frac{\mathbf{u}^{n+1}-\mathbf{u}^{*}}{\Delta t}=-\frac{1}{\rho}\nabla p^{n+1}, \tag{15}$$

where $\mathbf{u}^{*}$ is the intermediate velocity, and the superscript $n$ and $n+1$ represent the time step. The above equations are discretized by finite element method. The basis function is denoted by $\phi(x,y,z)$, and each variable adopts equal-order linear interpolation. The discretized equations are of the form:

- Discrete form of step 1

$$\frac{1}{\Delta t}\mathbf{M}\tilde{u}^{*}=\frac{1}{\Delta t}\mathbf{M}\tilde{u}^{n}-\mathbf{C}^{n}\tilde{u}^{n}-\frac{1}{\rho}\mathbf{K}_{u}^{n}\tilde{u}^{n}, \tag{16}$$

$$\frac{1}{\Delta t}\mathbf{M}\tilde{v}^{*}=\frac{1}{\Delta t}\mathbf{M}\tilde{v}^{n}-\mathbf{C}^{n}\tilde{v}^{n}-\frac{1}{\rho}\mathbf{K}_{u}^{n}\tilde{v}^{n}, \tag{17}$$

$$\frac{1}{\Delta t}\mathbf{M}\tilde{w}^{*}=\frac{1}{\Delta t}\mathbf{M}\tilde{w}^{n}-\mathbf{C}^{n}\tilde{w}^{n}-\frac{1}{\rho}\mathbf{K}_{u}^{n}\tilde{w}^{n}, \tag{18}$$

- Discrete form of step 2

$$\mathbf{K}_{p}\tilde{p}^{n+1}=-\frac{\rho}{\Delta t}\left(\mathbf{Q}_{u}\tilde{u}^{*}+\mathbf{Q}_{v}\tilde{v}^{*}+\mathbf{Q}_{w}\tilde{w}^{*}\right), \tag{19}$$

- Discrete form of step 3

$$\mathbf{M}\tilde{u}^{n+1}=\mathbf{M}\tilde{u}^{*}-\frac{\Delta t}{\rho}\mathbf{Q}_{u}\tilde{p}^{n+1}, \tag{20}$$

$$\mathbf{M}\tilde{v}^{n+1}=\mathbf{M}\tilde{v}^{*}-\frac{\Delta t}{\rho}\mathbf{Q}_{v}\tilde{p}^{n+1}, \tag{21}$$

$$\mathbf{M}\tilde{w}^{n+1}=\mathbf{M}\tilde{w}^{*}-\frac{\Delta t}{\rho}\mathbf{Q}_{w}\tilde{p}^{n+1}, \tag{22}$$

where $\tilde{u}$, $\tilde{v}$, $\tilde{w}$ and $\tilde{p}$ are node values of each physical quantity. The computation form of each matrix is as follows:

$$\mathbf{M}_{ij}=\left(\phi_i,\ \phi_j\right),\ \left(\mathbf{C}_{u}^{n+1}\right)_{ij}=\left(\phi_i,\ u^{n}\frac{\partial\phi_j}{\partial x}+v^{n}\frac{\partial\phi_j}{\partial y}+w^{n}\frac{\partial\phi_j}{\partial z}\right),$$

$$\left(\mathbf{K}_{u}^{n}\right)_{ij}=\left(\frac{\partial\phi_i}{\partial x},\ \eta^{n}\frac{\partial\phi_j}{\partial x}\right)+\left(\frac{\partial\phi_i}{\partial y},\ \eta^{n}\frac{\partial\phi_j}{\partial y}\right)+\left(\frac{\partial\phi_i}{\partial z},\ \eta^{n}\frac{\partial\phi_j}{\partial z}\right),$$

$$\left(\mathbf{K}_{p}\right)_{ij}=\left(\frac{\partial\phi_i}{\partial x},\ \frac{\partial\phi_j}{\partial x}\right)+\left(\frac{\partial\phi_i}{\partial y},\ \frac{\partial\phi_j}{\partial y}\right)+\left(\frac{\partial\phi_i}{\partial z},\ \frac{\partial\phi_j}{\partial z}\right),$$

$$\left(\mathbf{Q}_{u}\right)_{ij}=\left(\phi_i,\ \frac{\partial\phi_j}{\partial x}\right),\ \left(\mathbf{Q}_{v}\right)_{ij}=\left(\phi_i,\ \frac{\partial\phi_j}{\partial y}\right),\ \left(\mathbf{Q}_{w}\right)_{ij}=\left(\phi_i,\ \frac{\partial\phi_j}{\partial z}\right),$$

where $(f,g)=\int_{\Omega}f\cdot g\mathrm{d}\mathbf{x}$. In order to improve the computational efficiency, mass lump method is used in the first and third steps, and conjugate gradient method is used in the second step.

## 4. Simulation Result

### *4.1. Die Geometry*

Consider a trapezoidal double cavity die with equal section as shown in figure 1. The geometric parameters are list in table 1.

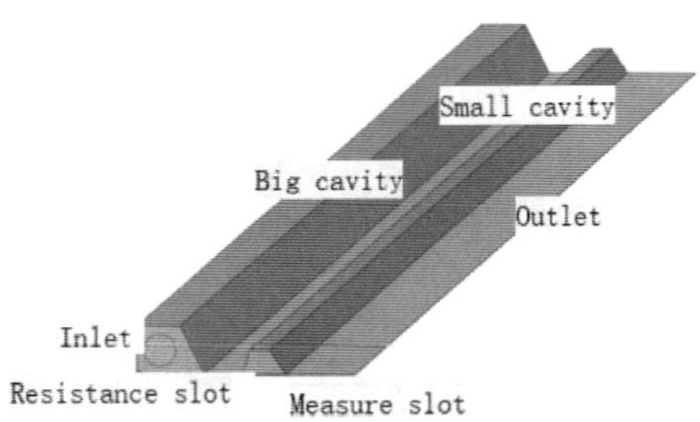


**Figure 1**. Schematic diagram of the trapezoidal double cavity die with uniform cross section.

**Table 1**. Geometric parameters of the trapezoidal double cavity die with uniform cross section.

| Parameter | Value | Parameter | Value |
|---|---|---|---|
| Upper width of big cavity | 15 mm | Transverse length of the die | 1000 mm |
| Bottom width of big cavity | 25 mm | Width of the resistance slot | 14 mm |
| Height of big cavity | 19 mm | Width of the measure slot | 25 mm |
| Upper width of small cavity | 5 mm | Height of the slot | 1 mm |
| Bottom width of small cavity | 16 mm | Diameter of the inlet | 12 mm |
| Height of small cavity | 10 mm | | |

*4.2. Constitutive Equation of Coating Liquid*

The coating liquid used in this paper consists of PVA, water and active agent. Both the Ostwald and Carreau models are used to fit the viscosity of the coating liquid. By comparison, the Ostwald model gives better results and the fitting error is less than 1.75%. The obtained constitutive equation is

$$\eta(\dot{\gamma})=\begin{cases}\dfrac{8.6491}{1-0.0456\dot{\gamma}^{-2.5793}} & \dot{\gamma}\le 10\text{s}^{-1}\\ 8.7282\dot{\gamma}^{0.9897-1} & \dot{\gamma}>10\text{s}^{-1}\end{cases}. \tag{23}$$

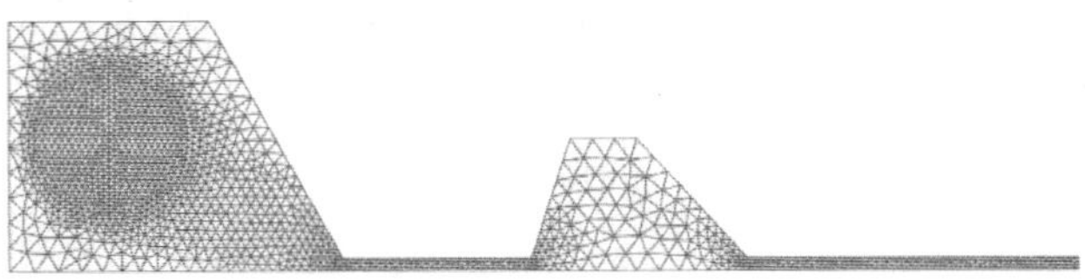

**Figure 2.** Lateral view of the mesh of the trapezoidal double cavity die with uniform cross section.

*4.3. Simulation Results of Flow Field in the Die*

In simulation, the die is divided into tetrahedral grids. Figure 2 shows mesh, which contains 4,756,915 tetrahedral elements. The coating density is set as $1000\text{kg/m}^3$, and the inlet flow rate is set as $0.5\text{kg/min}$. Figure 3 shows the simulation results, and the corresponding coating velocity distribution on the center line of the slot outlet is given in figure 4. It can be seen that the velocity near the entrance side is 13.25 mm/s, while the velocity far from the entrance side is 12.46 mm/s. The velocity distribution inhomogeneity along the transverse is about 6.34%, which is close to the experimental measurement value.

*4.4. Effect of Measure Slot Width on Transverse Homogeneity*

Next, we discuss the effect of measure slot width on transverse homogeneity. In the simulations, the inlet flow rate is set as $1.0\text{kg/min}$. Five cases of measure slot widths of 10mm, 14mm, 20mm, 25mm and 30mm are considered here. The numbers of tetrahedral grid cells used in the simulations are 3397876, 3753300, 4265063, 4756915 and 5104859, respectively. Figure 5 shows the coating velocity distribution on the center line of the slot outlet under various conditions and the trend of transverse inhomogeneity with respect to the slot width. From these results, it can be concluded that increasing the width of the measure slot can improve the transverse homogeneity of the outlet velocity. This result is consistent with the theoretical expectation. More specifically, as can be seen from figure 5(f), when the measure slot width is greater than 25 mm, the effect of increasing the measure slot width on improving the transverse homogeneity is no longer significant, which is consistent with the experimental results. In applications, measure slot width is usually set as 25 mm according to experience. Therefore, our simulation results can provide theoretical basis for practical applications.

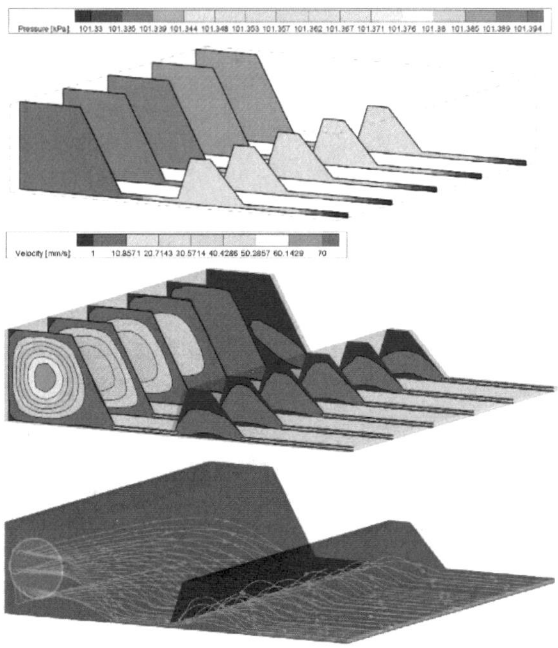


**Figure 3.** The simulation results of flow field in the die. From top to bottom are pressure, velocity and streamlines, respectively.

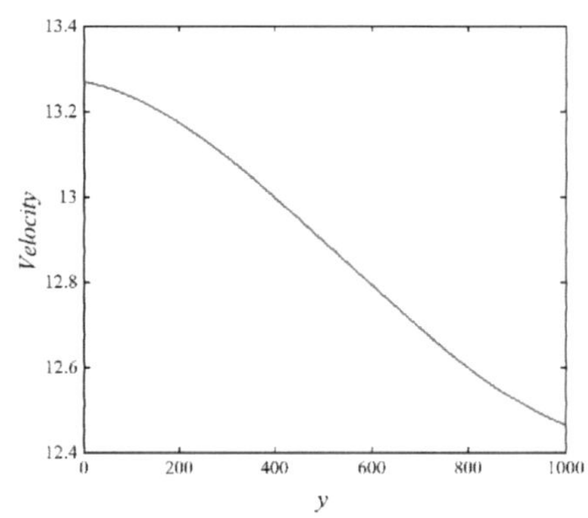


**Figure 4.** Coating velocity distribution on center line of the die exit.

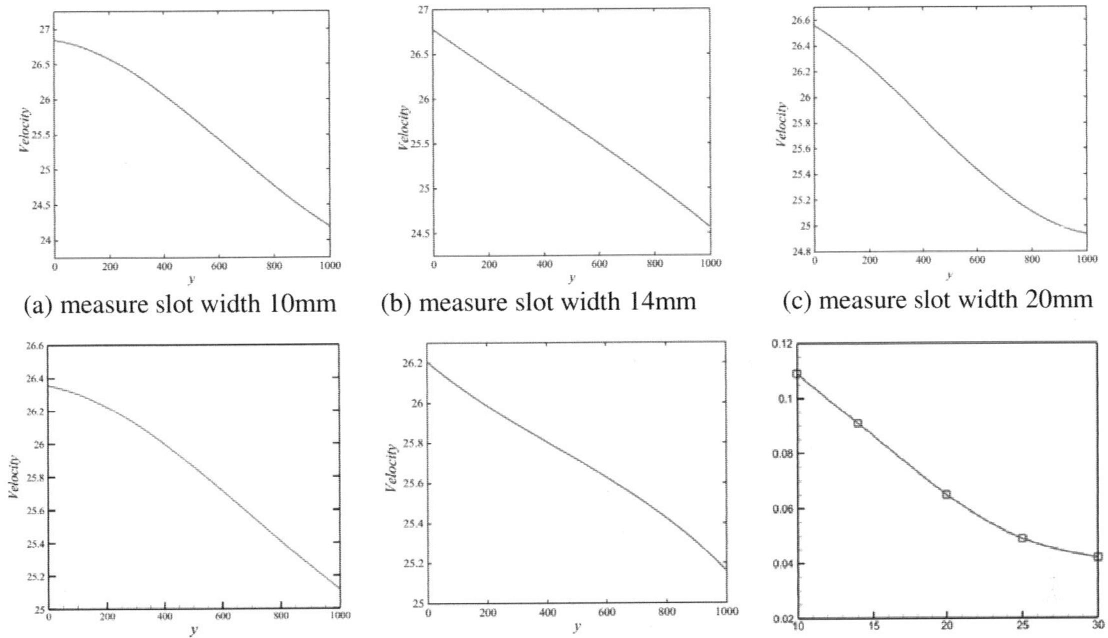

(a) measure slot width 10mm (b) measure slot width 14mm (c) measure slot width 20mm

(d) measure slot width 25mm (e) measure slot width 30mm (f) transverse inhomogeneity

**Figure 5**. Coating velocity distribution on center line of the die outlet under different measuring slot widths and the trend of transverse inhomogeneity with respect to the slot width.

## 5. Conclusions

We have established a mathematical model of flow field in slotted coating die. Incompressible Navier-Stokes equations were used to describe fluid flow, and generalized Newtonian fluid model was used to describe the rheological behaviour of coatings. Based on the finite element method, a numerical method for solving the model is proposed. To improve the efficiency, the Navier-Stokes equations were decoupled by time splitting algorithm, the first-order explicit scheme was used for time discrete, linear finite element method was used for space discrete, mass lump method was used for velocity equation, and conjugate gradient method was used for pressure equation. To verify the effectiveness of the model and algorithm, they were applied to simulate the internal flow field of a trapezoidal constant section double cavity die. The cavity was divided by tetrahedral mesh, and the constitutive relation of coating liquid was obtained by fitting the experimental data. The flow field distribution information inside the die was obtained by simulation, and the coating velocity distribution on the centre line of the die outlet was further analysed. The simulation results show that the transverse inhomogeneity is about 6.34%, which is close to the experimental measurement, thus verifying the validity of the model. Further on, the influence of measure slot width on transverse homogeneity is studied. The results show that the transverse inhomogeneity can be improved by increasing the width of the measure slot width, which is consistent with the theoretical expectation. When the measure slot width is larger than 25mm, the improvement is no longer significant, which is consistent with the experimental results. Therefore, our model can provide theoretical basis for practical applications.

## References

[1] Verma A, Martineau D, et al. 2020 *J. Mater. Chem. C* **8(18)** 6124-6135.
[2] Patidar R, Burkitt D, et al. 2020 *Mater. Today Commun.* **22** 100808.
[3] Kasischke M, Hartmann S, et al. 2021 *arXiv preprint*, arXiv:2101.01614.
[4] Ji HS, Ahn W G, et al. 2016 *Chem. Eng. Sci.* **143** 122-129.

[5] Lee Y W, Ahn W G, et al. 2017 *Rheol. Acta* **56(9)** 707-717.
[6] Jang I, Song S 2013 *Int. J. Heat Fluid FL.* **40** 180-185.
[7] Lin C F, Hill Wong D S, et al. 2010 *J. Polym. Process. Inst.* **29(1)** 31-44.
[8] Lee S H, Koh H J, et al. 2011 *Chem. Eng. Sci.* **66(21)** 4953-4959.
[9] Lee S, Nam J. 2015 *AIChE J.* **61(5)** 1745-1758.
[10] Siqueira IR, Rebouças RB, Carvalho MS 2017 *AIChE J.* **63(7)** 3187-3198.
[11] Siqueira IR, Carvalho MS 2019 *J. Coat. Tech. Res.* **16(6)** 1619-1628.
[12] Campana DM, Valdez Silva LD, Carvalho MS 2017 *AIChE J.* **63(3)** 1122-1131.
[13] Rebouças RB, Siqueira IR, Carvalho MS. 2018 *J. Non-Newton. Fluid* **258** 22-31.
[14] Kim S, Lee J, Lee C 2019 *J. Adv. Manuf. Tech.* **104** 2991-2997.
[15] Tan P, Diao S, et al. 2020 *Chem. Eng. Sci.* **222** 115716.

# Numerical Simulation of Flow Past Circular Cylinder Based on Overset Grid Method

**Liwei Song**

School of science, University of Science and Technology Liaoning, Anshan, Liaoning, China
Email: ur135@qq.com

**Abstract**. Based on overset grid method, the problems about flow past circular cylinder were numerically studied, which include the flow around a single circular cylinder at Reynolds number Re=100 and Reynolds number Re=200, the flow past two tandem circular cylinders at Reynolds number Re=200 and dimensionless distance $g^*$=4, and the flow around a lateral oscillating circular cylinder at Reynolds number Re=200. Overset grid method is effective in dealing with the problems of flow over multi-body and flow with moving boundaries. This grid method couple overlapping regions in an arbitrary manner through flow field information updating over acceptor cells in one region suing its donor cells in another region. The computed drag coefficient and lift coefficients of the circular cylinder, vortical structure in the flow wake and vortex shedding P+S mode are agreed well with the results in previous experiment investigations and numerical simulations.
**Keywords.** Overset grid method, oscillating circular cylinder, vortex shedding mode.

## 1. Introduction

The flow around a circular cylinder is a basic problem in computational fluid dynamics [1]. It is a typical example for the study of flow mechanism and the verification of numerical methods. Overset grid method [2] is different from body conformal grid. It divides different sub domain grids, and obtains the overall computing grid through the overlap between sub domain grids. This method is usually used for multi-body boundary and moving boundary problems, and has been widely used because of its convenient grid generation.

In this paper, the overset grid method is adopted in the simulations of flow over a single cylinder, flow over two tandem circular cylinders and and flow over an oscillating circular cylinder to obtain the lift coefficient, drag coefficient and the vortical structure in the wake flow field.

## 2. Numerical Simulation

### 2.1. Mathematical Model

The Navier-Stokes equations govern the two-dimensional unsteady incompressible fluids and are given in primitive variable form as:

$$\frac{\partial \boldsymbol{u}}{\partial t} + \boldsymbol{u} \cdot \nabla \boldsymbol{u} = -\nabla p + \frac{1}{\mathrm{Re}} \nabla^2 \boldsymbol{u} \tag{1}$$

$$\nabla \cdot \boldsymbol{u} = 0 \tag{2}$$

where $p$ is the fluid kinetic pressure, $\boldsymbol{u}$ is the fluid velocity vector, and Re represent the Reynolds number.

### 2.2. Solving Method

In the present work, SIMPLE [3] method is used to solve the incompressible Navier-Stokes equations. The Navier-Stokes equations are solved iteratively for the flow variables. This approach is also called pressure speed amending method. Pressure is obtained by solving Poisson equation for pressure-correction. Velocity coupling pressure is solved with the SIMPLE where the continuity equation constraint on momentum equation can be achieved by solving a pressure-correction equation:

$$p^{n+1} = p^n + \omega p' \quad (3)$$

$$\boldsymbol{v}_p^{n+1} = \boldsymbol{v}_p^* - \frac{V\nabla p'}{a'^{v}_{p}} \quad (4)$$

where $\omega$ is the relaxation factor, $p'$ is the pressure corrections, $a'^{v}_{p}$ is the central coefficient vector, and V is the volume.

## 3. Overset Grid Method

Overset grid method [4] is employed to divide the computational domain into several overlapping regions in an arbitrary manner and the computational grid of each region is constructed separately. They are most useful in problem with moving body or multi-body where a body can be enclosed in a single region and set to different velocity and position.

A typical overset grid has a background mesh covering all computational domain and one overlapping region that surrounds a body such as a cylinder in figure 1(a). The outer boundary of the overlapping region is defined as an overset boundary. The overset hole-cutting procedure is essentially a cell marking procedure as shown in figure 1(b). The hole in the background region is virtual because no cells are actually removed. They are deactivated and have no part in the solution process.

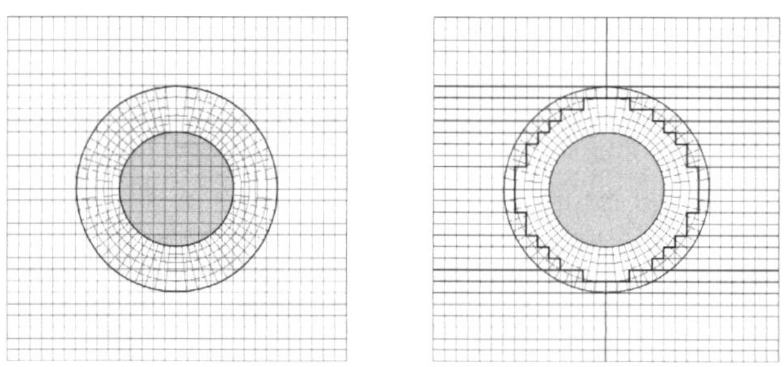

(a) (b)

**Figure 1**. Schematic view of overset grid method for flow around cylinder.

The overset grid method couple regions through flow field information update of acceptor cells in one region from its donor cells in another region are as shown in figure 2. The information at the acceptor cell is obtained by

$$\varphi_a = f(\varphi_1, \varphi_2, \cdots, \varphi_N) \quad (5)$$

in which $\varphi_a$ is flow variable, $N$ is the number of its donor cells, and $f$ is linear interpolation, bilinear interpolation or other interpolation function that can be used to calculate the flow variables in the overset boundary to achieve the data exchange between different regions.

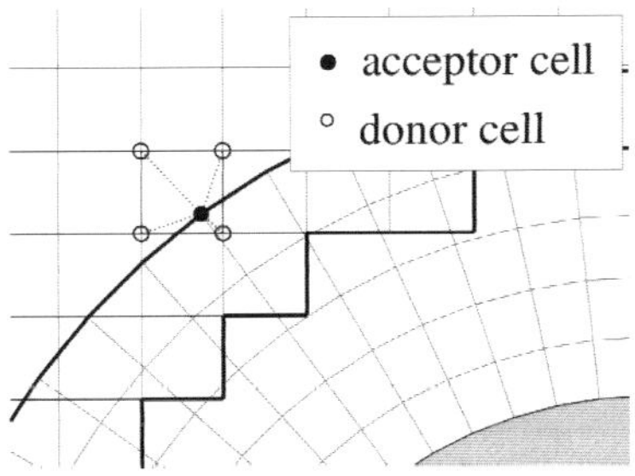


**Figure 2**. An acceptor cell and its donor cells in overset grid method.

## 4. Results

*4.1. Flow around Cylinder*

An unsteady flow around a cylinder is simulated to verify the accuracy of overset grid method. A rectangular computational domain is defined as background grid and discretized into non-uniform Cartesian grid. Another body conformal structural grid is established near the cylinder as shown in figure 1, which will be overlapped on the background grid. Figure 3 shows the computed vortical structure at Re=100 and Re=200. It can be visualized that the vortexes shed up and down alternatively and lead to Karman vortex shedding in the wake.

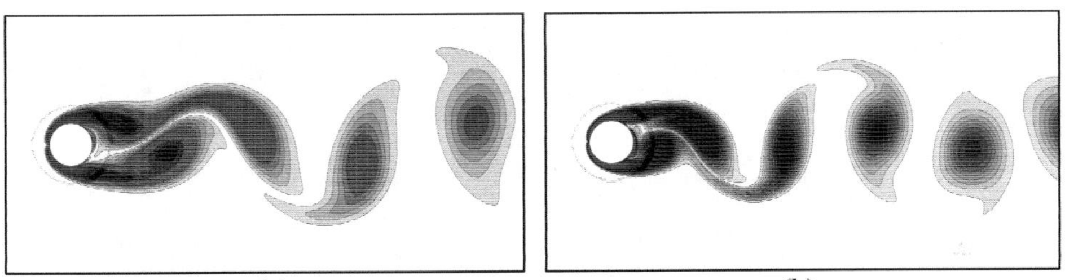

(a) (b)

**Figure 3**. Vorticity contours for (a) Re=100 and (b) Re=200.

The computed lift coefficient $C_L$, drag coefficient $C_D$ and vortex shedding Strouhal number St are shown in table 1. For comparison, the results of other numerical method are also listed in the table.

**Table 1.** Comparison of lift coefficients amplitude, mean drag coefficients and Strouhal numbers for different Reynolds numbers.

| | Re=100 | | | Re=200 | | |
|---|---|---|---|---|---|---|
| | $C_D$ | $C_L$ | St | $C_D$ | $C_L$ | St |
| Present | 1.39 | 0.340 | 0.169 | 1.37 | 0.70 | 0.200 |
| Chiu [5] | 1.35 | 0.303 | 0.167 | 1.37 | 0.71 | 0.198 |
| Xu [6] | 1.42 | 0.34 | 0.171 | 1.42 | 0.66 | 0.202 |
| Liu [7] | 1.35 | 0.339 | 0.164 | 1.31 | 0.69 | 0.192 |

*4.2. Flow around Two Tandem Cylinders*

The flow around two tandem cylinders is simulated to study the influence of proximity on vortex shedding. The dimensionless number $g^*=G/D$ is defined as the dimensionless distance between two cylinders, where D is the diameter and G is the minimum distance between the surfaces of two circular cylinders.

Figure 4 displays the vortical structure in the wake and figure 5 shows the time histories of the lift coefficient and drag coefficient at Re=200 and $g^*=4$.

**Figure 4**. Vorticity contours in the weak.

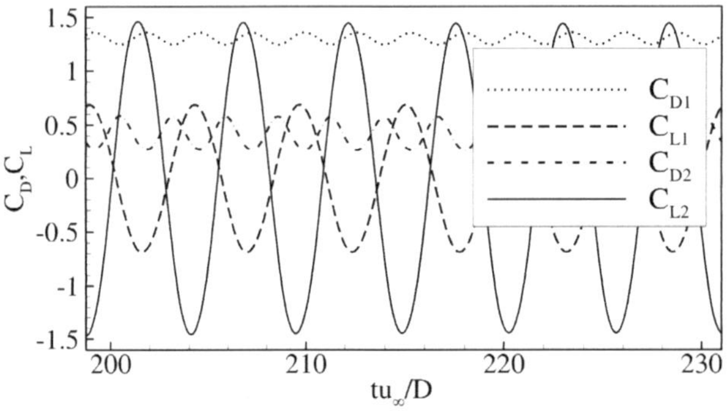


**Figure 5**. Lift coefficients and drag coefficients at Re=200 and $g^*=4$.

In table 2, the computed lift coefficient amplitude, mean drag coefficient and Strouhal number are compared with other numerical results. The identical Strouhal numbers for two circular cylinders show that the vortex shedding is synchronized, which is conform to experimental result [1].

**Table 2.** Lift coefficients amplitude, mean drag coefficients and Strouhal numbers at Re=200 and $g^*=4$.

| | $C_{D1}$ | $C_{L1}$ | $St_1$ | $C_{D2}$ | $C_{L2}$ | $St_2$ |
|---|---|---|---|---|---|---|
| Present | 1.30 | 0.70 | 0.185 | 0.42 | 1.44 | 0.185 |
| Chiu [5] | 1.27 | 0.65 | 0.180 | 0.35 | 1.23 | 0.180 |
| Farrant [8] | 1.25 | 0.71 | 0.179 | 0.38 | 1.59 | 0.179 |

### *4.3. Flow around Oscillating Cylinder*

The simulation of flow over an oscillating circular cylinder is used to study the mode of vortex shedding in the wake. The coordinate of the center position of the cylinder is denoted by $y=0.65D\sin(2\pi f_{ex}t)$, where $f_{ex}$ is the frequency of external forced oscillation. $f_{st}$ is the frequency of vortex shedding. When $f_{ex}/f_{st}=0.8$ the vortex shedding is P+S mode. A vortex pair and a single vortex shed from both sides of the cylinder surface respectively in a periodic. The current simulation is a long-distance oscillation, which can illustrate the ability of overset grid method to deal with moving boundary problem.

The lift coefficient and drag coefficient at Re=200 and $f_{ex}/f_{st}=0.8$ are displayed in figure 6. The asymmetric variation of the lift coefficient is caused by the vortex shedding P+S mode. Figure 7 shows the computed vorticity contours and the result in experiment [9].

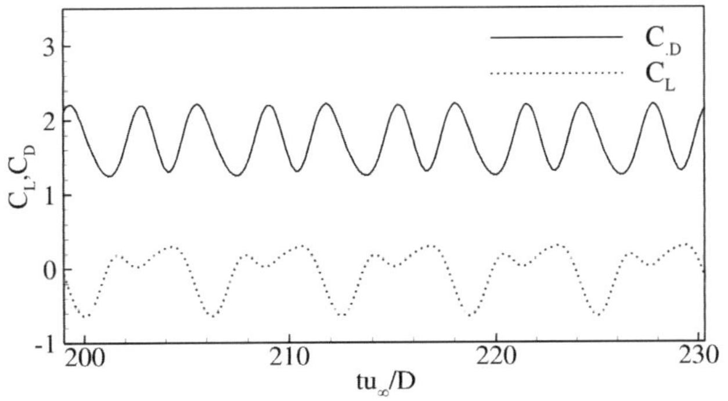


**Figure 6**. Lift and drag coefficient at Re=200 and $f_{ex}/f_{st}$=0.8.

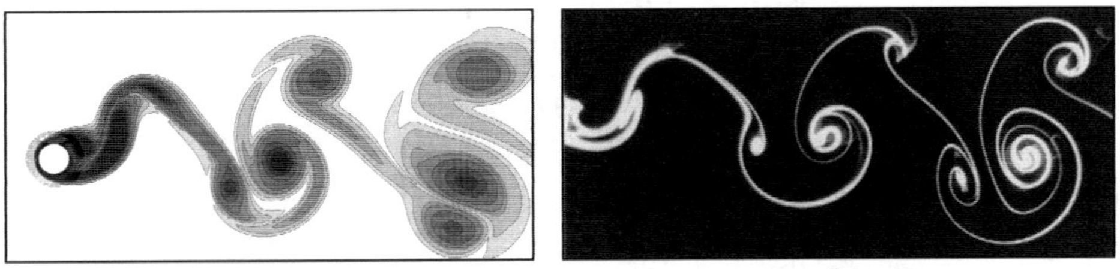

(a) (b)

**Figure 7**. Vorticity contours: (a) computed and (b) experiment.

## 5. Conclusion

The overset grid method is adopted in the simulations of flow over circular cylinder. All the computed results coincide with the experiment and other numerical results. It is proved that the method is effective in dealing with the flow over multi-body and the flow with moving boundaries.

## References

[1] Zdravkovich M M 1987 The effects of interference between circular cyllinders in cross flow *Jounnal of Fluids and Structures* **1**(2) 239–61

[2] Steger J L, Dougherty F C and Benek J A 1982 A Chimera grid scheme *ASME Mini-Symposium on Advances in Grid Generation* (Houston, TX) 59–69

[3] Patankar S V and Spalding D B 1972 A calculation procedure for heat, mass and momentum transfer in three-dimensional parabolic fows *International Journal of Heat and Mass Transfer* **15**(10) 1787–806

[4] Chen H, Qian L, Ma Z, Bai W, Li Y, Causon D and Mingham C 2019 Application of an overset mesh based numerical wave tank for modelling realistic free-surface hydrodynamic problems *Ocean Engineering* **176**(15) 97–117

[5] Chiu P H, Lin R K and Sheu T W H 2010 A differentially interpolated direct forcing immersed boundary method for predicting incompressible Navier-Stokes equations in time-varying complex geometries *Journal of Computational Physics* **229**(12) 4476–500

[6] Xu S and Wang Z J 2006 An immersed interface method for simulating the interaction of a fluid with moving boundaries *Journal of Computational Physics* **216**(2) 454–93

[7] Liu C, Zheng X and Sung C H 1998 Proconditioned multigrid methods for unsteady incompressible flows *Journal of Computational Physics* **139**(1) 35-57

[8] Farrant T, Tan M and Price W G 2001 A cell boundary element method applied to laminar vortex shedding from circular cylinders *Computers & Fluids* **30**(2) 211–36

[9] Williamson C H K and Govardhan R 2004 Vortex-induced vibrations *Annual Review of Fluid Mechanics* **36**(1) 413–55

# Study the Effect of SEN Depth on Jet Flow and Level Fluctuation under Permanent Magnet Controlling in Continuous Casting Mould by Numerical Simulation

**Zefeng Han [1, 2], Engang Wang [1, 2*] and Lin Xu[3]**

[1] School of Metallurgy, Northeastern University, Shenyang, 110004, P.R. China
[2] Key Laboratory of Electromagnetic Processing of Materials (Ministry of Education), Northeastern University, Shenyang, 110004, P.R. China
[3] Key Laboratory of Liaoning Province for Clean Combustion Power Generation and Heating Technology, Shenyang Institute of Engineering, Shenyang, 110136, P.R. China
Email: egawang@mail.neu.edu.cn

**Abstract.** In this paper, a new method named permanent magnet controlling mould to get majorization flow pattern and suppress fluctuation of free surface was proposed. The effect of permanent magnet controlling on the molten steel in the continuous casting process is simulated by using *k-ε* turbulence model. The features of permanent magnetic field and flow field inside a 1450 mm × 230 mm mould with combined permanent magnets have been calculated. The calculated results showed that the magnetic field generated by combined permanent magnets could take over both meniscus region and jet flow impingement region. Additionally, Lorentz force generated by the combined permanent magnets can effectively control the upper vortex and suppress the level fluctuation. With the SEN depth increasing, the fluctuation of free surface becomes more stable. However, the lower vortex core gets into the deeper of the slab.
**Keywords.** Permanent magnet, Level fluctuation, Continuous casting, SEN depth

## 1. Introduction

The crude steel produced by continuous casting (CC) process accounts for 96.6% of the total steel production [1]. How to ameliorate the quality of CC slab is the research focus of many scholars and Technologists. The quality of CC slab is affected by many factors, such as SEN depth [2], casting speed [3], argon injection [4], SEN port angle [5], and electromagnetic force [6]. During the CC process, SEN depth has a direct effect on the flow behaviour of liquid steel, fluctuation of free surface and solidification behaviour of continuous casting shell in the mould. Mallikarjuna et. al. [7] establishes a 1/2 scale water model to study the fluctuation of free surface in mould between different SEN depth and casting speed, and a safe operating matrix for avoiding mould slag entrapment was found. To obtain high quality slab, the electromagnetic braking (EMBr) techniques are widely used in CC process. The electromagnetic braking device consumes a lot of power resources when it's working. This paper proposes a type of permanent magnet controlling mould, which can control the jet flow and level fluctuation.

In this research, the VOF model is used to treat the steel phase and slag phase, and the *k-ε* turbulence model is used to study the flow behaviour in a CC slab mould. This paper is follow with interest the effect of the permanent magnets filed on the fluctuation of free surface and the flow behaviour in the mould for different SEN depths.

## 2. Description of Permanent Magnets Controlling Mould

The permanent magnets arranged by Halbach array are inlayed in the mould wall, which are 20mm to the inner wall of mold. One upper side of combined permanent magnets are equal to the free surface; the other one upper side of combined permanent magnets are 275mm to the free surface. The section of the slab is 1450 mm × 230 mm, and the calculation length of simulation is 800 mm (see figure 1).

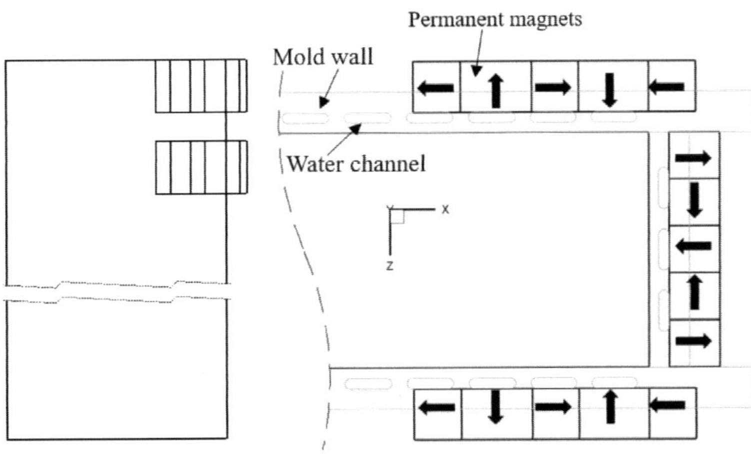


**Figure 1.** Schematic of the permanent magnets controlling mould.

## 3. Mathematical Model

### *3.1. Basic Assumption*

(1) Molten steel and slag are liquid Newtonian fluids.

(2) The permanent magnet has no individual performance difference.

(3) Heat transfer and solidification are ignored.

(4) The additional magnetic field of induced current is ignored, which is generated by molten steel after it passes through the stable magnetic field in the mould.

### *3.2. Equation of Flow Field*

(a) Continuity equation

$$\nabla \mathrm{u} = 0 \tag{1}$$

(b) Momentum equation

$$\partial_t u + (u \cdot \nabla) u + \frac{\nabla p}{\rho} = v \nabla^2 u + \frac{1}{\rho}(F_m + F_g) \tag{2}$$

where $u$, $\nabla p$, $v$, $F_m$ and $F_g$ are velocity, pressure gradient, kinematic viscosity, the electromagnetic forces and gravity, respectively.

(c) Standard $k$- $\varepsilon$ equation

$k$ equation:

$$\rho \frac{\partial (v_i k)}{\partial x_i} = \frac{\partial}{\partial x_i}\left[\left(\mu_l + \frac{\mu_t}{\sigma_k}\right)\frac{\partial k}{\partial x_i}\right] + G - \rho\varepsilon \tag{3}$$

$\varepsilon$ equation:

$$\rho\frac{\partial(v_i\varepsilon)}{\partial x_i}=\frac{\partial}{\partial x_i}\left[\left(\mu_l+\frac{\mu_t}{\sigma_\varepsilon}\right)\frac{\partial\varepsilon}{\partial x_i}\right]+c_l\frac{\varepsilon}{k}G-c_2\rho\frac{\varepsilon^2}{k} \tag{4}$$

where,

$$G=\mu_t\frac{\partial v_i}{\partial x_j}\left(\frac{\partial v_i}{\partial x_j}+\frac{\partial v_j}{\partial x_i}\right) \tag{5}$$

$$\mu_t=\rho c_\mu\frac{k^2}{\varepsilon} \tag{6}$$

where, $\mu_{\text{eff}}$ is effective viscosity coefficient, $\mu_t$ is kinetic eddy viscosity, $\mu_l$ is laminar viscosity coefficient; $k$ is turbulent kinetic energy, $\varepsilon$ is turbulent fluctuating kinetic energy dissipation rate. $c_1$, $c_2$, and $c_\mu$ is empirical constant, 1.44, 1.92, 0.09, 1.0, 1.3, respectively.

(d) MHD model

$$\boldsymbol{F}_{\text{m}}=\boldsymbol{j}\times\boldsymbol{B}_0 \tag{7}$$

where, $\boldsymbol{j}$ is the induced current, $\boldsymbol{B}_0$ is the magnetic flux density.

(e) VOF model

$$\frac{\partial\alpha_{st}}{\partial t}+\boldsymbol{v}\cdot\nabla\alpha_{st}=0 \tag{8}$$

where, $\alpha_{\text{st}}$=1 stands for the liquid molten steel; $\alpha_{\text{sl}}$=0 stands for the mould flux; and between 0 and 1, $\alpha_{\text{st}}$ represents the steel/slag interface.

(f) Computational domain

In order to exhibit the lower vortex flow effectively, the computational length was extended to 4 meters. The physical properties of permanent magnet are shown in table 1. The material properties and simulation parameters are shown in table 2.

**Table 1.** The physical properties of permanent magnet.

| Dimensions(mm) | Relative permeability | Magnetic Coercivity(kA) | Remanence(T) |
|---|---|---|---|
| 69×50×175, 46×50×175 | 1.099 | 890 | 1.4 |

**Table 2.** The material properties and simulation parameters.

| Parameter | Value | Parameter | Value |
|---|---|---|---|
| Casting speed ($m\cdot min^{-1}$) | 1.6 | Depth of SEN (mm) | 170, 190, 210 |
| Interface tension ($N\cdot m^{-1}$) | 1.2 | Magnetic permeability ($H\cdot m^{-1}$) | $1.257\times10^{-6}$ |
| Slab section (mm) | 1450×230 | Liquid steel viscosity ($kg\cdot m^{-1}\cdot s^{-1}$) | 0.0062 |
| Angle of nozzle port (°) | -15 | Electric conductivity ($S\cdot m^{-1}$) | 714000 |
| Molten steel density ($kg\cdot m^{-3}$) | 7020 | Liquid slag viscosity ($kg\cdot m^{-1}\cdot s^{-1}$) | 0.2664 |
| SEN outlet Size (mm) | 80×80 | Slag density ($kg\cdot m^{-3}$) | 3500 |
| Computational length (mm) | 4000 | | |

## 4. Results and Discussion

Figures 2 and figures 3 show the contour of magnetic flux density in height centre plane of permanent magnets and centre plane of width face of the mould. It can been seen that the maximum value of magnetic field is appeared in the corner of mould. The shape of magnetic field distribution seems "M" in height centre plane of permanent magnets. The magnetic flux intensity is sharply lower in deflection corner area, which is caused by permanent magnet arrangement.

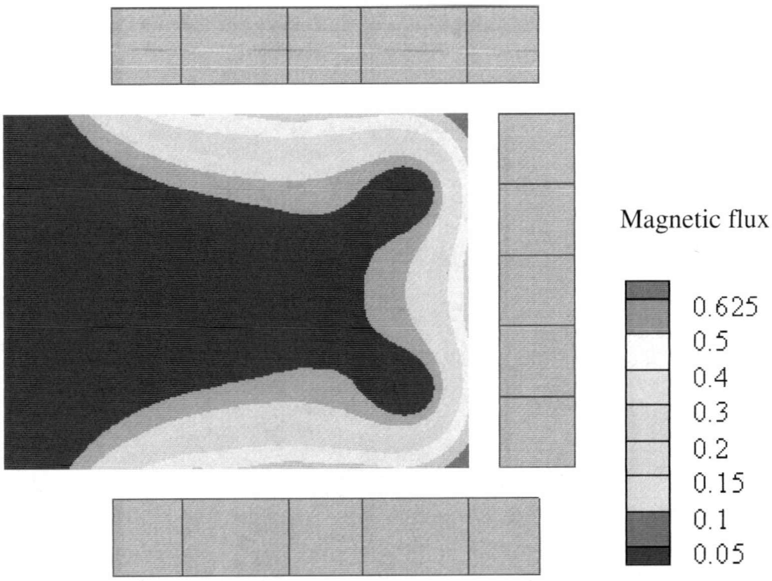


**Figure 2.** Contour of magnetic flux density in the height centre of permanent magnet.

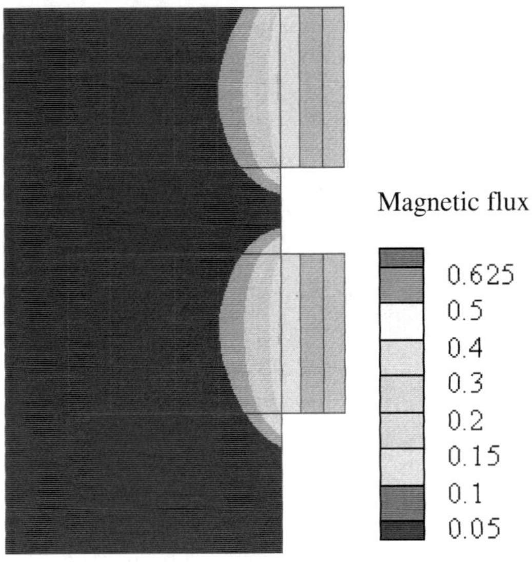


**Figures 3.** Contour of magnetic flux density in the centre of width side of mould.

Figure 4 presents the velocity distribution in the mid-plane at the mould wide face with different SEN depths. It is shown that the lower vortex centre is deeper with the increase of SEN depth.

However, the lower vortex centre is in the same position with the utilization of permanent magnets controlling mould. Moreover, the velocity of upper flow is also controlled.

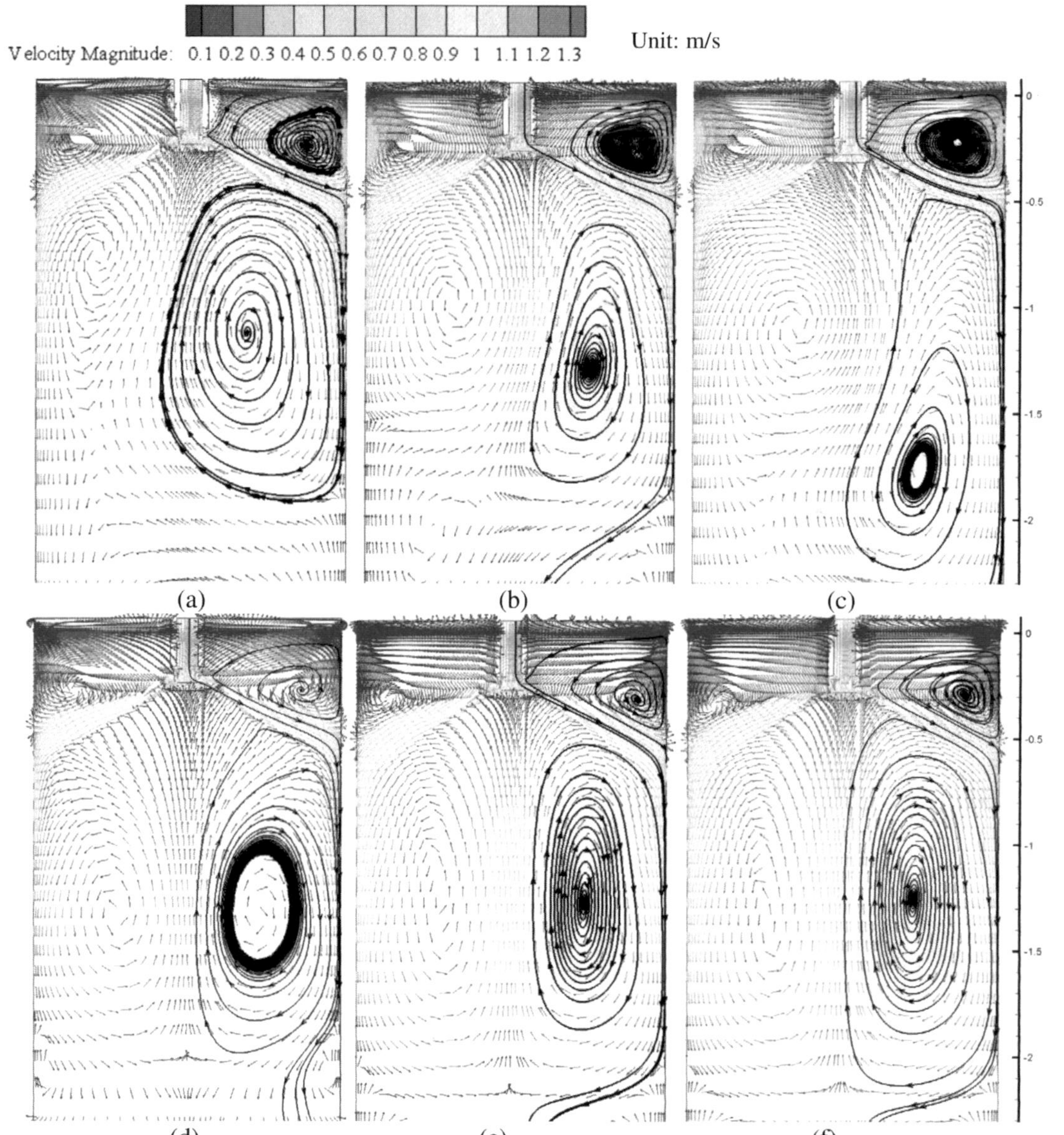


**Figure 4.** Distribution of velocity vector and streamline with different SEN immersion depth in the mid-plane with or without magnetic field: without magnetic field (a) 170mm, (b) 190mm, (c) 210mm; with magnetic field (d) 170mm, (e) 190mm, (f) 210mm.

The fluctuation of free surface and surface velocity with different SEN depths with magnetic field are illustrated in figure 5. When the permanent magnets controlling mould is used, the fluctuation and the maximum velocity is reduced. The fluctuation height is uneven at both sides of SEN, which means that the molten steel flow is swing left and right always. This situation has changed after the permanent magnets controlling mould is used.

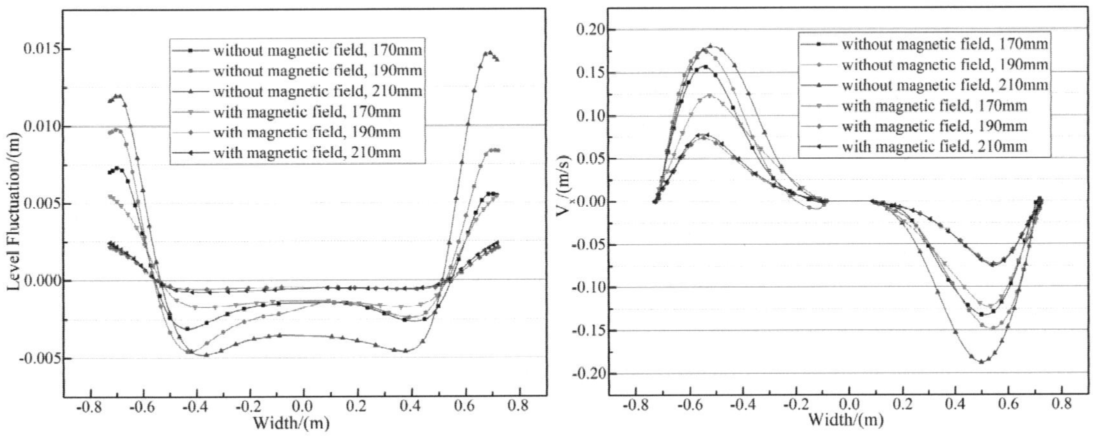


**Figure 5.** The fluctuation of free surface and surface velocity with different SEN immersion depth with or without magnetic field.

## 5. Conclusions

The permanent magnets controlling mould has two combined permanent magnets, which are arranged at both the meniscus region and jet flow impingement region near the thickness side of the mould. With the permanent magnets, the surface velocity and the fluctuation of free surface could be controlled effectively. Simultaneously, the steady flow in the mould can be obtained.

## Acknowledgments

The present work is supported financially by the National Natural Science Foundation of China (Grant No. 51574083 and Grant No. U1760206) and the 111 Project (2.0) of China (No. BP0719037).

## References

[1] Word Steel Association 2021 *World Steel in Figures* 2021 pp 11
[2] Ni P, Ersson M, Jonsson L, Zhang T, and Jönsson P 2018 *Metals* **8(11)** 910.
[3] Fei P, Yi M, Liu C, and Jiang M 2019 *International Journal of Minerals, Metallurgy and Materials* **26(2)** 186-193.
[4] Wu Y, Liu Z, Wang F, Li B, and Gan Y 2021 *Powder Technology* **387** 325-35.
[5] Lee W and Yi K 2020 *Metals and Materials International* **27(10)** 4168-181.
[6] Thomas B and Cho S 2018 *IOP Conference Series. Materials Science and Engineering* **424(1)** 12027.
[7] Rao P, Chougale V, Satish K, Rajendra T and Balachandran G 2021 *Metallurgical & Materials Engineering* **27(1)** 557-73

# CFD Simulation of the Cavitation Behaviour in Single Jet Process via Outlet Pressure

**Hao Zhang, Xinjie Chai, Shihao He, Facheng Qiu* and Zhiliang Cheng**

School of Chemistry and Chemical Engineering, Chongqing University of Technology, Chongqing 400054, China
Email: qfcandly@cqut.edu.cn

**Abstract.** Cavitation effect is the most common phenomenon in the process of jet. In this work, the method of CFD numerical simulation is used to study the cavitation behaviour in the process of single hole jet. In order to obtain the mixing situation and change process of gaseous water and liquid water produced in the nozzle, the Euler multiphase flow model and the realizable k-epsilon model in the turbulence model are used. The simulation results show that the degree and frequency of liquid column breakage of cavitation water jet are far higher than those of ordinary water jet. And the structure of the nozzle also has the most ideal situation. When the nozzle inlet size is 20 mm, the nozzle diameter is 1 mm, and the nozzle length is 5 mm, which is most conducive to the growth and diffusion of cavitation bubbles. The cavitation effect increases with the increase of the inlet jet velocity, but the promotion effect does not increase significantly when the inlet velocity is greater than 2.5 m/s. From the perspective of energy consumption, the optimum velocity at this time is 2.5 m/s. And the pressure environment at the nozzle outlet also affects the cavitation phenomenon, mainly as follows: positive pressure can inhibit the cavitation effect, negative pressure can promote the cavitation effect, but the negative pressure has a limited effect on the degree of cavitation.
**Keywords.** CFD, jet cavitation, nozzle inlet, jet velocity, outlet pressure.

## 1. Introduction

The jet breaking and atomization of liquid is a complex nonlinear dynamic evolution process. The so-called jet breaking refers to the phenomenon that some liquid falls off from the jet liquid column due to instability when the liquid is emitted from a relatively narrow aperture (nozzle). The phenomenon of jet fragmentation also exists in all aspects of today's society, such as faucet, shower head, fountain, fire gun, etc [1]. Since jet fragmentation was concerned by scholars, a lot of scientific research has been carried out at home and abroad [2]. For different application scenarios, many scholars have studied a variety of jet fragmentation forms and the main influencing factors behind them, such as liquid free jet, jet atomization in air flow and jet impact atomization.

Actually, the published research shows that the cavitation phenomenon of jet widely exists in various fields and has a significant impact on the jet fragmentation of fluid [3]. Sou et al. [4] examined the effect of cavitation on the atomization of liquid in the nozzle and jet. It was found that the state of the jet changed from the jet to the spray jet from the weak to the strong. The atomization of the liquid jet near the nozzle outlet depended on the cavitation state, indicating that the high intensity cavitation increased the effect of jet breakage. This confirms that the cavitation phenomenon in the injector will have a positive impact on the jet breaking and its primary breaking process. As for the effect of jet nozzle, Kumar et al. [5] and others believe that the different geometry of the nozzle outlet can change the flow mode before the jet occurs, which is very important for the instability of the jet. Etzold [6] et al. Conducted experiments with nozzles with small aspect ratio, and the data conclusions are inconsistent with the existing theoretical research on the rupture length of liquid jet. It is found that the round hole jet is a better choice for the formation of jet fragmentation in a non-special physical environment. Because the jet fragmentation and atomization of liquid is a complex and nonlinear dynamic evolution process, most scholars consider using the method of numerical simulation to explore the jet cavitation and achieve good results [7]. Thus, in our present work, the conditions of single circular hole jet are considered with the CFD simulation method.

In order to obtain a more specific research result of jet cavitation, different condition parameters, such as nozzle size, jet velocity and outlet pressure, were investigated. As for the mixing situation and change process of gaseous water and liquid water produced in the nozzle, the Euler multiphase flow model and the realizable k-epsilon model in the turbulence model are used. This research can provide support for the development of jet cavitation.

## 2. Numerical Simulation Condition Section

### *2.1. Cavitation Simulation Model*

The calculation model adopted is a simplified two-dimensional model to save calculation resources. As shown in the figure below: the liquid enters the small hole nozzle from the left to form a jet, and the right part is the jet development area. Firstly, the observation object is the jet development area, and the main research purpose is to explore the difference between ordinary jet and jet with cavitation effect, that is, cavitation jet development behaviour. Secondly, while the jet is formed, the influence mechanism of jet cavitation effect under various conditions is explored by mastering the gas-liquid two-phase evolution in the small hole nozzle and the cavitation bubble content parameters in the nozzle pipe.

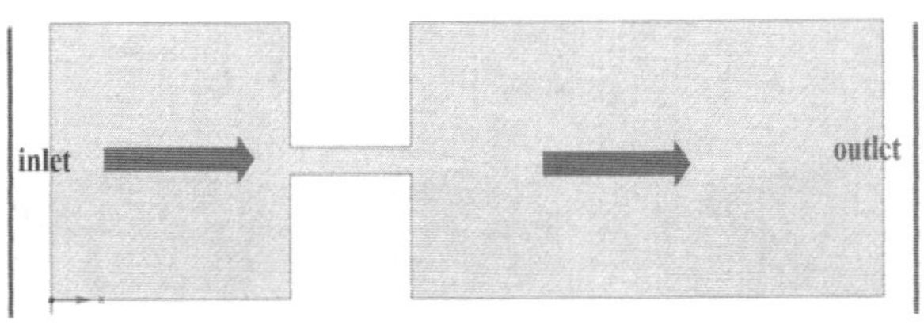


**Figure 1.** Schematic diagram of cavitation simulation model.

### *2.2. Mesh Generation and Independence Verification*

The main parameter size of the whole calculation domain is that the inlet size is 20 mm, the nozzle diameter is 1 mm and the nozzle length is 5 mm. In order to ensure the calculation accuracy, a structured grid division is adopted globally. Moreover, the nozzle area, especially the throat, needs to be properly densified according to the experimental needs. The number of grids selected by the model in the calculation is 65243, 142136 and 232564. Taking the volume fraction of internally generated cavitation as the physical reference, the calculation deviation of three grid numbers is compared. Among them, the calculation deviation of grid number is less than 5% in the range of 142136 to 232564. On the premise of improving the calculation accuracy and efficiency, the mesh grid number of 142136 is selected for calculation.

### *2.3. Numerical Simulation Method*

For the generation of cavitation bubbles and its effect on the jet process, the process of liquid column breaking involves complex multiphase flow and strong interphase movement. Liquid water, gaseous water and air are selected as the three-phase simulation medium. In order to observe the mixing of gaseous water and liquid water and its change process, the Euler model in multiphase flow and the standard k-epsilon model in turbulence model are used to accurately describe the changes of the surface crushing process of liquid column [8]. The simple calculation method and the first-order upwind difference scheme are adopted, and the velocity inlet and pressure outlet are selected as the boundary conditions. The overall calculation process is transient calculation, the time step is $10^{-5}$, and the maximum number of iterations is 20.

The simulation mainly involves the two-phase mixing of water vapor and liquid water, and the Euler model has high accuracy [9]. The Euler model adopts the concept of phase volume fraction. The volume fraction represents the space occupied by each phase, and each phase independently satisfies the laws of mass conservation and momentum conservation. The conservation equation is derived from the overall average or hybrid theory method of local instantaneous equilibrium of each phase. The governing equation of Euler die is shown in literature [10].

## 3. Results and Discussion

### *3.1. Variation Behaviour of Water Jet Cavitation*

Figure 2 shows the evolution process of cavitation bubbles in pipe diameter. When the inlet velocity of liquid water is 2.0 m/s, the liquid enters the nozzle from the inlet, and the moving section shrinks suddenly. When it enters the nozzle to form a jet, it also produces a local low pressure lower than its liquid saturated vapor pressure, resulting in cavitation. At this time, the jet is a cavitation water jet. It can be seen from the following cloud figure that cavitation bubbles are generated from the pipe wall of the nozzle and then move to the middle of the nozzle to form a cavitation water jet full of bubbles. After passing through the nozzle, the bubble expansion process is completed until rupture occurs. The energy generated by rupture makes droplets and even filaments fall off from the liquid column, resulting in the defect of the liquid column and strengthening the jet crushing effect.

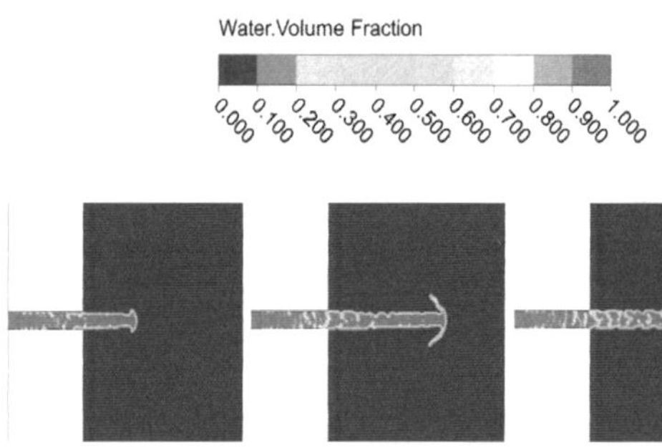


**Figure 2.** Schematic diagram of cavitation simulation model.

As shown in figure 3, the different jet motion development behaviour is shown. When the inlet velocity of liquid water is 0.5 m/s, there is no cavitation effect in the nozzle. At this time, the jet is an ordinary water jet. When the inlet velocity of liquid water is 1.5 m/s, cavitation occurs in the nozzle, and the jet behaves as cavitating water jet. By comparing the motion state cloud pictures of ordinary water jet and cavitating water jet, it is found that the ordinary water jet only depends on the jet instability caused by the surface tension of the liquid column itself, resulting in the slight deformation of the liquid column. During this period, only sporadic small droplets peel off from the liquid column, and the deformation degree of the jet column at adjacent time intervals in the process of jet development is weak. And there is no separation of large liquid filaments. The cavitating water jet is full of cavitation bubbles. During its movement, the first liquid filament falls off at 0.4 ms and the second liquid filament falls off at 1.0 ms, accompanied by a large number of small droplets. It shows that cavitating jet is not only better than ordinary water jet in instability, that is, the degree of jet fragmentation, but also the frequency of liquid column fragmentation is much higher than ordinary jet.

In the same time, the gas-liquid two-phase contact area is greatly improved, so as to strengthen the mass transfer efficiency between the two phases.

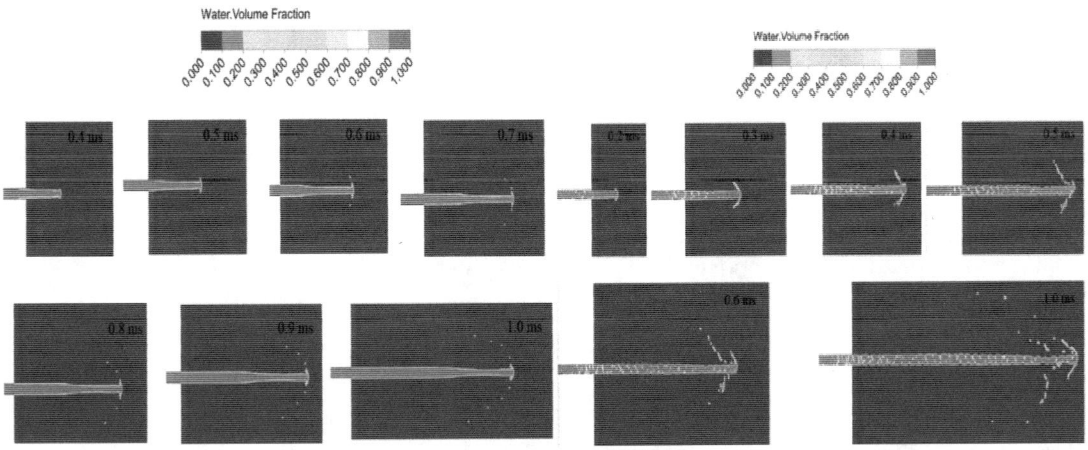


(a) (b)

**Figure 3.** Development behaviour of liquid column under two jet modes.

### *3.2. Effect of Nozzle Diameter*

Figure 4 shown the influence of nozzle diameter on cavitation under the conditions of inlet speed v= 1.5 m/s, nozzle length = 5 mm and outlet pressure P = 0 Pa. It can be seen from the figure that when the nozzle diameter is d=3 mm, there is no cavitation effect, and when d=2 mm, there is weak cavitation. The reason is that when the nozzle diameter is large, the local low pressure is insufficient, the cavitation effect is weak, and the degree of spatio-temporal effect increases significantly when d = 1 mm. Then, when the diameter is reduced to d = 0.8 mm, it is found that there is no obvious change in the cavitation area inside the pipe diameter, but at this time, the peak volume fraction of gaseous water produced by cavitation decreases from 0.76 to 0.60. When d=0.5 mm, the cavitation area inside the pipe diameter also decreases significantly with the decrease of the peak value of gaseous water to 0.23. When the pipe diameter is too small, it hinders the generation of cavitation and limits the development of cavitation effect. To sum up, in terms of nozzle diameter, the pipe diameter of 1mm is a more suitable choice.

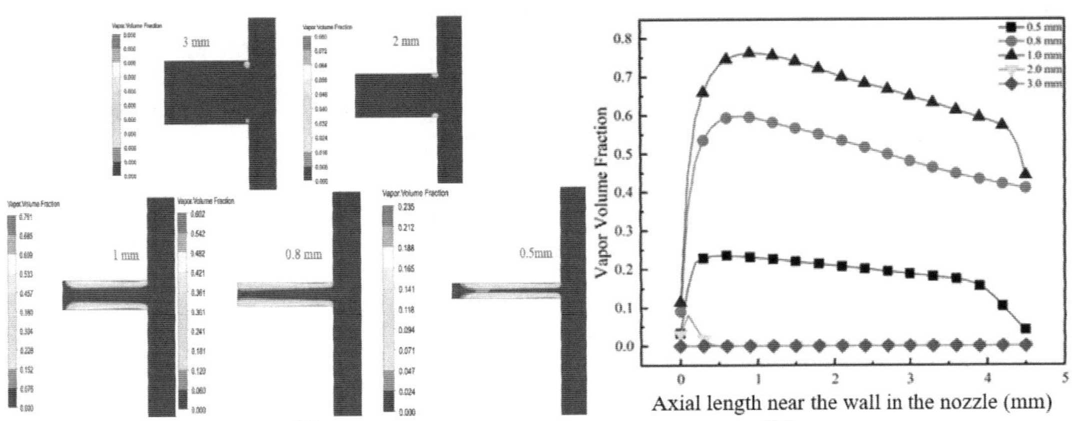


(a) (b)

**Figure 4.** Distribution of gaseous water in different nozzle diameters.

### 3.3. Effect of Nozzle Length

Figure 5 shows the discussion on the length scheme of the nozzle under the conditions of inlet speed v=1.5 m/s, nozzle diameter d=1 mm and outlet pressure P=0 Pa. The results showed that when the nozzle length increased from L=1 mm to L=9 mm, the peak value of gaseous water volume fraction was 0.71 and was relatively stable without large fluctuation. The region where gaseous water exists is always widely distributed in the pore diameter, that is, with the increase of pore diameter, the total amount of gaseous water increases, which promotes the cavitation phenomenon. However, through the further processing of the simulation results, it is found that the volume fraction of gaseous water is small when the nozzle length is L=9 mm. Then, through the analysis of the cavitation efficiency of multiple lengths, it is found that 44% of the areas with L=9 mm have low cavitation efficiency, while the areas with low efficiency cavitation are the least when l=3 mm and L=5 mm. In conclusion, considering the total amount and efficiency of gaseous water, it is a more suitable scheme when the nozzle length is L=5 mm.

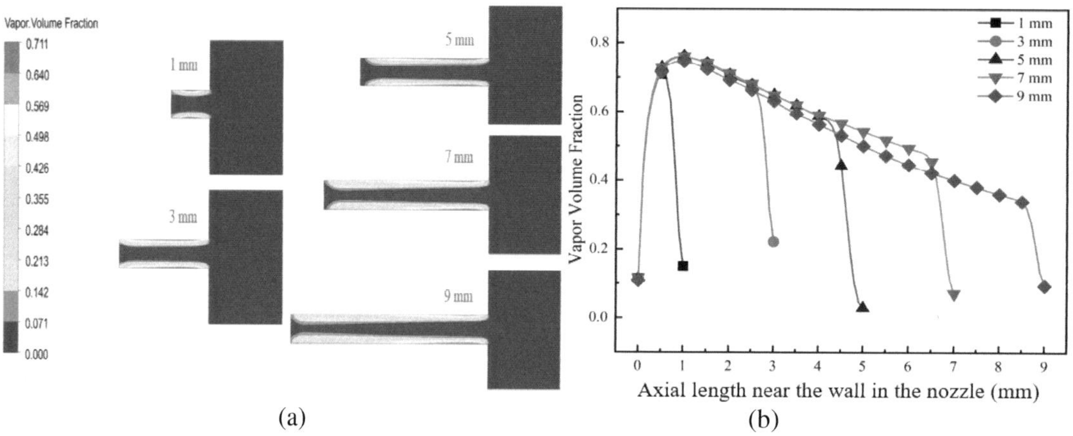


**Figure 5.** Distribution of gaseous water in different nozzle diameters.

### 3.4. Effect of Outlet Pressure

In this section, the nozzle length L=5 mm, nozzle diameter d=1mm and inlet speed 1.5m/s are selected to explore the cavitation degree from the pressure of physical environment. As shown in figure 6, through cloud image and data processing, it is found that the pressure at the nozzle outlet has a certain correlation with the cavitation effect. When the outlet pressure P is in the range of 0 pa-1500 pa, the cavitation effect area and the peak value of gaseous water do not change. Therefore, from the analysis of simulation results, the pressure value at the outlet within this pressure range has little effect on the cavitation effect inside the nozzle. Then increase the outlet pressure value to 0.1 MPa, and the peak value of gaseous water volume fraction does not change, but the spatio-temporal cavitation area with P = 0.1 MPa is less than half of the normal cavitation area. Then the cavitation effect inside the nozzle shows a significant inhibition effect, and the inhibition effect becomes stronger with the increase of outlet pressure.

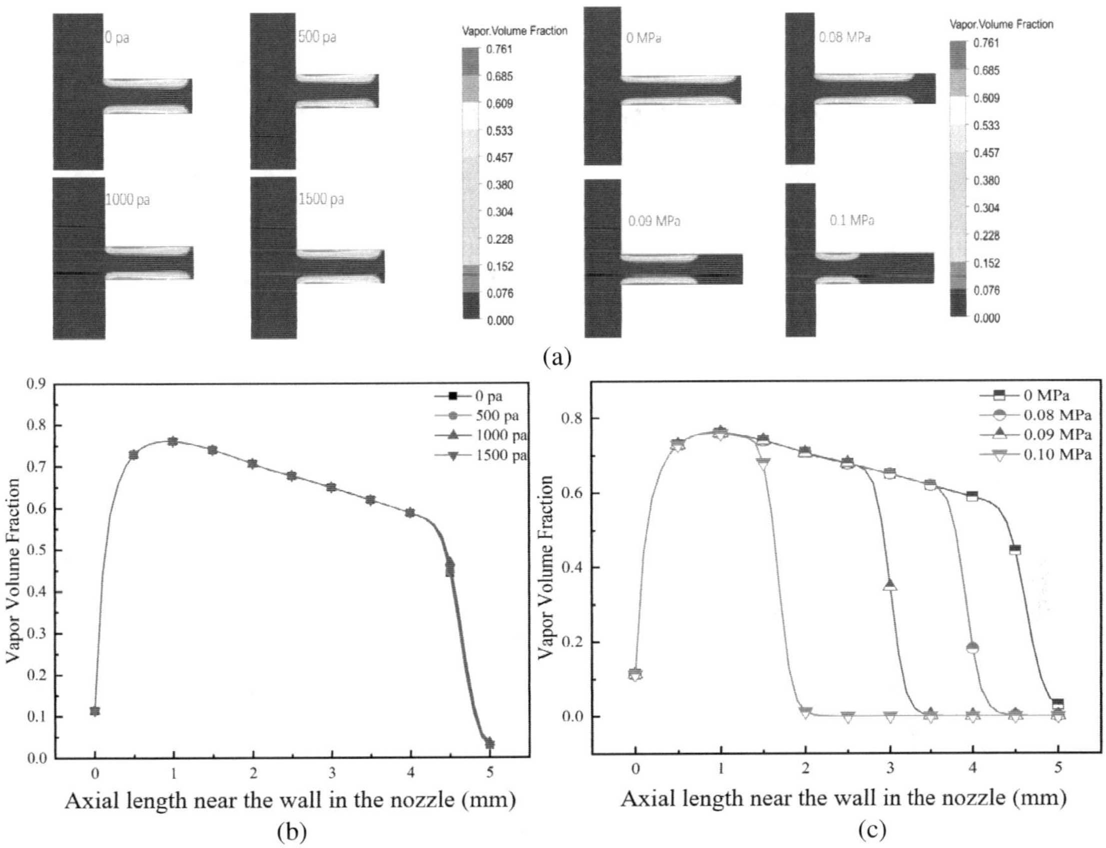


**Figure 6.** Variation law of cavitation effect under different pressures.

### *3.5. Effect of Inlet Velocity*

As one of the main physical parameters, inlet velocity is the key factor affecting the pressure change inside the nozzle diameter, and it is very important to the effect of cavitation. As shown in figure 7, the conditions selected in this section are nozzle diameter d=1 mm, length L=5 mm and outlet pressure P = 0 Pa. The influence on its speed is studied. When the speed is 0.5 m/s, the internal pressure of the nozzle does not meet the requirements of liquid cavitation. When v = 1.0 m/s, cavitation effect appears. The cavitation efficiency of 1 mm pipe diameter in front of the nozzle is consistent and reaches the peak at various inlet speeds. The difference is mainly reflected in the distribution of cavitation area. With the increase of inlet velocity, the area of cavitation distribution and the integral value of gaseous water body show a stable and gradual increasing trend. When the inlet speed is in the range of 2.5 m/s to 3.0 m / s, the increasing trend of cavitation degree in the pipe diameter remains unchanged, but the distribution between them and the numerical difference of gaseous water volume fraction are very small, that is, from the perspective of energy consumption, the optimal value of inlet speed under current conditions should be 2.5 m/s. Hence, when other conditions such as nozzle size and pressure are determined, the relationship between cavitation degree and inlet velocity variables is positively correlated, and there is an upper limit of cavitation degree. Therefore, reasonable inlet speed design is helpful to achieve the purpose of saving energy consumption and improving cavitation efficiency according to the demand in practical application.

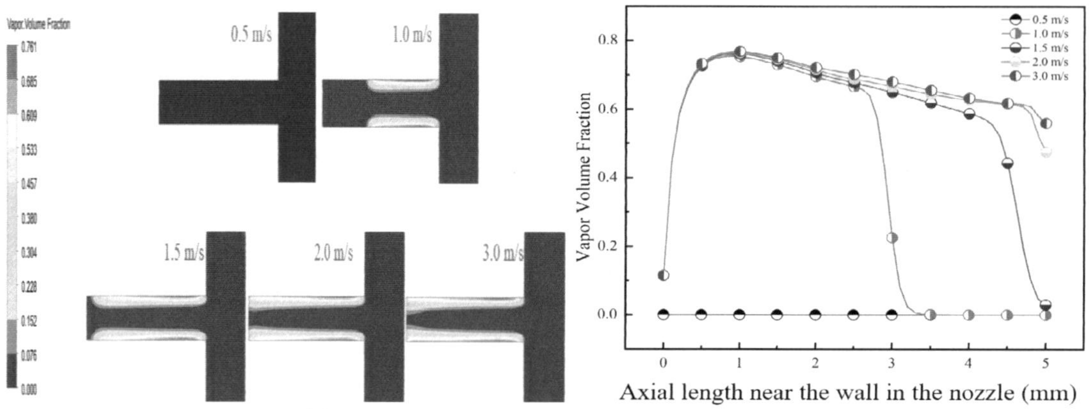


(a) (b)

**Figure 7.** Distribution of cavitation bubbles at different inlet velocities.

## 4. Conclusion

By establishing a two-dimensional model and using the method of simulation calculation, the main influencing factors of jet cavitation are explored

(1) Through the simulation of the development behaviour of ordinary jet and cavitation jet, it is found that the fragmentation degree and frequency of cavitation jet are far higher than those of ordinary jet, which greatly increases the contact area of gas-liquid two-phase and strengthens the mass transfer between the two phases.

(2) Based on the simulation study of the cavitation bubble distribution in the nozzle, it is found that there are the best conditions to promote the cavitation effect in the structure of the nozzle when the inlet size is 20 mm. The results show that under this condition, considering the total amount of cavitation and the efficiency of cavitation, the combination of nozzle diameter of 1 mm and nozzle length of 5 mm is the most conducive to the generation of cavitation effect.

(3) In terms of physical parameters, when the positive pressure at the nozzle outlet reaches 0.1 MPa, it has a violent inhibitory effect on the degree of internal cavitation. Compared with positive pressure, when the pressure environment at the outlet is negative pressure, it can promote the cavitation phenomenon in the nozzle. Then increase the negative pressure intensity, and its promotion has not changed significantly. At the same time, as the main influencing factor, the influence mechanism of the inlet speed is that the cavitation degree intensifies with the increase of the inlet speed, but the promotion effect on the cavitation phenomenon is not significantly strengthened when the inlet speed is greater than 2.5 m/s, indicating that there is an optimal inlet speed from the perspective of saving energy consumption, and the optimal value of the inlet speed is 2.5 m/s.

## Acknowledgments

The study was supported by project of Scientific Research Foundation of Chongqing University of Technology, Science and Technology Research Program of Chongqing Municipal Education Commission of China (KJQN202001112) and Natural Science Foundation of Chongqing, China (cstc2021jcyj-msxmX0184).

## References

[1] Lešnik L, Kegl B, Bombek G, Marko H and Ignacijo B 2018 The influence of in-nozzle cavitation on flow char-acteristics and spray break-up *Fuel* **222** 550-560.

[2] Abderrezzak B and Huang Y 2016 A contribution to the understanding of cavitation effects on droplet formation through a quantitative observation on breakup of liquid jet *Int. J. Hydrogen Energ.* **41**(35) 15821-15828.

[3] Piscaglia F, Giussani F, Hèlie J, Lamarque N and Aithal S M 2021 Vortex flow and cavitation in liquid injection: a comparison between high-fidelity CFD simulations and experimental visualizations on transparent nozzle replicas *Int. J. Multiphas. Flow* **138** 103605.

[4] Sou A, Hosokawa S and Tomiyama A 2007 Effects of cavitation in a nozzle on liquid jet atomization *Int. J. Heat and Mass Tran.* **50**(17-18) 3575-3582.

[5] Kumar A and Sahu S 2020 Influence of nozzle geometry on primary and large-scale instabilities in coaxial injectors *Chem. Eng. Sci.* **221** 115694.

[6] Etzold M, Deswal A, Chen L and Durst F 2018 Break-up length of liquid jets produced by short nozzles *Int. J. Multiphas. Flow* **99** 397-407.

[7] Tadjfar M and Jaberi A 2019 Effects of aspect ratio on the flow development of rectangular liquid jets issued into stagnant air *Int. J. Multiphas. Flow* **115** 144-157.

[8] Huang J and Zhao X 2019 Numerical simulations of atomization and evaporation in liquid jet flows *Int. J. Multiphas. Flow* **119** 180-193.

[9] Hu R, Li Q, Li C and Li C 2019 Effects of an accompanied gas jet on transverse liquid injection in a supersonic crossflow *Acta Astronaut.* **159** 440-451.

[10] Nambu T and Mizobuchi Y 2021 Detailed numerical simulation of primary atomization by crossflow under gas turbine engine combustor conditions *P. Combust. Insti.* **38**(2) 3213-3221.

# Particle Number Measurement and Its Numerical Simulation of the Airborne Particle Counter by 'Post-Collection-Check'

**Z L Gao[1], G L Zeng[2*], Q D Cheng[1], R J Wang[1], W H Zhang[1], F Q Wang[2], J Chen[3], J P Wu[1] and J Guo[2]**

[1] Beijing Orient Institute of Measurement and Test, Haidian District, Beijing, 100094, China.
[2] The First Branch of Beijing Hang Tian He Science Technology Development Co., Ltd., Changping District, Beijing, 102205, China.
[3] Suzhou Sujing Automation Equipment Corporation, Suzhou, 215000, Jiangsu, China.
Email: gzl514cast@126.com

**Abstract.** The measurement results of the light-scattering airborne particle counter can effectively evaluate the concentration of airborne particles in the clean environment and confined space, and the accuracy improvement of the calibration measurement results has attracted more and more attention. This paper introduces the 'post-collection-check' way of the airborne particle concentration measurement method of the OPC (Optical Particle Counter), based on FESEM (Field Emission Scanning Electron Microscopy) and AAO (Anodic Aluminum Oxide) template, the probability distribution of the collected PSL (Polystyrene Latex) particles on the surface of AAO template, and the pressure parameters of AAO template under the different particle over current areas are both simulated. The numerical simulation results will help to improve the accuracy of the inference statistics of the collected PSL particle number on the membrane filter, and to give guidance on the optimal design of the PSL particle collection tube and the membrane filter, in order to provide an important reference for further improving the OPC particle concentration calibration and traceability method based on statistical theory.
**Keywords.** Airborne particle counter, post-collection-check, numerical simulation, non-comparative calibration method, aerodynamics.

## 1. Introduction

As people pay more and more attention to particulate pollution in the air, such as digital new infrastructure, aerospace, medicine and chemical industry, precision instrument manufacturing, motor vehicle exhaust emissions, virus aerosol transmission [1-3], so the light-scattering airborne particle counter (also known as Optical Particle Counter, OPC) is often used to measure the airborne particle concentration. Traceability and its accuracy improvement of the airborne particle concentration measurement method are more and more valuable [4-6].

At present, the comparative test method is widely used to calibrate OPC particle concentration in the world, however, the measured objects of the standard particles passing through the calibrated OPC cannot be completely consistent in theory, so that there is a 'principle drawback' of 'uncertainty' in the 'true value' of the particle concentration.

On the basis of referring to the standards for measuring the air cleanliness by the way of filter (filter membrane) collection and electron microscope observation, such as IEST-RP-CC003.3:2003,

GB 50591-2010, QJ 2214-1991, the research team from Beijing Oriental Institute of Measurement and Test (BOIMT) tried to propose a new calibration and traceability scheme of the airborne particle concentration based on FESEM and AAO template, which is different from the existing international common comparison test calibration method with the 'principle drawback', built a 'post-collection-check' statistical calibration system of the 0.5μm PSL particle concentration and carried out theoretical and experimental verification [7-9].

## 2. Analysis of Measurement Method of Airborne Particle Concentration by 'Post-Collection-Check'

In order to ensure that the OPCs has the same measured gas path, the same gas flow rate and the same PSL standard particle subject, the airborne particle concentration calibration method based on FESEM and AAO template is to install two filter membranes after the calibrated OPC. The first one is installed behind the light-scattering chamber and used to collect the PSL particles passing through the particle counter, and the second one is installed behind the first one to confirm all the collected particles (see figure 1). Double-pass AAO template with 300 nm aperture for the selected monolayer membrane, and cold field emission model S-4700 for the selected FESEM (see figure 2).

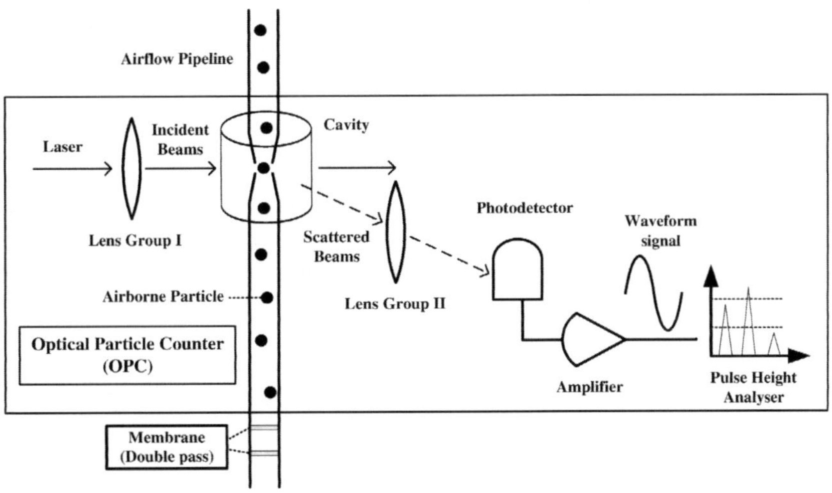


**Figure 1.** Scheme of standard particle number measurement by 'post-collection-check'.

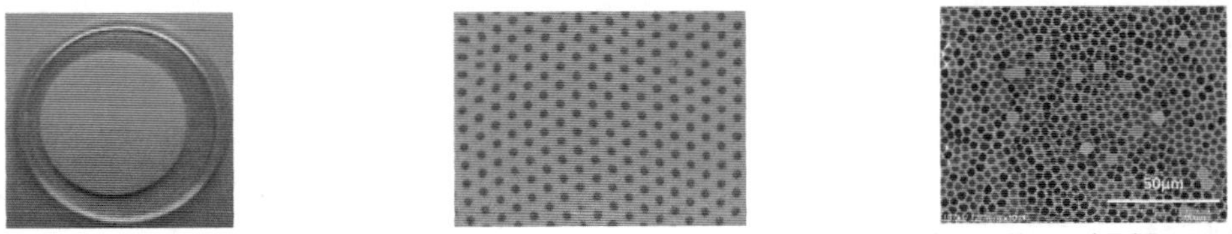

(a) AAO template (b) AAO template under FESEM (c) The collected PSL particles by AAO template under FESEM

**Figure 2.** AAO template and the collected PSL particles under FESEM.

The air outlet structure of the calibrated OPC should be modified (see figure 3), and it needs to increase 10cm gas straight pipe behind the outlet to buffer particle flow, and allocate special connection fixture of suitable size, with internal installation of monolayer membrane. At the same time, we keep the light scattering chamber outlet and the straight pipe consistent with the membrane effective diameter.

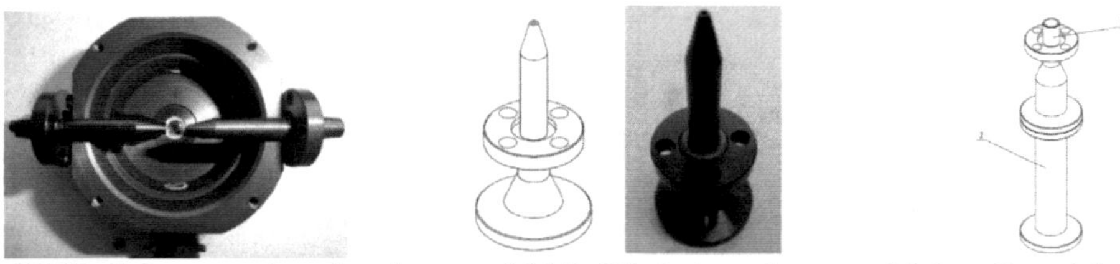

(a) OPC's light-scattering chamber (b) Modified air outlet (c) Install straight tube after modification of air outlet

**Figure 3.** Renovation components of the calibrated OPC.

Although the 'post-collection-check' calibration method of the airborne particle concentration is theoretically feasible, however, the PSL standard particle loss control scheme needs to be further optimized (such as reducing the influence of electrostatic adsorption), the AAO template working gas flow needs to be further expanded through numerical simulation (for example, the working flow of an AAO template is 0.59 L/min, but the standard working flow of the OPC under test is 2.83 L/min), and the measurement accuracy improvement by using the theories of aerodynamics and inferential statistics needs to be further improved [10, 11].

## 3. Simulation of the Calibrated OPC's PSL Standard Particle Flow by 'Post-Collection-Check'

### *3.1. Simulation Conditions and Analysis of the PSL Particle Stress*

After the gas flow rate and saturation concentration of 0.5 μm PSL particles are given, the model is imported into the ANSYS Fluent software for computational fluid dynamics (CFD) simulation. The particle velocity and PSL particle distribution (AAO template) and resistance are obtained by coupling the standard K-ε model with the discrete phase model (DPM). The initial and the boundary conditions of this simulation are shown in table 1 and table 2.

**Table 1.** Initial conditions of the PSL standard particle flow simulation.

| Serial number | Parameter category | Concrete value |
|---|---|---|
| 1 | Density | 1050 kg/m$^3$ |
| 2 | Specific Heat Capacity | 1330 J/kg·K |
| 3 | Thermal Conductivity | 0.116 W/m·K |
| 4 | Particle Size | 0.5 μm |
| 5 | Saturation concentration of PSL particles in airflow | 5500/2.83 L |
| 6 | Total Particle Flow | 1.9e$^{-18}$ kg/s |

**Table 2.** Boundary conditions of the PSL standard particle flow simulation.

| Serial number | Air Flow Rate | Corresponding Flow Velocity |
|---|---|---|
| 1 | 0.59 L/min (0.0354 m$^3$/h) | 0.142 m/s |
| 2 | 2.83 L/min (0.1698 m$^3$/h) | 0.68 m/s |
| 3 | 28.3 L/min (1.698 m$^3$/h) | 6.8 m/s |

During this simulation, PSL standard particles are mainly affected by drag force and lift force, and the trajectory and velocity of the discrete phase particles are predicted through the balance and offset the relationship between the forces. Therefore, all the forces acting on the particle are equal to the inertia of the particle generated by them [12], which is formulated as follows:

$$\frac{d\overrightarrow{u_p}}{dt}=\frac{\vec{u}-\overrightarrow{u_p}}{\tau_r}+\frac{\vec{g}[\rho_p-\rho]}{\rho_p}+\vec{F} \quad (1)$$

where, $\vec{F}$ is the additional acceleration (force/unit particle mass), $\frac{\vec{u}-\overrightarrow{u_p}}{\tau_r}$ is the drag force exerted by a single particle, where:

$$\tau_r=\frac{\rho_p d_p^2}{18\mu}\frac{24}{C_d Re} \quad (2)$$

where, $\tau_r$ is the particle relaxation time, $\vec{u}$ is the airflow velocity, $\overrightarrow{u_p}$ is the particle velocity, μ is the molecular viscosity of the airflow, $\rho$ is the gas density, $\rho_p$ is the particle density, $d_p$ is the particle size, $Re$ is Raynaud's constant, and $Re$ is calculated as follows:

$$Re=\frac{\rho d_p|\overrightarrow{u_p}-\vec{u}|}{\mu} \quad (3)$$

$\vec{F}$ in Formula (3) is mainly affected by the virtual mass force, which is the force accelerating the airflow around particles. The formula is as follows:

$$\vec{F}=C_{vm}\frac{\rho}{\rho_p}\left[\overrightarrow{u_p}\nabla\vec{u}-\frac{d\overrightarrow{u_p}}{dt}\right] \quad (4)$$

where, $C_{vm}$ is the virtual mass force factor, and the default value is 0.5.
Since the pressure gradient in the flow also generates additional forces, the formula is as follows:

$$\vec{F}=\frac{\rho}{\rho_p}\overrightarrow{u_p}\nabla\vec{u} \quad (5)$$

When the flow density is much lower than the particle density, the virtual mass force and pressure gradient force will weaken. When the density of PSL particles is similar to that of airflow, the effect of virtual mass force and pressure gradient force will be obvious. In this simulation, there is little difference between PSL particles and airflow density, so the effect of virtual mass force and pressure gradient force should be considered.

### *3.2. Simulation of the Probability Distribution of the PSL Particle Position Collected on AAO Template*

The renewed air outlet with a straight tube of the calibrated OPC in figure 3(b) is selected. The simulation is carried out under the three-dimensional structure (variable diameter cylinder) according to the real size. The modeled structure is imported into Gambit software for grid division, choosing the unsteady calculation and setting the time step as 0.0001. The resistance distribution and the flow velocity distribution on the X-axis pipeline profile are shown in figure 4.

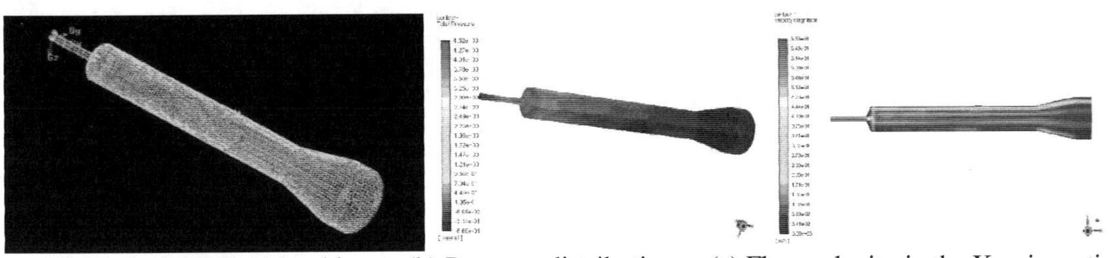

(a) 3D mathematical meshing (b) Pressure distribution (c) Flow velocity in the X-axis section

**Figure 4.** 3D modelling, pressure and flow velocity simulation results.

The simulation lasts for 1.75 s, and about 158,000 PSL particles enter the pipeline with the airflow. The particle distribution over time is shown in figure 5(a). The position distribution of particles with different numbers at the end of calculation is shown in figure 5(b).

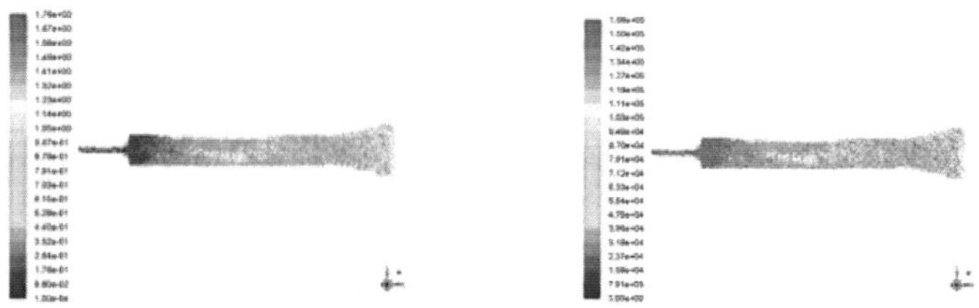

(a) Particle distribution over time (b) Different particle distribution

**Figure 5.** Simulation results of particles in pipeline.

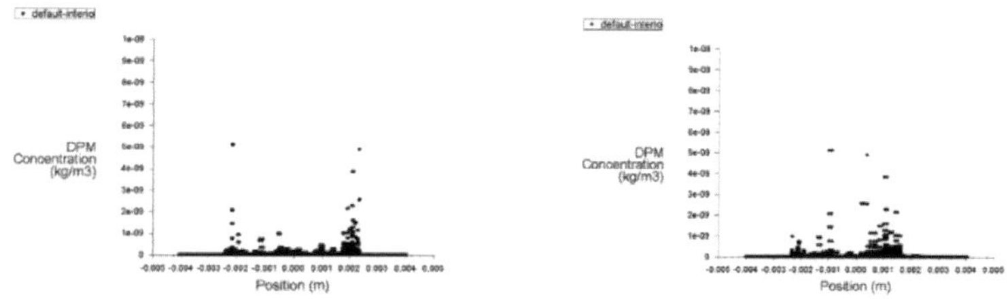


(a) Y-axis plane (b) Z-axis plane

**Figure 6.** Simulation results of particle distribution on Y-axis and Z-axis surfaces in pipelines.

*3.3. Simulation of Pressure Parameters of AAO Template under Different Particle Overcurrent Areas*

The structure No. 2 in figure 3(c) is selected as the simulation model. The simulated inlet condition is set as velocity inlet, and setting the surface jet mode as the initial speed, setting the outlet condition as pressure-outlet, setting the wall conditions as non-slip and particle rebound, setting pressure and velocity coupling as Phase Coupled SIMPLE algorithm, setting momentum, turbulent kinetic energy and turbulent dissipation rate as first order upwind schemes.

Figure 7 shows that the boundary effect decreases with the increase of flow velocity, and the peak motion state in the inner core flow area becomes more intense.

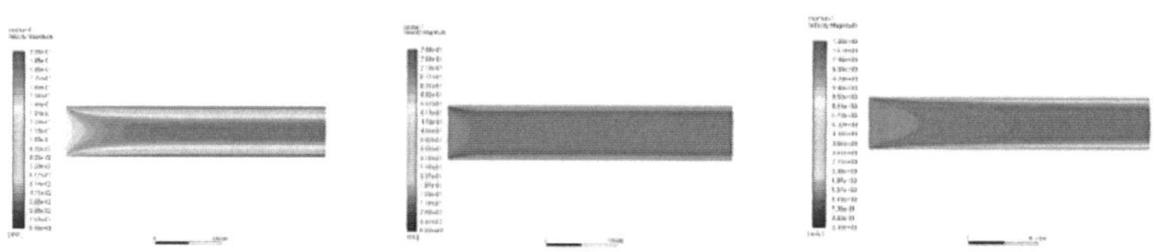

(a) 0.142 m/s (b) 0.68 m/s (c) 6.8 m/s

**Figure 7.** Flow velocity distribution diagram inside the pipeline.

Figure 8 shows that the high-pressure area gradually extends to the outlet with the increase of the flow velocity, and the pressure increases rapidly with the increase of 3~5 times of the flow velocity.

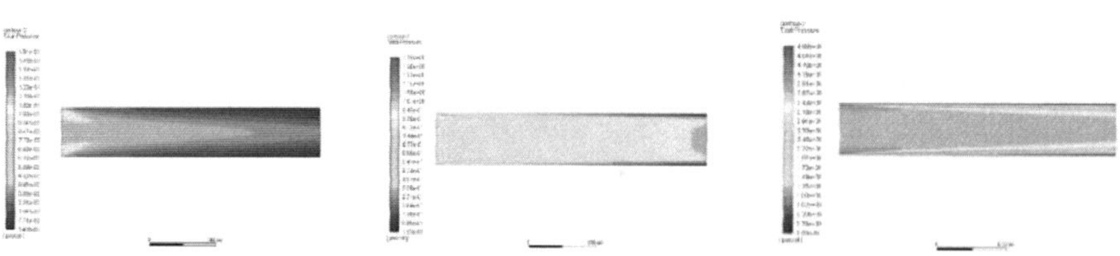

(a) 0.142 m/s (b) 0.68 m/s (c) 6.8 m/s

**Figure 8.** Diagram of pressure distribution inside the pipeline.

Figure 9 shows that, on the premise that the pipeline structure does not change, the change of air velocity has little impact on the flow pattern in the core area of particle movement, while the boundary area shows an unstable trend when the velocity increases due to the presence of wall action.

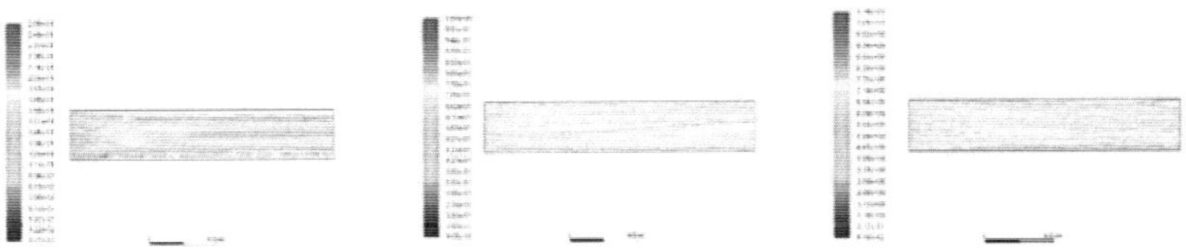

(a) 0.142 m/s (b) 0.68 m/s (c) 6.8 m/s

**Figure 9.** Particle flow diagram inside the pipeline.

Because the filter membrane itself is composed of thousands of very fine double-pass pores, and the pores among them are non-channel structures, the flow area of different filter membranes or in the process of particle collection is different, so the flow areas of three filter membranes (90%, 80% and 70% of the cross-sectional area of the channel respectively) are assumed.

Figures 10-12 show that, with the continuous decrease of filtration membrane flow area, the influence on the flow in the pipeline increases. With the increase of airflow velocity, the influence of membrane flow zone on airflow will increase. Due to the rapid reduction of the airflow area in the filtration membrane flow area, the airflow in this area presents a local high-speed zone, and a local low-speed swirl zone appears at the back end of the filtration membrane flow area. The higher the velocity is, the more obvious the high-speed zone and swirl zone are.

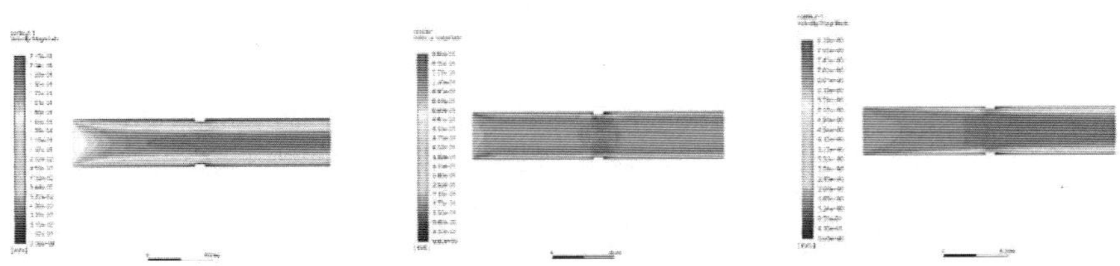

(a) 0.142 m/s (b) 0.68 m/s (c) 6.8 m/s

**Figure 10.** Flow velocity distribution diagram in the pipeline when the 90% membrane flow area.

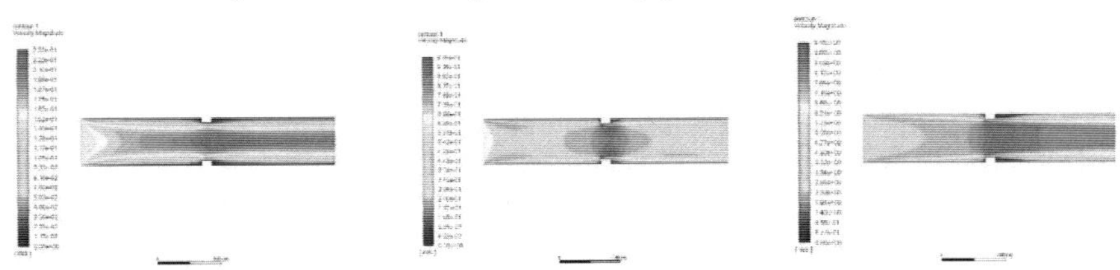

(a) 0.142 m/s (b) 0.68 m/s (c) 6.8 m/s

**Figure 11.** Flow velocity distribution diagram in the pipeline when the 80% membrane flow area.

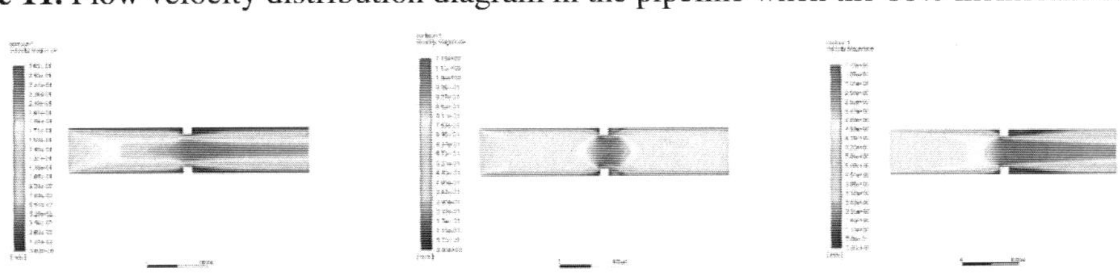

(a) 0.142 m/s (b) 0.68 m/s (c) 6.8 m/s

**Figure 12.** Flow velocity distribution diagram in the pipeline when the 70% membrane flow area.

Figures 13-15 shows the pressure distribution in the pipeline under three different flow rates under different filter membrane flow areas, it can be seen that, when the air flow rate is low, with the continuous decrease of the filter membrane flow area, the area of the low-pressure area on both sides of the back end of the filter membrane flow area decreases, but the maximum pressure inside the pipeline is basically the same. When the flow rate is high, with the decrease of the flow area of the filter membrane, the area of the low-pressure area on both sides of the back end of the filter membrane flow area increases, and the maximum pressure inside the pipeline also increases. When the membrane flow area is unchanged, the pressure inside the pipeline increases rapidly with the increase of the air flow rate, and at the same time, the pressure gradient and pressure difference appear obviously before and after the membrane flow area. When the flow rate is constant, the high-pressure area inside the pipeline gradually extends to the outlet with the decrease of the filtration membrane flow area.

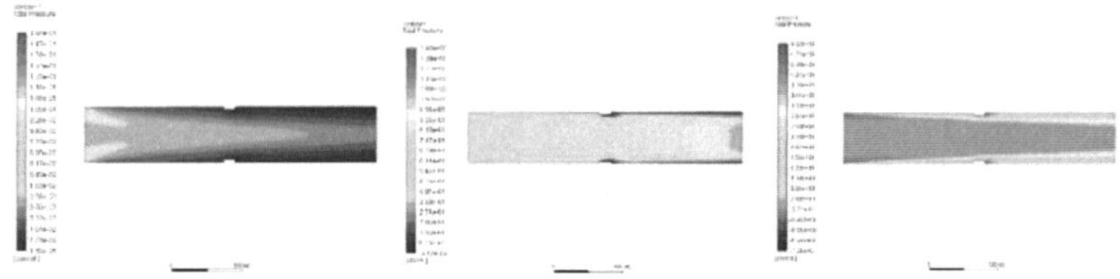

(a) 0.142 m/s (b) 0.68 m/s (c) 6.8 m/s

**Figure 13.** Diagram of pressure distribution inside the pipeline when the 90% membrane flow area.

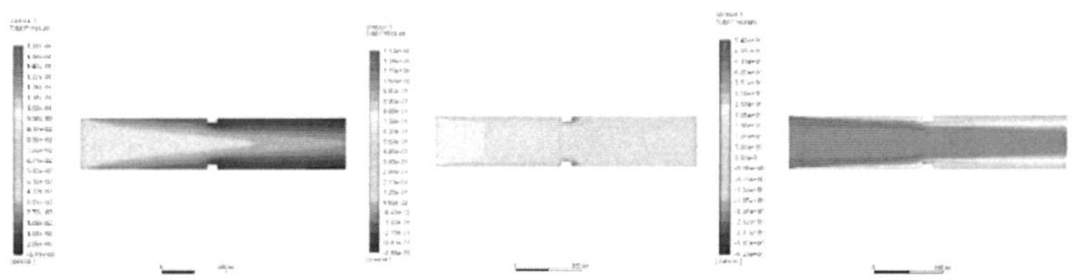

(a) 0.142 m/s (b) 0.68 m/s (c) 6.8 m/s

**Figure 14.** Diagram of pressure distribution inside the pipeline when the 80% membrane flow area.

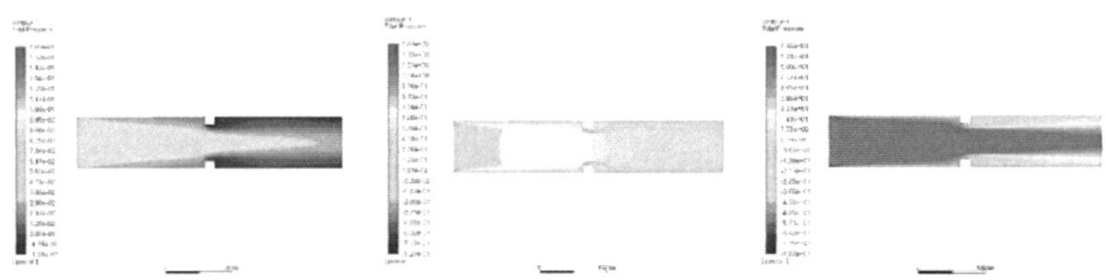

(a) 0.142 m/s (b) 0.68 m/s (c) 6.8 m/s

**Figure 15.** Diagram of pressure distribution inside the pipeline when the 70% membrane flow area.

Figure 16 shows that the reduction of filter membrane flow area will significantly strengthen the flow disorder on the inside side of the pipeline, accelerate the flow velocity in the second half of the pipeline, and weaken the stability of air distribution.

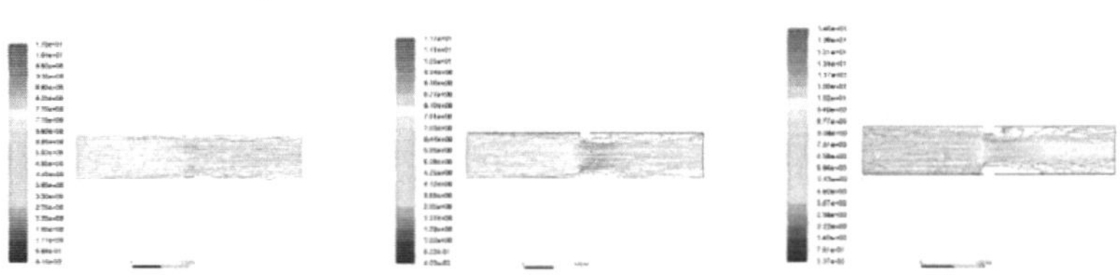

(a) 90% (b) 80% (c) 70%

**Figure 16.** Particle flow diagram inside the pipeline with the flow area of three filters.

## 4. Conclusion and Prospect

The calibration improvement of the OPCs is one of the key ways to obtain higher measurement accuracy of airborne particle concentration. The 'post-collection-check' measurement method can locate the particle distribution of the standard particle flow at each position on the outlet surface of the pipeline more clearly by numerical simulation. The total number of the simulated standard particles can be increased by perfecting the simulation method, and the probability distribution of collected standard particles on the filter membrane can be more and more "real". It can also master the characteristics of more motion states of the airflow with different flow rates under the same pipeline structure, verify the influence caused by different filter membrane flow area, and provide guidance for further development of pipeline and filter membrane morphology design, in order to further improve the calibration and traceability method of the airborne particle concentration of the sub-micron particle size (0.5 μm) based on statistical theory.

## Acknowledgments

This research is supported by the Outstanding Young Talents Fund of China Academy of Space Technology (CAST).

## References

[1] Chau J P C, Liu X, Lo S H S, Chien W T and Wan X 2020 Effects of environmental cleaning bundles on reducing healthcare-associated Clostridioides difficile infection: a systematic review and meta-analysis *Journal of Hospital Infection* **106** 734-44

[2] Nowak N, Scheiber K, Pfeil J, Meyer J, Dittler A, Koch T and Kasper G 2020 Sampling and conditioning of engine blow-by aerosols for representative measurements by optical particle counters *Journal of Aerosol Science* **148** 105612

[3] Pillai S G, Haldorai K, Seo W S and Kim W G 2021 COVID-19 and hospitality 5.0: Redefining hospitality operations *International Journal of Hospitality Management* **94** 102869

[4] Peng J C, Zhou Z M and Li X H 2020 Analysis on the Quantity Value Traceability of Airborne Particle Counter *Metrology & Measurement Technique* **47** 5-6, 11

[5] Wlasits P J, Stolzenburg D, Tauber C, Brilke S, Schmitt S H, Winkler P M and Wimmer D 2020 Counting on chemistry: laboratory evaluation of seed-material-dependent detection efficiencies of ultrafine condensation particle counters *Atmospheric Measurement Techniques* **13** 3787-98

[6] Liu J J, Qi T Y, Wang J W and Liu T 2018 Calibration Technology and Traceability System of Particle Counter (PNC) *Proceedings of the 22nd Symposium on Air Pollution Prevention and Control Technology* 153-9

[7] Gao Z L, Ji Q Z, Chen J, Li G F, Wang R J, Wu J P, Zeng G L, Cheng Q D, Hu W Z and Tang X 2020 Metrological traceability methods for the light scattering airborne particle concentration *The International Society for Optical Engineering: 11th International Symposium on Precision Engineering Measurements and Instrumentation (ISPEMI 2020)* 37

[8] Gao Z L, Ji Q Z, Chen J, Zhang X B, Zhang W H, Tan J G and Wang C Y 2018 Statistical measurement method of the standard particles through airborne particle counter based on FESEM *Proceedings of SPIE (EI: 20191106638812) - The International Society for Optical Engineering: 10th International Symposium on Precision Engineering Measurements and Instrumentation (ISPEMI 2018)* **110530** 110530I-1-8

[9] Ji Q Z, Gao Z L, Zhang X B and Chen J 2014 Discuss on traceability method of light-scattering airborne particle counter's counting performance *Proceedings of SPIE (EI: 20151200656569) - The International Society for Optical Engineering: 9th International Symposium on Precision Engineering Measurements and Instrumentation (ISPEMI 2014)* **9446** 94464C-1-6

[10] Gao Z L, Cheng Q D, Zeng G L, Wen Y, Li G F, Chen J, Dong Y B and Ji Q Z J 2021 Review of Calibration and Improvement Methods of Light-Scattering Airborne Particle

Concentration *Journal of Physics (EI: 20214911291306), IOP Publishing-2021 International Conference on Fluid and Chemical Engineering (ICFCE 2021)* **2097** 012008

[11] Wei L B, Li D H, Chen Z G, Sun M Y and Zhu X S 2017 Numerical simulation of force and separation on particles in pulsing airflow *Journal of China University of Mining & Technology* **46** 162-168, 176

[12] Li J R, Li X, Chen J H, Dong X X and Xin M Y 2018 Numerical Simulation of Clean Room Air Distribution for Typical Characteristics *Journal of Harbin University of Science and Technology* **23** 24-28

# Simulation Analysis of Double-effect Seawater Desalination System Using Compression Heat Pump

**Yanping Xiao[1], Hao Yuan[2], Jiubin Shen[1*] and Siyuan Yan[1]**

[1]School of Energy and Power, Jiangsu University of Science and Technology, Zhenjiang 212003, China.
[2]Shanghai Marine Diesel Engine Research Institute (SMDERI), Shanghai 201100, China
Email: shenjiubing@just.edu.cn

**Abstract.** The lack of freshwater resources has become the biggest obstacle to the sustainable development of the world economy with the expansion of population and the impact of climate change. Seawater desalination is recognized as the most effective means to alleviate water scarcity. A double-effect seawater desalination system using compression heat pump is proposed by combining a compression heat pump and a double-effect desalination system. Performances of the proposed system are simulated using Aspen simulation software. Based on the simulation results, the influence of heat pump evaporating/condensing temperature, temperature difference between the two effects and the brine circulation ratio on the system Performance Ratio (PR) and freshwater production are analysed. It is concluded that the value of PR increases but the freshwater production reduces with the increase of evaporating temperature of the heat pump. Both the value of PR and the freshwater production decreases with the condensing temperature of the heat pump. An optimal temperature difference between the two effects exits to get the highest PR and freshwater production for a constant heat pump operation condition. The freshwater production can increase slightly with the increase of the brine circulation ratio. When the mass flow rate of feed seawater is 80kg/h and the evaporation temperature of primary effect evaporator is 40°C, the PR of the system reaches 40.07 with an evaporating/condensing temperature of 20°C/50°C. The freshwater production is 39.76kg/h with an evaporating/condensing temperature of 10°C/50°C.
**Keywords.** Heat pump, seawater desalination, double-effect evaporation, Aspen Plus.

## 1. Introduction

Water scarcity has become a global problem, and providing adequate quantities of qualified freshwater has become a challenging work. The reserve of seawater accounts for 97.4% of the total water resources of the earth. If the seawater is fully utilized and converted into freshwater available to human beings, the shortage of freshwater resources can be greatly alleviated. The development of seawater desalination is a crucial mean to increase the supply of water resource and optimize the water supply structure. It is of great significance to relieve the water scarcity in coastal areas and offshore islands.

In the contemporary, the seawater desalination technology is mainly based on distillation [1], which is to evaporate seawater to produce steam and condensate the steam to obtain high purity freshwater. Hao et al. used Aspen Plus software to simulate the four-effect low-temperature multi-effect seawater desalination plant of SIDEM Company in France and the seawater desalination plant in the circular economy project of Beijiang Power Plant [2]. The credibility of simulation results of

Aspen Plus software were verified. Ma et al. established three simulation models for low-temperature multi-effect desalination systems by using Aspen Plus software, which includes the parallel flow, forward flow and backward flow processes. Performance analyses contained the gained output ratio, motive steam flow and heat transfer area were conducted as a function of the process type, number of effects, heating steam temperature and the effect NO. for thermal vapor compression ejection [3]. Angelica et al. proposed a small-scale multi-effect distillation plant by renewable energy or waste heat sources. The feed mass flow in each effect was varied to maximize the recovery ratio by imposing the maximum salt concentration in the brine related to calcium sulphate precipitation. The configurations implementing seawater preheating increased the performance ratio up to 10% [4].

As a new type of high-efficiency and energy-saving technology, the heat pump has been widely used in many fields. The application of heat pumps to seawater desalination can realize the perfect coupling of the two technologies, consume a small amount of high-grade energy (mechanical energy, electric energy, etc.) and convert low-grade heat energy into high-grade heat energy [5], which has advantages of high energy utilization rate, low cost of water production, no pollution to the environment and good market prospect. In 1984, Reali M applied the compression heat pump technology to seawater desalination for the first time, using the heat pump condenser/evaporator to heat seawater and condensing the steam [6]. Slesarenko proposed that the seawater desalination technology using compression heat pump was conducive to the efficient use of thermal energy in desalination small-capacity plant. The thermodynamic and economic indexes of the desalination process by 2-3 times [7]. Zhou et al. applied the evaporative condenser to the heat pump seawater desalination device to study the influence of inlet temperature, spray volume and wet bulb temperature at the side outlet of the evaporative condenser on freshwater production, providing reference for the design of a novel household heat pump seawater desalination device [8]. Zhao designed a small heat pump seawater desalination device, and analysed the appropriate heat exchanger area, heat transfer coefficient and capillary length of the system under different condensing and evaporating temperatures [9]. Ahmed adopted a heat pump seawater desalination system with water as working medium combined with passive vacuum system [10]. In order to solve the traditional condensting temperature of the heat pump was lower than the evaporating temperature of seawater, reduce the evaporation temperature of seawater by raising the distillation tank to maintain the vacuum pressure inside the tank. In addition, some scholars have combined other types of heat pump with multi-effect distillation for research [11-13].

Under the background of increasingly stringent environmental requirements and energy crisis, this paper established a double-effect seawater desalination system using compression heat pump to reduce the energy consumption of the seawater desalination system and reduce the cost of water production. Analysis of the system can provide theoretical basis for the research and development of an efficient seawater desalination device.

## 2. System Description

Diagram of the double-effect seawater desalination system using compression heat pump is shown in figure 1, which is combined of a heat pump system and a double-effect evaporation system. The heat pump evaporator is used to recover the condensation heat of the final-effect low-temperature steam. The heat released by the condenser is used for the evaporation of seawater in the primary effect.

The working process of the system is as follows: the feed seawater flows through preheater 1 and preheater 2 sequentially, and then enters the condenser of heat pump which is also the primary effect evaporator of the desalination system. By the heating effect, primary steam is produced in the primary effect evaporator, using the primary steam as the heat source of the secondary effect evaporator to yield the secondary steam. The generated secondary steam dissipates some heat in the auxiliary condenser and then enters into the evaporator of heat pump system where it is condensed. The condensed water from the secondary effect evaporator and heat pump evaporator is stored in the freshwater tank.

The brine out of the primary effect evaporator preheats feed seawater in the preheater 2. After that, it is mixed with the circulating brine and enters into the secondary effect evaporator. The concentrated brine discharged from the bottom of the secondary effect evaporator preheats feed seawater in preheater 1 and then is divided into two streams. Part of the brine enters back into the secondary effect evaporator as circulating seawater. The other part is discharged. The pressure of the secondary effect evaporator is lower than the pressure of the heat pump condenser. Accordingly, the evaporation temperature of seawater in each effect evaporator is different. An auxiliary condenser is mainly used to dissipate the heat difference between the condenser and evaporator of the heat pump.

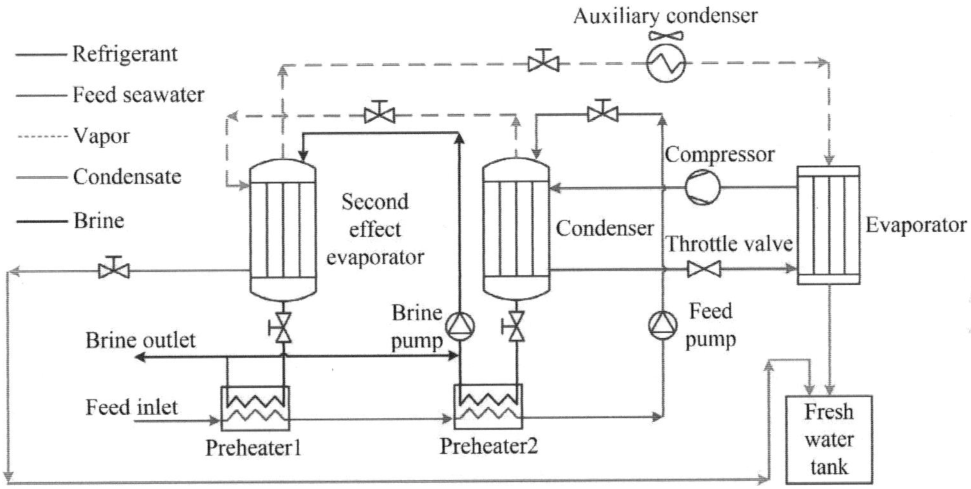


**Figure 1.** Flow chart of double-effect seawater desalination system using compression heat pump.

## 3. System Simulation

### *3.1. Establish the System Simulation Model*

Uisng the HeatX, Valve2, Aircooler, FSsplit and Mixer modules to establish the system simulation model as shown in figure 2. To simplify the simulation, the following assumptions are made:

1. System operates in steady state;
2. The heat loss and fluid flow resistance of each component are ignored;
3. The internal heat transfer process of each heat exchanger is carried out under constant negative pressure;
4. The condensate produced is pure water;
5. Ignore the influence of non-condensable gas.

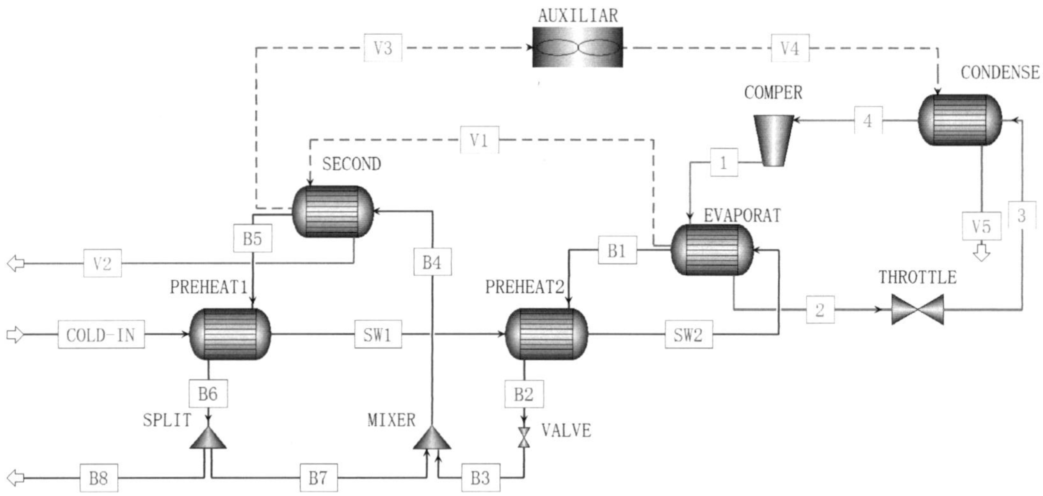


**Figure 2.** Simplified diagram of double-effect seawater desalination system using compression heat pump model.

### *3.2. Boiling Point Elevation of Seawater*

The relationship between Boiling Point Elevation (BPE) of seawater and salinity is calculated by equation (1) according to the literature [14]:

$$BPE = 0.528764S + 0.826030\times10^{-2}St - 0.315082\times10^{-6}St^2 + 0.320553\times10^{-1}S^2$$
$$-0.144367\times10^{-3}S^2t + 0.184416\times10^{-5}S^2t^2 \quad (1)$$

where $S$ is salinity and $t$ is seawater temperature.

The BPE of seawater with a salinity of 35‰ and temperature of 40°C increases to 0.35°C. The higher salinity and temperature of the seawater, the higher the BPE value. The actual freshwater production is less than the target freshwater production due to the influence of BPE.

### *3.3. Heat Transfer Coefficient*

The heat transfer coefficient of heat pump condenser and heat transfer coefficient of heat pump evaporator are calculated by equation (2) and equation (3) according to the literature [15]:

$$U_e = 1.9394 + 1.40562\times10^{-3}T_i - 2.0752\times10^{-4}T_i^2 + 2.3186\times10^{-6}T_i^3 \quad (2)$$

$$U_c = 1.6175 + 0.1537\times10^{-3}T_v + 0.1825\times10^{-3}T_v^2 - 8.026\times10^{-8}T_v^3 \quad (3)$$

where $T_i$ is the brine boiling temperature and $T_v$ is vapor saturation temperature.

As the pressure of evaporation of seawater in the heat pump condenser is constant, the boiling temperature of seawater at this time is 40°C. Based on the above formula, the heat transfer coefficient of the heat pump condenser is 1.81.

### *3.4. System Performance*

Performance Ratio (PR) is defined as the amount of distillate produce per 2326 kJ of heat input or number of pounds of distillate produce per 1000 Btu of heat [16]. PR is calculated by equation (4) according to the literature [16]:

$$PR = \frac{2326m_d}{Q_{in}} \quad (4)$$

$$Q_{in} = W_{comp} \quad (5)$$

where $m_d$ is freshwater production, kg/h; $Q_{in}$ is the input work of the system, kW; $W_{comp}$ is the compressor power, kW.

### *3.5. Parameter Setup*

The basic parameters required for simulation input are summarized in table 1. The working medium of the heat pump in this system is refrigerant R134a, and the mass flow is 320 kg/h. The pressure of the primary effect evaporator is 7.24 kPa, and the corresponding evaporation temperature of seawater is 40°C. The pressure of the secondary effect evaporator is affected by the temperature difference between the two effects.

**Table 1.** The basic parameters.

| Name | Numerical | Unit |
|---|---|---|
| Feed seawater temperature | 20 | °C |
| Feed seawater salinity | 35 | ‰ |
| Feed seawater flow | 80 | kg/h |
| Target freshwater production | 40 | kg/h |
| Evaporation temperature of the primary effect evaporator | 40 | °C |
| Degree of superheat or supercooling | 5 | °C |

Taking the defined standard seawater salinity of 35‰ as the basic parameter of feed seawater, the NaCl solution with the same salinity was simulated. The selection of physical properties is the basis for Aspen Plus to calculate thermodynamic and transfer properties, and proper selection is crucial to the accuracy and reliability of process simulation. ELECNRTL physical property method is used in simulation because salt water belongs to inorganic electrolyte.

## 4. Results and Analysis

### *4.1. Influence of Heat Pump Evaporating/Condensing Temperature*

The heat pump evaporating/condensting temperature determine the energy consumption of the refrigeration compressor. Accordingly, the compressor power is an important element to affecting the PR value of the double-effect evaporation system. The brine circulation ratio is 60% and the temperature difference between the two effects of the system is 5°C are fix parameters. The relationship between PR and freshwater production with the evaporating temperature and condensing temperature is shown in figure 3.

The figure 3 is shown that the freshwater production reduces with the increase of evaporating temperature, and the decline amplitude is very small. In the range of evaporating temperature from 10 to 20°C, the freshwater production only decreases by 0.5 kg/h, while the PR value increases with the rising of the evaporating temperature. The freshwater production and PR reduce with the increase of condensting temperature when the evaporating temperature is constant. The difference in freshwater production between 50°C and 60°C of condensting temperature is 3.39kg/h. The PR of the system reaches 40.07 with an evaporating/condensing temperature of 20°C/50°C. The freshwater production is 39.76kg/h with an evaporating/condensing temperature of 10°C/50°C.

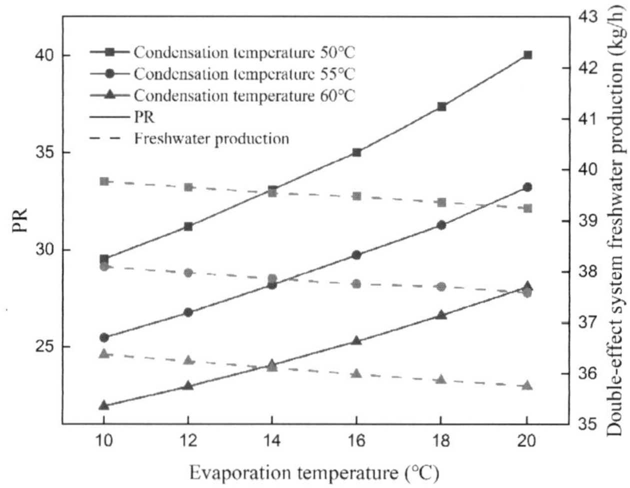


**Figure 3.** The relationship between PR and freshwater production with the evaporation temperature and condensting temperature of the heat pump.

### *4.2. Influence of Temperature Difference between the Two Effects*

The difference between the primary effect evaporation temperature of seawater and the secondary evaporation temperature of seawater is called the temperature difference between the two effects in the double-effect system. Taking heat pump evaporating temperature of 50°C, heat pump condensting temperature of 10°C and brine circulation ratio of 60% as fix parameters, the relationship between the temperature difference between the two effects and primary effect steam production and secondary effect steam production and freshwater production is as shown in figure 4.

When the primary effect evaporation temperature is constant, the secondary evaporation temperature of seawater decreases with the rising of the temperature difference between the two effects. Therefore, the BPE also decreases and the secondary effect steam production increases a little with the increase of the temperature difference between the two effects. The heat release of brine in preheater 2 will increase with the increase of temperature difference between the two effects, and then the temperature of feed seawater will rise. However, the temperature of secondary steam in the secondary effect evaporator decreases and the heat release of the auxiliary condenser increases, which leads to the trend of steam production in the primary effect evaporator increasing and then decreasing. The maximum freshwater production is 39.78 kg/h when the temperature difference between the two effects is 5°C. Based on the formula for PR, the change of temperature difference between the two effects does not influence on the energy consumption of the heat pump compressor. Therefore, the value of PR increases and then decreases with the increase of the temperature difference between the two effects.

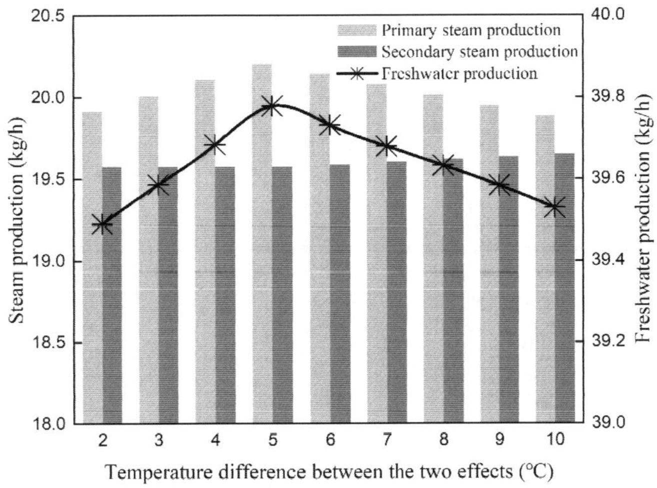


**Figure 4.** The relationship between the temperature difference between the two effects and primary effect steam production, secondary effect steam production, freshwater production.

*4.3. Influence of Brine Circulation Ratio*

The brine circulation ratio is one of the key parameters affecting the double-effect system operation. It is defined as the ratio of brine re-entering into the secondary effect evaporator to the total brine flowing out of the secondary effect evaporator. When the evaporating/condensting temperature of the heat pump is 50°C/10°C and the temperature difference between the two effects is 5°C, the relationship between the brine circulation ratio and primary effect steam production and secondary effects steam production and freshwater production is shown in figure 5.

As displayed in figure 5, the steam production of the primary effect evaporator increases with the increase of brine circulation ratio, while the steam production of the secondary effect evaporator decreases and the total freshwater production increased slightly. The difference in freshwater production between 0% and 60% of brine circulation ratio is only 0.46 kg/h. The reason is the increase of brine circulation ratio increases the seawater flow and salinity into the secondary effect evaporator. With the increase of concentrated seawater flow in preheater 1 increases the temperature of feed seawater into the primary effect evaporator and the BPE value of seawater in the secondary effect evaporator. Due to the heat pump operation condition is unchanged and the constant energy consumption of the heat pump compressor, and the brine circulation ratio has little influence on the PR when the total freshwater production changes slightly. Although increasing the brine circulation ratio can increase freshwater production, it will also increase the energy consumption of the circulating pump and improve the risk of corrosion of the secondary effect evaporator. In actual system operation, the brine circulation ratio should not exceed 60%.

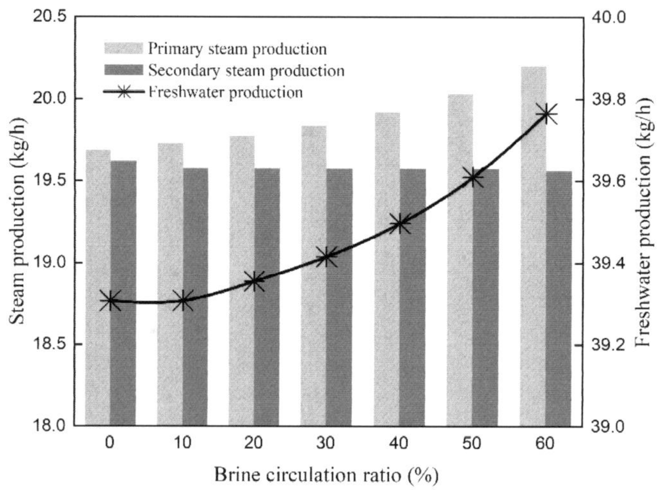


**Figure 5.** The relationship between brine circulation ratio and primary effect steam production, secondary effect steam production, freshwater production.

## 5. Conclusion

- The value of PR increases but the freshwater production reduces with the rising of evaporating temperature of the heat pump. Both the value of PR and the freshwater production decreases with the condensing temperature of the heat pump;
- When the pressure in the primary effect evaporator or seawater evaporating pressure is constant, the total freshwater production increases and then reduces with the increase of the temperature difference between the two effects. The freshwater production of the double-effect evaporation system is maximum when the temperature difference between the two effects is 5°C;
- With the increase of the brine circulation ratio of the secondary effect evaporator, the steam production of the primary effect evaporator increases, the steam production in the secondary effect evaporator decreases and the total freshwater production increases slightly;
- The double-effect seawater desalination system using compression heat pump is not only limited to use for seawater desalination, but also can be used for purification treatment of other solutions. The biggest advantage of this system is that it can reduce the concentration of the treated solution in the primary effect evaporator. The concentrated solution can be evaporated and concentrated in the secondary effect evaporator at a lower temperature, which increases the stability of the equipment.

## Acknowledgments

This work was supported by the National Natural Science Foundation of China (Grant No.52006092).

## References

[1] Zhang Y S 2021 The status and development trend of seawater desalination tech industry *Industrial Water Treatment* **41** 26-30

[2] Hao D Q, Sha Z L, Wang Y F, Shu C, Zeng X D, Zhao X Y and Xiong C 2011 Establish LT-MED simulation system with Aspen Plus *Journal of Salt and Chemical Industry* **40** 37–39

[3] Ma Z Q, Huo S H and Su M 2013 Simulation with Aspen Plus and performance analysis of LT-MED seawater desalination system *Applied Mechanics and Materials* **397–400** 948–956

[4] Liponi A, Wieland C and Baccioli A 2020 Multi-effect distillation plants for small-scale seawater desalination: thermodynamic and economic improvement *Energy Conversion and Management* **205** 112337

[5] Hu B, Wang W Y, Wang K and Cao F 2011 The application of high-temperature heat pump in industrial refrigeration field *Journal of Refrigeration* **32** 1–5
[6] Reali M 1984 A refrigerator-heat-pump desalination scheme for fresh-water and salt recovery *Pergamon* **9** 583–588
[7] Slesarenko V V 2001 Heat pumps as a source of heat energy for desalination of seawater *Desalination* **139** 405–410
[8] Zhou J F, Yuan G F and Zhang H F 2005 Experimental study on heat pump seawater desalination system *Tianjin Chemical Industry* **02** 22–23+30
[9] Zhao L 2005 Analysis of variable working conditions of small heat pump system for seawater desalination *Chinese Journal of Mechanical Engineering* **05** 225–229
[10] Attia A A A 2016 Heat pump seawater distillation system using passive vacuum generation system *Desalination* **397** 151–156
[11] López-Zavala R, Velázquez N, González-Uribe L A, Quezada-Espinoza K M, Aguilar-Jiménez J A, Islas S, Nakasima-López M and González E 2019 Absorption cooling and desalination system with a novel internal energetic and mass integration that increases capacity and efficiency *Desalination* **471** 114144
[12] Evyatar S, Wang Z X, Akshay D, Vasiliki K and Menachem E 2020 Membrane distillation assisted by heat pump for improved desalination energy efficiency *Desalination* **496** 114694
[13] Elbassoussi M H, Ahmed M A, Zubair S M and Qasem N A A 2021 On a thermodynamically-balanced humidification-dehumidification desalination system driven by a vapor-absorption heat pump *Energy Conversion and Management* **238** 114142
[14] Gao C J and Chen G H 2004 *Desalination Design and Engineering Manual* (Beijing: Chemical Industry Press) p 16
[15] El-Dessouky H T, Ettouney H M and Al-Juwayhel F 2000 Multiple effect evaporation—vapour compression desalination processes *Chemical Engineering Research and Design* **78** 662–676
[16] Amin Z M and Hawlader M N A 2015 Analysis of solar desalination system using heat pump *Renewable Energy* **74** 116–123

# Numerical Calculation of Solid-Liquid Two-Phase Flow in Open-Design Vortex Pump

**Xiongfa Gao [1,2], Weidong Shi [2,3,a*], Junhua Wang [1,b*], Hongzhong Lu [1,c*], Jiabin Wang [4], Wanli Meng [1] and Jiamin Jin [1]**

[1] Shanghai Kaiquan Pump (Group) CO. LTD., Shanghai 201804, China;
[2] National Research Center of Pumps, Jiangsu University, Zhenjiang 212013, China;
[3] School of Mechanical Engineering, Nantong University, Nantong 226000, China;
[4] Shandong shuanglun CO. LTD., Weihai 264203, China.
Email: [a] wdshi@ujs.edu.cn, [b] wangjunhua@kaiquan.com.cn;
[c] luhongzhong@kaiquan.com.cn.

**Abstract.** This paper mainly based on the CFD platform, numerical simulation of the internal flow field in the open-design vortex pump under different solid particle concentration and particle size is carried out to explore the flow trajectory characteristics of solid particles in the impeller channel and the lateral cavity, as well as the distribution characteristics of solid particles, and calculate the sliding velocity of particles in the impeller channel. The results show that at different solid particle concentrations, the solid volume fraction *f* in the front and rear cover plate area of the blade is high and that in the middle area is low; The solid volume fraction *f* decreases gradually from the working face to the back, and the solid particles are widely distributed in the inlet area. Due to the asymmetry of the volute structure, the particle concentration in the inlet area corresponding to the casing tongue is significantly higher than that in other areas; Due to the existence of circulating flow and through flow, the solid particles are distributed in a circular direction in the lateral cavity, and the particle concentration in the middle region is large; In the impeller area corresponding to the casing tongue, the sliding velocity of solid particles decreases with the increase of particle size, and the sliding velocity values of 0.5mm and 1.0mm particle sizes are close.
**Keywords.** Open-design vortex pump; Numerical simulation; Through flow; solid-liquid two-phase flow.

## 1. Introduction

Compared with conveying one-phase medium, the performance of open-design vortex pump changes greatly when conveying medium containing solid particles, and the solid-liquid two-phase flow in one-phase medium is complex and special, and it is easy to be blocked when conveying medium containing large particles. Therefore, many scholars at home and abroad have revealed the law of solid-phase flow in open-design vortex pump through theoretical analysis, experimental research and CFD analysis.

In 2002, Daniel Baun [1] used the method of combining experimental analysis and CFD calculation to study the velocity at different sections in the impeller, the stress at the working face and back of the blade, and the static pressure at the impeller outlet. It was concluded that the numerical calculation results were roughly similar to the experimental results, indicating the feasibility of the numerical simulation method. In 2007, Mehta M and Sastry S [2] used PIV technology to measure the velocity of solid particles in the pump. In 1998, wuyulin [3] and others used *K-ε-* Ap turbulence model and SIMPLEC algorithm simulate the distribution of solid-liquid two-phase medium in the flow passage components. It is found that when the solid volume concentration is low, the influence of solid particles on the liquid phase velocity is very small. Yang Minguan [4-6] and others used the PDPA technology to measure the liquid-solid two-phase flow field inside the self-designed swirl model pump, and obtained the three-dimensional velocity distribution of each phase inside the vaneless cavity and impeller. Then they used the advanced laser Doppler anemometry (LDA) to measure the flow field inside the vaneless cavity and impeller, and obtained the circumferential velocity, axial velocity Radial velocity and its corresponding RMS fluctuating velocity distribution. Gao Bo [7, 8] also calculated the particle concentration distribution in the impeller and lateral cavity through CFD, and tested the vaneless cavity with phase Doppler particle velocity field analyzer (PDA) to explore the turbulent fluctuation characteristics of salting-out particles in the pump and its impact on the liquid flow field. Zhao Wanyong [9-12] conducted experimental research on vortex pump from different solid particle sizes and concentrations. Wang Congli [13], Li Yisong [14], Xu Haiming [15] and others analyzed the distribution characteristics of solid-liquid two-phase flow in the flow field in the vortex pump based on CFD, and obtained the distribution characteristics of solid particles in the vortex pump.

## 2. Simulation Model and Meshing

The vortex pump studied in this paper is SCP200-400 type vortex pump in Jiangsu Industry-University-Research Cooperation Project. The performance parameters of the vortex pump are: flow rate $Q= 500m^3/h$, head $H = 28m$, rotating speed $n = 1485r/min$, specific speed $n_s= 166$. Main structural parameters: impeller outside diameter $D_2 = 350mm$, blade width $b = 60mm$, blade outlet angle $\beta_2 = 45°$, blade inlet angle $\beta_1=60°$. The whole simulation domain of this study is mainly divided into five parts, namely, the inlet section, the lateral cavity, the impeller, the volute and the outlet section. ICEM software is used to divide structured grid of the model water bodies.

## 3. Numerical Simulation Method and Boundary Conditions

In this paper, based on ANSYS-CFX commercial numerical simulation software platform, the multi-reference coordinate system method was used to conduct the dynamic and static coupling between the impeller and the volute, and the time-averaged N-S equation was used as the basic control equation. It is assumed that the shape of the solid particles was unchanged during the operation, and the flow of the fluid medium in the pump was a steady flow. During the simulation, the surface roughness was set to 0.05mm.

The channel flow field was set as incompressible steady viscous flow, using Standard k-ε turbulence model. The irrotational flow was set at the impeller inlet and the pressure at the inlet was set zero as the reference point. The flow at the outlet was assumed fully developed outflow. The near-wall flow was treated by the standard wall function and assuming no slip solid wall, the pressure and speed were coupled by SIMPLE-C algorithm.

## 4. Analysis of Numerical Simulation Results

In order to correspond to the rapeseed particles used in the experiment, the particle distribution in the impeller and the lateral cavity area is analyzed when the solid particle diameter *D* is 0.5mm and the solid volume concentration cv is 5%. Figure 1 shows the code sketch and cross section diagram of the impeller blades (where "B" represents the blade, "B-P" represents the pressure side of blade, "B-S" represents the suction side of blade, and "S" represents the cross section).

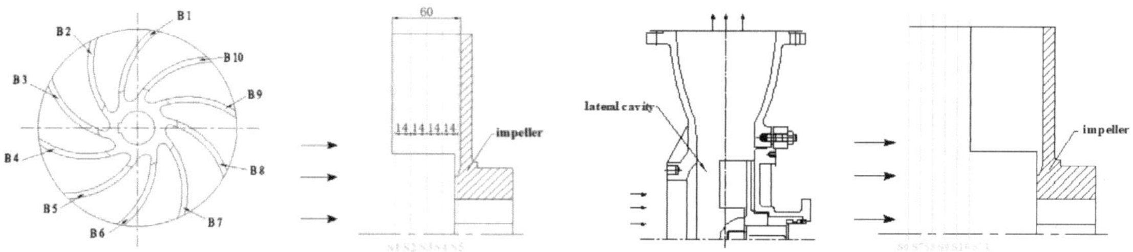


**(a)** **(b)** **(c)** **(d)**

**Figure 1.** Cross-section and the impeller blade code sketch: (**a**) impeller blade code; (**b**) cross-section of impeller; (**c**) lateral cavity; (**d**)cross section of lateral cavity.

### *4.1. Solid-Liquid Two-Phase Analysis of Impeller Blade Wall*

In the previous section, the influence of particles on the internal flow field under a particle diameter and solid particle concentration is analyzed. The following discussion will be conducted under the rated flow condition, with the solid particle diameter *d* being 0.5mm, and the solid volume concentration $c_v$ respectively 1%, 5% and 10%, analyze the influence of particle diameter on the flow field of the vortex pump. Due to the large number of blades, only blades B1 and B10 were selected for comparative analysis in this analysis, as the cross section, only the middle section of impeller back shroud and vaneless cavity were selected for analysis, and the cross section on B1 and B9 is selected for research.

Figure 2 and figure 3 show the contrast nephogram of the solid volume fraction distribution in blade pressure side of B1 and B10. It can be seen from the analysis of this figure that the distribution of particles in the blade wall area is greatly affected by solid-phase particle concentration. From figure 2, it can be seen that when the solid-phase volume concentration is 1%, the distribution of particles in blade pressure side of B1 is significantly less than that of 5% and 10%. When the solid volume concentration is 5%, solid particles accumulate more on the blade surface, not only a large number of particles accumulate in the shroud ends area and back shroud area of the blade, but also more particles exist on the blade wall. When the solid volume concentration is 10%, the distribution of particles in blade pressure side increases, and even the solid-phase in some areas fills the blade wall. The situation of blade B10 in figure 3 is the same as that of blade B1 in figure 2. The three solid volume concentrations show that the solid volume fraction *f* in front and back shroud area of the blade is high and that in the middle area is low.

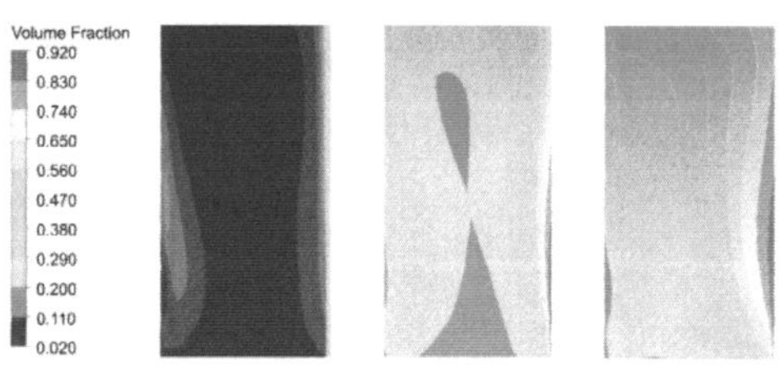


(a)$c_v$=1% (b)$c_v$=5% (c)$c_v$=10%

**Figure 2.**The particle concentration distribution nephogram in blade pressure side of B1.

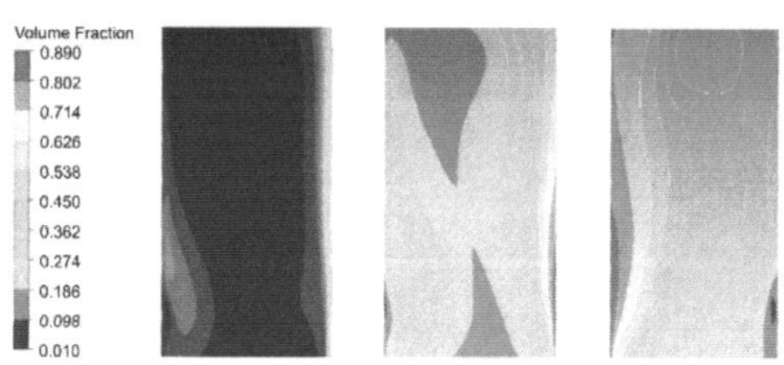


(a)$c_v$=1% (b)$c_v$=5% (c)$c_v$=10%

**Figure 3.** The particle concentration distribution nephogram in blade pressure side of B10.

### *4.2. Influence of Solid Particle Concentration Change on Internal Flow Field Distribution*

Figure 4 shows the contrast nephogram of the solid volume fraction distribution of the impeller back shroud. Analysis of the diagram shows that the larger the concentration of solid particles is given, the more particles accumulate in back shroud area, and the larger the solid volume fraction *f* of the working face is. The solid volume fraction *f* gradually decreases from the working face to the back, and the solid particles in the inlet area are widely distributed. Due to the asymmetry of the volute structure, the particle concentration in the inlet area corresponding to the tongue is significantly higher than that in other areas.

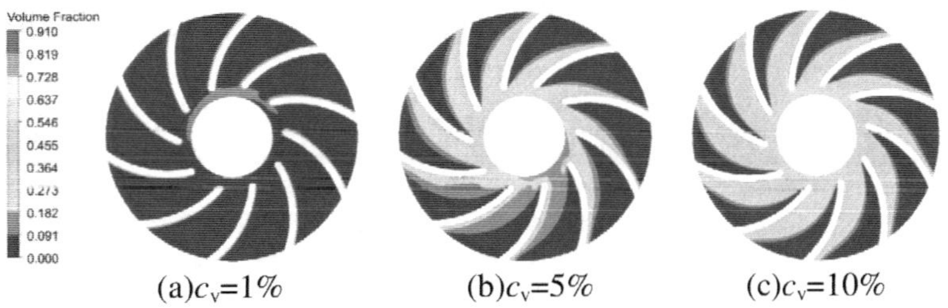


**Figure 4.** The particle concentration distribution nephogram in back shroud.

Figure 5 shows the contrast nephogram of the solid volume fraction distribution in the middle section of the vaneless cavity. It can be clearly seen from the figure that the solid particles have circumferential distribution in the vaneless cavity and are divided into three-ring distribution. When the concentration of solid particles is 1%, there are more solid particles in the area corresponding to the tongue and the aspect ratio is about 0.4, and then decrease in an anti-clockwise spiral manner. When the solid particle concentration reaches 10%, the maximum solid volume fraction shifts to the area where the aspect ratio is about 1 below the right of the vaneless cavity. The annular distribution is obvious, there are two annular distribution characteristics, and there are more particles accumulated in the central area.

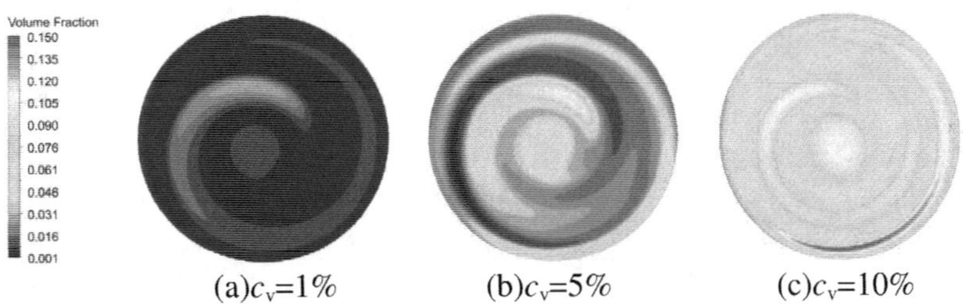


**Figure 5.** The particle concentration distribution nephogram in vaneless cavity middle section.

Figure 6 shows the curve of solid volume fraction and velocity in front streamline of blade. It can be seen from figure 6 (a) that when the diameter of solid particles is 0.1mm, the value of solid volume fraction in front streamline of the blade B1 fluctuates gently, and the solid volume fraction from the inlet to the outlet of the blade is maintained at about 0.1. With the increase of the solid particle diameter, the fluctuation of the solid volume fraction is relatively obvious. When the diameter of solid particles is 0.5mm and 1.0mm, the fluctuation trend of solid volume fraction is consistent and the value is close. The corresponding solid-phase slip velocity $v$ in front streamline is consistent with the increasing trend of solid-phase particle concentration. In the inlet area, the solid-phase slip velocity $v$ of 0.1mm solid particles is slightly larger than that of 0.5mm and 1.0mm solid particles. When the aspect ratio is greater than 0.55, the solid-phase slip velocity $v$ is significantly smaller than the values when the solid particle diameters are 0.5mm and 1.0mm. It can be seen from figure 6 (b) that when the diameter of solid particles is 0.1mm, the fluctuation of solid volume fraction in front streamline on the back of B1 blade is small. When the diameter of solid particles is 0.5mm and 1.0mm, the fluctuation amplitude is large, but the difference of solid volume fraction at corresponding points between them is relatively small. From the trend of solid-phase slip velocity $v$, the solid-phase slip velocity $v$ in front streamline of the blade back does not change significantly with the increase of solid particle diameter. It can be seen from figure 6 (c) that the solid volume fraction in front streamline of the blade B9 working face changes greatly with the increase of the solid particle diameter. Especially when the aspect ratio is about 0.8, the solid volume fraction increases from 0.1 to nearly 0.7 when the diameter of solid particles increases from 0.1mm to 1.0mm. And the corresponding position of the slip velocity is relatively large, the wear of particles in this area is more serious. In this area, when the solid particle diameter is 1.0mm, the solid volume fraction is significantly higher than that when the solid particle diameter is 0.5mm. The variation trend of the solid-phase slip velocity $v$ of the blade is similar to that of the B1 blade with the increase of the solid-phase particle diameter. It can be seen from figure 6(d) that the variation trend of the solid volume fraction in front streamline on the back of the blade B9 with the increase of the solid particle diameter is close to that of the blade B1, but the solid-phase slip velocity $v$ is different. When the diameter of solid particles is 0.5mm and 1.0mm, the solid-phase slip velocity $v$ at the corresponding point of the blade is very close. When the diameter of solid particles is 0.1mm, the solid-phase slip velocity $v$ in front streamline with aspect ratio less than 0.9 is higher than that when the diameter of solid particles is 0.5mm and 1.0mm.

The changes of solid volume fraction and slip velocity of the two blades B1 and B9 are different, mainly due to the different positions of the blades B1 and B9 in the pump cavity. The position of blade B1 corresponds to the tongue-separated area, which has tongue-separated blocking flow and small cross-sectional area. The blade B9 is located near the eighth section of the volute, where the medium flows out of the volute after the impeller does work on the medium. The medium flow in these two areas is complex. It has a great influence on solid volume fraction $f$ and solid-phase slip velocity $v$.

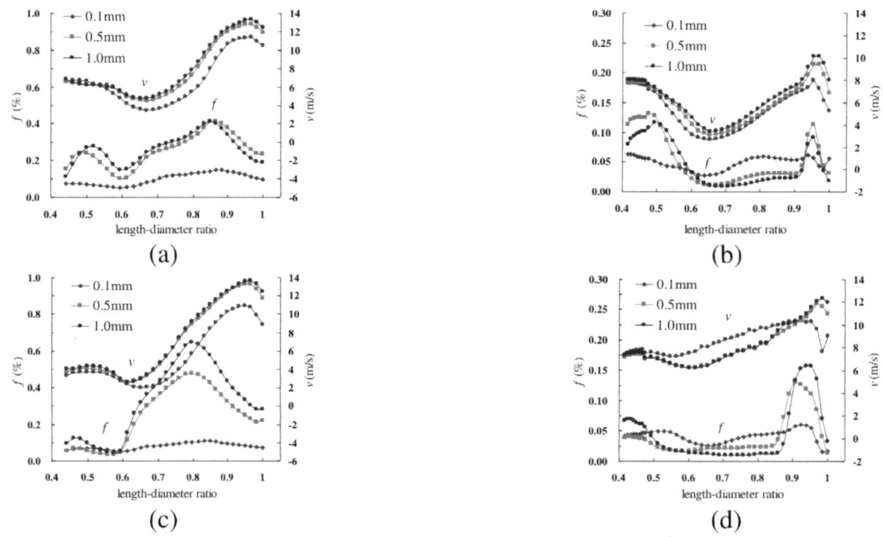


**Figure 6.** The solid volume fraction and velocity in front streamline of blade: (**a**) Pressure side of B1; (**b**) Suction side of B1; (**c**) Pressure side of B9; (**d**) Suction side of B9.

Figure 7 shows the curve of solid volume fraction and velocity in centre streamline of blade. It can be seen from figure 7(a) and figure 7(c) that when the particle diameter is 0.1mm, the solid volume fraction $f$ in streamlines of blade B1 and B9 working faces is very gentle. However, when the particle diameter increases to 0.5mm, the solid volume fraction $f$ increases.

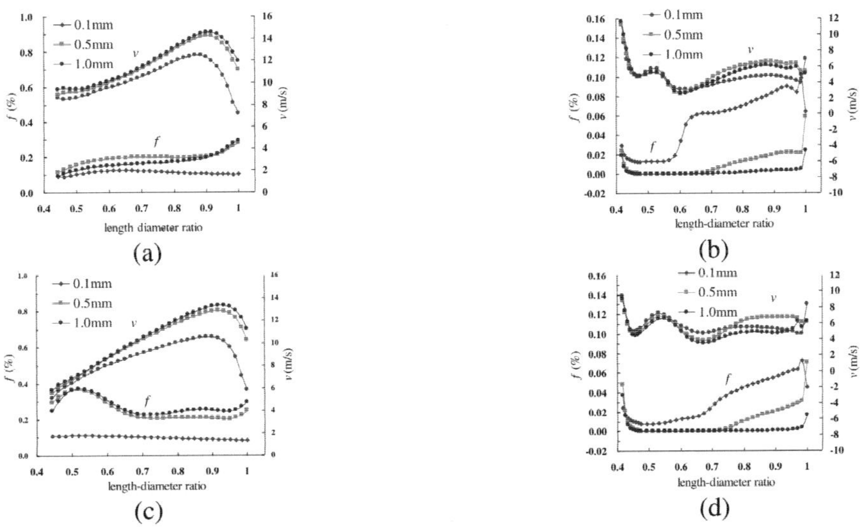


**Figure 7.** The solid volume fraction and velocity in center streamline of blade: (**a**) Pressure side of B1; (**b**) Suction side of B1; (**c**) Pressure side of B9; (**d**) Suction side of B9.

The trend of solid-phase slip velocity of these two blades is consistent. The streamline area at the inlet of blade B1 does not increase with the increase of particle diameter, and the wear is small in this area. Figure 7(b) and figure 7(d) show that when the particle diameter is 0.1mm, the solid volume fraction *f* in centre streamline of blade back is larger than that of 0.5mm and 1.0mm, and increases from inlet to outlet. When the diameter increases to 1.0mm, the solid volume fraction *f* tends to zero in the other areas of the blade, except for a small part of the particle distribution in the inlet and outlet areas. However, the solid-phase slip velocity *v* does not change significantly with the increase of particle diameter.

## 5. Conclusion

(1) At different solid particle concentrations, the solid volume fraction f in the front and rear cover plate area of the blade is high and that in the middle area is low; The solid volume fraction f decreases gradually from the working face to the back, and the solid particles are widely distributed in the inlet area. Due to the asymmetry of the volute structure, the particle concentration in the inlet area corresponding to the casing tongue is significantly higher than that in other areas;

(2) In the lateral cavity, due to the existence of circulating flow, the solid particles in the lateral cavity show the characteristics of circular distribution, and the particle concentration in the middle region is large, and the particles show the characteristics of stratified circular distribution; At higher concentration, more particles gather in the fifth and sixth cross sections of the volute.

(3) In the impeller area corresponding to casting tongue, the slip velocity of solid particles decreases with the increase of particle size; when the particle size is larger than 0.5 mm, the change of concentration and slip velocity on the front face of the impeller blade in the corresponding area of casting tongue is not obvious; In the impeller area corresponding to the eighth section of volute, when the particle size is 0.1 mm, the particle concentration on the working surface of the impeller blade is obvious, and the particle size is less than 0.5 mm and 1.0 mm. On the back of the impeller blade, the particle concentration under the particle size of 0.1 mm is much greater than that of 0.5mm and 1.0 mm.

**Acknowledgments**
This research was funded by the Special funds for Taishan Industrial Leading Talents Project of China (Grant No. tscy20200225), Enterprise Postdoctoral Program of China (Grant No. 8421440002).

## Reference

[1] Daniel B 2002 *Hydrodynamic Force in Centrifugal Pump and Compressor Impellers in Volue Casings: Measurements Using Magnetic Bearing and CFD Simulation* (Virginia: University of Virginia) 3057473

[2] Mehta M, Kadambi J R, Sastry S, Sankovic J M, Wernet M P, Addie G and Visintainer R 2007 Particle velocities in the rotating impeller of a slurry pump *The 5thJoint ASME.JSME Fluids Engineering Conference* p 369-378

[3] Wu Y L, Cao S L, Ge L, Wang L, Wang X M and Sun Z X 1998 Solid-liquid two-phase turbulent flow calculation and measurement through centrifugal slurry pump impellers *Journal of Tsinghua University Science and Technology* **38(1)** 71-74.

[4] Yang M G, Gao B, Gu H F and Li H 2008 Measurement on 3D turbulent flow field in vortex pump *Drainage and Irrigation Machinery* **26(1)** 60-64 (in Chinese)

[5] Yang M G, Gao B, Liu D, Gu H F and Li H 2008 Experimental Investigation of Salt-out Two-phase Flow in a Vortex Pump by PDPA Measurements *Journal of Engineering Thermophysics* **29(2)** 237-240 (in Chinese)

[6] Yang M G, Gao B, Liu D, Gun H F and Li H 2007 Analysis on Liquid-solid Two-phase Flow Field in a Vortex Pump by PDPA Measurement *Transactions of the Chinese Society for Agricultural Machinery* **38(12)** 53-57 (in Chinese)

[7] Yang M G and Gao B 2009 Particle concentration distribution and its effect on salt-out features in a vortex pump *Journal of Engineering Thermophysics* **30(12)** 275-278.

[8] Yang M G and Gao B 2010 Research on turbulent velocity fluctuations of salt-out particles in a vortex pump volute *Journal of Engineering Thermophysics* **31(2)** 2031-2033.

[9] Zhao W Y, Liu T B, Li Y S, Zhang L and Wang Z 2007 Flow mechanism in votex-pump *Drainage and Irrigation Machinery* **25(6)** 52-55 (in Chinese)

[10] Zhao W Y, Li Y S, Wang Z and Zhang L 2007 Inquiry into equation of particles motion in vortex-pump *Journal of Lanzhou University of Technology* **33(6)** 47-49 (in Chinese)

[11] Zhao W Y, Wang Z, Jing Y, Li Y S and Zhang L 2009 Numerical analysis of the influence of different flow parameters on the performance of vortex pumps *Journal of Xihua University* **28(4)** 89-92.

[12] Zhao W Y, Jing Y, Wang Z, Zhang L and Zhang F 2009 Numerical analysis of influences on vortex pump exterior characteristics for solid particles diameter and concentration *Oil Field Equipment* **38(10)** 5-10 (in Chinese)

[13] Wang L H, Song P Y and Zhao L 2008 The numerical simulation research on distribution of solid particle on solid-liquid vortex pump *Machinery* **35(12)** 9-12.

[14] Li Y S and Wei J F 2010 Numerical simulation of interior flow in vortex-pump for liquid-solid two-phase medium with small size particles *Fluid Machinery* **38(7)** 20-23.

[15] Xu H M, Zhuang W J, Tai J and Li H L 2011 Numerical simulation research on solid-liquid two-phase on vortex pumps *China Rural Water and Hydropower* **6** 86-88.

# Fault Analysis and Elimination of Hydraulic Driving System for Rubber Rig in Coal Mine

**Fenglei Chen***

Xi'an Research Institute, CCTEG, Xi'an, 710077, Shaanxi, China
Email: 282979472@qq.com

**Abstract.** Aiming at the failure of the rubber wheel drilling rig with unstable walking speed during use, the principle of the walking hydraulic system is introduced, the cause of the failure is analyzed, and the principle of first easy and then difficult, first observation and then dismantling and inspection is followed. According to the output pressure of the differential valve block and the output characteristics of the hydraulic pump, combined with the principle of the hydraulic travel system, it is finally determined that the internal leakage of the differential valve block is the main cause of the failure. After replacing the differential valve block, the fault was eliminated, and the traveling system of the whole machine returned to normal.
**Keywords.** Rubber Wheel Rig; Hydraulic Closed System; Instable speed; Fault Diagnosis; Hydraulic System Optimization.

## 1. Introduction

At present, coal mine underground gas extraction drilling rigs are mainly full hydraulic crawler type, generally composed of crawler chassis, main engine, pump station, console and other parts. Compared with traditional split drilling rigs, crawler drilling rigs have shorter movement and stabilization time, which reduces the number of workers [1-5]. However, the long-distance underground relocation of crawler drilling rigs still adopts the methods of crawler self-propelled, integral relocation, and split relocation and reassembly, especially in mines with trackless rubber-wheeled transportation and auxiliary relocation conditions. , which restricts the production and transportation efficiency of the mine.

Trackless auxiliary transportation equipment developed relatively late in China. It was not until the early 1990s that trackless rubber tire auxiliary transportation equipment was introduced [6-7]. The auxiliary transportation equipment that is more suitable for domestic mines not only has a short production cycle, but also has a relatively low price [8-9].

As the infrastructure construction of trackless transportation increases with the depth of coal mining, the state pays more attention to the safe and efficient production of coal mines, and the contradiction between drilling and excavation becomes more and more prominent. Requirements [10-12]. In response to this demand, a rubber wheel type full hydraulic drilling rig was developed.

The drilling rig is mainly hydraulically controlled, including main systems such as drive system, steering system, outrigger system and drilling construction. The article analyzes and provides solutions for the failures of the walking system in using.

## 2. The Fault Phenomenon Description and Principle Analysis

The designed maximum speed of the equipment is 20 km/h. During testing and use, the vehicle speed may be unstable, and the maximum speed can only reach about 12km/h. At the same time, when the accelerator pedal is released, the whole machine suddenly moves forward. The walking hydraulic closed system adopts a pump-controlled motor for speed regulation. The driver manipulates the pilot handle to change the displacement of the pump, thereby controlling the motor speed to change the vehicle speed. The diesel engine is used as the power source, which transmits the torque to the hydraulic pump, and the high-pressure oil in the pump It is connected with the motor to drive the wheels to rotate and walk [13-15]. The system principle of the drilling rig is shown in figure 1. The main circuit is composed of 2 A4VG series pumps and 2 motors. The direction handle controls the forward and backward, and the differential valve group is used to balance the speed of the wheels on both sides during the turning process. According to the fault phenomenon, in the case of ensuring the correct pipeline connection, check the oil charge pressure, brake valve, hydraulic pump DA valve directional handle, and differential valve group.

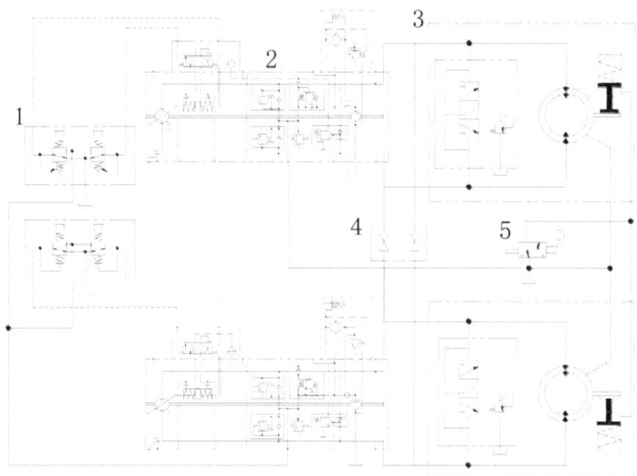


1-Pilot handle for walking; 2-Hydraulic pump; 3-Hydraulic motor;
4-Differential valve group; 5-Parking valve

**Figure 1.** The schematic of hydraulic system.

## 3. The Analysis and Troubleshoot of Failure

The design, manufacture, installation and debugging of the hydraulic system are closely linked, especially in the debugging stage. The analysis and elimination of faults are related to the success of the entire system. Now through the analysis of the failure situation of the hydraulic system during the debugging process, and propose a scheme.

### *3.1. The Detection of Fuel Charge and Parking Pressure*

First, check the fuel supply pressure and the output pressure of the parking valve. After the engine starts normally, the fuel supply pressure and the output pressure of the parking valve group are about 27bar, which means that the fuel supply system and the braking system are working normally, and the parking brake is normally opened. It is shown in figure 2. Rule out the possibility that the parking gear is not fully opened.

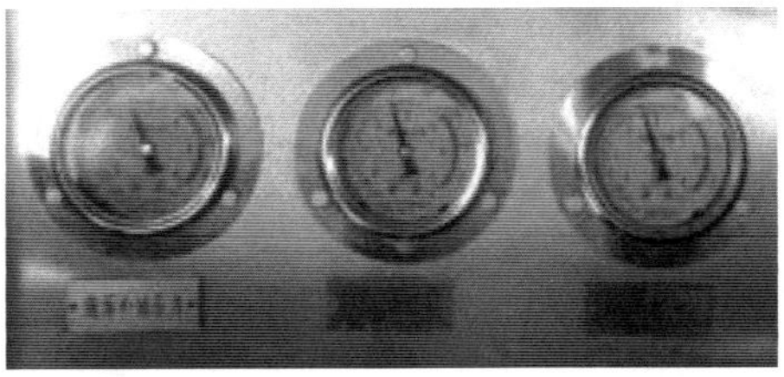

**Figure 2.** The detection of fuel charge and parking pressure.

### *3.2. The Detection of DA Valve in Hydraulic Pump*

The system uses Rexroth A4VG series pumps with DA valve control function, which are suitable for the static pressure transmission mechanism and closed circuit system of mining machinery, which can maintain the best match between the output of the hydraulic pump and the output of the engine, and prevent the mechanical system from being overloaded by the engine. It is large and stalled, and improves the efficiency of use and the operability of the driver [16]. The operator disassembled and cleaned the DA valve, measured the size of the orifice to meet the output requirements, and ensured that the maximum displacement of the pump could meet the requirements for use. It is shown in figure 3.

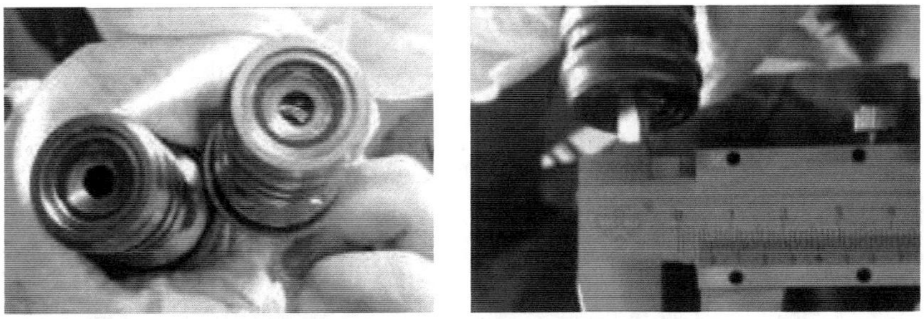

**Figure 3.** The detection of DA valve.

### *3.3. The Output Detection of Pilot Handle*

The pilot handle is used to control the maximum output of the hydraulic pump and the output direction of high-pressure oil, and its output curve is shown in figure 4. When the pilot control element is in the neutral position, the displacement of the hydraulic pump is zero and the running is stopped; using the proportional variable pump controlled by DA, after pulling the control handle in any direction, the speed of the vehicle increases with the increase of the engine speed. , the vehicle walks forward or backward; the combination of the pilot control element and the DA control can better control the low-speed stability of the vehicle. The operator checks the pilot output pressure, the output pressure range is normal and meets the requirements for use.

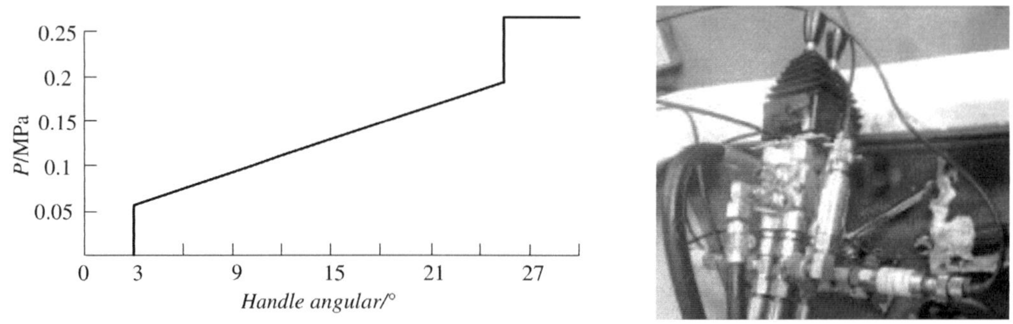


**Figure 4.** The output detection of pilot handle.

### *3.4. The Joint Debugging Test of Whole Machine*

In the single inspection, after excluding the oil charge and parking pressure, the output pressure of the hydraulic pump DA valve and the pilot valve, the traveling system is jointly adjusted and tested, including parameters such as engine speed, system pressure, wheel speed and system oil temperature. The field connection test is shown in figure 5.

After the system is connected, start the engine, and after the system oil temperature is normal, conduct the hydraulic oil temperature, engine speed, wheel speed, A port (high pressure), B port (low pressure) and Ps port (control) of the two hydraulic pumps. During the test, the hydraulic oil temperature was normal, the engine speed conformed to the characteristics of the output curve, and the wheel speed was converted into a vehicle speed of about 12 km/h. At this time, the pressure curves of the A/B port and the Ps port of the hydraulic pump are shown in figure 6.

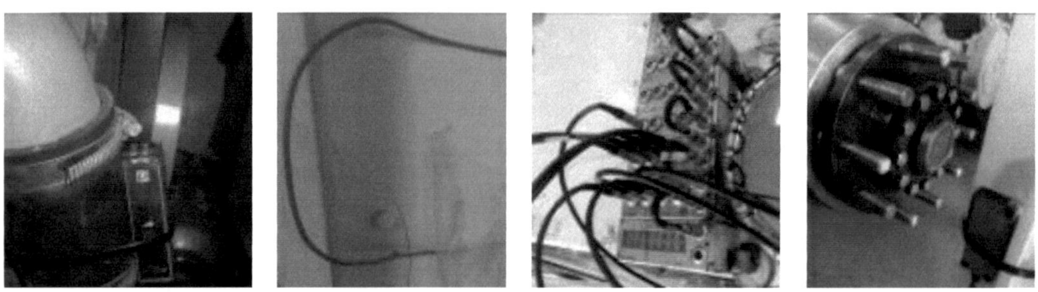

**Figure 5.** The detection of engine speed, system pressure, wheel speed and oil temperature.

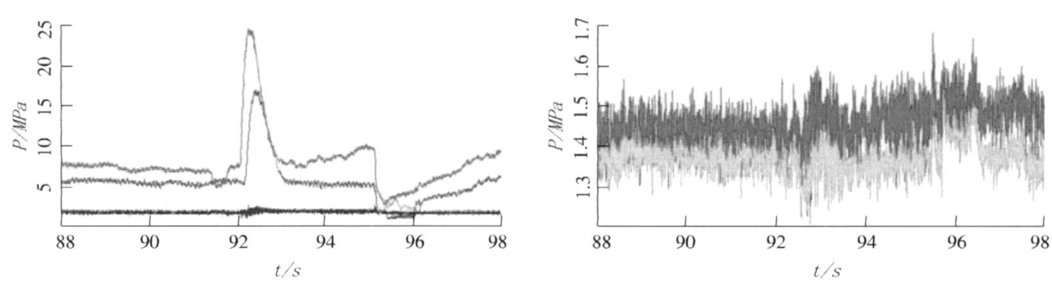


a) The curve 1 of hydraulic pump A/B b) The curve 1 of hydraulic Ps

**Figure 6.** The curves 1 of hydraulic pumps.

It can be seen from the pressure curve that the high pressure ports of the two pumps have a step at about 92 s and suddenly decrease at about 95 s. By analyzing the schematic diagram of the hydraulic system and combining the above preliminary fault judgment results, the debugging personnel initially identified the fault point. It may be the cause of the failure that differential valve block.

After the differential valve group was removed on site, the test was re-tested. The wheel speed was equivalent to about 19 km/h. At this time, the pressure curves of the A/B port and the Ps port of the hydraulic pump were shown in figure 7.

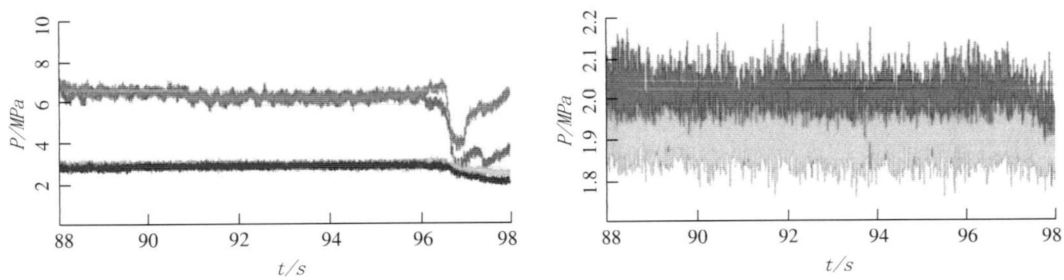


a) The curve 2 of hydraulic pump A/B    b) The curve 2 of hydraulic Ps

**Figure 7.** The curves 2 of hydraulic pumps.

The differential valve group was repeatedly loaded and unloaded on site for testing, and the situation was basically consistent with the two test curves. After replacing the new differential valve group to test, the travel speed test is normal. At this time, the cause of the failure is basically found out. This kind of failure phenomenon is relatively rare in the travel drive system. It may be that the magazine causes the spool of the differential valve block to be blocked. Carry out comprehensive cleaning and testing, restore the system pipeline and original parts, and the whole system is running normally. In the future, the reliability and functionality of each component should be checked before assembly, which will be of great help and practicability to eliminate potential safety hazards and hidden troubles.

## 4. Conclusions

The article expounds the instability of vehicle speed and vehicle shifting faults in the system debugging and operation of the rubber-wheeled drilling rig. Through the principle analysis of the system and components, the fault points are found and eliminated. In recent years, due to the development of computer technology, detection technology, information technology and intelligent technology, the development of hydraulic system fault detection and diagnosis technology has been greatly promoted. The system performs real-time detection to improve the reliability of the system.

## References

[1] Wu D, Yao K and Chen L 2014 Application and development of track-mounted drilling rig with acrossing transportation belt *Coal Science and Technology* **42** (7) 81-83, 29.

[2] Tian H L 2008 Manufacture and application on track-mounted rig for gas drainage *Coal Geology and Exploration* **36**(5) 77-80.

[3] Wu D, Zhang R and Wang S 2017 Development of the ZDY3600LQ All-dip drilling rig *Metal Mine* (12) 99-102.

[4] Yao N P, Sun R J and Ye G F 2008 Equipment and technology of borehole construction for gas drainage in china underground mines *Coal Science and Technology* **36** (3) 13-16.

[5] Wu D 2016 Design and application of ZDY4300LF(A) split crawler drilling rig *Coal Geology and Exploration* **44** (5) 160-163,167.

[6] Li Z N 2019 Application of trackless rubber tyred vehicle in mine auxiliary transportation system *Colliery Mechanical and Electrical Technology* **40** (2) 66-69.

[7] Jin W W 2019 Application status and development direction of auxiliary transportation system for mine trackless rubber wheeled vehicles *Mechanical and Electrical Engineering Technology* **48**(3) 188-190.
[8] Zhu J J and Li Y H. 2000 Developing trend of the auxiliary transport facilities at coal mine in our country *Coal Mine Machinery* (1) 1-3.
[9] Fan Y P and Li C Y 2005 Design on TY100FB transport vehicle of double drum coal shearer *Coal Science and Technology* **33**(1) 33-35.
[10] Shen B H, Liu J Z and Zhang H 2007 The technical measures of gas control in china coal mines *Journal of China Coal Society* **32** (7) 673-679.
[11] Yao Y F, Wang H J and Fan D 2008 Development of the ZDY1200L track-mounted hydraulic tunnel drilling rig *Coal Geology and Exploration* **36** (3) 75-77.
[12] Li W N, Shao J J and Guo X P 2018 Application on ZDY750L drilling rig in anchoring of floor in xinyuan coal mine *Coal Geology and Exploration* **46** (2) 196-200.
[13] Liu B T, Yao Y and Jiang P 2018 Closed hydrostatic walking system of raspberry picking machine *Chinese Hydraulics and Pneumatics* (9) 108-113.
[14] Zhou Y H, Zhao J and Liu N N 2017 Design of closed loop hydraulic driving control system for construction vehicle *Rock Drilling Machinery & Pneumatic Tools* (2) 17-20.
[15] Lou T X, Ding H G and Dong J 2019 Design and simulation analysis of electro-hydraulic proportional close loop drive systems in closed loop control for coal mine hydraulic winch *Chinese Hydraulics and Pneumatics* (6) 63-68.
[16] Liang G P and He X H 2010 Analysis and application of control principle for A4V the hydraulic pump with DA valve *Machinery Design and Manufacture* (12) 105-107.

# Research on Momentum Parameters of Tailings Dam-break Mortar

**Hongkang Zhou[1], Yanzhang Liu[1,2*], Liang Shan[1], Wenjie Wang[1,2] and Huabing Cao[3]**

[1] School of Resources and Environmental Engineering, Wuhan University of Science and Technology, Wuhan, Hubei 430081, China;
[2] Hubei Key Laboratory for Efficient Utilization and Agglomeration of Metallurgic Mineral Resources, Wuhan University of Science and Technology Wuhan, Hubei 430081, China;
[3] Chengchao Mining Co. Ltd of Wuhan Resources Group, Ezhou, Hubei 436050, China
Email: 756089103@qq.com

**Abstract.** The tailings dam-break mortar with high potential energy will cause impact damage to the downstream buildings, and the momentum parameters of mortar can effectively reflect its impact on buildings. Taking the distribution status of downstream geomorphology and production and living facilities of Yangjiawan tailing pond as the calculation background, Numbering 4 sensitive analysis points from 1 to 4 in line with the distance between the key protected structures and the breach of the tailing pond from the near to the distant, The three-dimensional fluid simulation software FLOW 3D was used to simulate the evolution process of dam-break mortar under the full-break condition of the accumulation dam above the initial dam, and the variation characteristics of the flow mortar momentum at the key protection structures in the evolution process were analyzed. The results show that the momentum of dam-break mortar at the sensitive points decreases from the near to the distant, Compared with instantaneous momentum, cumulative momentum could reflect the cumulative impact effect of mortar on buildings, and the distinction of impact effect was more obvious.; The peak inundation depth of dam-break mortar at each sensitive point was 1, 3, 2 and 4 sensitive points in descending order, the momentum and submergence depth were inconsistent, The momentum could clearly distinguish the difference of mortar impact effect, and the research results could provide some reference for dam - breaking mortar disaster assessment of tailing pond
**Keywords.** Momentum parameter, Dam-break mortar, Numerical simulation, Tailing pond.

## 1. Introduction

Tailings pond is used to store the residual waste residue generated during ore mining, i.e. tailings sand [1-3]. Once the dam breaks, the mortar with huge energy will impact or even destroy the downstream buildings [4-6]. The impact effect of flowing mortar on buildings can be reflected by the momentum value it carries. The impact effect of mortar on buildings can be studied through momentum analysis.

In this paper, the dam-break mortar of Yangjiawan tailings pond is taken as the research object, and the three-dimensional numerical simulation software FLOW3D is used to simulate the evolution process of the mortar in the downstream under the condition that the tailings above the initial dam are completely broken. The variation law of the momentum of the dam-break mortar with the evolution time and the distance from the initial dam of the tailings pond is analyzed. The research is helpful to further accurately analyze the impact effect of dam-break mortar on key protected buildings and structures, and is of great significance to predict the disaster consequences of tailings dam break and do a good job in disaster reduction and prevention.

## 2. Mortar Momentum Calculation Based on FLOW3D

FLOW3D, a three-dimensional dynamic simulation software based on finite difference method, can track the change of free surface of flowing mortar and simulate the evolution process of dam-break mortar in tailings pond downstream to obtain the momentum parameters of mortar. After the dam break of tailings pond, the impact damage of downstream buildings and structures is related to the size and duration of the impact force of flowing mortar. The size and duration of the impact force of mortar are reflected by the momentum carried by it. Equation (1) is the momentum theorem of flowing mortar:

$$\int Fdt = \int d\ (mv) = \int dp \tag{1}$$

where $F$ is the impact force of flowing mortar, unit: N; t is the action time of impact force of flowing mortar, unit: s; m is the mass of flowing mortar, unit: kg; $v$ is the flow velocity of flowing mortar, unit: m/s; $p$ is the instantaneous momentum of the flowing mortar, unit: n·s.

FLOW3D divides the surface of the downstream computing area into several equal area grids. The bottom area of each grid is s, and the unit is $m^2$. The momentum of mortar in the grid is calculated by taking the prism above the grid as the object. The density of mortar in the grid is ρ, Unit: $kg/m^3$; Instantaneous velocity is $v$, unit: m/s; The submergence depth is $h$, and the unit is m. Then the instantaneous momentum $p$ carried by the dam-break mortar is calculated according to equation (2):.

$$p = mv = \rho\ s\ h \cdot \tag{2}$$

Considering the cumulative effect of continuous impact of mortar on buildings and structures during the evolution process, the cumulative momentum can be calculated, as shown in equation (3).

$$M = \sum_{i=1}^{n} p_i \tag{3}$$

where M is the cumulative momentum of dam-break mortar at the nth moment, n · s; Pi is the instantaneous momentum of mortar at the ith moment, i=1, 2, 3 … n, n·s.

## 3. Overview of Tailings Pond and Downstream Landform

Yangjiawan tailings pond is located in Yangjiawan ditch, 1.0KM north of Chengchao Iron Mine, Ezhou City, Hubei Province. The elevation of tailings dam crest is 95.0m, the elevation of initial dam bottom is 32.6m, and the elevation of dam crest is 50.0m. The side slope ratio of the sub dam of the accumulation dam is 1:3. The effective storage capacity is 17.113 million $m^3$, and the storage capacity below the initial dam is 1.687 million $m^3$.

The topographic plan of the downstream of the tailings pond is shown in figure 1. The distance between each sensitive point and the initial dam of tailings pond, terrain elevation and building elevation are shown in table 1.

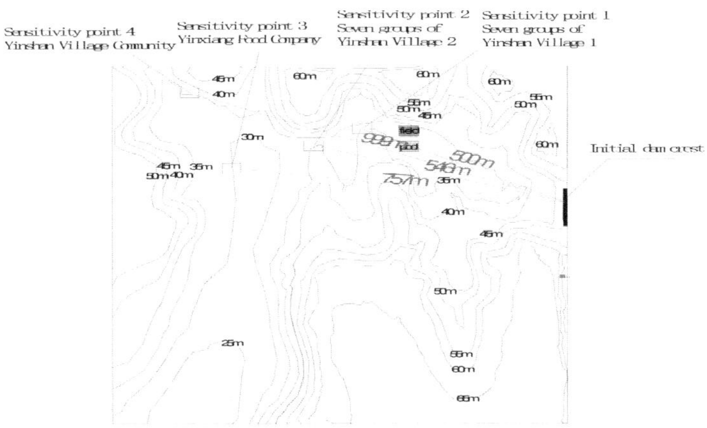


**Figure 1.** Topography plan of Yangjiawan tailings pond downstream.

**Table 1.** Sensitive point spatial position parameter table.

| Downstream sensitive poin | Distance from initial dam of tailings pond $L$/m | Topographic elevation $H$/m | Building elevation $D$/m |
|---|---|---|---|
| Sensitive point 1 | 257.0 | 34.6 | 44.6 |
| Sensitive point 2 | 420.0 | 35.1 | 45.1 |
| Sensitive point 3 | 550.0 | 30.2 | 45.2 |
| Sensitive point 4 | 900.0 | 40.8 | 49.8 |

## 4. Numerical Simulation of Dam-Break Mortar Evolution

### *4.1. Numerical Simulation Calculation Condition*

(1) Calculation condition

The most dangerous situation is often considered in the dam break disaster assessment of tailings pond. Since the initial dam is not easy to break, the instantaneous total failure of the accumulation dam above the initial dam is selected as the most dangerous calculation condition, and the total amount of tailings discharged is 15.426 million m$^3$.

(2) Mortar fluid parameters

According to the total amount of sand discharge and hydrogeological data of Yangjiawan tailings pond under the most dangerous working condition, the mass concentration of dam-break mortar under this working condition is 75%. According to the relevant tailings mortar test report, the dynamic viscosity of the tailings mortar fluid at 75% concentration is 93.0pa·s, and the relative density is 1611.0kg/m$^3$.

(3) Boundary condition

In this numerical simulation, the dam crest boundary of the initial dam is set as the inflow boundary, and the empirical formula method is often used to calculate the inflow boundary flow. Generally, the burst flow hydrograph is generalized as a quartic parabola. Input the flow based on the flow hydrograph in figure 2.

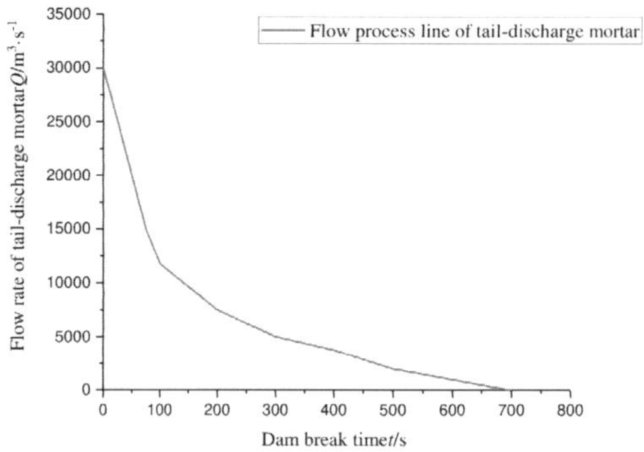


**Figure 2.** Flow process line of tail-discharge mortar.

The downstream outlet is set as the outflow boundary (free boundary), the top is the pressure boundary, the size is the standard atmospheric pressure, and other boundaries are all constrained wall boundaries.

*4.2. Analysis of Numerical Simulation Results*

After 4000s of dam break, the mortar at the downstream outflow boundary presents a slow and uniform flow state, and the movement state parameters of mortar at each sensitive point are in a relatively stable state after the peak. Subsequently, the numerical simulation results from 0 to 4000s are selected for analysis.

*4.2.1. Momentum of Dam Break Mortar at Sensitive Point.* Combined with the change of mortar inundation depth with time at the sensitive points after the breach of the dam obtained by equation (2), the change of mortar mass with time at each sensitive point is shown in figure 3(a), the change of mortar flow rate with time at each sensitive point obtained by numerical simulation is shown in figure 3(b), and the change of mortar instantaneous momentum with time at each sensitive point obtained by figure 3 is shown in figure 4. In figure 4, the difference between the magnitude of mortar momentum at sensitive point No. 4 and other sensitive points is large, so the mortar momentum curve at sensitive point No. 4 in the figure is not obvious.

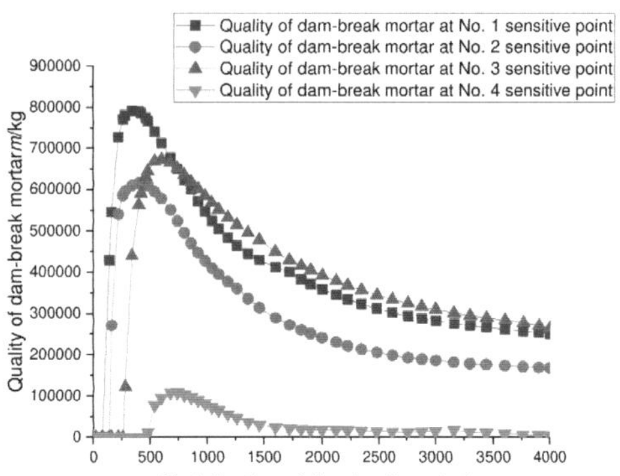


(a) Curve of dam-break mortar quality changing with time

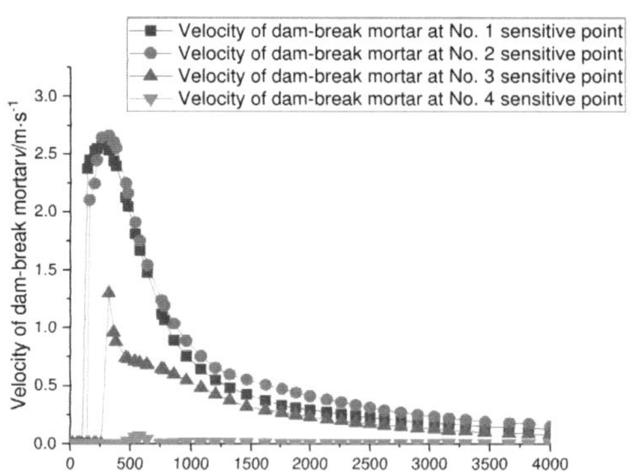


(b) Curve of flow velocity of dam-break mortar changing with time

**Figure 3.** Variation curve of dam-break mortar quality and velocity changing with time at sensitive point.

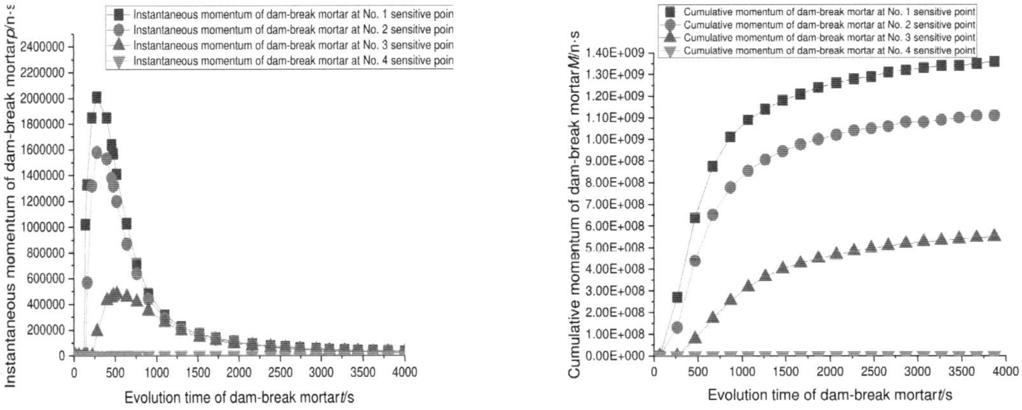


(a) Curve of instantaneous momentum of mortar changing with time

(b) Curve of cumulative momentum of mortar changing with time

**Figure 4.** Variation curve of dam-break mortar momentum with time at sensitive points.

Table 2 shows the momentum and evolution process parameters of dam-break mortar at each sensitive point according to figure 3~4.

**Table 2.** Comparison table of numerical simulation results of dam-break mortar at all sensitive points.

| Downstream sensitive poin | Instantaneous momentum peak $p$/n·s | Instantaneous momentum peak time $t$/s | Peak submergence depth $h$/m | Peak time of submergence depth $t$/s | Submerged state of buildings and structures | Time of arrival $t$/s | Accumulated momentum at 4000s $M$/n·s |
|---|---|---|---|---|---|---|---|
| Sensitive point 1 | 2010695.10 | 302 | 19.79 | 356 | been submerged | 124 | $1.36\times10^9$ |
| Sensitive point 2 | 1616943.50 | 320 | 15.41 | 416 | been submerged | 154 | $1.11\times10^9$ |
| Sensitive point 3 | 477087.10 | 522 | 16.83 | 594 | been submerged | 276 | $5.53\times10^8$ |
| Sensitive point 4 | 6280.40 | 586 | 2.70 | 716 | been submerged | 464 | $1.64\times10^6$ |

It can be seen from table 2 and figure 4 (a) that the momentum of dam-break mortar decreases with the increase of the distance from the initial dam. This is because when the dam just breaks, the terrain near the break is high. When the dam-break mortar reaches the initial dam crest, it will convert high potential energy into high velocity and reach the early peak. After that, the mortar collided with the ground, the direction of velocity changed, some energy was lost, and the instantaneous momentum decreased sharply. Then, under the combined action of cohesion and surface friction, the instantaneous momentum decreases gradually.

It can be seen from figure 4 (a) and (b) that the instantaneous momentum of dam-break mortar at each sensitive point first rises rapidly with the dam break time, and then slowly dissipates after reaching the peak. The cumulative momentum of dam-break mortar in the whole analysis period rises rapidly at first, and then gradually tends to be stable, which is more in line with the actual impact damage of buildings and structures by mortar in dam break. Therefore, compared with the instantaneous momentum, the cumulative momentum can reflect the cumulative impact effect of mortar on buildings and structures, and the distinction of impact effect is more obvious, which can better reflect the actual impact disaster process of flowing mortar.

*4.2.2. Comparison of Mortar Momentum, Submergence Depth and State at Sensitive Point.* At present, domestic and foreign scholars study the disaster caused by dam-break mortar by analyzing the submerged state and depth of dam-break mortar in the downstream. In this paper, the momentum characteristics of dam-break mortar are analyzed by comparing the motion state parameters of dam-break mortar at sensitive points. In order to compare the momentum of dam-break mortar with the submerged state, it is necessary to obtain the arrival time of dam-break mortar. The evolution diagram of dam-break mortar at several typical moments obtained by numerical simulation is shown in figure 5.

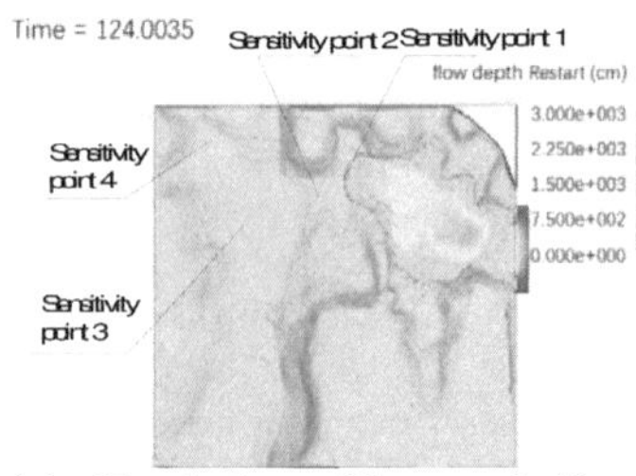


(a) Evolution Chart of Dam Break Mortar(124s)

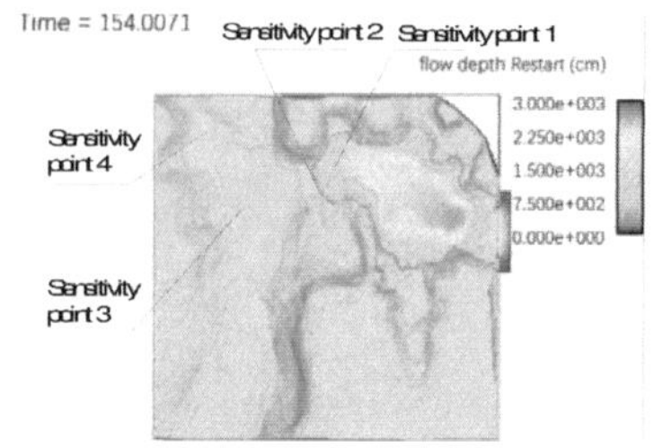


(b) Evolution Chart of Dam Break Mortar(154s)

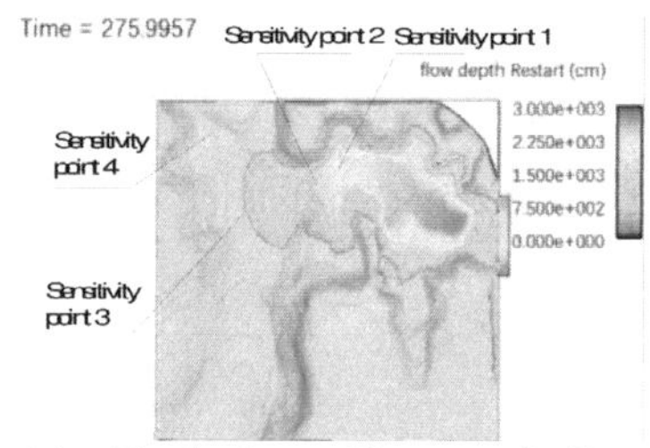


(c) Evolution Chart of Dam Break Mortar(276s)

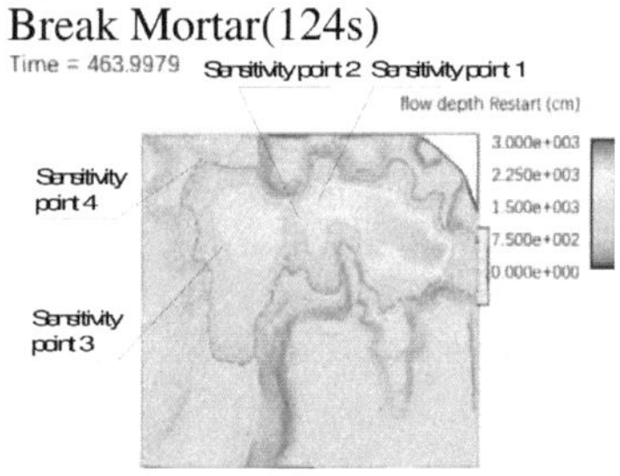


(d) Evolution Chart of Dam Break Mortar(464s)

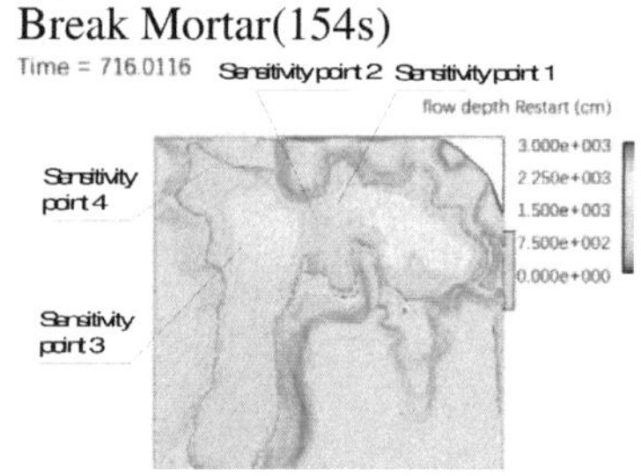


(e) Evolution Chart of Dam Break Mortar(716s)

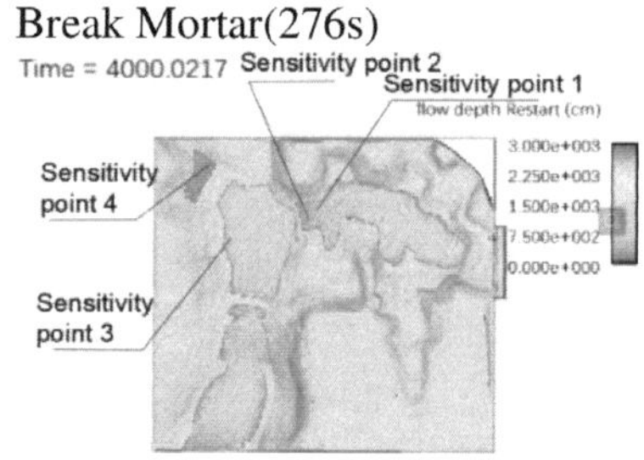


(f) Evolution Chart of Dam Break Mortar(4000s)

**Figure 5.** Evolution process diagram of dam breaking mortar at typical moment.

The instantaneous momentum peak value of mortar at the sensitive point decreases from near to far, and the submergence depth peak value from large to small is No. 1, 3, 2 and 4 sensitive points. The instantaneous momentum and submergence depth are not consistent. The reason is that the flow velocity of the mortar from sensitive point 2 to sensitive point 3 gradually decreases under the action of its own cohesion and surface friction. Although the terrain at sensitive point 3 is lower than that at sensitive point 2, the mortar submerged depth is large, but the ground potential elevation difference at the bottom of the downstream ditch of Yangjiawan tailings pond is relatively small, and the reduction of the flow velocity is greater than the increase of the submerged depth. Combining the flow velocity with the submerged depth, the mortar momentum at No. 3 sensitive point is smaller. Therefore, the momentum parameter can clearly distinguish the impact effect on the key protected buildings and structures, and provide a certain reference for the dam-break mortar disaster assessment of tailings pond.

## 5. Conclusion

In this paper, the finite difference calculation software FLOW3D is used to simulate the downstream evolution process of dam-break mortar of Yangjiawan tailings pond, and the variation law of momentum impact on key protected buildings with time and distance from the initial dam of tailings pond is analyzed. The conclusions are as follows:

(1) Numerical simulation can better simulate the evolution process of dam-break mortar in the downstream of tailings pond, obtain the motion state parameters of dam-break mortar in the downstream evolution process, extract the submergence depth and velocity to obtain the momentum parameters, and analyze its momentum characteristics.

(2) The momentum of dam-break mortar decreases with the increase of the distance from the initial dam. Compared with the instantaneous momentum, the cumulative momentum can reflect the continuous impact effect of mortar on buildings and structures, and can better reflect the actual impact disaster process of flowing mortar, which can provide a reference for the dam break disaster assessment of tailings pond.

(3) The submergence depth and impact momentum of mortar on buildings and structures are not consistent. When predicting the disaster caused by dam-break mortar in tailings pond, the momentum value can combine the submergence depth with flow velocity to help more clearly understand the difference of impact effect of dam-break mortar on downstream buildings and structures.

**Reference**

[1] Zhang L T 201 *Journal of Hydraulic Engineering* **44** 594-600
[2] Jiang Q H, Hu L M and Lin H 2017 *Advances in Science and Technology of Water Resources* **37** 77-86
[3] Yang C He, Zhang C and Li Q M 2021 *Rock and Soil Mechanics* **42** 1-17
[4] Lee S J and Jae B 2017 *Journal of Mechanical Science & Technology* **31** 207-214
[5] Deng Z, Chen S S and Ming Z Q 2017 *Chinese Journal of Geotechnical Engineering* **39** 932-938.
[6] Wu S F, Yan J and Cai H 2020 *Chinese Journal of Geotechnical Engineering* **42** 219-225.

# Numerical Simulation of Low Frequency Oscillation in an Aeroacoustic Wind Tunnel

**Ruixuan Ma[1,2,*], Hu Li[1], Shuaibin Han[1], Conghai Wu[1], Shuhai Zhang[1] and Xunnian Wang[2]**

[1]State Key Laboratory of Aerodynamics, China Aerodynamics Research and Development Center, Mianyang 621000, China
[2]Key Laboratory of Aerodynamics Noise Control, China Aerodynamics Research and Development Center, Mianyang 621000, China
Email: my200638@qq.com

**Abstract.** The low frequency pressure oscillation in an aeroacoustic wind-tunnel seriously affects the flow quality and acoustic measurement. At present, there are many measures to suppress pressure oscillation, but due to the lack of in-depth understanding of its generation mechanism, these methods are not of general significance. In this paper, based on the open-source software OpenFOAM, the SA-DDES method is used for the unsteady computations for the two-dimensional simplified wind-tunnel model. The steady and unsteady fields inside and outside the jet flow are showed. The computational results show that the nonlinear interaction between the large scale vortices rolled up by the instabilities of the shear layer and the front wall of the collector of the wind-tunnel is the main reason responsible for the low frequency oscillation in an aeroacoustic wind-tunnel. In addition, it is concluded that in the feedback loop path of the excitation-feedback effect, the pressure waves produced by the large scale vortex structures inside the shear layer impinging on the front wall of the collector are reflected by the rear wall of the plenum chamber. The end of the path that the pressure waves travel upstream is where vortex shedding happens, not the nozzle exit.
**Keywords.** Pressure oscillation, wind-tunnel, generation mechanism, shear layer.

## 1. Introduction

A general phenomenon in low-speed free-jet wind-tunnels is low frequency pressure oscillation in the open test section. Since the flutter frequency of the test section in most low-speed wind-tunnels is below 20 Hz, it is called low frequency pressure oscillation [1]. The existence of the low frequency oscillation phenomenon destroys the flow quality of test section, which seriously affects the validity of the wind-tunnel testing measurement. For acoustic wind-tunnels, low frequency oscillations also bring a threat to the safety of the anechoic chamber. In order to obtain a relatively stable and safe testing environment, it is particularly important to understand the flow mechanism of low frequency pressure oscillation and to control the low frequency oscillation of the wind-tunnel [2].

In order to reduce the low frequency pressure oscillation in the wind-tunnel, researchers and technicians in related fields have done a lot of work. Sellers et al. reduced the low frequency pressure oscillation in the open wind-tunnel jet by adding a spoiler device at the nozzle of the test section and gave the optimal spoiler configuration [3]. However, the method of nozzle turbulence decreased the effective test area of the open free jet wind-tunnel. Manuel et al. at Langley Company of America provided another method to reduce the low frequency pressure oscillation in the wind-tunnel, which was optimizing the shape of the collector. Experimental tests were carried out with six different configurations of collectors. The testing results also showed that the interval between the diffuser section and the collector helps reduce turbulence oscillation [4]. Researchers have also done some experimental studies on low frequency oscillation in the DNW wind-tunnel [2], and obtained many valuable results. Mark Rennie et al. studied the generation mechanism of the low frequency oscillation and investigated the influence of the crucial physical parameters on the low frequency oscillation in a 3/4 aeroacoustic wind-tunnel, and proposed that the excitation-feedback feedback is the main cause of low frequency pressure oscillation, and gave the explanation that the interval between the diffuser and the collector can reduce the low frequency oscillation [5]. In 2004, Rennie et al. in South Korea's Hyundai Company studied the generation mechanism of low frequency pressure oscillation in a 3/4 aeroacoustic wind-tunnel. It was concluded that low frequency pressure oscillation would affect the results of the steady force test, and a correction method was given in their paper [6]. In 2009, Kudo et al. in Mitsubishi Corporation of Japan investigated the suppression measures of low frequency pressure oscillation in a 7m × 4.5m automobile wind-tunnel. Their study concluded that the low frequency oscillation of the wind-tunnel mainly propagates in the loop (the standing wave mode can be clearly measured along the loop), changing the acoustic boundary of the loop can reduce the oscillation, and finally, the amplitude of the low frequency oscillation is reduced by the method of loop opening [7]. Recently, researchers of German DLR studied the generation mechanism of low frequency pressure oscillation in the DNW-NWB wind-tunnel, and concluded that the pressure oscillation was related to the propagation mode of the wind-tunnel pipeline. Through the transformation of the wind-tunnel structure, a sudden change in the cross-sectional area of the wind-tunnel is formed between the cooling section and the fan section, which makes the acoustic impedance of the pipeline generate a step, thereby suppressing the propagation of low frequency pressure oscillation [8]. In China, Jia et al. at Tongji University have done a lot of experimental research in a 3/4 open-jet wind automobile tunnel. The mechanism of low frequency pressure oscillation was explained by investigating the sound pressure spectrum, pressure oscillation coefficient, and the migration of the main frequency of the low frequency oscillation with the wind speed [9, 10].

However, most of the research focuses on the method and mechanism of reducing low frequency oscillation for specific wind-tunnels, which do not have general significance. In addition, the experimental research where the testing is conducted only at a limited number of pulsating pressure measuring points inside and outside the core flows in the test section of the wind-tunnel lacks the resolution on the details of the flow field, which is not enough to fully understand the flow mechanism responsible for the low frequency pressure oscillation. Therefore, a two-dimensional computation is performed to investigate the pressure oscillation in a low-velocity free-jet wind-tunnel with a conventional structure in this paper.

## 2. Methodology

In this section, we introduce the physical model and the numerical method used in the present work.

### *2.1. Physical Model*

In this paper, as shown in figure 1, a recirculation-type aeroacoustic wind-tunnel is adopted. The focus of our research is on the test section with an anechoic chamber. As the physical model used in the current study, a horizontal plane across the jet centerline of the simplified test section with an anechoic chamber is simulated numerically. $D$ denotes the length scale of the nozzle. P1 and P2 are testing points in the experiments and simulations. The complete test section is surrounded by an anechoic chamber with the size of $5D$. The collector has an expansion angle of 8 degrees. Figure 2 shows the grid used in this paper.

### *2.2. Computational Settings*

Before the unsteady computational process begins, a steady computation is conducted to achieve a statistical steady state as an initial field of the unsteady computations, then the unsteady computations are solved over hundreds of time periods of the pressure oscillation.

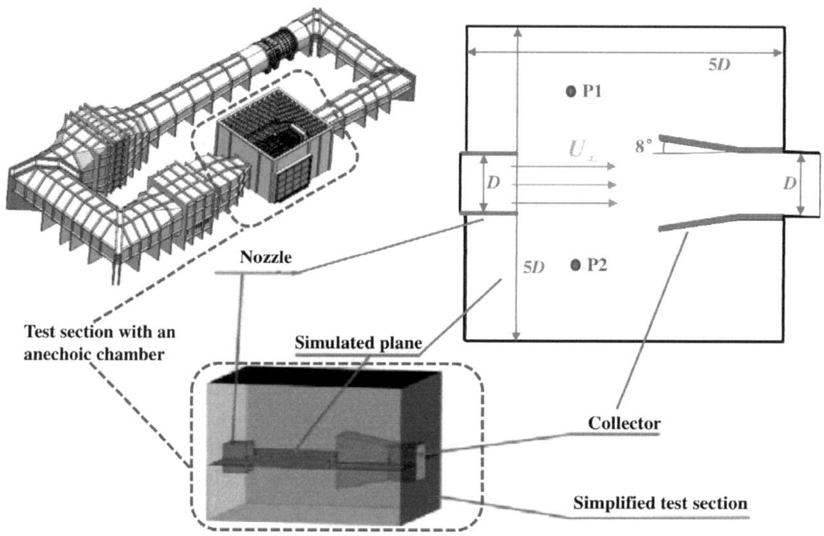


**Figure 1.** Simplified physical model.

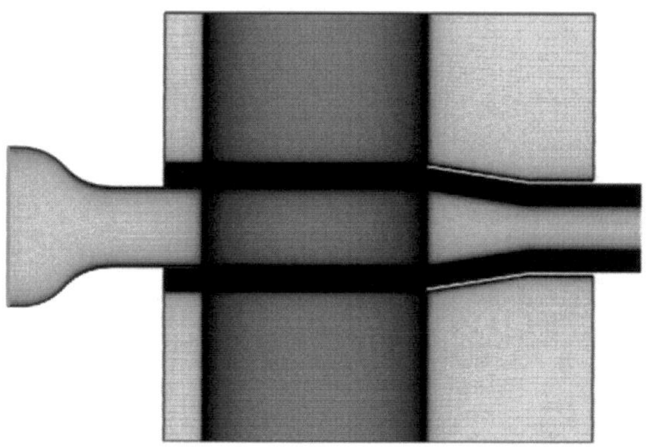

**Figure 2.** Computational mesh.

## 3. Results and Discussions

In this section, we give the typical computational results and corresponding discussions to reveal the generation mechanism of low frequency oscillation in an aeroacoustic wind-tunnel.

### *3.1. Experimental Validation*

Before investigating the generation mechanism of low frequency pressure oscillation, it is necessary to verify our numerical results by comparing them with the experimental results. Figure 3 provides the comparison of pressure oscillation in a far field outside the jet flows between the experimental and numerical data. Our computational results are in a quality agreement with experimental data. The calculation captures the approximate single-frequency pressure oscillation signal measured in the experiment. This single frequency is close to that of the experiment, and the amplitude is about half of that of the experiment.

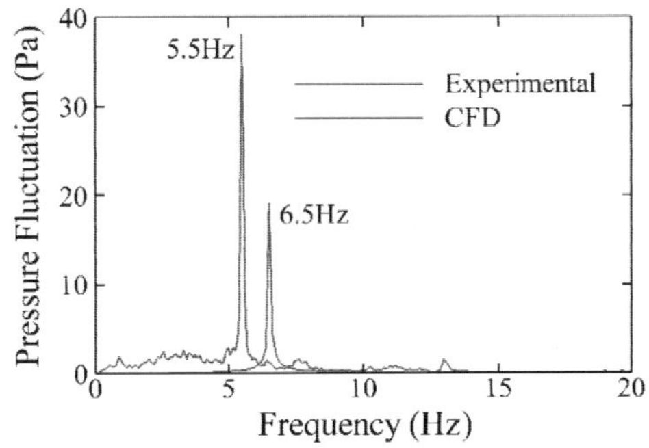


**Figure 3.** Comparison of pressure oscillation between the numerical and the experimental results.

*3.2. Steady Field*

Figure 4 illustrates steady-state calculation results. With the development of the jet, the downward diffusion of the lower edge of the shear layer is smaller than the upward diffusion of the upper edge, so that the center of the shear layer moves upward continuously. The thickness of the shear layer increases continuously along the flow direction, but the growth rate first increases and then decreases. The turbulence kinetic energy (TKE) in the shear layer of the jet flows is higher than that outside the shear layer, and its value reaches the maximum in the central region of the shear layer of the jet flows.

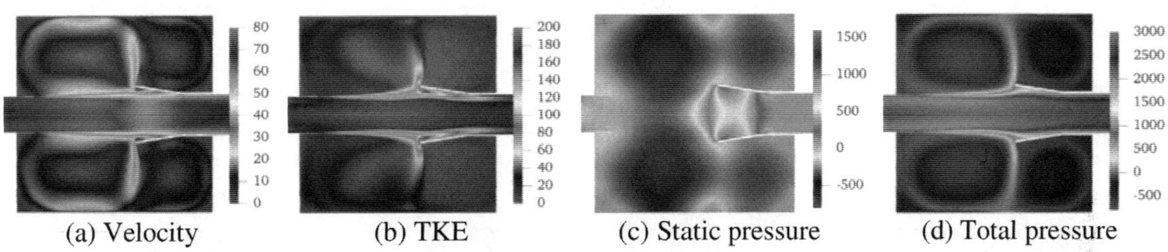


**Figure 4.** Steady flow field.

Figure 5 illustrates the steady flow field structure. There are multiple vortex structures with different scales outside the jet flow in the test section, including larger-scale circulation vortex and smaller-scale corner vortex vortices. In addition, at the back of the impact point, where the shear layer of the jet flows hits the collector, two small-scale vortices appeared at the junction of the front and rear larger-scale circulation vortices.

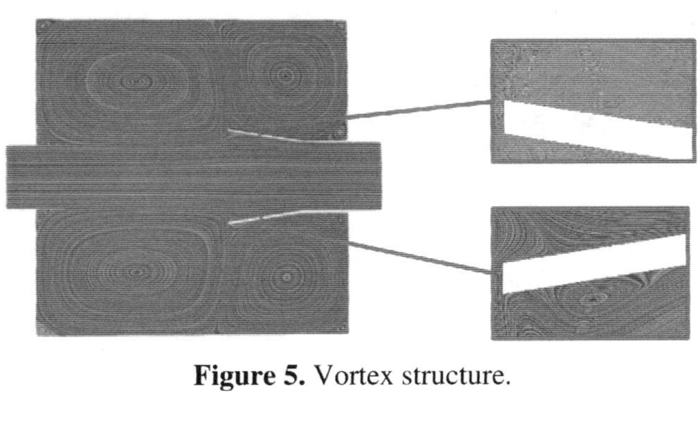

**Figure 5.** Vortex structure.

### *3.3. Unsteady Field*

Figure 6 shows the oscillation pressure far from the jet flows. As shown in figure 1, P1 and P2 are located on both sides of the jet flow, and symmetrical about the centerline of the wind-tunnel. The results obtained by wind-tunnel testing and numerical computations are consistent. The amplitude of the pressure oscillation at two points that are symmetrical about the axis of the wind-tunnel is equal and the phase is opposite.

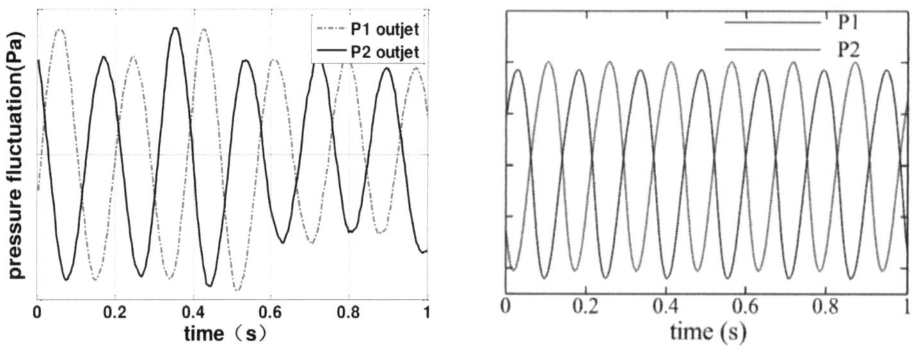


(a) Experimental (b) Computational

**Figure 6.** Oscillation pressure outside the jet flow.

The time evolution of pressure and vorticity is shown in figure 7 and figure 8, respectively. In simplicity, the generation mechanism of low frequency oscillation can be well recognized as the feedback mechanism between the instabilities of jet shear layer and the interaction of the vortex and the collector. The cyclic mechanism could be described as three critical steps: (1) generation of feedback waves related to the interaction of large scale vortex structures and the front wall of the collector (2) propagation of feedback pressure waves to upstream (3) excitation of unstable shear-layer near the nozzle, which leads to further vortex shedding and large scale structures.

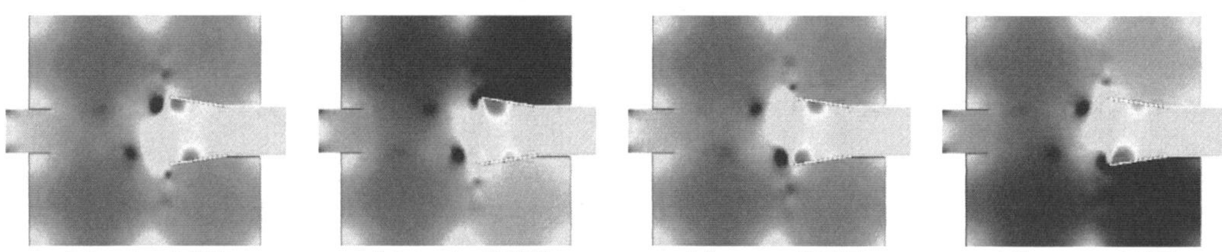

(a) $t=t_0+1/4T$ (b) $t= t_0+2/4T$ (c) $t= t_0+3/4T$ (d) $t= t_0+4/4T$

**Figure 7.** Pressure in a feedback loop.

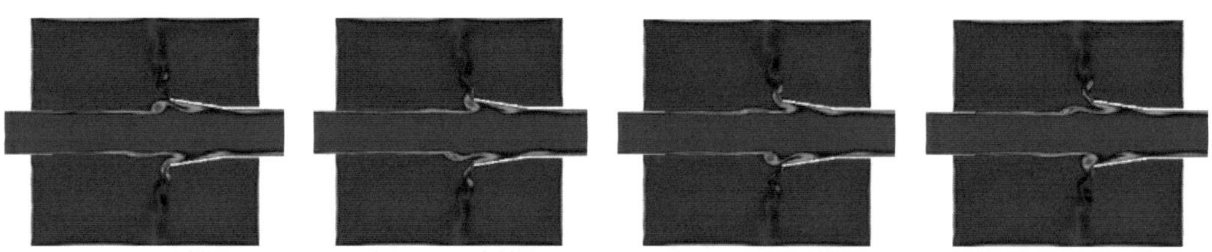

(a) $t=t_0+1/4T$ (b) $t= t_0+2/4T$ (c) $t= t_0+3/4T$ (d) $t= t_0+4/4T$

**Figure 8.** Vorticity in a feedback loop.

According to the above analysis, the frequency of pressure oscillation in an aeroacoustic wind-tunnel can be predicted as

$$f = \frac{1}{\frac{1}{m}\frac{L_1}{0.65U_\infty} + \frac{L_1+2L_2}{c}}. \quad (1)$$

Here $L_1$ denotes the distance between the position where the large scale vortex structures generate due to the instabilities of the shear layer and the front wall of the collector. $L_2$ denotes the distance between the front wall of the collector of the back wall of the plenum chamber. $m$ is an integer variable that represents the mode of vortices $c$ denotes the velocity of sound propagation. It should be noted that vortex structure is generated near the nozzle and propagates downstream along the shear layer at speed of $0.65U_\infty$ until it impinges upon the collector, as shown in figure 9. Moreover, in the feedback loop, the pressure waves produced by the large scale vortex structures inside the shear layer impinging on the front wall of the collector are reflected by the back wall of the plenum chamber, propagate upstream along the outer region of the shear layer, and excite a new instability vortex. The end of the path that the pressure waves travel upstream is where vortex shedding happens, not the nozzle exit.

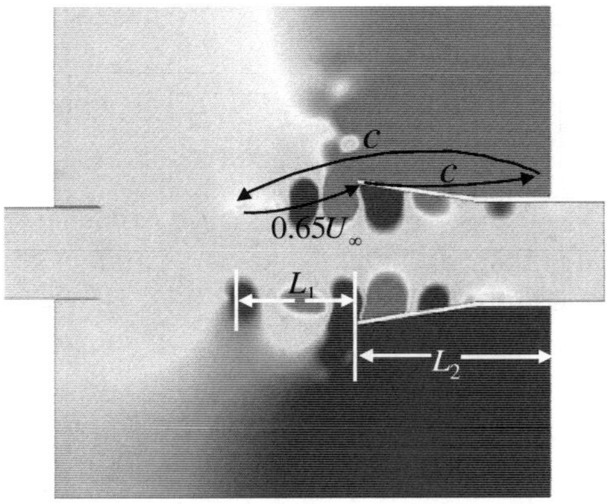


**Figure 9.** Schematic of a feedback loop with pressure oscillation.

## 4. Conclusions

Based on the SA-DDES turbulence model, the low frequency oscillation phenomenon in an aeroacoustic wind-tunnel was investigated numerically. The typical results are in good agreement with the experiments, which indicated that our methods and models used in this paper can capture the main features of pressure oscillation in an aeroacoustic wind-tunnel.

The generation mechanism responsible for the low frequency oscillation is discussed. It can be well recognized as the cyclic loop between the instabilities of the shear layer and the interaction of the vortex and the collector.

The feedback loop path is specified. The pressure waves are reflected by the back wall of the plenum chamber. The end of the path that the pressure waves travel upstream is where vortex shedding happens.

**Acknowledgments**

The authors gratefully acknowledge the support for this research provided by National Natural Science Foundation of China (Grant Nos. 11732016 and 12172374), Sichuan Science and Technology Program (Grant No. 2018JZ0076), CARDC Fundamental and Frontier Technology Research Fund (Grant No. PJD20200185), and National Numerical Windtunnel project.

## References

[1] Zheng Z, Wang Y and Yang Z 2009 Buffeting phenomenon of open-jet automotive aero-acoustic wind-tunnel *Tongji Daxue Xuebao* **37** 253-257
[2] Holthusen H and KooiModel J W 1997 Model and full scale investigations of the low frequency vibration phenomena of the DNW open jet *AGARD Conference Proceedings* pp 26.1-26.8
[3] Sellers W L, Applin Z T and Molloy J K 1984 Effect of Jet exit vanes on flow oscillations in an aeroacoustic wind-tunnel *NASA TM-86299*
[4] Manuel G S, Moller J M and Barba P S 1992 Effect of collector configuration on test section turbulence levels in an aeroacoustic wind-tunnel *NASA TM-4333*
[5] Rennie M 2000 Effect of jet length on pressure fluctuations in ¾ aeroacoustic wind-tunnels *SAE 2000 World Congress and Exhibition*, Session: Vehicle aerodynamics-wind-tunnel technology and correlation
[6] Rennie M and Kim M S 2004 Suppression of open-jet pressure fluctuations in the hyundai aeroacoustic wind-tunnel *SAE 2004 World Congress and Exhibition*, Session: Vehicle aerodynamics-wind-tunnel technology and correlation.
[7] Kudo K, Komatsu Y, Maeda K and Nishimura M 2009 Techniques for reducing low frequency fluctuations in aero-acoustic wind-tunnels *Journal of Environment and Engineering* **4** 289-301.
[8] Pottpollenske M, Heesen W V and Bergmann A 2012 Acoustical pre-examination work and characterization of the low noise wind-tunnel DNW-NWB *Aiaa/ceas Aeroacoustics Conference* **3**
[9] Jia Q, Zhu Y, Bao D, Rashidi M M and Yang Z 2019 On the low frequency pressure fluctuation in a 3/4 open jet automotive wind-tunnel *Journal of Applied Fluid Mechanics* **12** 1359-1369
[10] Jia Q, Zhu Y, Bao D, Rashidi M M and Yang Z 2021 Experimental research of active control optimization on a 3/4 aeroacoustic wind-tunnel's jet section *AEJ - Alexandria Engineering Journal* **60** 2265-2278

# Efficient Shock Capturing Based on Parallel Adaptive Mesh Refinement Framework

**Weihao Li and Jian Xia***

College of Aerospace Engineering, Nanjing University of Aeronautics and Astronautics, Nanjing 210016, China

Email: jxia@nuaa.edu.cn

**Abstract.** Based on AMReX, a software framework for massively parallel, block-structured adaptive mesh refinement (AMR) applications and in combination with finite difference weighted essential non-oscillatory (WENO) method, a numerical procedure is developed to provide universal discontinuity capturing capability for inviscid, compressible flow. Test cases of one-dimensional and two-dimensional flows containing shock waves, contact discontinuities, flow instabilities, and their interactions are considered to validate the high resolution of AMR for characteristic flow structure under unsteady conditions. The current elaborate AMR-based code is proven to have superior efficiency relative to global refinement via experiments including different initial mesh intervals, refinement ratios at all levels, and. Evidence in executions with multiple processes and domain division sizes indicates its high parallel scalability. Thus, the application built on AMReX is promising for simulations of phenomena undergoing rapid local variation such as compressible turbulence and multiphase interactions.

**Keywords.** Shock capturing, adaptive mesh refinement, parallel computing.

## 1. Introduction

The existence of discontinuities such as shockwaves in compressible flow simulations has long been a challenge to numerical methods aimed at high accuracy. Conventional high order schemes derived from Taylor series expansion are based on the assumption of local smoothness, which no longer holds near a discontinuity. Improper employment of such methods will suffer from nonphysical oscillations and lead to numerical instability or divergence. Special treatments to overcome these difficulties can be traced back to the 1960s when Gino Moretti established the shock-fitting method and made remarkable contributions to its development in his lifetime. In a shock-fitting method, discontinuities in the flow field are located at first. They are treated as boundaries of different smooth flow regions in which a general scheme can be applied. Next, analytical Rankine-Hugoniot jump conditions are enforced to solve for flow variables at both sides of the boundary. These solutions serve as boundary conditions for updating variables in smooth regions as well as the location of the discontinuity. To respect the physical nature that shock waves and other wave patterns arise from the hyperbolic inheritance of Euler equations governing inviscid compressible, the upwind difference is introduced [1] to follow the characteristic direction. The advantage of a shock-fitting method is its accuracy in smooth regions and saving of computational cost. On the other hand, the location of a discontinuity in three-dimensional flows containing complex interactions is not easy to find which restricts the application of shock-fitting methods [2].

The succeeding shock-capturing methods, on the other hand, use a unifying solver for every location in a flow field without any prior knowledge of discontinuity. Due to their simplicity for implementation, shock-capturing methods have become widely used and have almost taken the place of their predecessor in modern computational fluid dynamics codes. But since the thickness of a shock wave is equivalent to the free path of fluid particles in order of magnitude, the defect of shock-capturing methods manifests as the misrepresentation of discontinuities on a mesh with finite spacing, where a shock-fitting routine could help [3]. In the regime of finite difference method, Brehm et al. [4] compared three classes of higher-order shock capturing methods: (1) central finite-difference schemes with explicit artificial dissipation, (2) a compact centered finite-difference scheme with localized artificial diffusivity (LAD), and (3) weighted essentially non-oscillatory (WENO) schemes in both explicit and compact finite difference forms. It is concluded that not a single class of methods does the best in all aspects, while WENO schemes are more robust but less accurate and efficient than central schemes with LAD. A number of WENO schemes with improved spectral resolution is also available, including ones with different smooth indicators [5] and optimized weights [6]. Fast reconstruction can be achieved with approximate characteristic transformation [7] or unified reconstruction coefficients for all components of the flux vector [8]. But as suggested by [9], no version of modified WENO schemes is guaranteed to perform well in all situations, so the classical fifth-order WENO scheme proposed by Jiang and Shu [10] is selected in the current work for its proven stability and fidelity in various applications.

To alleviate the influence of WENO on flow resolution and computational performance, apart from improvement in the shock-capturing algorithm, the adaptive mesh refinement (AMR) technique, however, is adopted as an alternative approach. In physical problems of large scale and where disparities between structures are significant, AMR allows for coarser mesh in smooth regions to reduce computational cost and finer mesh in regions containing rapid local variation in space and time, such as chemical reactions or discontinuity interfaces, to reach higher accuracy, hence achieving an overall balance between simulation quality and time. The AMR approach was initially employed to solve hyperbolic systems of conservation laws followed by extensions to parabolic and elliptical equations. In areas such as astrophysics and aerospace engineering, simulation with adaptive mesh refinement has been a routine practice, enabling users to obtain data of high quality without much experience in problem modeling and mesh generation [11]. Therefore, in modern numerical simulation frameworks for research and industry, such as Launch Ascent and Vehicle Aerodynamics (LAVA) [12] and AMReX [13] used in current work, parallel adaptive mesh refinement (PAMR) has been incorporated as an infrastructure to help the development of various efficient solvers. It should be noted that, though implementation details of PAMR can be transparent to applications built on it, it's the developer's responsibility to design algorithms and data structures to fully exploit the potential for performance improvement. Last but not least, the consistence of physical meaning between refined levels should be maintained carefully like conservations on fine/coarse mesh interfaces [14].

The rest of the current article is organized as follows: Section 2 shows the governing equation and basic space and time discretization on a uniform cartesian mesh. Section 3 briefly illustrates the procedure of AMReX and deals with the key part of the design implemented herein. Section 4 demonstrates the accuracy and efficiency improvement of AMR as well as parallel scalability via canonical one-dimensional and two-dimensional test cases. Conclusions and prospects for further applications are included in Section 5.

## 2. Numerical Method

### *2.1. Governing Equations*

The differential Euler equation governing compressible inviscid flows in terms of vector variables can be written as

$$\frac{\partial W}{\partial t}+\nabla\cdot F=0 \text{ with } F=\begin{pmatrix}F_1 & F_2 & F_3\end{pmatrix} \tag{1}$$

In three-dimensional space, the conservative state variable vector is $W=\begin{pmatrix}\rho & \rho u & \rho v & \rho w & \rho E\end{pmatrix}^T$ and convective flux vectors are

$$F_1=\begin{bmatrix}\rho u\\ \rho uu+p\\ \rho uv\\ \rho uw\\ \rho uH\end{bmatrix},\quad F_2=\begin{bmatrix}\rho v\\ \rho vu\\ \rho vu+p\\ \rho vw\\ \rho vH\end{bmatrix},\quad F_3=\begin{bmatrix}\rho w\\ \rho wu\\ \rho wv\\ \rho ww+p\\ \rho wH\end{bmatrix} \tag{2}$$

For a perfect gas with specific heat ratio $\gamma=C_p/C_v$ and gas constant $R=C_p-C_v$, total enthalpy is defined as $H=C_pT+u_iu_i/2$ together with total energy $E=C_vT+u_iu_i/2$ per unit mass. In the rest of the article, $\gamma$ remains constant with a value of 1.4 for simplicity. Finally, the equation of state is necessary to enclose the equation.

$$p=\rho RT \tag{3}$$

### 2.2. Van-Leer Flux Vector Split

To reflect the fact that compressible flow is dominated by wave propagation phenomena, a convective flux variable is split into upwind and downwind parts which are discretized forward and backward in space respectively. A component-wise flux vector split of the Van-Leer type is employed here [15]. The forward split flux is a smooth function of the local Mach number

$$F^+=\begin{cases}F, & M>1\\ F^{+}, & |M|<1\\ 0, & M<-1\end{cases} \tag{4}$$

For example, the forward x-flux with $M=u/c$ and $f^+_{mass}=\rho c\left\{\frac{1}{2}(M+1)\right\}^2$ in the subsonic region is

$$F_1^+=\begin{bmatrix} f^+_{mass}\\ f^+_{mass}\cdot\{(\gamma-1)u+2c\}/\gamma\\ f^+_{mass}\cdot v\\ f^+_{mass}\cdot w\\ f^+_{mass}\cdot\left[\{(\gamma-1)u+2c\}^2/\{2(\gamma^2-1)\}+\frac{1}{2}(v^2+w^2)\right]\end{bmatrix} \tag{5}$$

Next, the backward flux is calculated by subtraction $F^-=F-F^+$.

### 2.3. Weighted Essential Non-Oscillation Method (WENO)

The basic idea of the WENO scheme is to first evaluate fluxes reconstructed from several sub-stencils to the target location and then assemble them using nonlinear weights into the final result. For a sub-stencil, if the flow variable is detected smooth inside, the nonlinear weight is close to the linear (ideal) weight. Superposition of flux using linear weights will recover a high order reconstruction on the full stencil. Thus, the scheme will reduce order near discontinuities to enhance stability and preserves high approximation accuracy in smooth regions. The original WENO consists of five stencil points with three sub-stencils. For forward flux on cell interface $I_{i+\frac{1}{2}}$, the points in sub stencils are

$$\begin{aligned} S_1 &= \{x_{i-2}, x_{i-2}, x_{i-2}\} \\ S_2 &= \{x_{i-1}, x_i, x_{i+1}\} \\ S_3 &= \{x_i, x_{i+1}, x_{i+2}\} \end{aligned} \tag{6}$$

Reconstructed fluxes and coefficients on sub stencils are

$$\begin{aligned} f_0 &= \frac{1}{3} f_{i-2} - \frac{7}{6} f_{i-1} + \frac{11}{6} f_i \\ f_1 &= -\frac{1}{6} f_{i-1} + \frac{5}{6} f_i + \frac{1}{3} f_{i+1} \\ f_2 &= \frac{1}{3} f_i + \frac{5}{6} f_{i+1} - \frac{1}{6} f_{i+2} \end{aligned} \tag{7}$$

Linear weights are obtained by comparison with coefficients on the full stencil

$$\begin{aligned} c_0 f_0 + c_1 f_1 + c_2 f_2 &= \frac{1}{30} f_{i-2} - \frac{13}{60} f_{i-1} + \frac{47}{60} f_i + \frac{9}{20} f_{i+1} - \frac{1}{20} f_{i+2} \\ \Rightarrow c_0 &= \frac{1}{10},\ c_1 = \frac{6}{10},\ c_2 = \frac{3}{10} \end{aligned} \tag{8}$$

Smooth indicators are defined as

$$\begin{aligned} \beta_0 &= \frac{13}{12}\left(f_{i-2} - 2f_{i-1} + f_i\right)^2 + \frac{1}{4}\left(3f_{i-2} - 4f_{i-1} + f_i\right)^2 \\ \beta_1 &= \frac{13}{12}\left(f_{i-1} - 2f_i + f_{i+1}\right)^2 + \frac{1}{4}\left(3f_{i-1} - f_{i+1}\right)^2 \\ \beta_2 &= \frac{13}{12}\left(f_i - 2f_{i+1} + f_{i+2}\right)^2 + \frac{1}{4}\left(f_i - 4f_{i+1} + f_{i+2}\right)^2 \end{aligned} \tag{9}$$

Nonlinear weights are computed based on smooth indicators and

$$\omega_j = \frac{\omega_j'}{\sum_i \omega_i'}, \quad \omega_j' = \frac{c_j}{\left(\varepsilon + \beta_j\right)^2} \quad \text{for } j = 0,1,2 \tag{10}$$

Assembled numerical flux approximated by WENO is

$$f_{i+1/2} = \sum_j \omega_j f_j \tag{11}$$

The stencil of backward flux is symmetric about the cell $i$. The final flux difference is given by

$$\frac{\partial F_i}{\partial x_i}(x_i) = \frac{f_{i+1/2} - f_{i-1/2}}{\Delta x_i} + O\left(\Delta x_i\right) \tag{12}$$

*2.4. Time Marching*

A three-stage Runge-Kutta total variation diminishing (TVD) method is used here. From time level n to n+1

$$
\begin{aligned}
&W^{(0)} = W^{(n)} \\
&W^{(1)} = W^{(0)} + \Delta t \cdot L(W^{(0)}) \\
&W^{(2)} = \frac{3}{4}W^{(0)} + \frac{1}{4}W^{(1)} + \frac{1}{4}\Delta t \cdot L(W^{(1)}) \\
&W^{(3)} = \frac{1}{3}W^{(0)} + \frac{2}{3}W^{(2)} + \frac{2}{3}\Delta t \cdot L(W^{(2)}) \\
&W^{(n+1)} = W^{(3)}
\end{aligned} \tag{13}
$$

The function $L$ is the right-hand side of the semi-discretization equation

$$\frac{\partial W}{\partial t} = L(W) = -\sum_i \frac{\partial F_i(W)}{\partial x_i} \tag{14}$$

## 3. Parallel Adaptive Mesh Refinement

This section deals with the mesh technique. A general description of AMReX and some important aspects of the current implementation is presented. A flow chart of a single physical time step on the coarsest mesh (level 0) in AMReX is depicted in figure 1.

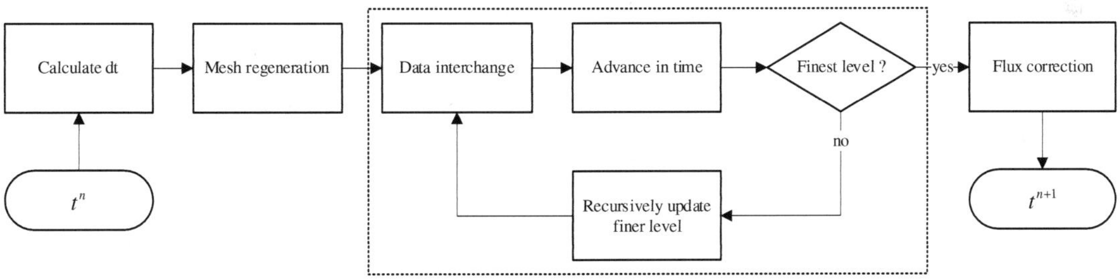


**Figure 1.** A time step on the coarsest mesh.

*Remark 1:* Steps in the dotted box in figure 1 are part of a recursive procedure as described in figure 2. Two sub-steps are used here because all levels are expected to march in time with a unifying CFL number. In general, the time step at a level is nearly proportional to its mesh interval but not exact. Therefore, step sizes at all levels are first scaled by their refinement ratio, and the smallest one is taken as the step size at level 0 denoted by $\Delta t$. Then step size at each level is determined via dividing $\Delta t$ by refinement ratios to ensure strict satisfaction of CFL and time synchronization.

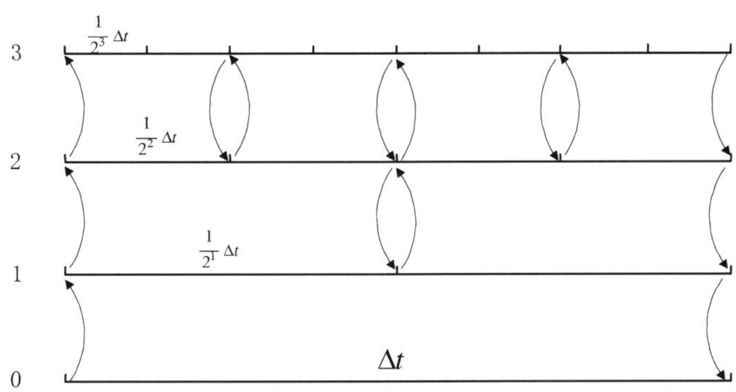


**Figure 2.** The sketch of recursive sub-steps.

*Remark 2:* Other important operations in figure 2 include the date prolongation (up arrow) to the finer level and restriction (down arrow) to the coarser level. Since the conservative variables representing volume averaged quantities are stored in a cell-centered form, the data interchange in the overlapped region is straightforward by doing a volume-weighted sum in a restriction operation and simple duplication in a prolongation operation (figure 3(a)).

(a) (b)

**Figure 3.** Data interchange operations between levels including (a) Prolongation and restriction and (b) flux correction.

Also, flux correction is required at the coarse/fine interface as shown in figure 3(b). The conservative variable in the cell of the coarser level near the interface should be modified by $\Delta\phi$ to compensate for the difference in flow rate between cell faces:

$$\Delta\phi = \frac{\sum A_i f_i - F\sum A_i}{\Omega}\Delta t \tag{15}$$

where $\Omega$ is the volume of the coarser cell and $A_i$ the face area of a neighboring finer cell.

*Remark 3:* Data parallelization in AMReX is achieved by dividing the structured mesh index domain into blocks. Domain blocks can be distributed to different processes in the context of Message Passing Interface (MPI). More fine-grained parallelization is achieved by scattering cell iterations over CPU threads or GPU cores. The dimension of a block must be the multiple of a power of 2 to help load balancing across processes.

Though details of the numerical methods have been given in section 3, the underlying data structure must also be tuned carefully to maximize parallelization. In current implementation values of split flux at all cell centers are evaluated before WENO reconstruction, thus avoiding any repetitive computation between stencils and sub stencils. It also decouples the flux vector splitting method and reconstruction method, leading to a more flexible architecture available on heterogeneous machines.

Data interchange between blocks of a level is achieved via ghost cells. Besides, the imposition of non-periodic boundary conditions is realized by filling ghost cells using data out of the physical domain specified for level 0, or that from the underlying level for finer levels. Effects of the number of processes and the size of blocks are shown in section 4.

*Remark 4:* AMReX must be informed of whether a cell should be refined. This step is called tagging and is the key part of mesh regeneration. To capture discontinuity, a tagger based on the local jump threshold of density is employed in the current code. As will be shown in section 4, a proper threshold is critical to instruct mesh refinement.

## 4. Test Cases

### *4.1. Sod Shock Tube*

The canonical test case is introduced by Sod [16]. The solution can be calculated analytically which may contain different wave patterns depending on the initial condition on both sides. The range of the computational field in the x-direction is $[-0.5, 0.5]$. The initial condition employed here is producing a rarefaction wave, a contact discontinuity, and a shockwave from left to right.

$$(\rho,\ u,\ p)=\begin{cases}(1,\ 0,\ 1) & x<0 \\ (0.125,\ 0,\ 0.1) & x\geq 0\end{cases} \tag{16}$$

In order to validate the accuracy and efficiency of current AMR code, different configurations of initial mesh and refinement ratios are used as described in table 1.

**Table 1.** AMR configurations of Sod shock tube.

| Configuration | Initial cell interval | Max level | Refinement ratios | Line color | Line style |
|---|---|---|---|---|---|
| 1a | 1/256 | 3 | 2, 2, 2 | red | solid |
| 1b | 1/256 | 2 | 2, 4 | red | dashed |
| 1c | 1/256 | 2 | 4, 2 | red | dash-dot |
| 2a | 1/512 | 2 | 2, 2 | green | solid |
| 2b | 1/512 | 1 | 4 | green | dashed |
| 3 | 1/1024 | 1 | 2 | blue | solid |
| 4 | 1/2048 | 0 | N/A | black | solid |

Note that the current numerical experiment is actually conducted in a two-dimensional space with the same number of cells in the y-direction, so the impact of mesh refinement in all directions on performance is included. All experiments are carried out with $t=0.2$, $CFL=0.6$, the finest cell interval of 1/2048, and all parallel techniques are turned off to focus on the AMR algorithm itself.

The distribution of final cell density is depicted in figure 4(a). Since the difference between results of configurations is small, the distribution of absolute error between numerical results and analytic solution for density is presented instead in figure 4(b).

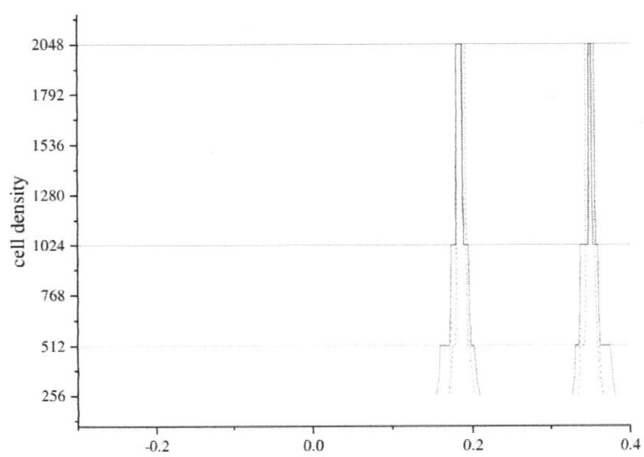


(a)

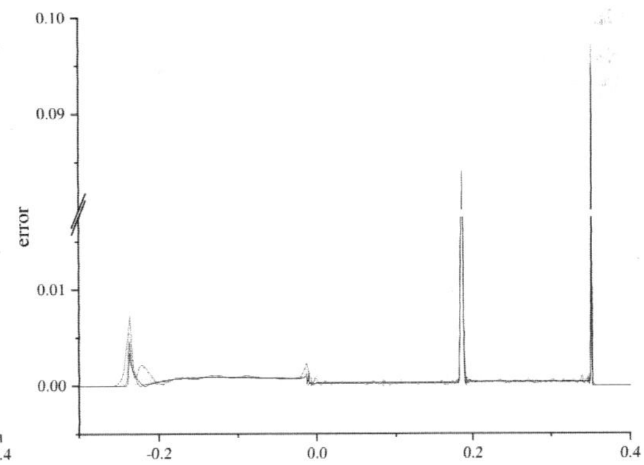


(b)

**Figure 4.** Distribution of (a) cell density and (b) absolute error of density.

It can be noticed in figure 4(a) that refinement is not triggered at the boundaries of the rarefaction wave because there is no discontinuity in analytic function but in its derivative and the jump of variable is less than the threshold. In this case, the mesh interval is positively correlated to local error as usual. Also, a larger refinement ratio leads to fewer total cells with the same initial cell size. For contact and shock, refinement is correctly enabled. Two extreme points in figure 4(b) emerge from the fact that a numerical method on a mesh of finite space smears the exact location of discontinuity which is inevitable. Errors in the approximation of analytic solution functions are quantified via the cell average of the error in form of vector norm as shown in table 2.

**Table 2.** Errors in the approximation of analytic solution functions.

| | | 1a | 1b | 1c | 2a | 2b | 3 | 4 |
|---|---|---|---|---|---|---|---|---|
| $\rho$ | $L_2$ | 3.586E-03 | 3.538E-03 | 3.481E-03 | 3.478E-03 | 3.493E-03 | 3.456E-03 | 3.521E-03 |
| | $L_\infty$ | 9.709E-02 | 9.495E-02 | 9.632E-02 | 9.554E-02 | 9.542E-02 | 9.475E-02 | 9.514E-02 |
| $u$ | $L_2$ | 1.698E-02 | 1.680E-02 | 1.673E-02 | 1.663E-02 | 1.670E-02 | 1.656E-02 | 1.661E-02 |
| | $L_\infty$ | 6.902E-01 | 7.152E-01 | 7.135E-01 | 7.088E-01 | 7.125E-01 | 7.044E-01 | 7.074E-01 |
| $p$ | $L_2$ | 3.178E-03 | 3.162E-03 | 3.073E-03 | 3.051E-03 | 3.045E-03 | 3.033E-03 | 3.079E-03 |
| | $L_\infty$ | 1.331E-01 | 1.323E-01 | 1.342E-01 | 1.331E-01 | 1.330E-01 | 1.322E-01 | 1.331E-01 |

Enlarged plots of shock region for all variables and contact region for density are presented in figure 5.

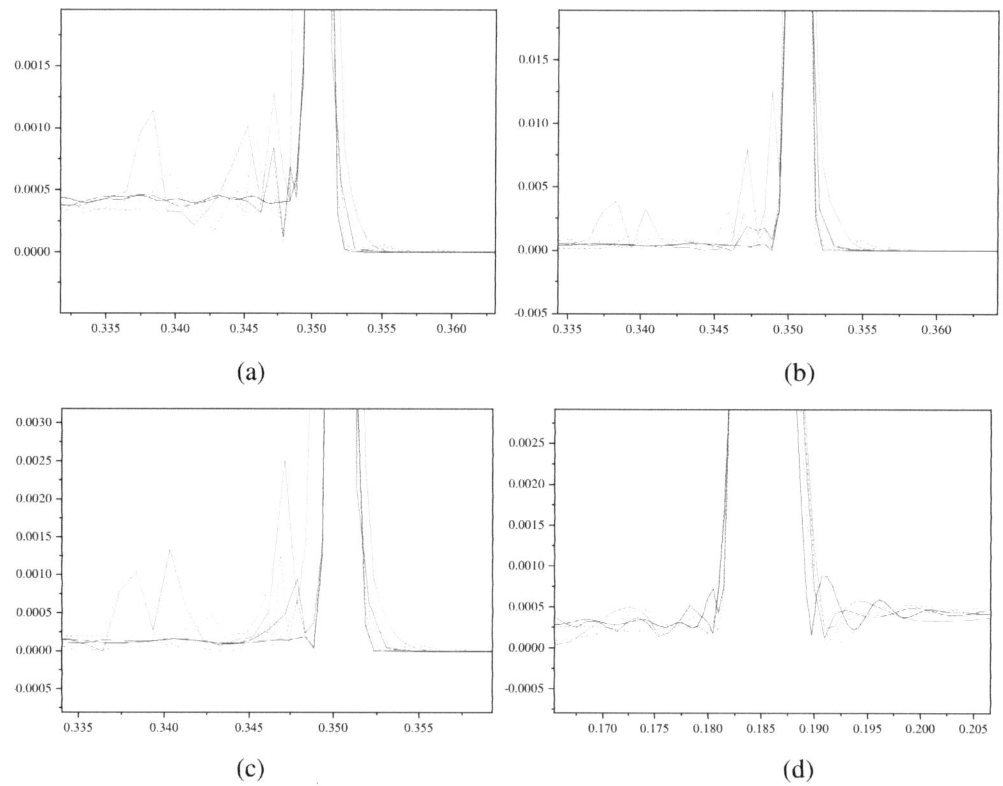


**Figure 5.** Enlarged plots for (a) density, (b) velocity, (c) pressure near shock and (d) density near contact discontinuity.

Further, the efficiency of a configuration is indicated through total CPU time $t_{cpu}$, the number of iterations $N$ , and CPU time per iteration per cell $t_{cell}$, as listed in table 3.

**Table 3.** Efficiency of AMR configurations.

| | 1a | 1b | 1c | 2a | 2b | 3 | 4 |
|---|---|---|---|---|---|---|---|
| $t_{cpu}/s$ | 10.8092 | 11.8079 | 12.0001 | 8.82543 | 9.141 | 17.0033 | 54.6053 |
| $N$ | 187 | 187 | 187 | 374 | 374 | 748 | 1497 |
| $t_{cell}/s$ | 1.74E-05 | 1.90E-05 | 1.73E-05 | 4.67E-06 | 4.34E-06 | 2.65E-06 | 2.23E-06 |

By comparison of 1a, 2a, and 3, it is observed that in general larger initial mesh interval leads to more error overall. However, errors of AMR in 2 and 3 are all smaller than that of global refinement case 4. Moreover, on a coarse mesh, a larger refinement ratio can gain some compensation for accuracy with additional computational cost via looking into 1b and 1c in contrast to 1a, which does not hold for 2b with respect to 2a.

The number of iterations is determined by the time step on the coarsest mesh since a fixed CFL and total time is enforced. Though total CPU time is significantly reduced as expected due to mesh refinement in 1 and 2 compared with 3 and 4, it is interesting to find that case 1 is less efficient than 2. It could be the consequence of more data interchange between levels, including data transferring and flux correction especially in two- or three-dimensional space. Thus, moderate initial mesh size and refinement ratios are necessary to reduce communication costs for an efficient simulation with AMR.

*4.2. Double Mach Reflection of a Strong Shock*

The case is proposed by Woodward and Colella [17] to study the capability of numerical methods in simulation flows containing strong shock. The range of the computational field is $[0,4]\times[0,1]$. A strong shock is moving to the right at Mach 10 relative to the undisturbed area before the wave front and has an angle of 60 degrees relative to the plate. The reflecting flat plate has a leading edge located at $x=0.16$ and coincides with the lower boundary to avoid any treatment for an irregular boundary. Flow conditions on the upper boundary are set to represent the accurate movement of the shock wave. In all following simulations, physical time is $0.2s$, $CFL=0.6$, block max level is 2 with refinement ratio $(4,2)$. The initial condition of primitive state variables is

$$(\rho,\ u,\ v,\ p)=\begin{cases}(1.4,\ 0,\ 0,\ 1) & \text{pre shock}\\(8,\ 7.145,\ -4.125,\ 116.5) & \text{post shock}\end{cases} \tag{17}$$

Settings of initial boundary conditions are presented in figure 6.

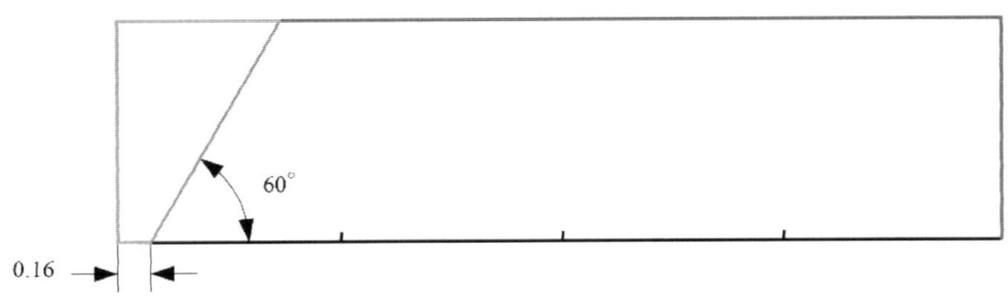


**Figure 6.** Settings of initial boundary condition, including initial shock front (blue), post-shock state(green), pre-shock state(red) and reflecting wall (black).

The first group of experiments is to test the parallel scalability of the current AMR code. The tendency of total CPU time concerning the number of MPI processes on two coarsest mesh intervals is presented in figure 7(a) and (b). Block size is small enough so the number of blocks is always more than that of running processes to take full advantage of computational resources. Note that the y-axis is transformed to be the reciprocal of time, and the dashed line denotes the ideal linear speedup for reference.

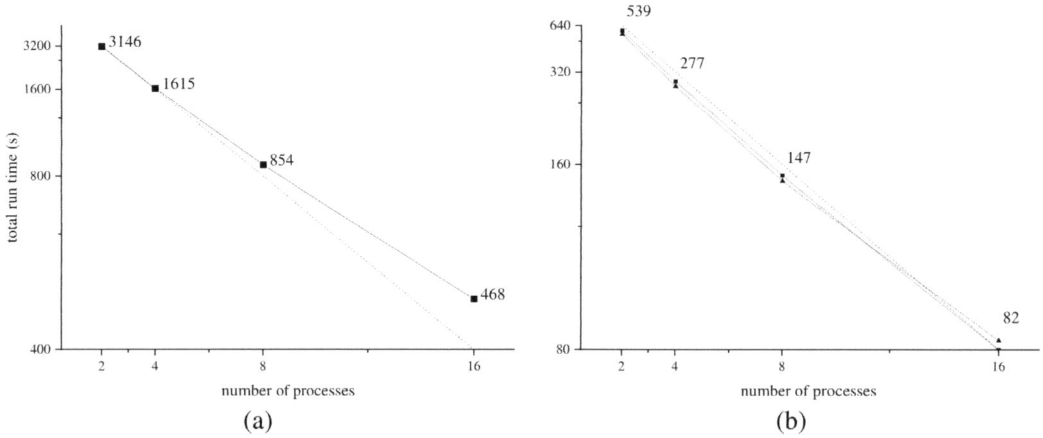


**Figure 7.** Parallel performance with the different number of processes with coarsest mesh interval of (a) 1/128 with block size 64 and (b) 1/64 with block size 32 (rectangle) and 16 (triangle).

The effect of parallel acceleration declines with more processes as expected due to communications between processes, but significant improvement in performance has demonstrated the scalability of the current AMR code. Another noticeable fact is that too small block size does not contribute to speedup especially when the number of processes is much less than that of processes by looking into the two lines in figure 7(b).

The second group is designed to show the importance of an appropriate refinement tagging threshold. Experiments using two jump thresholds: $10\Delta x$ and $20\Delta x$ for detecting density discontinuity are carried out. Outlines of AMR blocks at all levels and enlarged contours for density in the reflection area are presented in figure 8 and figure 9 respectively. Characteristic flow structures, such as incident shock ( $I$ ), reflected shocks ( $R$ and $R'$ ), Mach stems ( $M$ and $M'$ ), slip lines ( $S$ and $S'$ ), and two triple points ( $T$ and $T'$ ) can be identified. No rarefaction wave exists following theoretical results [18]. Note that the second slip line $S'$ attached to the triple point $T'$ is not recognized when using a larger jump threshold in figure 9(b).

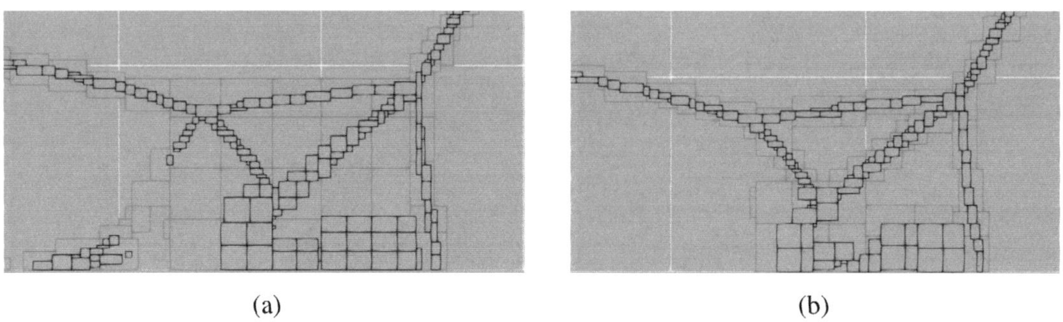


**Figure 8.** Outlines of AMR blocks at level 0 (white), 1 (grey), and 2 (black) with tagging thresholds (a) $10\Delta x$ and (b) $20\Delta x$ .

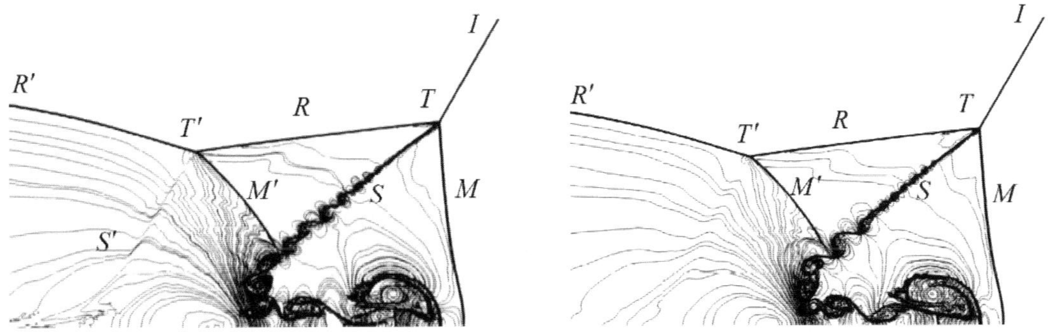


**Figure 9.** Enlarged contours of density for double Mach reflection area with tagging thresholds (a) $10\Delta x$ and (b) $20\Delta x$ .

## 5. Conclusions

In the current work, a set of parallel adaptive mesh refinement (PAMR) codes is built on the modern AMReX numerical computation framework. Through incorporating the finite difference WENO method, robust capturing of shockwaves and other discontinuities in inviscid compressible flows is achieved on a uniform Cartesian mesh. Improved resolution of characteristic flow structures is validated through canonical test cases. High parallel scalability is demonstrated through experiments on single and multiple CPU processes. In addition, appropriate settings, including initial cell interval, refinement ratios, tagging thresholds, and block division sizes should be selected to fully utilize the power of AMR to increase not only accuracy but also efficiency. Thus, the application of AMReX is promising for simulations containing rapid local variation, such as compressible turbulence, chemical reactions, and multiphase interactions.

## References

[1] Moretti G 1979 The λ-scheme *Computers & Fluids* **7** 191–205
[2] Bonfiglioli A, Paciorri R, Nasuti F and Onofri M 2016 Moretti's Shock-Fitting Methods on Structured and Unstructured Meshes *Handbook of Numerical Analysis* vol 17 (Elsevier) pp 403–39
[3] Pirozzoli S 2011 Numerical Methods for High-Speed Flows *Annu. Rev. Fluid Mech.* **43** 163–94
[4] Brehm C, Barad M F, Housman J A and Kiris C C 2015 A comparison of higher-order finite-difference shock capturing schemes *Computers & Fluids* **122** 184–208
[5] Borges R, Carmona M, Costa B and Don W S 2008 An improved weighted essentially non-oscillatory scheme for hyperbolic conservation laws *Journal of Computational Physics* **227** 3191–211
[6] Martín M P, Taylor E M, Wu M and Weirs V G 2006 A bandwidth-optimized WENO scheme for the effective direct numerical simulation of compressible turbulence *Journal of Computational Physics* **220** 270–89
[7] Su X, Sasaki D and Nakahashi K 2013 On the efficient application of weighted essentially nonoscillatory scheme *International Journal for Numerical Methods in Fluids* **71** 185–207
[8] Peng J, Zhai C, Ni G, Yong H and Shen Y 2019 An adaptive characteristic-wise reconstruction WENO-Z scheme for gas dynamic euler equations *Computers & Fluids* **179** 34–51
[9] Shu C-W 2020 Essentially non-oscillatory and weighted essentially non-oscillatory schemes *Acta Numerica* **29** 701–62
[10] Jiang G-S and Shu C-W 1996 Efficient Implementation of Weighted ENO Schemes *Journal of Computational Physics* **126** 202–28

[11] Plewa T, Linde T J and Weirs V G 2005 *Adaptive mesh refinement, theory and applications: proceedings of the Chicago Workshop on Adaptive Mesh Refinement Methods, Sept. 3-5, 2003* (Berlin ; New York: Springer) preface pp III-VI
[12] Kiris C C, Housman J A, Barad M F, Brehm C, Sozer E and Moini-Yekta S 2016 Computational framework for Launch, Ascent, and Vehicle Aerodynamics (LAVA) *Aerospace Science and Technology* **55** 189–219
[13] Zhang W, Almgren A, Beckner V, Bell J, Blaschke J, Chan C, Day M, Friesen B, Gott K, Graves D, Katz M P, Myers A, Nguyen T, Nonaka A, Rosso M, Williams S and Zingale M 2019 AMReX: a framework for block-structured adaptive mesh refinement *Journal of Open Source Software* **4** 1370
[14] Berger M J 1987 On Conservation at Grid Interfaces *SIAM J. Numer. Anal.* **24** 967–84
[15] van Leer B 1997 Flux-Vector Splitting for the Euler Equation *Upwind and High-Resolution Schemes* ed M Y Hussaini, B van Leer and J Van Rosendale (Berlin, Heidelberg: Springer Berlin Heidelberg) pp 80–9
[16] Sod G A 1978 A survey of several finite difference methods for systems of nonlinear hyperbolic conservation laws *Journal of Computational Physics* **27** 1–31
[17] Woodward P and Colella P 1984 The numerical simulation of two-dimensional fluid flow with strong shocks *Journal of Computational Physics* **54** 115–73
[18] Anon 1985 A numerical study of oblique shock-wave reflections with experimental comparisons *Proc. R. Soc. Lond. A* **398** 117–40

# Study on Impeller Configuration Strategy of Dredger Pump with Different Conveying Distance

**Haifei Zhuang***, **Mingming Liu and Yongding Wu**

CCCC National Engineering Research Center of Dredging Technology and Equipment CO., TD, Room 2902, 1088 Yangshupu Road, Shanghai, China
Email: zhuanghaifei@cccc-drc.com

**Abstract.** For DN850 dredge pumps of 3,500 $m^3/h$ series cutter suction dredger, on the premise of not changing the impeller size, by changing the parameters of impeller blade to reduce the head of dredge pump, special new impellers for inboard pump and submerged pump are designed; The flow field of the dredge pump is analyzed by numerical simulation. There is no vortex in the impeller channel and near volute tongue, which indicates that the hydraulic performance is excellent; At the flow rate of 10,000 $m^3/h$, compared with the original impeller, the head of the two pumps in series with new impellers is reduced by 23 m, and the efficiency is close to that before modification, reaching more than 85%; This paper analyzes the matching between the dredger pump and the pipeline under three kinds of common sandy mud. By comparing the unit power consumption and productivity, it obtains the suitable conveying distance range of special new impellers of dredger pumps, and forms the impeller configuration scheme of 3,500 $m^3/h$ series cutter suction dredger with different conveying distance, which provides guidance for dredging engineering.
**Keywords.** Dredge pump, impeller, numerical simulation, working condition matching, energy saving and efficiency increasing.

## 1. Introduction

As the key equipment of dredger, the dredge pump is widely used in cutter suction dredger, trailing suction dredger. The performance of dredge pump affects the construction efficiency of dredger directly. When the length of the dredging pipeline changes, especially under the condition of short conveying distance, the flow of the dredge pump is too large [1]. At present, most of dredge pumps are driven by diesel engines through the gear reducer, the speed adjustment range is limited [2], and the diesel engine can not be operated at low speed for a long time. The throttle tube of pipeline is often used to reduce the flow of the dredge pump under the condition of short conveying distance [3]. Although this method is simple and easy, however, by increasing the pipe resistance consumption, the throttle tube is equivalent to an "energy dissipator", which wastes lots of energy.

**Table 1.** Throttle tube under short conveying distance.

| Dredger name | Project location | Conveying distance (km) | Throttle tube diameter (mm) |
|---|---|---|---|
| Xin Hai Kun | Putian, Fujian | 1.5 | 600 |
| Xin Hai Li | Meizhou Bay, Fujian | 2.6 | 550 |
| Xin Hai E | Gulei, Fujian | 0.8 | 550 |

At present, 3,500 $m^3/h$ series cutter suction dredgers reduce the flow rate through throttle tube when the pipeline length is 1-2.5 km and the inner diameter of the pipeline is 850 mm (table 1). If the flow rate can be controlled by replacing the impeller of dredge pumps with throttle tube flow regulation, the energy consumption can be reduced and the dredging cost can be saved.

Under special conditions, the dredger shall be equipped with targeted special equipment to improve the dredging efficiency. Some dredging companies attach great importance to the analysis of various influencing factors in the dredging process, look for the corresponding relationship between the impact of dredging equipment, and conduct quantitative research on the relationship of them. At the same time, many R&D institutions carry out forward-looking research. Many cutter suction dredgers are equipped with different dredge pumps to meet the requirements of construction conditions, which is more targeted, Bugdayci. H. H. et al. [4] developed a curved impeller for the rock special pump with larger ball channel and specially used for heavy rock operation; Ukhin BV [5] studied the influence of the number and shape of impeller blades on the parameters of dredge pump through experiments. Some scholars have studied the matching performance of the dredge pump and pipeline: Wu Yongbin et al. [6] have analyzed the spherical channel size of the dredge pump impeller, the impact of impeller cutting on the performance of the dredge pump, the calculation of blade cutting length and impeller cutting operation, and adjusted the diameter of the impeller to adapt to the complex working conditions of rocky. Han Haiqiang et al. [7] realized the matching between the dredge pump and the short conveying distance pipeline by adjusting the speed of the dredge pump and using a small diameter impeller; In the dredge pump and pipeline system, Wang Bin et al. [8] constructed the dredge pump pipeline model according to the similarity law, and studied the influence of different soil and mud concentration on the head of the dredge pump; Yang Zhengjun et al. [9] has developed a dredge pump impeller with a suction diameter of 800 mm suitable for short conveying distance, and the water efficiency has reached 84%; Zhai Qiwei [1] analyzed the countermeasures for the construction of cutter suction dredger under the conditions of long and short conveying distance, and improved the construction efficiency by appropriately changing the speed of dredger pump and arranging pipelines reasonably. The twisted impeller with three blades of the inboard pump of the cutter suction dredger with a suction diameter of 900 mm is developed [10], which avoids the use of pipeline throttle tube and improves the dredging efficiency. Combined with the methods of numerical simulation and model test, the dredge pump of the dredger is optimized and the efficiency of the dredge pump is improved. The commonly used "small impeller" in China cuts the impeller of dredge pump [11-12]. Due to the change of blade profile, the efficiency of pump is reduced more, which is not conducive to energy saving.

By changing the wrap angle of blade, inlet and outlet placement angle of blade, blade profile and number of blades of the dredge pump impeller, without changing the overall dimension of the impeller, change the head of the dredge pump, optimize the parameters of the impeller and improve the efficiency of the dredge pump. It will not affect the normal dredging and improve the dredging capacity of the cutter suction dredger under different conveying distances. The dredge pump impeller is a vulnerable part, then the dredge pump impeller needs to be replaced in the form of spare parts without any modification to the structure of the dredge pump [10].

## 2. Research and Development of New Impeller of Dredge Pump

### *2.1. Impeller Design Parameters*

The new impeller is designed, which keep the overall dimension of the impeller unchanged, appropriately reduce the blade diameter (the impeller wall diameter remains unchanged), design the new blade, adjust the blade inlet angle, outlet angle and blade wrap angle, repeatedly optimize the blade profile, reduce the head of the two pumps by 20-25 m, and the efficiency reaches more than 86%. As the impeller of high-efficiency pump has three blades, the blade number can not be reduce; But the efficiency is reduced too much by directly cutting the blades. Therefore, the method of appropriately reducing the blade diameter and combining the blade profile optimization is adopted to ensure the design goal of the head and the efficiency. The reduction of blade diameter shall be controlled within 10%. The outer diameter of original impeller of DN850 inboard pump is 2,210 mm, and the outer diameter of new impeller is 2,010 mm, which is reduced by 9.0%; The outer diameter of original impeller of DN850 submerged pump is 1,740 mm, and the outer diameter of new impeller is 1,650 mm, which is reduced by 5.2%. Meanwhile the blade profile is redesigned. Combined with the numerical method, the hydraulic design and performance optimization of the inboard pump and submerged pump under the condition of short conveying distance are carried out, and the performance prediction of the final design scheme is carried out.

*2.2. Numerical Simulation Method and Boundary Conditions*
Computational fluid dynamics (CFD) technology is used for internal flow analysis and performance prediction of fluid machinery. It has the advantages of low cost, high efficiency, convenience, rapidity and less time. The method of using numerical simulation to simulate the full 3-D flow field of the internal flow of fluid machinery, and then to predict the performance of fluid machinery is more and more widely used by researchers [13].

The setting of boundary conditions mainly involves three conditions: inlet boundary, outlet boundary and wall boundary. The inlet boundary condition is usually set as velocity inlet or pressure inlet. The velocity inlet boundary condition is suitable for incompressible flow problems, and the magnitude, direction or velocity components of the velocity need to be specified. The outlet boundary condition is usually set as free flow or pressure outlet. The free outflow boundary condition is applicable to the case where the pressure or velocity at the outlet is unknown. The free outflow boundary condition is used to ensure that the flow is a fully developed turbulence, and the diffusion flux of all variables at the outlet is zero. No slip boundary condition is adopted at the fixed wall. The moving wall represented by the impeller adopts the rotating reference coordinate system, and it is assumed that the rotating speed of the impeller relative to the fluid in its channel is zero; The fixed wall represented by volute adopts static absolute coordinates.

The inlet and outlet extension sections of inboard pump and submerged pump are added as the calculation model respectively (figure 1), and each part of the calculation model is meshed by ICEM CFD 17.2. The calculation condition is the same as that of the original pump. The design speed of the inboard pump is 325 rpm, the speed of the submerged pump is 330 rpm, and the calculation fluid medium is clean water.

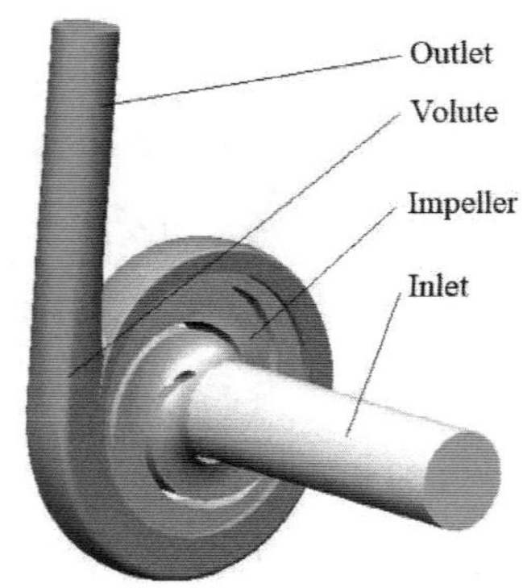


**Figure 1.** Schematic diagram of computational domain.

*2.3. Flow Field Analysis*
The figure 2 and the figure 3 show the pressure distribution when the flow rate is 10,000 $m^3/h$. From the impeller center to the radial direction, the pressure of the water section increases, the low-pressure area is concentrated in the impeller center, and the pressure gradient conforms to the law of fluid motion.

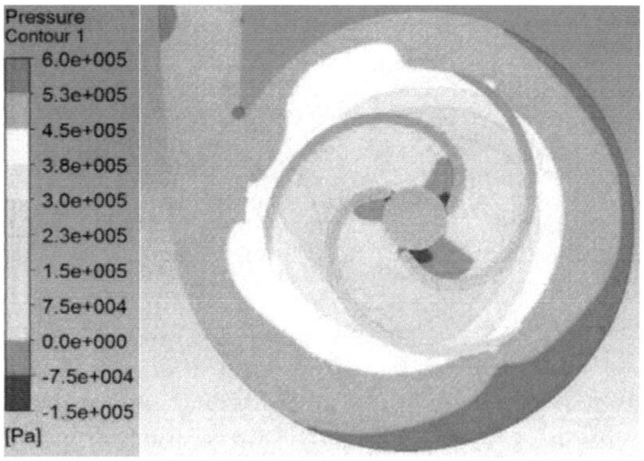


**Figure 2.** Middle section of inboard pump.

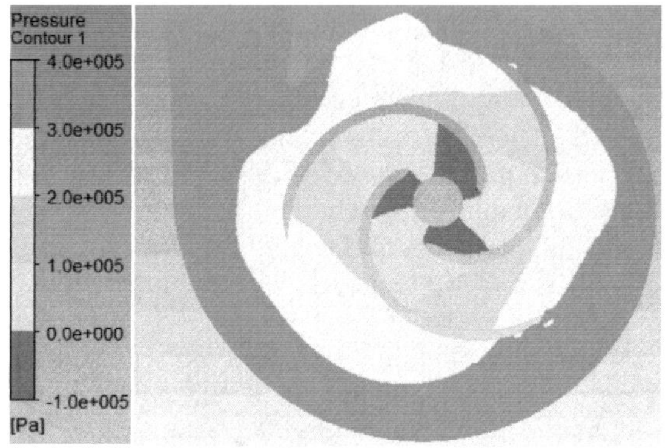


**Figure 3.** Middle section of submerged pump.

The figure 4 and the figure 5 show the relative velocity distribution of the section in the impeller and volute channel at 10,000 $m^3/h$. The relative velocity direction in the impeller channel is consistent. Along the blade direction, there is no obvious vortex and reflux, and there is no vortex in the relative velocity near the volute tongue, indicating that the hydraulic design is excellent. At the same time, the abrasion of the slurry on the channel wall is reduced and the wear resistance of the pump is improved.

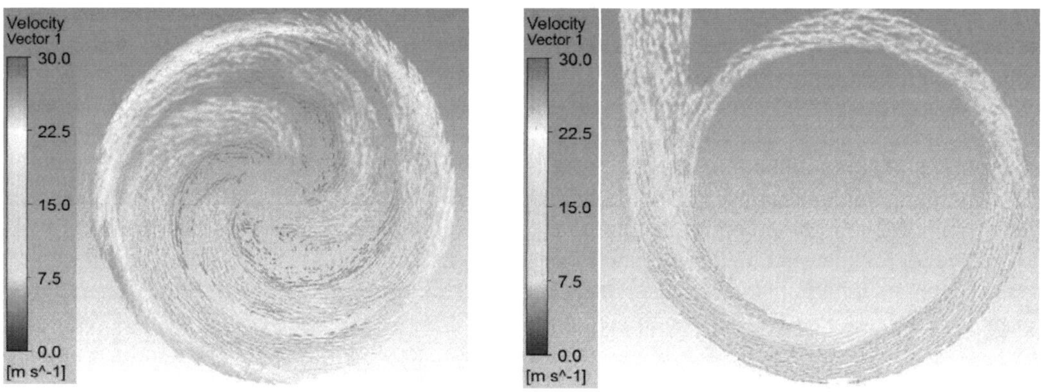


(a) Relative velocity of impeller (b) Relative velocity of volute

**Figure 4.** Relative velocity of middle section of inboard pump.

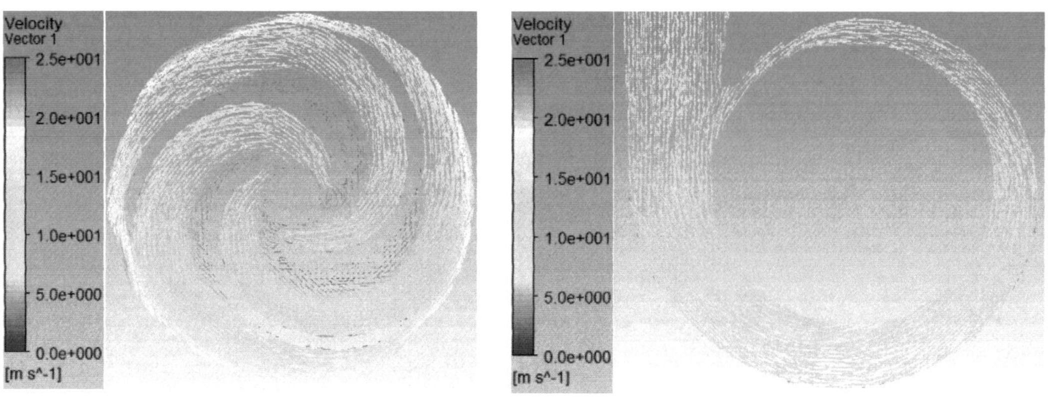


(a) Relative velocity of impeller (b) Relative velocity of volute

**Figure 5.** Relative velocity of middle section of submerged pump.

### *2.4. Predicted Performance Curve*

According to the results of numerical simulation, the head and the power at different flow points are calculated, and then the hydraulic efficiency of dredge pump is calculated. Finally, the flow--head, flow--power and flow--efficiency data are obtained.

(1) Performance of inboard pump with original impeller and new impeller

Table 2 shows the characteristic data of inboard pump. With a flow 8,000-12,000 $m^3/h$, the efficiency of the inboard dredge pump with the new impeller of is equivalent to that of the original impeller, the high efficiency area coincides, the maximum efficiency is 86.4%, the head is reduced by 18-19.5 m and the power is reduced by 500-750 kW.

**Table 2.** Comparison of performance data of inboard pump with two impellers.

| | Original impeller | | | Difference of new impeller | | |
|---|---|---|---|---|---|---|
| Flow $m^3/h$ | Head m | Power kW | Efficiency % | Head difference m | Power difference kW | Efficiency difference % |
| 8,000 | 87.5 | 2,351.4 | 81.5 | -18.0 | -500.0 | 0.0 |
| 9,000 | 85.5 | 2,502.3 | 83.8 | -18.5 | -549.7 | 0.4 |
| 10,000 | 82.8 | 2,648.7 | 85.2 | -18.8 | -610.0 | 0.6 |
| 11,000 | 80.0 | 2,779.5 | 86.4 | -19.0 | -671.2 | 0.2 |
| 12,000 | 77.0 | 2,923.1 | 86.2 | -19.5 | -750.0 | -0.2 |

(2) Performance of submerged pump with original impeller and new impeller

Table 3 shows the characteristic data of submerged pump. With a flow of 8,000-12,000 $m^3/h$, the efficiency of the submerged dredge pump with the new impeller is 1.5 percentage points higher than that of the original impeller, the maximum efficiency is 86.1%, the head is reduced by 3.3-4.9 m and the power is reduced by 126-182 kW.

**Table 3.** Comparison of performance data of submerged pump with two impellers.

| | Original impeller | | | Difference of new impeller | | |
|---|---|---|---|---|---|---|
| Flow $m^3/h$ | Head m | Power kW | Efficiency % | Head difference m | Power difference kW | Efficiency difference % |
| 8,000 | 45.9 | 1,249.0 | 80.0 | -3.3 | -125.9 | 2.6 |
| 9,000 | 44.5 | 1,313.5 | 82.9 | -3.4 | -137.8 | 2.7 |
| 10,000 | 43.0 | 1,378.0 | 84.9 | -4.0 | -153.0 | 1.7 |
| 11,000 | 41.2 | 1,435.0 | 85.9 | -4.3 | -160.7 | 0.8 |
| 12,000 | 39.3 | 1,492.0 | 86.1 | -4.9 | -182.1 | -0.3 |

## 3. Condition Matching Analysis

### *3.1. Operating Point Analysis*

Based on the power and parameter configuration submerged pump and inboard pump of 3,500 $m^3/h$ cutter suction dredger, high-efficiency dredger adopts original impeller and new impeller respectively. The two pumps of cutter suction dredger are constructed in series. The production rate and efficiency of three slurry types ($d_{50}$:0.05 mm silty sand, 0.2 mm fine sand and 1 mm coarse sand) and different conveying distance are predicted and compared. In order to determine the suitable conveying distance range of different impellers [14], and provide a theoretical basis for impeller configuration and construction productivity prediction. Comparison description: In order to unify the comparison

Journal of Physics: Conference Series **2329** (2022) 012019 doi:10.1088/1742-6596/2329/1/012019

conditions of original impeller and new impeller, it is assumed that the dredge pump operates at the rated speed. When the power of the dredge pump with original impeller at the intersection of short conveying distance exceeds the maximum power provided by the engine, adjust the dredge pump to the full load working point through pipeline throttle tube.

Using the calculation method of dredging specification JTS 181-5-2012 [15], calculate the intersection of head curve and pipeline consumption pipe resistance curve for different mud properties, and obtain the power consumption per kilometer, total power consumption of two pumps and productivity of original impellers and new impellers with different conveying distances. Figures 6-9 show the comparison curve of unit power consumption, the power and the productivity of pumps with fine sand. The comparison curve of silty sand and coarse sand can be obtained by the same method, which will not be repeated here.

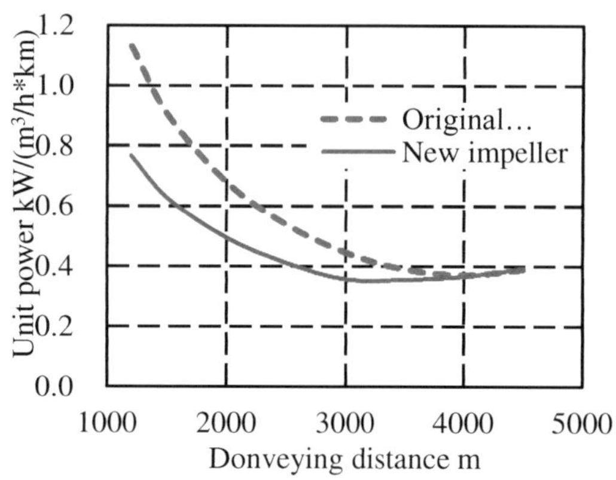


**Figure 6.** Comparison of unit power consumption.

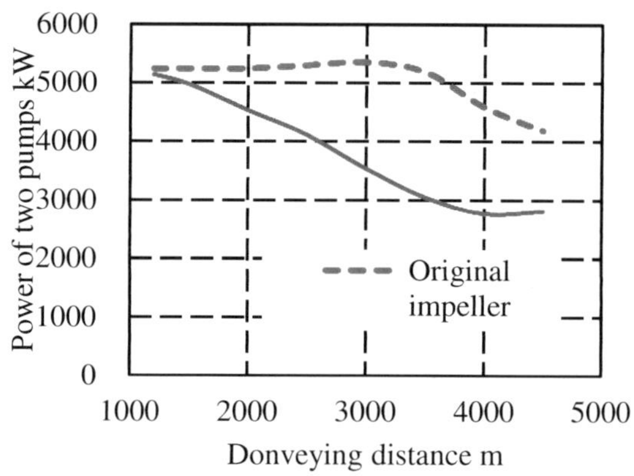


**Figure 7.** comparison of power impeller.

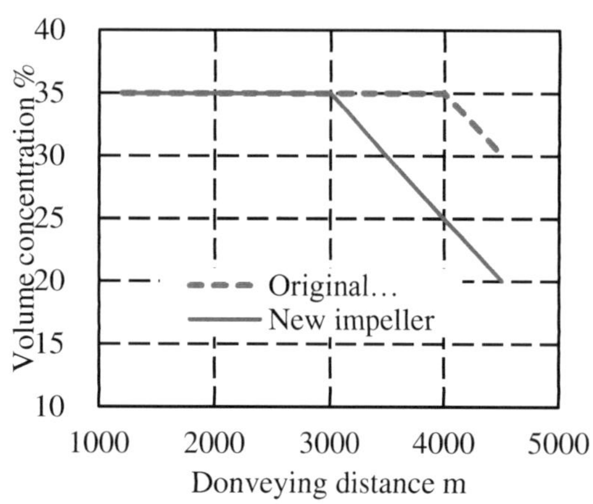


**Figure 8.** Volume concentration.

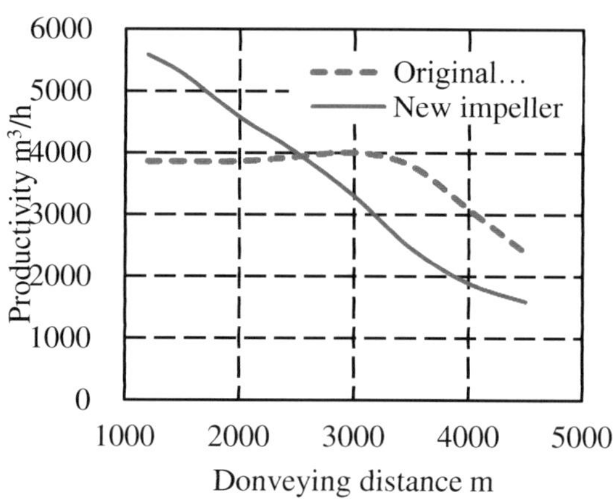


**Figure 9.** Comparison of productivity.

### *3.2. Impeller Configuration Strategy*

By comparing the unit power consumption, the power and the productivity of dredge pumps, the suitable pipeline distance range of the new impeller is obtained. When the conveying distance is less than which in the table 4, it is recommended to construct a single submerged pump with original impeller. When it is greater than the distance range in the table 4, it is recommended to construct the inboard dredge pump and submerged pump with original impeller in series.

**Table 4.** Energy conservation and productivity increase of new impellers.

| | Adaptive distance km | Average unit energy saving % | Average productivity increase % |
|---|---|---|---|
| Silty sand | 1.2-4.0 | 28 | 19 |
| Fine sand | 1.2-2.5 | 28 | 25 |
| Coarse sand | 0.8-1.2 | 25 | 14 |

(1) Silty sand, in the range of 1.2-4.0 km, the shorter the distance, the greater the effect of productivity improvement. When the distance is at 3.5 km, the productivity of new impeller is equivalent to that of original impeller, but the unit power consumption of the new impeller is reduced by more than 20%. At this time, it only has the advantage of energy saving. When the conveying distance is greater than the value in table 4, although the new impeller still has the advantage of energy saving, the productivity also decreases. For example, the productivity of dredge pumps decreases by 11.5% with a distance of 6-8 km. It can be judged to use the new impeller or the original impeller according to the needs. When the conveying distance is less than 1.2 km, it is recommended to construct a single submerged pump with the original impeller. When the conveying distance is greater than 4 km, it is recommended to construct original impellers with inboard pump and submerged pump in series.

(2) Fine sand, in the range of 1.2-2.5 km, the shorter the conveying distance, the greater the effect of productivity improvement. When the distance is at 2.5 km, the productivity of new impeller is equivalent to that of original impeller, but the unit power consumption of new impeller is reduced by about 20%. When the distance is greater than the value in table 4, the unit power consumption and dredge pump power of the new impeller are lower than that of the original impeller, but the productivity also decreases. For example, with a distance of 3.5 km, the unit power consumption of dredge pumps with new impellers decreases by 9.1% and the productivity also decreases by 35.3%. It can be judged to use the new impeller or original impeller according to the needs. When the distance is less than 1.2 km, it is recommended to construct a single submerged pump with original impeller. When the distance is greater than 2.5 km, it is recommended to construct the inboard pump and submerged pump with original impeller in series

(3) Coarse sand, in the range of 0.8-1.2 km, the new impeller has some advantages. When the distance is at 1.2 km, the productivity of new impeller is equivalent to that of original impeller, but the unit power consumption is reduced by about 20%. When the distance is greater than the above value in table 4, the power of dredge pumps with new impellers is lower than that of original impellers, but the unit power consumption increases and the productivity also decreases. For example, with a 2 km distance, the unit power consumption of dredge pumps increases by 4.4% and the productivity also decreases by 33.6%. Different from silty sand and fine sand, the suitable distance range of new impeller is narrow in coarse sand. When it is larger than the suitable distance range, the unit power consumption of dredge pump with new impeller increases and the productivity decreases. It is recommended to use original impeller in coarse sand, that is, the applicability of new impeller is poor.

## 4. Conclusion

- Keeping the overall dimension of the impeller unchanged, the new impellers of inboard pump and submerged pump are designed by appropriately reducing the blade diameter and redesigning the blade profile. The flow field is numerically simulated and analyzed. The pressure gradient conforms to the fluid law, and the relative velocity has no vortex along the impeller channel;
- The flow is 8,000-12,000 $m^3/h$, the maximum efficiency of the inboard pump with the new impeller is 86%, the head is reduced by 18-19.5 m, and the power is reduced by 500-750 kW; The maximum efficiency of the submerged pump with the new impeller is more than 86%, the head is reduced by 3.3-4.9 m, and the power is reduced by 126-182 kW;

- The power consumption and productivity of dredge pump with new impeller and original impeller are compared and analyzed, the suitable conveying distance range is obtained, the energy-saving and efficiency increasing effect of new impeller is predicted, and the impeller configuration strategy for different distance is formed.
- The research provides equipment support for the fine construction of dredgers. Whether it is applied to the energy-saving and efficiency improvement of dredgers or the design of new dredge pump, it is conducive to improve the dredging efficiency of dredgers.

**References**

[1] Zhai Q W 2014 Working mechanism and countermeasures of cutter suction dredger under special conveying distance *China Water Transport* **14**(6) 21-24
[2] Xu Y H and Yue W F 2016 Analysis of transport process of cutter-suction dredger *China Harbour Engineering* **36**(9) 62-65
[3] Jin X M 2019 Comparative study on two optimization construction methods of cutter suction dredger in short conveying distance *Pearl River Water Transport* (8) 29-30
[4] Bugdayci H H, Munts E and Grinwis H 2013 Latest developments in dredge pump technology *20th World Dredging Congress and Exhibition (WODCON XX)*. Brussels, (Belgium: Central Dredging Association) **2013** 260-277
[5] Ukhin B V 2002 Effect of the quantity and shape of impeller blades on dredge pump parameters *Power Technology & Engineering* **36**(3) 168-172.
[6] Wu Y B, Feng C, Li J F and Yang Z J 2019 Performance adjustment of small flow dredge pump for trailing suction hopper dredger under multi-rock sediment *Port Waterway Engineering* (02) 22-27
[7] Han H Q and Wang C 2020 Analysis on matching of mud pump and short discharge pipe for "Junjiang"ship *Port Waterway Engineering* (z1) 47-50
[8] Wang B, Fan S D, Wen Q, Liu B H and Zhang M 2020 Digital Simulation Matching Calculation of Suction System of Cutter Suction Dredger *Ship Engineering* **42**(06) 61-67.
[9] Yang Z J, Zheng X B and Guo Z Y 2020 Development and application of dredging pump with inlet pipe diameter 800 mm impeller *Port Waterway Engineering* (z1) 81-85
[10] Zhuang H F, Liu M M and Jin B X 2020 Research and application of centrifugal pump impeller modification for cutter-suction dredger *Fluid Machinery* **48**(4) 42-45
[11] Zhuang H F, Wang W K, Zheng L Z, Yang C and Hu J Z 2015 Optimization design of dredge pump based on CFD and model experiment *China Rural Water and Hydropower* (10) 130-132
[12] Zhuang H F, Yang C and Hu J Z 2017 Research and application of dredge Pump for 4500m$^3$ trailing suction hopper dredger *China Harbour Engineering* **37**(1) 63-66
[13] Wang F J 2004 *Computational Fluid Dynamics Analysis* (Beijing: Tsinghua University Press) chapter 1 p 2-3
[14] Li M Z, He Y P, Zhuge W and Huang C 2015 Working range and operation optimal of slurry transportation system in dredger *Port Waterway Engineering* (4) 201-204, 213
[15] JTS 181-5-2012 2012 *Design Code for Dredging and Reclamation Works* (Beijing: China Communications Press) p 50-59

# Numerical Simulation of Solid-Liquid Two-Phase Flow in Non-Clog Pump

**Xiongfa Gao [1,2], Junhua Wang [1,a,*], Hongzhong Lu [1,b,*], Weidong Shi [2,3], Junli Duan [4], Wanli Meng [1] and Jiamin Jin [1]**

[1] Shanghai Kaiquan Pump (Group) CO. LTD., Shanghai 201804, China;
[2] National Research Center of Pumps, Jiangsu University, Zhenjiang 212013, China;
[3] School of Mechanical Engineering, Nantong University, Nantong 226000, China;
[4] Shimge Pump Industry (Jiangsu) Co., Ltd, Suqian 223800, China.
Email: [a] wangjunhua@kaiquan.com.cn; [b] luhongzhong@kaiquan.com.cn.

**Abstract.** This paper mainly based on the CFD numerical simulation method to analyze the solid-liquid two-phase flow of the vortex pump. Three kinds of solid-phase volume concentration and three solid-phase particle diameters were given respectively to carry out the numerical simulation of solid-liquid two-phase flow in the internal flow field of the pump. The distribution of solid particles in the impeller and the lateral cavity were analyzed to study the influence of different solid-phase particle diameters and solid-phase particle concentrations on the pump performance. The results showed that there was no obvious concentration distribution at the junction area between blades and hub. The solid phase concentration on the back is lower than that on the working face. Due to the characteristics of circulating flow in the vortex pump, some particles enter the lateral cavity from the blade working face, and some particles enter the blade back face from the lateral cavity under the action of low pressure on the back and circulating flow. Solid particles are mainly distributed in the center of the cavity and are circularly distributed. The overall fluctuation of the solid phase slip velocity on the working face is not large, while the magnitude of the solid phase slip velocity on the back is relatively disordered.
**Key words.** Vortex pump; numerical simulation; circulating flow; solid-liquid two-phase flow.

## 1. Introduction

The structure of the vortex pump is special and the internal flow is complex. The structural feature of this pump is generally that the impeller retracts in the back pump cavity of the volute, and the lateral cavity has a spacious space. Since there is a large gap between the impeller and the volute, when the pump is running, the energy of the liquid is increased by the centrifugal force, and the liquid entering the blade moves with the impeller under the impetus of the rotating blades [1-3]. The liquid near the top of the impeller outlet forms a through flow due to the larger centrifugal force, while the liquid near the middle of the impeller forms a circulating flow due to the smaller centrifugal force. This kind of pump has good non-clogging performance, but low efficiency and high energy consumption, therefore, it is urgent to improve performance and reduce energy consumption.

Japanese scholar Akira OTA [4] used graphic method to give the movement patterns of different solid particles in the pump impeller and the changes of solid particle trajectory. After that, with the continuous development of test methods and test equipment, high-speed photography, particle imaging velocimetry (PIV), finite element analysis and other technologies have been continuously applied to the research of solid-liquid two-phase flow pump. Daniel baun [5] studied the velocity at different sections of the impeller, the stress on the working face and back of the blade, as well as the static pressure at the outlet of the impeller by combining experimental analysis and CFD simulation. The numerical simulation results were roughly close to the experimental research results, which showed the feasibility of the numerical simulation method. Mehta m and Sastry s [6] measured the velocity of solid particles in the pump by PIV technology. Yang Minguan, Gao Bo et al. [7-9] used PDPA technology to measure the liquid-solid two-phase flow field in the self-designed vortex model pump, and obtained the three-dimensional velocity distribution of each phase in the lateral cavity and impeller. Later, they measured the flow field in the lateral cavity and impeller with advanced laser doppler anemometry (LDA), and obtained important information such as circumferential velocity, axial velocity, radial velocity and its corresponding root mean square fluctuation velocity distribution. Zhao Wanyong [10-13] carried out experimental research on vortex pump from different solid particle sizes and concentrations.

In view of the test limitations, through the mature CFD numerical prediction software platform, this paper analyzes the distribution of solid particles on the blade surface and lateral cavity of the vortex pump, and explores the blockage and wear characteristics of it when transporting the medium containing solid particles, which has a certain reference significance in theoretical analysis and engineering practical application.

## 2. Simulation Model and Meshing

The vortex pump studied in this paper is SCP200-400 type vortex pump in Jiangsu Industry-University-Research Cooperation Project. The performance parameters of the vortex pump are: flow rate $Q = 500m^3/h$, head $H = 28m$, rotating speed $n = 1485r/min$, specific speed $n_s = 166$. Main structural parameters: impeller outside diameter $D_2 = 350mm$, blade width $b = 60mm$, blade outlet angle $\beta_2 = 45$ °, blade inlet angle $\beta_1 = 60°$. The whole simulation domain of this study is mainly divided into five parts, namely, the inlet section, the lateral cavity, the impeller, the volute and the outlet section. ICEM software is used to divide structured grid of the model water bodies.

## 3. Numerical Simulation Method and Boundary Conditions

In this paper, based on ANSYS-CFX commercial numerical simulation software platform, the multi-reference coordinate system method was used to conduct the dynamic and static coupling between the impeller and the volute, and the time-averaged N-S equation was used as the basic control equation. It is assumed that the shape of the solid particles was unchanged during the operation, and the flow of the fluid medium in the pump was a steady flow. During the simulation, the surface roughness was set to 0.05mm.

The channel flow field was set as incompressible steady viscous flow, using Standard k-ε turbulence model. The irrotational flow was set at the impeller inlet and the pressure at the inlet was set zero as the reference point. The flow at the outlet was assumed fully developed outflow. The near-wall flow was treated by the standard wall function and assuming no slip solid wall, the pressure and speed were coupled by SIMPLE-C algorithm.

## 4. Analysis of Numerical Simulation Results

In order to correspond to the rapeseed particles used in the experiment, the particle distribution in the impeller and the lateral cavity area is analyzed when the solid particle diameter $D$ is 0.5mm and the solid volume concentration cv is 5%. Figure 1 shows the code sketch and cross section diagram of the impeller blades (where "B" represents the blade, "B-P" represents the pressure side of blade, "B-S" represents the suction side of blade, and "S" represents the cross section). In order to facilitate the analysis, the blades are respectively projected onto the normal plane of the respective central axis.

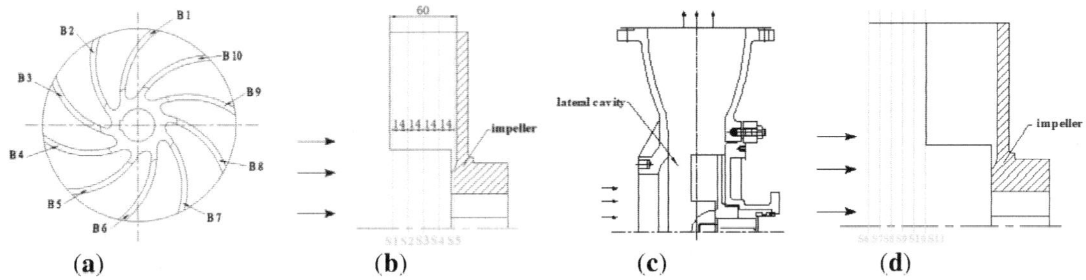


**Figure 1.** Cross-section and the impeller blade code sketch: (**a**) impeller blade code; (**b**) cross-section of impeller; (**c**) lateral cavity; (**d**) cross section of lateral cavity.

*4.1. Solid-Liquid Two-Phase Analysis of Different Cross-Sections of the Impeller*

Figure 2 shows the particle concentration distribution cloud diagram of the impeller section. The cutting method is to cut five surfaces in turn from the front shroud of the blade to the back shroud, as shown in figure 1(b). Section S1 is 2mm away from the face of the front shroud, and section S5 is 2mm away from the face of the back shroud. From the solid particle concentration distribution in the figure, the S5 section near the back shroud has a relatively high particle concentration. Except for the S5 section, from S1 to S4 section, the solid particle concentration in the impeller inlet area gradually decreases.

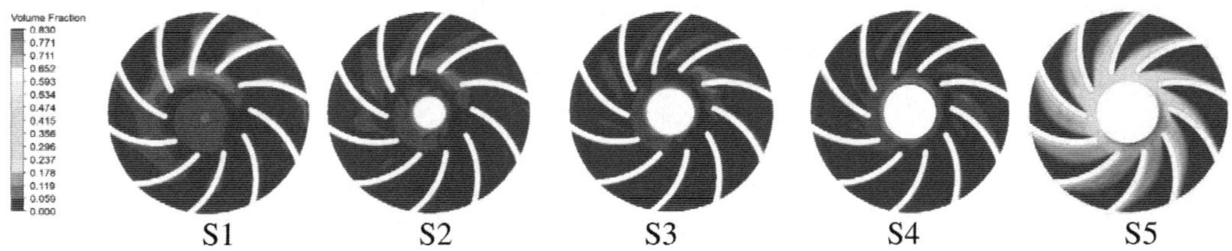


**Figure 2.** The particle concentration distribution cloud picture in impeller cross-section.

*4.2. Solid-Liquid Two-Phase Analysis of Different Cross-Sections of the Lateral Cavity*

Figure 3 shows the cloud picture of particle concentration distribution in the section of lateral cavity. The way of taking sections is to take an average of six sections from the inlet face of lateral cavity to the front face of impeller blade, and the section diagram is shown in figure 1 (d). S6 is the inlet wall of the lateral cavity, and the section S11 is 1mm from the blade front shroud. The maximum volume fraction of solid phase in the lateral cavity appears in the lower left area of the wall. From the perspective of the entire section, the concentration of solid phase particles in the middle part near the exit direction is larger, and the concentration values of the six sections in this area are close. It can be clearly seen from the figure that the solid phase particles follow the liquid phase in the lateral cavity in a circulating flow. It can be seen from the section S11 that this area is close to the impeller. Due to the effect of dynamic and static interference, the solid phase volume fraction distribution in this area is quite different from other sections.

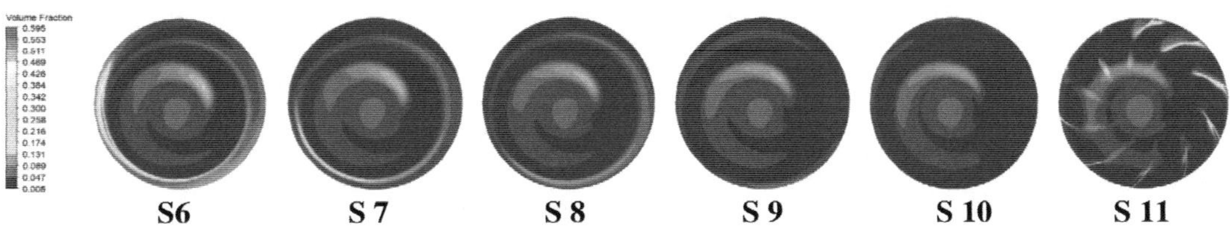


**Figure 3.** The particle concentration distribution cloud picture in lateral cavity cross-section.

### *4.3. Solid-Liquid Two-Phase Analysis of Different Impeller Streamlines*

Starting from the specific location, the following is a more subtle exploratory study of the sliding speed, distribution of solid particles and wear on the surface of the impeller. The following are mainly selected B1, B3, B5, B7 and B9 blades as the research objects, extract the intersection line between the blade and the front shroud, the middle section of the impeller and the back shroud, and divide the points, calculate the solid phase volume fraction $f$ and solid phase slip velocity $v$ at each node through numerical simulation, and the comparative analysis is carried out.

Figure 4 shows the curve of the solid phase volume fraction and the slip velocity of the front wall of the impeller blade. It can be clearly seen from figure 4(a) that before the B1 blade length-diameter ratio (the ratio of the distance from the point on the section line to the impeller axis and the outer diameter of the impeller) of 0.6, the solid phase volume fraction $f$ rises first and then decreases, while the solid phase volume fraction $f$ at the points taken by the other four intercept lines slowly increases; after the length-diameter ratio is greater than 0.6, the solid phase volume fraction $f$ of the B1 and B9 section lines increases significantly and then decreases, while the volume fraction $f$ of the solid phase on the B3 section line rises slowly. Except B3, the maximum slip velocity $v$ falls back about 2m / s when the aspect ratio is 1. When the length-diameter ratio of B3 section lines is between 0.8 and 0.9, the solid phase volume fraction $f$ tends to 0, and the corresponding solid phase slip velocity $v$ is also less than the slip velocity of other section lines. The rotating impeller blades do work on the medium. Theoretically, the speed should increase from the inlet to the outlet, but the figure shows that it first decreases and then increases. This may be due to the existence of circulating flow in the lateral cavity. In the area of the front shroud of the impeller blade inlet, there is a medium flowing into the impeller flow channel from the lateral cavity, and the incoming flow interacts with the medium flowing in the inner and outlet direction of the impeller.

Figure 4(b) shows the distribution of solid phase volume fraction and solid phase slip velocity at the intersection of the suction side and the front face of the blade. It can be seen from the figure that the solid phase volume fraction $f$ is significantly lower than the solid phase volume fraction $f$ at the intersection of the blade front face and the pressure side, and the solid phase slip velocity $v$ is not much different from the blade front face working face. Most of the solid phase volume fraction $f$ of the five intercept points are less than 0.5. While the B3 section line rises first in sinusoidal form, reaches the maximum when the length-diameter ratio is about 0.6, and then decreases; for the B9 section line, the solid volume fraction $f$ has little change before the length-diameter ratio is 0.85. The fluctuating trends of the solid-phase slip velocity $v$ of the five section lines are basically the same. The solid-phase slip velocity $v$ of the B1 section reaches the minimum when the length-diameter ratio is about 0.61, and then continues to increase, when the length-diameter ratio is about 0.95, it reaches the maximum value, and then the slip speed decreases to the exit. The solid-phase slip velocity $v$ of the B3, B5, and B7 sections decreases from the inlet first, reaches the lowest value when the length-diameter ratio is about 0.65, and then gradually increases.

Figure 5 shows the solid volume fraction and slip velocity curve of streamline in impeller blade, Figure 5(a) shows the solid phase volume fraction and solid phase slip velocity diagram on the midstream line of the pressure side, It can be seen from the figure that except for the large fluctuations in B3 and B9, the streamline fluctuations in the other three blades are relatively small, and the trend is basically the same; the solid phase volume fraction $f$ first gradually rises, reaching the peak when the length-diameter ratio is 0.7, then decreases, and rises again to the exit area; B3 is similar to the three middle streamlines, except that the solid phase volume fraction $f$ fluctuates greatly, and its maximum value $f$ is close to 0.4. The solid phase volume fraction $f$ of the B9 middle streamline first rises and then decreases at the inlet section, and the value of the latter half tends to be stable. The difference from the other four middle streamlines is that the inlet solid phase volume fraction $f$ value is much higher than other blade center streamline.

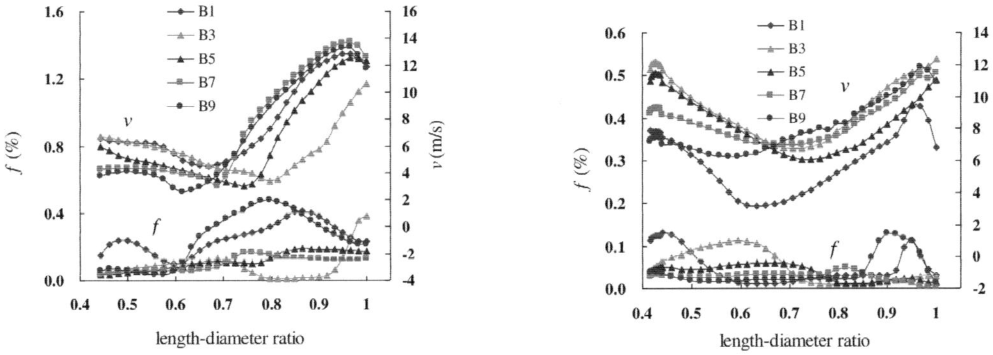


(a) (b)

**Figure 4.** The solid volume fraction and velocity in front shroud streamline of blade: **(a)** Intersecting line between pressure face and impeller front cover face; **(b)** Intersecting line between suction face and impeller front cover face.

Figure 5(b) shows the solid phase volume fraction and solid phase slip velocity curve on the back center streamline. It can be seen from the figure that the solid phase volume fraction *f* on the back center streamline is relatively small, and the value of *f* in the inlet is about 0.2, and then quickly drops to 0. The *f* value at the inlet of the streamline in B9 is higher than that of other blades. When the length-diameter ratio is between 0.42 and 0.68, the *f* values of the five back center streamline are all zero; when the length-diameter ratio is greater than 0.68, the *f* values of B1 and B9 increase slowly, while the *f* value of the other three center streamlines only increased after the aspect ratio was 0.83. From the perspective of the solid phase slip velocity *v* of the center streamline, its value fluctuates greatly. Especially for B3, B5 and B7, when the length-diameter ratio is between 0.41 and 0.5, it first drops rapidly, and then rises rapidly. When the length-diameter ratio is greater than 0.5, the fluctuation of the value becomes slower. The solid phase slip velocity *v* of the streamline in B1 is lower than the values of the other four blades. It reaches the lowest value when the length-diameter ratio is about 0.56, and then gradually increases, and starts to decrease when the length-diameter ratio is about 0.88. The fluctuation of the solid phase slip velocity *v* of B9 is similar to that of B1. Different from the streamline in the pressure side, the fluctuation of the solid phase slip velocity *v* in the outlet area of the blade is turbulent.

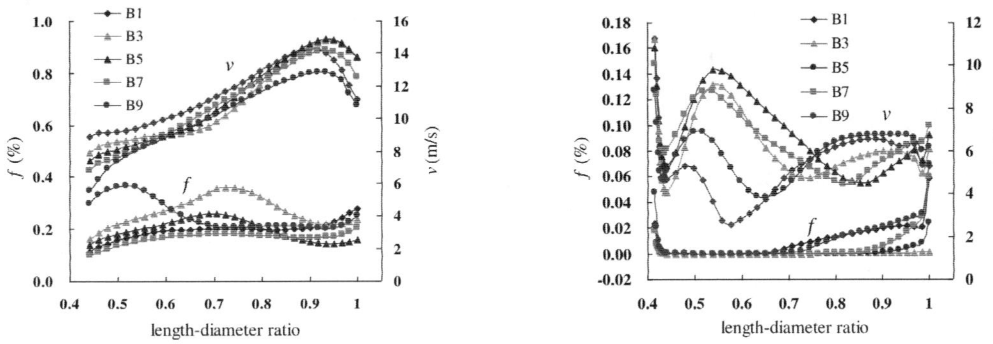


(a) (b)

**Figure 5.** The solid volume fraction and velocity in center streamline of blade: **(a)** Intersecting line between pressure face and impeller middle section; **(b)** Intersecting line between suction face and impeller middle section.

### *4.4. Effect of Particle Diameter and Concentration on the Performance of Vortex Pump*

Figure 6 shows the influence curve of solid particle diameter and solid particle concentration on pump performance. Figure 6 (a) shows the influence curve of solid particle diameter change on hydraulic performance when the solid particle concentration cv=5%. The analysis figure shows that when the concentration is constant, the hydraulic efficiency and head of the vortex pump decrease with the increase of solid particle diameter. When the diameter of solid particle is 0.1mm, the hydraulic efficiency is 48.57% and the head is 31.73m; When the diameter of solid particle is 0.5mm, the efficiency decreases by nearly 1 percentage point and the head decreases by nearly 1.5m. When the solid particle diameter continues to increase to 1.0mm, the hydraulic efficiency and head decrease slowly.

Figure 6(b) shows the influence curve of solid particle concentration change on hydraulic performance when solid particle diameter d=0.5mm. From the chart, we can analyze that when the solid particle concentration increases from 1% to 5%, the hydraulic efficiency decreases by more than 2 percentage points; When the solid particle concentration increases to 10%, the hydraulic efficiency decreases slowly. Different from figure 6(a), the head increases with the increase of solid particle concentration. When the solid particle concentration increases from 1% to 10%, the head increases by about 0.35m.

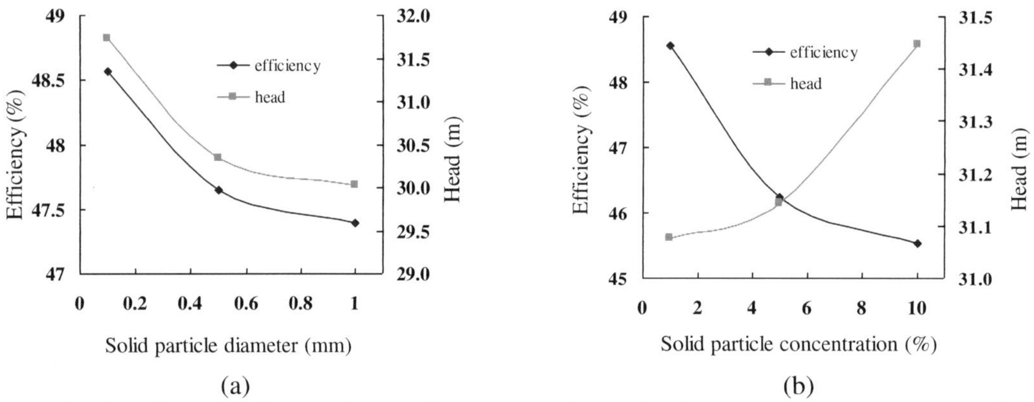


**Figure 6.** The diameter sizes and particle concentration affection on performance: (**a**) Effect of diameter on hydraulic performance (cv=5%); (**b**) Effect of concentration on hydraulic performance (d = 0.5mm).

## 5. Conclusion

(1) Due to the characteristics of circulating flow in the vortex pump, some particles enter the lateral cavity from the blade working face. Under the action of low pressure on the back and circulating flow, some particles enter the back of the blade from the lateral cavity.

(2) The solid particles in the lateral cavity are mainly distributed in the central region and distribute circumferentially. There are a variety of complex flow structures in the lateral cavity. Driven by the circulation flow, more particles are brought to the lateral cavity and solid particles do circulation flow with the liquid phase in the lateral cavity. At the outer ring of the lateral cavity, some particles flow circumferentially.

(3) The overall fluctuation of the solid-phase slip velocity on the working face is small, while the solid-phase slip velocity on the back is disordered. The solid volume fraction f on the back of the blade is significantly lower than that on the front-end face of blade working face, while the solid slip velocity v is almost the same as that on the front-end face of blade working face.

**Acknowledgments**
This research was funded by the National Natural Science Foundation of China (Grant No. 51909108), Enterprise Postdoctoral Program of China (Grant No. 8421440002).

**Reference**

[1] Li S H and Feng J 1987 A Review of the Development of Vortex Pump *Journal of Beijing Agricultural Engineering University* **(3)** 55-63. (in Chinese)
[2] Feng J and Li S H 1988 Study on flow in vaneless cavity of vortex pump *Pump technology* **(1)** 25228.
[3] Feng J 1988 *Study on Internal Flow Performance and Application of Vortex Pump* (Beijing: Beijing Agricultural Engineering University).
[4] Ohta A 1962 Investigation of sand pump theory *Collection of Mechanical Society* **4(66)** 91-100.
[5] Daniel B 2002 *Hydrodynamic Force in Centrifugal Pump and Compressor Impellers in Volue Casings: Measurements Using Magnetic Bearing and CFD Simulation* (Virginia: University of Virginia) 3057473
[6] Mehta M, Kadambi J R, Sastry S, Sankovic J M, Wernet M P, Addie G and Visintainer R 2007 Particle velocities in the rotating impeller of a slurry pump *The 5thJoint ASME.JSME Fluids Engineering Conference* p 369-378
[7] Yang M G, Gao B, Gu H F and Li H 2008 Measurement on 3D turbulent flow field in vortex pump *Drainage and Irrigation Machinery* **26(1)** 60-64.
[8] Yang M G, Gao B, Liu D, Gun H F and Li H 2008 Experimental investment of Salt-out two-phase flow in a vortex pump by PDPA measurements *Journal of Engineering Thermophysics* **29(2)** 237-240.
[9] Yang M G, Gao B, Liu D, Gun H F and Li H 2007 Analysis on Liquid-solid Two-phase Flow Field in a Vortex Pump by PDPA Measurement *Transactions of the Chinese Society for Agricultural Machinery* **38(12)** 53-57.
[10] Zhao W Y, Liu T B, Li Y S, Zhang L and Wang Z 2007 Flow mechanism in votex-pump *Drainage and Irrigation Machinery* **25(6)** 52-55.
[11] Zhao W Y, Li Y S, Wang Z and Zhang L 2007 Inquiry into equation of particles motion in vortex-pump *Journal of Lanzhou University of Technology* **33(6)** 47-49.
[12] Zhao W Y, Wang Z, Jing Y, Li Y S and Zhang L 2009 Numerical Analysis of the Influence of Different Flow Parameters on the Performance of Vortex Pumps *Journal of Xihua University* **28(4)** 89-92.
[13] Zhao W Y, Jing Y, Wang Z, Zhang L and Zhang F 2009 Numerical Analysis of Influences on Vortex Pump Exterior Characteristics for Solid Particles Diameter and Concentration *Oil Field Equipment* **38(10)** 5-10.

# Design of Hydraulic System of Microcomputer Governor for Small Turbine

**Minchao He[1], Huali Hu[1], Hongwei Quan[1], Qibo Liu[2] and Bingsen Chen[1*]**

[1] Guangxi Vocational College of Water Resources and Electric Power, Nanning, Guangxi, China.
[2] Nanning Jingneng Power Generation Equipment Co. LTD, Nanning, Guangxi, China
Email: 82412777@qq.com

**Abstract.** Taking a power station as an example and according to the technical and performance requirements of the turbine governor in small rural hydropower stations, a hydraulic station with high oil pressure is designed by adopting a highly reliable general PLC as the control device and using standardized hydraulic components; Taking into account requirements for regulating performance under the conditions of automatic regulation, manual regulation and emergency shutdown, and the reliability requirements, the hydraulic control circuit with automatic adjustment, manual adjustment and emergency stop is designed to avoid possible problems of hydraulic system by taking effective measures. The product based on the design have been developed and put into production to meet the requirements of industry standard and its application with good effect can be further promoted.
**Keywords.** Hydropower station, governor, hydraulic pressure, technical renovation.

## 1. Introduction

By the end of 2020, China had built 43,957 small rural hydropower stations (stations with installed capacity less than 25MW), with a total installed capacity of 81.3GW, accounting for 22 % of China's total installed hydropower capacity. The annual power generation is 242.3TWh, accounting for 17.9% of China's hydropower generation [1]. It can be seen that rural small hydropower stations play a very important role in China's power grid.

Most of the rural small hydropower stations in China built before 2000, and due to the design and manufacturing conditions at that time, most of them used mechanical hydraulic governors with rated oil pressure of 2.5MPa or electric speed control mechanisms to regulate and control the hydro-turbines. This kind of regulation control device, with slow response, low regulation performance and poor control accuracy, make it difficult for the power quality to meet the requirements of modern power grid. Therefore, technical transformation and upgrading is in urgent need.

With the development of computer control theory and technology, the use of computer devices and digital PID control laws to achieve turbine regulation and control has become an important application of computer monitoring technology for modern hydropower station,which not only contributes to developing towards full automation and intelligence, but also helps the turbine to obtain excellent static and dynamic regulation to improve the power quality, and enhance the safety and stability of hydropower station [2]. In addition, with the standardization of hydraulic components and the continuous progress of manufacturing technology, high oil pressure and standardized hydraulic components are applied to design the hydraulic system of turbine governor, effectively simplifying the system structure, reducing the size, and improving the operation and maintenance of the system.

Therefore, the design of high oil pressure governor of hydraulic turbine controlled by microcomputer and using standard hydraulic components is in line with trend of the technical transformation and upgrading of small hydropower stations in rural areas.

## 2. General Design Idea of Small Microcomputer Governor

According to the demand of rural small hydropower station for microcomputer governor of hydraulic turbine with high performance, the scheme designed is shown in figure 1.

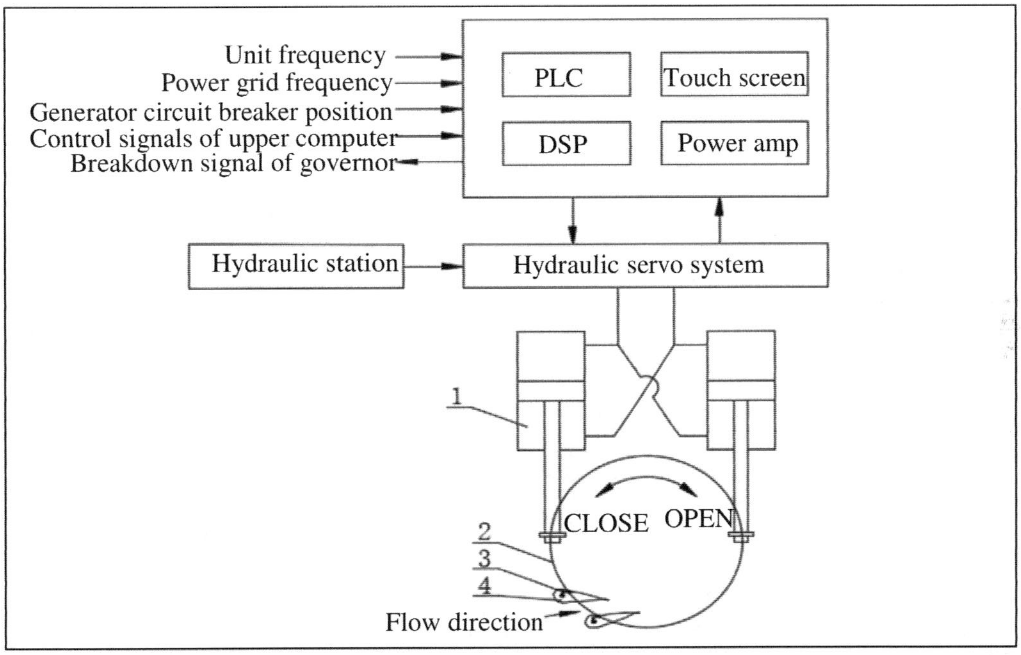


1. Hydraulic cylinder: 2. Water turbine control ring: 3. Movable water guide vane drive connecting rod: 4. Movable water guide blade

**Figure 1.** Design scheme of small microcomputer governor system.

PLC (Programmable Logic Controller), a highly reliable general-purpose computer control device, is applied in the governor to collect the unit frequency, power grid frequency, position of generator circuit breaker and control signals of upper computer (short for the upper level computer). Equipped with a touch screen as the human-computer interface, the function of input, output and display are realized. The hydraulic station is responsible for providing the high-pressure oil required for the hydraulic servo system. Signals are analyzed, judged and calculated by the program based on PID controlling law, and output results are amplified by power to control the hydraulic servo system so as to realize the rotation control of hydraulic cylinder 1 and hydraulic turbine control ring 2. When the control ring 2 rotates, through the transmission linkage 3, the governor realizes the control of the movable water guide vane 4, which evenly distributed around the whole circle, from fully closed to fully open at any angle, and finally completes the regulation and control of the flow of the turbine [3].

For a small hydropower station, with a rated power of 1250kW for axial flow fixed-blade turbine, a runner diameter of 1.6 m, a maximum head of 13.6 m, and a rated flow of 11.7m$^3$/s, the turbine should be equipped with a governor with speed regulation power of 13644Nm or more, according to the calculation formula of speed regulation power A of small turbine [4].

$$A = 250Q\sqrt{H_{\max} D_1} = 250 \times 11.7 \times \sqrt{13.6 \times 1.6} \approx 13644(\text{Nm}) \tag{1}$$

## 3. Design of Hydraulic System for Small Microcomputer Governor

According to the calculation results of formula (1) and the overall design of figure 1, the hydraulic system of the governor is shown in figure 2.

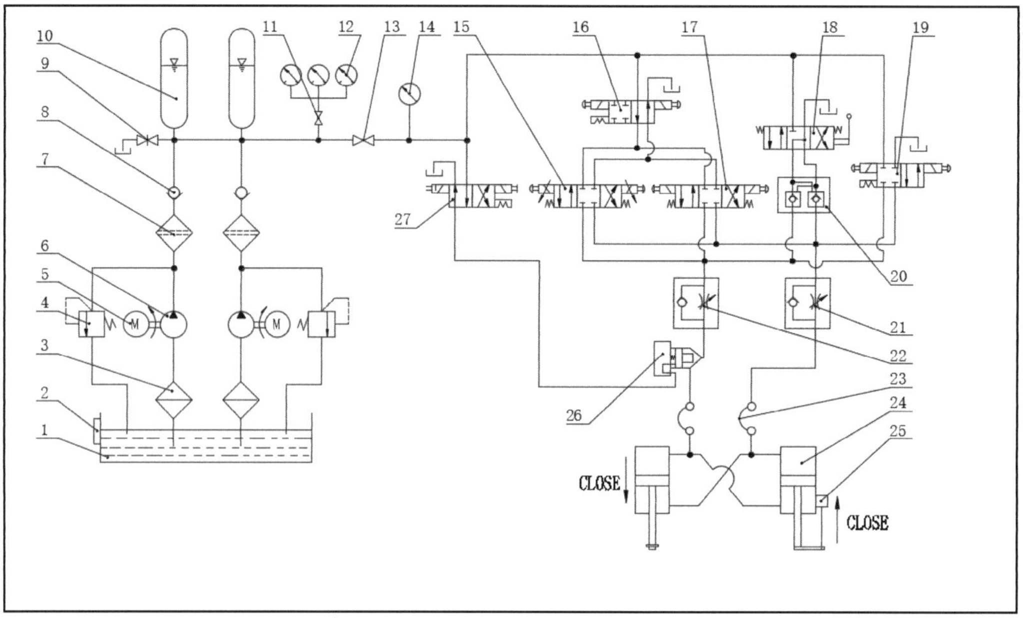


1. Fuel tank; 2. Liquid level gauge; 3. Oil suction filter; 4. Relief valve; 5. Motor; 6. Oil pump; 7. Oil filter; 8. One-way valve; 9. Ball valve; 10. Capsule accumulator; 11. Ball valve; 12. Electric contact pressure gauge; 13. Ball valve; 14. Pressure gauge; 15. Small fluctuation electromagnetic proportional valve; 16. Manual automatic switching solenoid valve; 17. Large fluctuation electromagnetic directional valve; 18. Manual reversing valve; 19 emergency stop solenoid valve; 20. Two-way hydraulic lock; 21. Throttle valve for the fastest shutdown time ; 22. Throttle valve for the fastest startup time; 23. High pressure hose; 24. Hydraulic cylinder; 25. Feedback device; 26. Hydraulic lock; 27. Lock-Up solenoid valve.

**Figure 2.** Hydraulic system of small high oil pressure microcomputer governor.

### *3.1. Hydraulic Station Design*

The rated oil pressure of the hydraulic station is 16.0MPa with the lowest rated oil pressure of 12.5MPa, and the total volume of tank 1 is 200L. Two oil pumps 6 (each with three working positions: working, standby and removal) are used to supply oil to two capsule accumulators with each volume of 40L successively through oil suction filter 3, outlet filter 7 and one-way valve 8. The accumulator is filled with nitrogen gas at a pressure of 11MPa, and the energy of the oil is stored in the gas because of the compressible nature of the gas. The operation values of the three electric-contact pressure gauges 12 are set according to table 1, which are respectively used for start/stop control of working oil pump and standby oil pump, as well as high oil pressure alarm and low oil pressure accident shutdown [5]. The function of relief valve 4 is to discharge part or all of the oil from the oil pump outlet to the tank when the oil pump cannot be stopped for any reason, so as to avoid damage caused by high oil pressure of the hydraulic station and hydraulic servo system. The function of one-way valve 8 is to prevent the oil in the accumulator from flowing back to the tank through the oil pump when the oil pump stops. Ball valve 13 (normally open) is the main oil supply valve of the governor hydraulic servo system, which plays a role in controlling the total oil source of the hydraulic servo system [6].

**Table 1.** Setting value of electrical contact pressure gauge in hydraulic station.

| NO. | Signal Contact | Pressure Setting Value | Signal name | Control function |
|---|---|---|---|---|
| 1 | Accident Reduction | ≤11MPa | Low oil pressure of accident | Emergency shutdown |
| 2 | Extra-Low | ≤13MPa | Under lower limit of normal oil pressure | Standby pump starting |
| 3 | Lower | ≤14MPa | lower limit of normal oil pressure | Working pump starting |
| 4 | Normal | ≥16MPa | Upper limit of normal oil pressure | Termination of pumping |
| 5 | Higher | ≥17MPa | Above upper limit of normal oil pressure | High oil pressure alarm |

### *3.2. Design of Hydraulic Servo System*

According to the functions of manual adjustment, automatic adjustment and emergency shutdown control required by the governor, the automatic adjustment circuit consisting of manual automatic switching solenoid valve 16, small fluctuation electromagnetic proportional valve 15, and large fluctuation electromagnetic directional valve 17, the manual adjustment circuit consisting of manual directional valve 18 and two-way hydraulic lock, the emergency stop control circuit consisting of emergency stop solenoid valve 19, the hydraulic cylinder lock circuit consisting of locking solenoid valve 27 and hydraulic lock valve 26, and the fastest on/off time control circuit consisting of the fastest shutdown time throttle valve 21 and the fastest startup time throttle valve 22 are designed. The specific design ideas and key points are as follows:

*3.2.1. An Automatic Adjustment Circuit is Used for Small Fluctuation Adjustment and Large Fluctuation Adjustment.* As is shown in figure 2, when the valve core of manual automatic switching solenoid valve 16 is in the automatic adjustment position, the pressure oil from the oil source flows through the solenoid valve 16 and reaches the automatic adjustment circuit consisting of small fluctuation electromagnetic proportional valve 15 and large fluctuation electromagnetic directional valve 17 in parallel. When small change needs to be adjusted, PLC controls electromagnetic proportional valve 15 by controlling hydraulic cylinder operation with continuous adjustable small flow, to achieve small fluctuations in micro-adjustment. When the frequency changes greatly, PLC controls the electromagnetic directional valve 17 by controlling the hydraulic cylinder with large flow, to achieve rapid large adjustment. The combination of small and large fluctuation regulation can effectively realize the rapidity of large regulation and stability requirements of small regulation, and effectively reduce the overshoot and the number of adjustments. Under 2Hz disturbance, the overshoot is ≤30% and the adjustment times are ≤2 times [7].

*3.2.2. Manual Adjustment Circuit Equipped with Two-Way Hydraulic Lock is Adopted.* When the manual automatic switching solenoid valve 16 is in the manual position, the oil path is cut off from the oil source to the small fluctuation electromagnetic proportional valve 15 and large fluctuation electromagnetic directional valve 17, with a result that PLC can not control the hydraulic cylinder through the automatic adjustment circuit. However, operating manual directional valve 18 and two-way hydraulic lock 20 can change the oil path on/off and reverse direction, to achieve manual control of the hydraulic cylinder. Manual adjustment control mode can control the turbine's start/stop and load adjustment in case of computer failure or electrical failure. It is not suggested to use manual adjustment control mode except in special circumstances because it doesn't have automatic adjustment function.

Two-way hydraulic lock composed of two hydraulic control check valves is applied to lock the oil path between the hydraulic cylinder and the manual valve when manual adjustment is not carried out, so as to avoid the hydraulic cylinder drifting in the open direction under the action of water thrust, resulting in the loss of control of the opening degree of water guide blade, and even serious accidents caused by the damage of hydro-generator set.

The working principle of two-way hydraulic lock is shown in figure 3. B1 is connected to the hydraulic cylinder shutdown chamber while B2 is connected to the hydraulic cylinder starting chamber. When the manual reversing valve is in the middle position, A1, A2, K1 and K2 of the two-way hydraulic lock are connected with T, the oil discharge port, and the oil circuit between the hydraulic cylinder and the reversing valve will be closed and locked as the two one-way valves are closed. When manual reversing valve operates in the close direction, pressure oil flows from P to A, A to A1 and K2, and two one-way valves are opened resulting in pressure oil from P to B1, B2 to T, which can achieve manual operation to close the opening of the guide vane. The principle of operating in opposite direction is the same.

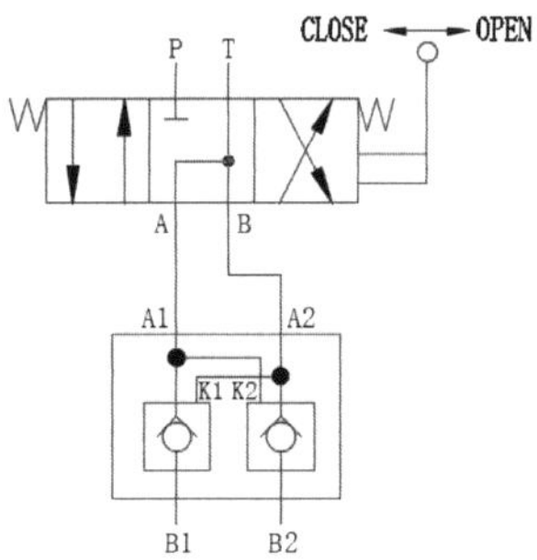


**Figure 3.** The working principle of two-way hydraulic lock.

*3.2.3. Setting the Circuit with the Fastest Open/Shut down Time by "Throttle Valve + Check Valve".* Different hydropower stations have different natural conditions, and the pipe diameter, length and pressure of the diversion system are also different. Water guide blade need to be closed quickly in an emergency, the shorter the closing time, the better to avoid unit speed rising too high, but the pressure of water hammer in water diversion system will be greater, which would cause the system pipe break or damage, directly affecting the safety of hydropower station. On the other hand, if the action of closing water guide blade is slower, the water hammer pressure will not rise much, but the rotational speed of the unit will rise greatly, and the rotating part of the unit may be damaged due to excessive centrifugal force. This will also compromise the safety of the hydropower station. Therefore, throttle valve 21 for the fastest shutdown time and throttle 22 for the fastest startup time are set to adjust the fastest startup/shutdown time to meet the safety requirements of each power station.

When accident happens in the unit, emergency stop solenoid valve 19 will work and pressure oil flows through solenoid valve 19, throttle valve 22 and its parallel one-way valve quickly into the hydraulic cylinder shutdown chamber. However, because oil can only be discharged through throttle valve 21 by the hydraulic cylinder starting chamber, the fastest speed of closing water guide blade is determined by throttle valve 21, whose opening degree can be changed according to the requirements of power station, and the fastest shutdown time can be obtained through the operation test. Similarly, the fastest startup time is adjusted by throttle valve 22.

*3.2.4. Reducing the Internal Pressure Pulsation in System by Hydraulic Hose.* For governor hydraulic system with high oil pressure, hydropower station requires seamless steel pipe connection. As solenoid valve 17 and emergency stop solenoid valve 19 working in switch mode, large pressure pulsation will be formed in the hydraulic system at the moment of operation, causing negative effect to hydraulic cylinder, pipeline and joint. Therefore, near the position close to the hydraulic cylinder and the hydraulic cylinder cross butt, high pressure hose 23 with pressure resistance of more than 20MPa is used for connection, which helps not only to meet the needs of the small angle swing for hydraulic cylinder when working, but also to reduce the pressure pulsation in the hydraulic system by taking advantage of the elasticity of high pressure hose [8].

*3.2.5. Realizing locking with Hydraulic Lock.* The circuit consisting of locking solenoid valve 27 and hydraulic lock 26 can be locked to fix the water guide blade in the fully closed position when the unit is shut down and the water guide blade is completely closed, so as to avoid serious accidents caused by turning on water guide blade by mistake due to hydraulic thrust action.

After the unit is shut down and the water guide blade is turn off, the upper computer controls locking electromagnetic valve 27, through which pressure oil flows directly into the right side of the hydraulic lock piston, pushing the piston to the right by differential force, to cut off the hydraulic cylinder shutdown chamber circuit, which prevents oil from entering and discharging, achieving the locking function by locking the hydraulic cylinder piston in the fully closed position of the water guide blade. When the locking is withdrawn, the hydraulic locking piston is connected to the right side to discharge the oil, and the locking piston is pushed to the leftmost end under the pressure of the hydraulic cylinder shutdown chamber pipeline. The pipeline is clear, and the operation of hydraulic cylinder is not affected.

## 4. Conclusion

Adopting a highly reliable general PLC as the control device and using the standardized hydraulic components, the microcomputer governor of small water turbine which is suitable for the technical transformation and upgrading of small rural hydropower stations is developed. The governor is simple in structure with fewer components, low in cost and easy to maintain. By adopting digital PID regulation law, different regulation parameters can be set under different operating conditions, so as to achieve excellent static and dynamic regulation performance, and the maximum overshoot and regulating times meet the requirements of industry standard under the influence of disturbance.

The governor developed according to this design has been put into production and used in many small rural hydropower stations, and the power quality and automation level of hydropower stations are greatly improved by combining PLC communication function with the computer monitoring system. The application can be further promoted in small rural hydropower stations.

## References

[1] *Analysis of the Current Situation of China's Rural Hydropower Construction: the Scale of Rural Hydropower Stations under Construction in 2020 is 3.482 Million kW* [EB/OL] https://www.chyxx.com/industry/202202/995019.html, 2022-02-10
[2] Cai Y S 2009 *Hydro-Turbine Governing* (Zhengzhou: Yellow River Water Conservancy Press) p166
[3] Song H P 2021 Study on abnormal oil temperature in operation of governor hydraulic system *Mechanical & Electrical Technique of Hydropower Station* (01) 61-63.
[4] Liang J H, Tong W Y, Chen B S, Li X L and Guo H M 2005 *Hydraulic Turbine Textbook Jian Qun* (Beijing: China WaterPower Press) p 216
[5] Eremin A V, Trubitsyn K V and Kudinov V A 2018 Designing of hydraulic systems on computer models.*IOP Conference Series: Earth and Environmental Science* **194**(7) 072002.
[6] Hong X and Tang X H 2013 *Automatic Operation and Monitoring of Hydropower Generating Units* ed S W Huang (Beijing: China Electric Power Press) p 132.
[7] Majeed Farah Ayad Abdul 2021 Steam Turbine Governor (STG) Controller Design Based on Particle Swarm Optimization (PSO) *IOP Conference Series: Materials Science and Engineering* **1105**(1) PP012050.

[8] Mateo Beus and Hrvoje Pandžić 2019 Application of an adaptive model predictive control algorithm on the Pelton turbine governor control *IET Renewable Power Generation* **14**(10) 1720-1727.

# Damage Characteristics of Roadway Wall under the Impact of Gas Explosion

**Zhenzhen Jia***, **Shujuan Li, Qing Ye, Wenxue Ding and Yangzhou Teng**

School of Resource, Environment and Safety Engineering, Hunan University of Science and Technology, Xiangtan 411201, China.
Email: jiazhenzhen1982@126.com

**Abstract.** In order to explore the damage process and characteristics of roadway wall under the impact of gas explosion, in this paper, the numerical model of gas explosion in roadway is established, and the explosion impact response of roadway wall and the stress change of gas explosion in roadway wall are analyzed. The results show that the impact produced by gas explosion is repeatedly reflected and superimposed on the wall, resulting in the high peak overpressure and duration growth in the explosion load process formed in the roadway. The shock wave propagation of gas explosion in wall is dynamic, and will produce tensile stress and compressive stress, causing damage to the interior of the wall. It will drive the wall structure of adjacent areas to move along the action direction of explosion load, resulting in the overall instability and damage of roadway wall structure. The research results can provide reference for gas explosion disaster prevention and explosion accident identification.
**Keywords.** Gas explosion; Wall damage; Impact stress; Numerical simulation.

## 1. Introduction

Gas explosion accident in coal mine will not only cause casualties of operators in the roadway, but also destroy the roadway wall structure and mine ventilation facilities, resulting in adverse social impact [1-3]. In order to effectively control the damage caused by gas explosion, many scholars have studied the propagation characteristics and damage mechanism of gas explosion, and have analyzed the influencing factors, propagation law, the destructive effect of gas explosion and formulated various measures to prevent and suppress explosion, and achieved fruitful results [4-13]. However, the damage characteristics of roadway wall under the impact of gas explosion need to be deeply analyzed.

Based on the analysis on the propagation law of shock wave of gas explosion in coal mine, the stress field of gas explosion acting on the roadway wall is analyzed, and the flow field analysis model of shock wave of gas explosion in the roadway is established. Through the simulation on damage characteristics of explosion impact with different loads in the roadway model, the explosion impact response of roadway wall and the change of explosion stress of roadway wall are analyzed. It is expected that the research results will be helpful to the prevention and control of gas explosion disaster in coal mine.

## 2. Methodology

### *2.1. Assumed Conditions and Mathematical Equation*

Gas explosion is an extremely rapid and complex chemical reaction. A large number of intermediate and instantaneous products are involved in the explosion process. In order to reduce the amount of calculation, some reasonable assumptions are made for the model: (1) The intermediates of chemical reaction are ignored. (2) At standard atmospheric pressure. (3)The process of gas explosion is one-way irreversible process, and the explosion process is adiabatic process. (4)The wall is an adiabatic rigid wall, and the heat transfer and radiation between the inner space of the roadway and the outside world are not considered. (5)The boundary is set to a non reflective boundary condition. (6)The gas mixture is evenly mixed at room temperature and the specific heat capacity follows the mixing rules. (7)There are no obstacles inside the roadway.

The damage process of gas explosion is the flow process of explosion shock wave. The Lagrangian description increment method is adopted. In the rectangular coordinate system of space, the conservation equations of the damage process of gas explosion are the equations of mass conservation, momentum conservation and energy conservation [14-16].

### 2.2. Physical Model

In order to facilitate the numerical simulation, according to the central symmetry of the roadway's geometric model, a quarter roadway model of gas explosion is established for numerical simulation. As shown in figure 1. The geometric parameters of the roadway are adopted in this numerical simulation as follows. The outer radius of the roadway is 1m, the inner radius of the roadway is 0.9m, the roadway length is 10m. Cylindrical and elliptical sealed explosion models are used in the simulation, and the explosion source is located at the front end. According to the axial symmetry of the model, the model is divided into three parts. The first part is the gas explosion unit in the middle, the second part is the air unit with large space, and the third part is the roadway wall.

Solid164 eight-node-hexahedral solid element is adopted in the model, the bottom and the other side are fixed constraints, the air boundary is set as non reflective boundary condition, Euler is used to divide the grid for air, and Lagrange is used to divide the grid for solid structure. The unified unit system (kg-m-s) is adopted in the material model, and the mapping grid division is adopted. The finite element model after grid division is shown in figure 2. The size of the setting unit is 0.05 m. The solution termination time is set to 0.05s and the time step is controlled to 0.67.

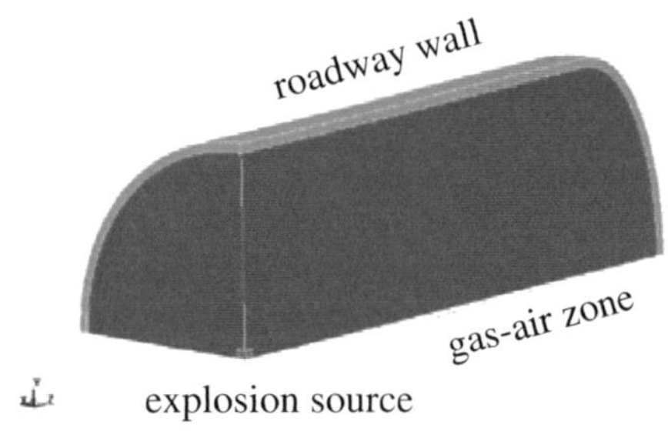


**Figure 1.** Physical model of roadway.

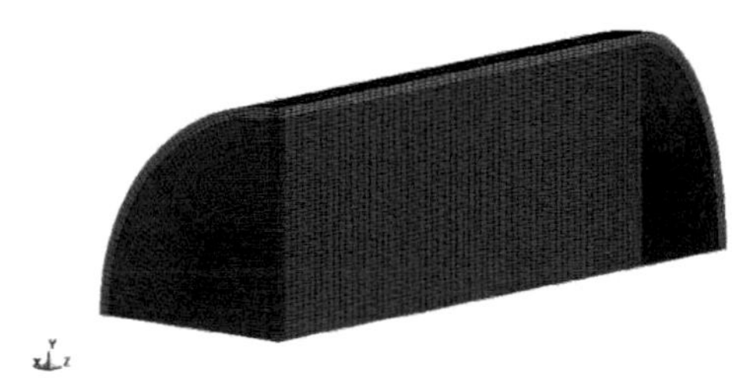

**Figure 2.** Finite element model of roadway.

### 2.3. Definition of Unit Type and Material Model

(1) Gas material model and state equation

High explosive material model (HIGH_EXPLOSIVE_Burn) is adopted for gas. Jones Wilkins Lee (JWL) state equation is used to describe the relationship between explosion pressure and volume. The JWL state equation is as follows

$$p = A\left(1-\frac{\omega}{A_1 V}\right)e^{-R_1 V} + B\left(1-\frac{\omega}{R_2 V}\right)e^{-R_2 V} + \frac{\omega E_0}{V} \tag{1}$$

where, $p$ is unit pressure, $V$ is relative volume, $E_0$ is initial internal energy density， Parameters A and B are material constants. $R_1$、$R_2$ is a dimensionless constant. Ω is Gruneisen constant, that is the change rate of pressure relative to internal energy under constant volume condition, and the parameters of JWL state equation are shown in table 1.

**Table 1.** JWL state equation parameters.

| Density (kg/m$^3$) | Velocity (m/s) | PCJ GPa | v0 | A GPa | B GPa | R1 | R2 | W | E0 GPa |
|---|---|---|---|---|---|---|---|---|---|
| 1.2×103 | 4.×103 | 5.15 | 1 | 293.9 | 21.73 | 6.366 | 2.152 | 0.2069 | 3.14 |

(2) Air material model and state equation

The *MAT_NULL is adopted in air material model, and is generally described by *EOS_LINEAR _ POLYNOMIAL state equation. A linear-polynomial state equation is as follows.

$$P = C_0 + C_1\mu + C_2\mu^2 + C_3\mu^3 + \left(C_4 + C_5\mu + C_6\mu^2\right)\rho_0 E \tag{2}$$

where, $P$ is the explosion pressure, $E$ is the internal energy per unit volume. $\mu = \rho / \rho_0 - 1$, $\rho$ is current density, $\rho_0$ is initial density, $\rho / \rho_0$ is the relative density, $E$ is the internal energy of the material, $C_0 \sim C_6$ is the parameter of the equation of state. The parameters of air material state equation are shown in table 2.

**Table 2.** Air material equation of state parameters.

| Material | ρ0 (kg/m$^3$) | C0 (Pa) | C1 | C2 | C3 | C4 | C5 | C6 | E (J/m$^3$) |
|---|---|---|---|---|---|---|---|---|---|
| Air | 1.29 | -1·105 | 0 | 0 | 0 | 0.4 | 0.4 | 0 | 2.5·105 |

(3) Material parameters of roadway wall

Under the action of shock wave of gas explosion, compression damage will occur on the wall near the explosion center, and tensile stress wave will be formed on the surface, resulting in the damage of tensile layer. Therefore, Holmquist-Johnson- cook (HJC) model is adopted. It is controlled by the degree of tensile damage. The J-C model in LS-DYNA is adopted for the roadway wall model, and the material parameters are shown in table 3, in which the relationship between stress and strain is as follows:

$$\sigma_S = \left[A + B\left(\varepsilon^p\right)^n\right]\left[1 + C\ln\frac{\dot{\varepsilon}^p}{\dot{\varepsilon}_1}\right] \tag{3}$$

where, $\sigma_S$ is yield strength, $\varepsilon^p$ is effective plastic strain, $\dot{\varepsilon}^p$ is strain rate, $\dot{\varepsilon}_1$ is the reference plastic strain rate, $A$, $B$, $n$ is constant. $C$, $p$ are Strain parameter of Cowper-Symonds. Therefore, the *mat_Rigid material model is adopted in the roadway wall material model, wall material parameters are shown in table 3.

**Table 3.** Parameters of wall material.

| *MAT_RIGID | | | | | | | |
|---|---|---|---|---|---|---|---|
| MID | RO | E | PR | N | COUPLE | M | ALIAS |
| 1 | 0.78E04 | 2.06E12 | 0.25 | 0.0 | 0.0 | 0.0 | 0.0 |
| CMO | CON1 | CON2 | | | | | |
| 1.0 | 7.0 | 7.0 | | | | | |
| LOC | A2 | A3 | V1 | V2 | V3 | | |
| 0.0 | 0.0 | 0.0 | 0.0 | 0.0 | 0.0 | | |

Journal of Physics: Conference Series **2329** (2022) 012022 doi:10.1088/1742-6596/2329/1/012022

## 3. Result Analysis and Discussion

### *3.1. Analysis on Explosion Impact Response of Roadway Wall*

Figure 3 shows the damage process of roadway wall structure under the action of gas explosion. After the gas explosion, the shock wave of gas explosion has experienced the of propagation process from the air to the roadway wall structure. In the process of stress propagation, the wall structure experienced the explosive damage process formed by stress impact, stress expansion, tensile damage and instability damage. The stress wave directly acts on the wall structure, which will produce tensile stress, resulting in huge compressive stress along the circumferential direction. The wall structure is still damaged under the continuous action of internal stress wave, which drives the wall structure in adjacent areas to move along the action direction of explosion load, resulting in the overall instability and damage of roadway wall structure.

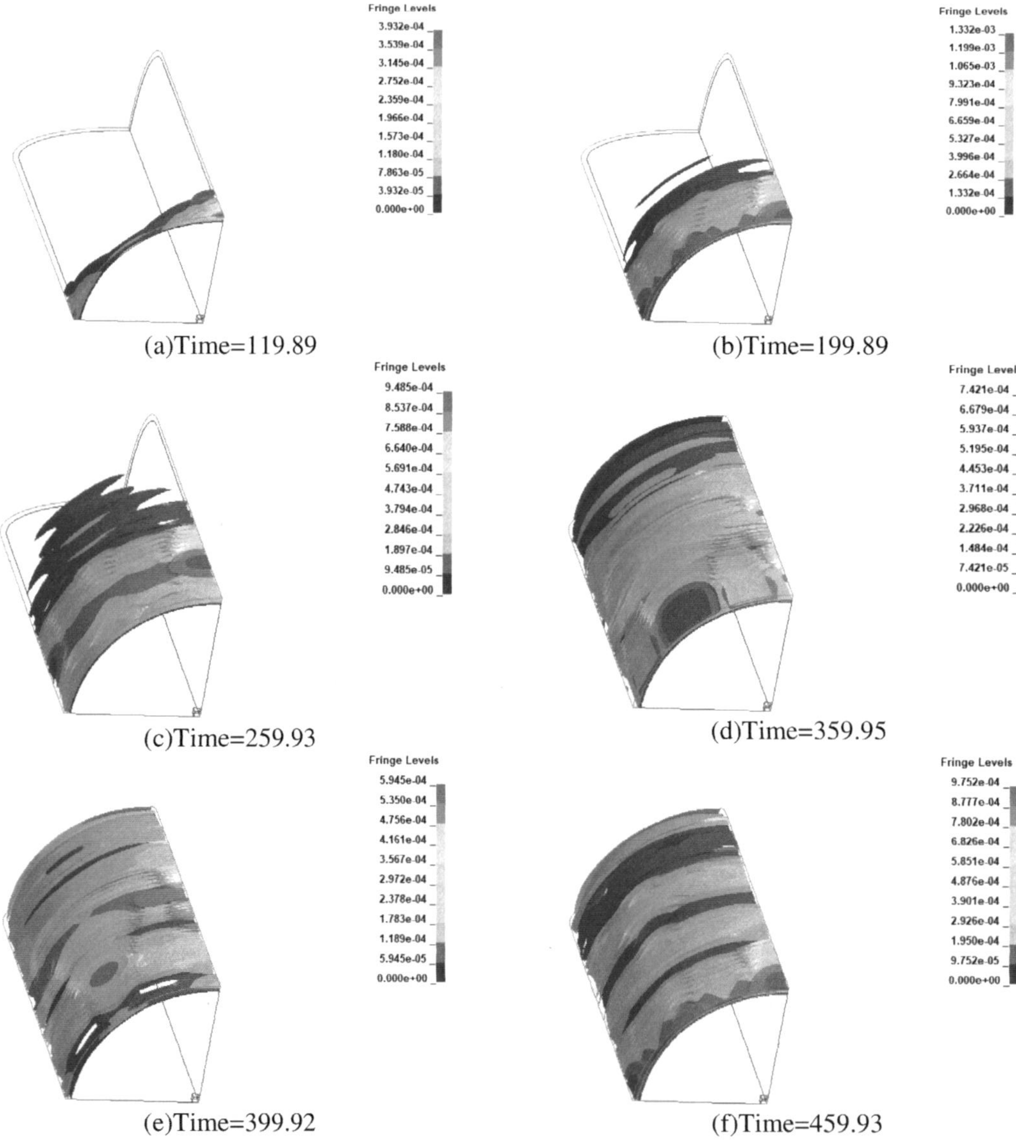


**Figure 3.** Damage process of roadway wall structure under gas explosion.

Under the action of gas explosion, the floor and the upper- lower corners of the roadway are damaged first and then the damage diffuses around. The upper and lower floor structure of the roadway is subjected to the tensile stress caused by the transverse explosion load and the outward expansion deformation of the side wall, resulting in the longitudinal stress along the roadway direction. Subsequently, the middle area of the side of the roadway structure is directly affected by the explosion impact load, resulting in the damage with different degrees. In the confined space of the roadway, the shock wave of gas explosion propagates along the radial direction of the roadway and cannot be released in a short time. The shock wave is reflected and superimposed in the roadway space, causing continuous damage to the wall structure. The impact produced by gas explosion is repeatedly reflected and superimposed on the wall, resulting in high peak overpressure and long duration of explosion load in the roadway.

### *3.2. Analysis of Explosion Stress Change of Roadway Wall*

Figure 4 is the variation process of wall stress. After the gas explosion, the wall stress is instantly loaded on the end wall, but the stress is not immediately transmitted to the whole wall. The transmission velocity in the wall is smaller than that of the shock wave. The stress of the wall is the result of the direct action of the explosion shock wave. The explosion stress wave is transmitted to the wall, resulting in high stress at the explosion facing surface. It causes the stress concentration at the explosion facing surface of the wall, and causes the strength damage of local materials, and then drives the wall in the adjacent area to move along the action direction of explosion load, finally causes cumulative damage. Then it is transmitted along the axial direction, and the whole wall will be affected by the stress wave. After that, the follow-up action of the stress wave causes the continuous damage of the wall, which makes the stress wave on the wall increase and decrease repeatedly until it decays gradually. Due to the repeated action of shock wave, the stress on the wall increases continuously, and greater stress damage is also produced at the position far away from the explosion source, causing greater damage to the wall. Due to the transmission of structural stress along the axial direction, the greater tensile stress is generated on the wall.

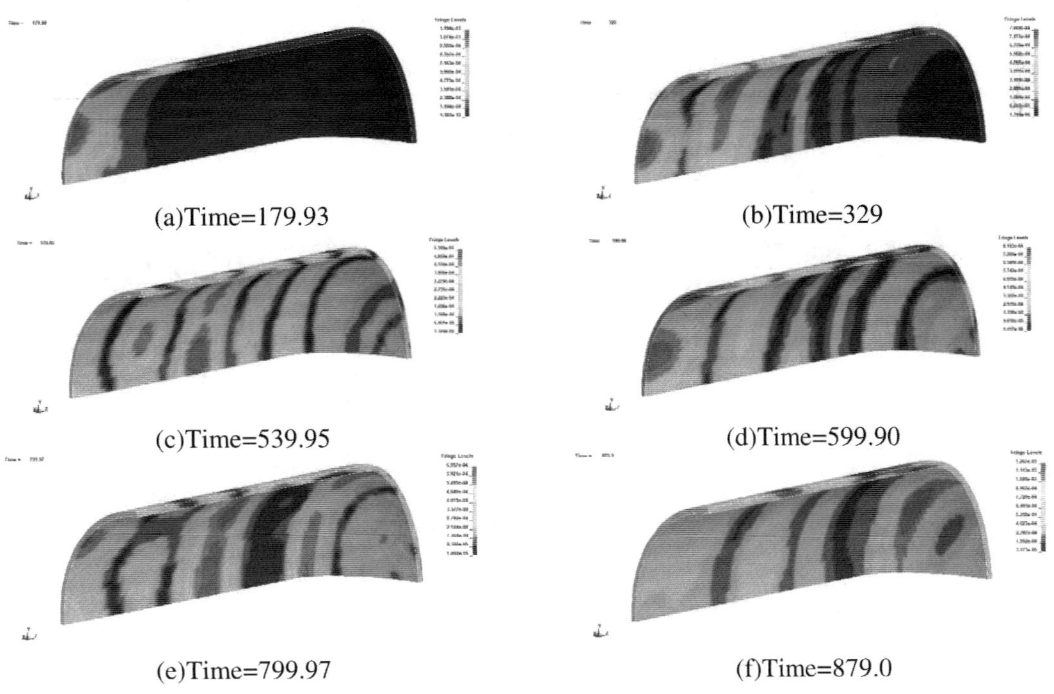

(a)Time=179.93 (b)Time=329

(c)Time=539.95 (d)Time=599.90

(e)Time=799.97 (f)Time=879.0

**Figure 4.** Diagram of wall stress change process.

Under the action of tensile stress, the wall structure will suffer direct or indirect tensile damage, the direct action is the deformation of the wall structure caused by the direct impact of the explosion wave; the indirect action is the damage caused by the displacement change of the internal structure caused by the stress of the wall structure.

## 4. Conclusions

In this paper, the gas explosion model of roadway is established by using the finite element software ANSYS / LS-DYNA, and the influence of gas explosion impact on the damage of roadway wall is analyzed. Among them, the explosion impact response of roadway wall and the change of explosion impact stress of roadway wall are analyzed, and the conclusions are obtained as follows.

1) From the initiation of gas explosion, the shock wave has experienced the propagation process in the air and the roadway wall structure. In the process of stress propagation, the wall structure experienced an explosion damage process formed by stress impact, stress expansion, tensile damage and instability damage.

2) The stress wave produces tensile stress in the wall structure, resulting in huge compressive stress along the circumferential direction of the wall. Under the continuous action of stress wave, the interior of the wall structure is damaged, which drives the wall structure in adjacent areas to move along the action direction of explosion load, resulting in the overall instability and damage of the roadway wall structure.

3) The impact produced by gas explosion is repeatedly reflected and superimposed on the wall, resulting in high peak overpressure and long duration of explosion load in the roadway.

4) The explosion stress wave is transmitted to the wall structure, resulting in high stress at the explosion facing structure, resulting in stress concentration on the explosion face of the wall and local material strength damage.

5) In the process of axial movement of stress, the greater tensile stress will be generated in the wall structure. Under the action of tensile stress, the wall structure will be subject to direct or indirect tensile damage. Therefore, the impact produced by the wall leads to the damage of the internal structure of the wall.

## Acknowledgements

This work is financially supported by the National Natural Science Foundation Project of China (52174177, 52174178) and project supported by Scientific Research Fund of Hunan Provincial Education Department (20B240). Those supports are greatly acknowledged.

## References

[1] Gao K, Li S N, Han R, Li R Z, Liu Z M and Qi Z P 2020 Liu Z Y Study on the propagation law of gas explosion in the space based on the goaf characteristic of coal mine *Safety Science* **127** 104693

[2] Wang C, Huang F L, Addai E K, et al. 2016 Effect of concentration and obstacles on flame velocity and overpressure of methane-air mixture *Journal of Loss Prevention in the Process Industries* **43** 302-310

[3] Liu D, Si R and Li R Z 2015 Ambient humidity influence on explosion characteristics of methane-air mixture *Chinese Journal of High Pressure Physics* **29**(4) 69 -74

[4] Zheng L G, Du DP Wang J, Dou Z G, Wang X, Jin H W and Wang Y 2019 Study on premixed flame dynamics of CH4/O2/CO2 mixtures *Fuel* **256** 115913

[5] Yang Z H, Ye Q, Jia Z Z and Li H 2020 Numerical Simulation of Pipeline-Pavement Damage Caused by Explosion of Leakage Gas in Buried PE Pipelines *Advances in Civil Engineering* **2020** 4913984

[6] Frazier C, Lamnaouer M, Divo E, Kassab A and Petersen E 2011 Effect of wall heat transfer on shock- tube test temperature at long times *Shock Waves* **21**(1) 1–17

[7] Ye Q, Jia Z Z, Wang H Z and Pi Y 2013 Characteristics and control technology of gas explosion in gob of coal mines *Disaster Advance* **6**(S3) 112-118

[8] Jiang B Y, Su M Q, Liu Z L Cai F, Yuan S J, Shi S L and Lin B Q 2016 Effects of changes in fuel volume on the explosion-proof distance and the multi parameter attenuation characteristics of methane-air explosions in a semi-confined pipe Journal of Loss Prevention in the *Process Industries* **39** 17-23

[9] Ye Q and Jia Z Z 2014 Effect of the bifurcating duct on the gas explosion propagation characteristics *Combustion, Explosion, and Shock Waves* **50**(4) 424-428

[10] Ye Q, Wang G G X, Jia Z Z and Zheng C S 2017 Experimental study on the influence of wall heat effect on gas explosion and its propagation *Applied Thermal Engineering* **118**(25) 392–397

[11] Ye Q, Jia Z Z and Zheng C S 2017 Study on hydraulic-controlled blasting technology for pressure relief and permeability improvement in a deep hole *Journal of Petroleum Science and Engineering* **159** 433–442

[12] Xu J D and Yang G Y 2004 Numerical simulation of the barricade encouraging effect in the process of gas explosion propagation *Journal of China Coal Society* (1) 53-56

[13] Deng J, Cheng F M, Luo Z M and Wang H 2008 Experimental study on methane explosion characteristics in turbulent state *China Safety Science Journal* **18**(8) 85-88

[14] Shao H, Wang C and Yu H K 2020 Effect of copper foam on explosion suppression at different positions in the pipe. Powder Technology 360: 695–703

[15] Gao K, Li S N, Liu Y J, Jia J Z and Wang X Q 2021 Effect of flexible obstacles on gas explosion characteristic in underground coal mine *Process Safety and Environmental Protection* **149** 362- 369

[16] Xiong X and Zhou L 2011 Approximation regression method in predicting blasting induced praticle velocity by constructing a vibration velocity function *Soil Engineering and Foundation* **25**(5) 62-65

# Analysis and Calculation of Hydrodynamic Pressure on Asphalt Pavement

**Xueying Zhao[1]* and Baofu Ma[2]**

[1] School of Civil Engineering, Qingdao University of Technology, 777 Jialingjiang Road, Huangdao District, Qingdao, 266520, China;
[2] China Communications Construction Company Highway Consultants Co., Ltd., 33 Dongsiqianchaomian Hutong, Dongcheng District, Beijing, China.
Email: zhaoxueying@qut.edu.cn

**Abstract.** Asphalt pavement has been used worldwide. However, exposed to the natural environment and under the application of repeated vehicle loads, a series of typical distresses develops especially water damage. In this paper, the hydrodynamic pressure of asphalt pavement was calculated based on momentum conservation laws and verified by the hydrodynamic pressure measured in the case study engineering Xinhe-Xinzhuangzi section of Rongwu highway in Shandong Province through univariate linear regression. It was found that calculation results of equations deduced in this paper and hydrodynamic pressure measured in the case study engineering follow univariate linear regression. Therefore, hydrodynamic pressure on asphalt pavement can be calculated through equations in this paper.
**Keywords.** Asphalt pavement, Hydrodynamic Pressure, Momentum conservation law, Calculation and verification

## 1. Introduction

Asphalt pavement has been widely used due to its characteristics of comfortable driving, simple construction, and fast maintenance, etc. The mileage of highway in China is up to 113, 000 km by the end of 2020 and 90% of them used asphalt pavement, according to statistics published by the Ministry of Transport of the People's Republic of China [1]. Exposed to the natural environment, asphalt pavement continuously withstands the influence of environmental factors and repeated vehicle loads, including temperature, humidity, and rainfall, etc. Therefore, a series of typical distresses such as looseness, potholes, cracking, and rutting may appear one after another in the service life and water damage is one of the most serious distresses.

At present, scholars have carried out a lot of research on the influence of water on asphalt pavement. In Chinese Standard Test Methods of Bitumen and Bituminous Mixtures for Highway Engineering [2], immersed and freeze-thaw Marshall tests are recommended to evaluate the ability of water damage resistance. However, the working conditions simulated by these two tests are quite different from that in reality, in which moving traffic loads and hydrodynamic pressure are applied on the asphalt pavement simultaneously, rather than applying water pressure first and then applying traffic loads. On the one hand, for the immersed Marshall test, the process of immersion is static, which is quite different from the process of hydrodynamic scouring. Water inside the pavement structure would be squeezed and suctioned repeatedly by the movement of wheel loads, which causes the spalling of asphalt from aggregate surface. Water pressure on asphalt mixture is not hydrostatic, but hydrodynamic. On the other hand, for the freeze-thaw Marshall test, which is considered to be a more rigorous evaluation of water sensitivity, it mainly focuses on the water damage process caused by ice expansion and melting. However, water damage happens all year, instead of only cold season. Hence it is also insufficient to simulate the process of the application of traffic loads and hydrodynamic pressure.

Therefore, to help prevent the distress caused by hydrodynamic pressure and guide pavement design, it is of vital importance to study the hydrodynamic pressure under the application of traffic loads. In this paper, the hydrodynamic pressure of asphalt pavement was calculated based on momentum conservation laws and the hydrodynamic pressure of the case study engineering Xinhe-Xinzhuangzi section of Rongwu highway in Shandong Province was analyzed and compared with the calculation results the justify its accuracy.

## 2. Calculation of Hydrodynamic Pressure on Asphalt Pavement

The hydrodynamic pressure on the road surface is closely related to the water film thickness, vehicle axle loads and driving speed based on Reference [3]. According to relevant research, when the water film thickness is small (< 3 mm), the water film between the wheel and the road surface is wedge-shaped; when the water film thickness is large (> 3 mm), the contact area between the water film and the tire is large. Even if the average pressure on the tire caused by water is less than the tire inflation pressure, the tire can still be lifted, which is referred to as water drift [4]. Since there will be a large gap of the hydrodynamic pressure between different water film thickness [5], in this paper, it assumes that the thickness of water film on the road surface is less than 3 mm.

The stress diagram of the tire driving on the watery road is shown in figure 1. The wheel is subjected to the upward force $\boldsymbol{W_K}$ and backward resistance $\boldsymbol{W_Z}$, which were caused by the front water film during traveling. Also, the wheel is subjected to the supporting force $\boldsymbol{N}$ of the pavement and its own gravity $\boldsymbol{G}$. It is assumed that water on the road surface is incompressible and remains in an ideal state during driving. At the same time, due to the existence of water, the resistance of the tire is ignored.

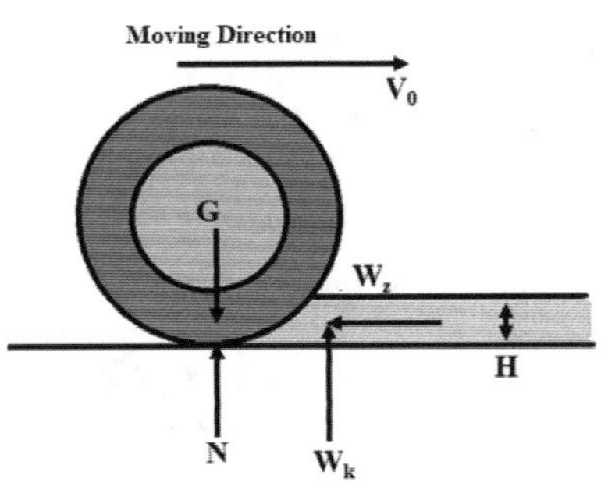


**Figure 1.** The stress diagram of the tire.

It is assumed that the variation of water film velocity follows the linear variation law within the tire grounding range [6], and hence water film velocity $\boldsymbol{V_x}$ is given by equation (1).

$$V_x = V_0 - (V_0/L)x \ \ (x \in [0, L]) \tag{1}$$

where $\boldsymbol{V_0}$ is the vehicle velocity, and $\boldsymbol{L}$ is the length of the tire grounding range.

According to Newton's third law, Force $\boldsymbol{F_x}$ of tire on water film in horizontal direction equals Force $\boldsymbol{W_Z}$ of water film resistance to tire backward, and hence based on momentum theorem there would be equation (2).

$$F_x \cdot t = W_z \cdot t = m(V_0 - V_x) \tag{2}$$

where $\boldsymbol{m}$ is the mass of water film, and $\boldsymbol{t}$ is impact time between water film and tire.

When equation (1) was substituted into equation (2), there would be equation (3).

$$W_z t = \rho H B x V_0^2 L^{-1} \tag{3}$$

where $\boldsymbol{H}$ is the thickness of water film, $\boldsymbol{B}$ is the wheel width and $\boldsymbol{\rho}$ is water density; in this study, $\boldsymbol{\rho} = 1.0\,\mathrm{g/cm^3}$.

It is assumed that the water flow range on the road surface is($\boldsymbol{x} \in [0, L]$), and it is integrated. Hence, the horizontal resistance force of water film to tire $\boldsymbol{W_z}$ is given by equation (4).

$$W_z = \int_0^L \left(\frac{\rho H B V_0^2 x^2}{L^2}\right) \mathrm{d}x = \frac{\rho H B L V_0^2}{3} \tag{4}$$

Besides, the grounding area $\boldsymbol{S}$ between road surface and tire is given by equation (5).

$$S = \frac{G}{P} \tag{5}$$

where $\boldsymbol{S}$ is the grounding area, $\boldsymbol{G}$ is tire gravity and $\boldsymbol{P}$ is tire pressure.

The length of the tire grounding range is given by equation (6).

$$L = \frac{S}{B} = \frac{G}{PB} \tag{6}$$

When substituting equation (6) into equation (4), the horizontal resistance force of water film to tire $\boldsymbol{W_z}$ is given by equation (7).

$$W_z = \frac{\rho G H V_0^2}{3P} \tag{7}$$

If the wedge angle was set as $\boldsymbol{\beta}$, upward force $\boldsymbol{W_K}$ is given by equation (8).

$$W_k = W_z \cot\beta = W_z \cdot \frac{L}{H} \tag{8}$$

Therefore, the hydrodynamic pressure of the water film on the tire is given by equation (9).

$$P_{\text{hydrodynamic pressure}} = \frac{W}{S} = \frac{\left(W_z^2 + W_k^2\right)^{1/2}}{S} = \frac{\rho V_0^2}{3\left(H^2 + L^2\right)^{1/2}} \tag{9}$$

Based on equation (9), hydrodynamic pressure is affected by driving speed, water film thickness and tire grounding length.

According to Reference [7], the pore on the surface of newly built asphalt pavement usually does not exceed 10mm, and most of them are 1-3mm. Therefore, the thickness of surface water film is assumed to be 2mm in this section. According to equation (9), the hydrodynamic pressure of asphalt pavement under different driving speeds and standard axle load can be calculated, as shown in figure 2.

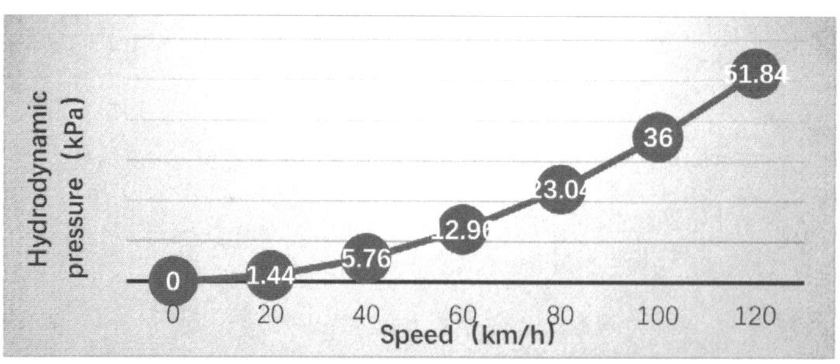


**Figure 2.** Calculation results of the hydrodynamic pressure.

## 3. Analysis and Verification of Calculation Results

In order to explore the accuracy of the calculation results of hydrodynamic pressure by equation (9) in Section 2, the theoretical calculation results are verified by the field measurement results. The hydrodynamic pressure of Xinhe Xinzhuangzi section of Rongwu Expressway in Shandong Province in Reference [8] was referred to.

The test vehicle has three axles and six wheels. The axle load of the front axle is about 50kN and that of the rear axle is about 100kN. The measured hydrodynamic pressure on road surface under different vehicle speeds is shown in table 1.

**Table 1.** The measured hydrodynamic pressure under different vehicle speeds.

| Vehicle speed (km/h) | 60 | 80 | 100 | 120 |
|---|---|---|---|---|
| Hydrodynamic pressure (kPa) | 11.84 | 24.77 | 32.04 | 47.93 |

To test whether the theoretical calculation results of hydrodynamic pressure $y_i'$ and measured hydrodynamic pressure $y_i$ follow univariate linear regression, it can be assumed as equation (10).

$$y_i' = \alpha + \beta \cdot y_i + \varepsilon \tag{10}$$

where $\alpha$ and $\beta$ are coefficients of univariate linear regression and $\varepsilon$ is subject to normal distribution $N(0,\sigma^2)$.

In this way, it can be turned into testing hypothesis $\mathbf{H_0}$: whether $\beta$=0 is established. If the assumption holds, it is considered that the linear regression is not significant; otherwise, linear regression is considered significant.

At a given significance level of 5%, it could be given by equation (11).

$$\mathrm{T} = \frac{\hat{\beta}\sqrt{\sum_{i=1}^{n}(y_i - \overline{y_i})^2}}{\sigma^*} \tag{11}$$

where $\sigma^*$ is sample standard deviation, $\sigma^* = \sqrt{\sigma^{*2}}$, $\sigma^{*2} = \frac{n}{n-2}\sigma^2$ and

$$\sigma^2 = \left(\frac{1}{n}\sum_{i=1}^{n} y_i'^2 - \overline{y_i'}^2\right) - \hat{\beta}^2\left(\frac{1}{n}\sum_{i=1}^{n} y_i^2 - \overline{y_i}^2\right);$$

$\hat{\beta}$ is the estimated value of $\beta$, $\hat{\beta} = \frac{\overline{y_i y_i'} - \overline{y_i}\cdot\overline{y_i'}}{\overline{y_i^2} - \overline{y_i}^2}$,

$\overline{y_i} = \frac{1}{n}\sum_{i=1}^{n} y_i$, $\overline{y_i'} = \frac{1}{n}\sum_{i=1}^{n} y_i'$, $\overline{y_i^2} = \frac{1}{n}\sum_{i=1}^{n} y_i^2$ and $\overline{y_i \cdot y_i'} = \frac{1}{n}\sum_{i=1}^{n} y_i \cdot y_i'$.

If the equation (11) is satisfied, $\mathbf{H_0}$ is rejected and the linear regression is considered significant; otherwise, it is considered that the linear regression is not significant.

$$|T| \geq t_{\frac{\alpha}{2}}(n-2) \tag{12}$$

Hence, verification results were shown in table 2.

**Table 2.** Verification results of calculated and measured hydrodynamic pressure.

| Parameter | $\hat{\beta}$ | $\sigma^*$ | $\|T\|$ | $n$ | $t_{\frac{\alpha}{2}}(n-2)$ | $H_0$ | Linear regression |
|---|---|---|---|---|---|---|---|
| Value | 1.09 | 2.8887 | 20.5969 | 6 | 2.1318 | Rejected | Significant |

According to the calculation results in table 2, it can be considered that the theoretical calculation of hydrodynamic pressure in Section 2 is in line with the reality. Therefore, the calculation results are correct, and equation (9) can be used to calculate hydrodynamic pressure.

## 4. Conclusion

In this paper, the hydrodynamic pressure of asphalt pavement was calculated based on momentum conservation laws and verified by the hydrodynamic pressure measured in the case study engineering Xinhe-Xinzhuangzi section of Rongwu highway in Shandong Province through univariate linear regression. It was found that calculation results of equation (9) deduced in Section 2 and hydrodynamic pressure measured in the case study engineering follow univariate linear regression. Therefore, equation (9) can be used to calculate the hydrodynamic pressure on asphalt pavement.

## Reference

[1] Ministry of Transport of the People's Republic of China 2021.5.19 Statistical bulletin on the development of transportation industry in 2020 http://www.gov.cn/xinwen/2021-05/19/content_5608523.htm

[2] Ministry of Transport of the People's Republic of China 2011 *Standard Test Methods of Bitumen and Bituminous Mixtures for Highway Engineering* (Beijing: China Communications Press)

[3] Teng X Q, Wang H F and Wen H 2016 Calculation and influence factors analysis for hydrodynamic pressure of asphalt pavement with large slope based on momentum theorem, *Chinese Journal of Computational Mechanics* **33** 257-262

[4] Li Q 2012 *Calculation and Research of Hydrodynamic Pressure of Road Surface* (Chongqing: Chongqing Jiaotong University)

[5] Wang W T 2020 *Evaluation on the Influence of Dynamic Pore Water Pressure on the Damage of Asphalt Pavement Considering Multi-Factor Coupling Effects* (Beijing: University of Science and Technology Beijing)

[6] Li J Q and Ma S Z 1983 *Fluid Mechanics Fundamentals* (Beijing: Science Press)

[7] Zhang Q 2014 *Microscopic Analysis of Asphalt Pavement Water Damage Based on Computational Fluid Dynamics* (Jilin University Changchun)

[8] Tang W Z, Ou J Q, Cui X Z, Lou J J, Xiao M, Zhang J, Huang D and Hou F 2015 Field test and research on vehicle load induced dynamic pore pressure in asphalt pavement *Journal of Shandong University (Engineering Science)* **45** 84-90

# Optimization of Brayton Cycle $CO_2$-Based Mixed Working Medium for Marine Low-Speed Diesel Engine Flue Gas Waste Heat Recovery

**Liangtao Xie[1] and Jianguo Yang[1,2,3*]**
[1] School of Naval Architecture, Ocean and Energy Power Engineering, Wuhan University of Technology, Wuhan 430063, China;
[2] Key Laboratory of Marine Power Engineering Technology Transportation Industry, Wuhan 430063, China;
[3] National Engineering Laboratory for Marine and Ocean Engineering Power System, Electronic Control Sub-laboratory for Low-speed Engine, Wuhan 430063, China.
Email: 969870955@qq.com

**Abstract.** In this paper, the S-$CO_2$ recompression Brayton cycle system of Marine Low-speed diesel engine flue gas waste heat recovery is built. By analyzing the physical characteristics of working medium and the law of the influence on the efficiency, five kinds of $CO_2$-based mixed working medium are selected to optimize the $CO_2$-based mixed working medium for Marine Low-speed diesel engine flue gas waste heat recovery from the aspects of total recovery energy, net recovery work and efficiency. The results show that the net recovery work of $CO_2$-based mixed working medium supplemented with 20.11% propane increases by 17.02kW compared with that of pure $CO_2$.The circulating efficiency of $CO_2$ based mixed medium with 32.03% propane is increased by 5.42%.
**Keywords.** Marine Low-speed diesel engine; Waste heat of flue gas; Recompression Brayton cycle; Working medium optimization.

## 1. Introduction

With the development of the global shipping industry and the increasingly stringent regulations of the International Maritime Organization (IMO), there is a general consensus for efficient operation and energy saving [1]. As one of the important ways of energy conservation and emission reduction, Waste Heat Recovery (WHR) of marine diesel engine can effectively improve the overall efficiency of engine and reduce carbon emission. Although the thermal efficiency of marine low-speed two-stroke diesel engine can reach 50%, most of the heat is still carried away by the flue gas and cooling medium, resulting in serious energy loss [2].

The temperature of marine low-speed engine flue gas can reach over 400°C, which is a high-grade heat source with recovery potential. Currently, the main methods of recovering flue gas waste heat from marine diesel engine [3] are Power Turbine (PT), Thermoelectric Generator (TEG), Organic Rankine Cycle (ORC) [4], and Kalina Cycle (KC). However, these methods are not suitable for waste heat recovery of marine high power and low speed engines due to the efficiency, heat source grade and power density. Supercritical Carbon Dioxide Brayton Cycle (SCBC) has the characteristics of high efficiency, high power density and small volume. It is suitable for high grade heat source and has a wide power application range. It can be used for flue gas waste heat recovery in marine low-speed engines.

Working medium screening is one of the important contents of SCBC. Some scholars have carried out SCBC working medium screening of internal combustion engine waste heat. Guo [5] used Brayton cycle in solar thermal power generation, and added inert gas xenon and krypton to the working medium $CO_2$, which increased the cycle efficiency by 1.44% and 3.04%. SANDIA Laboratory [6] tried to add a certain proportion of $SF_6$ and Ne to the $CO_2$ of pure compressed Brayton cycle, and measured the change of working medium density with temperature. The results showed that the critical point of mixed working medium changed. Few studies have been carried in the optimization of SCBC working medium screening for marine low speed diesel engine.

In this paper, the S-$CO_2$ recompression Brayton cycle (SCRBC) was constructed for the flue gas waste heat recovery of a marine low-speed engine. Five kinds of $CO_2$-based mixed working medium were selected to evaluate the $CO_2$-based mixed working medium of marine low-speed diesel engine flue gas waste heat recovery from the aspects of the flue gas energy, net recovery work and efficiency, and S-$CO_2$ Brayton cycle working medium was selected.

## 2. Brayton Cycle Layout and Model

The SCBC system of marine low-speed engine flue gas waste heat is shown in figure 1. Due to the limitation of the space, simple arrangement should be selected for SCBC and the device for waste heat recovery should be minimized in size and weight. Recuperative cycle is an effective method to improve the efficiency of the Brayton cycle, but the efficiency increase is limited. Intercooling, multistage heat recovery and reheat are combined on this basis [7]. The recompression cycle improved the regenerator efficiency and the system efficiency significantly due to it could reduce the heat released from the working medium to the cold source due to the shunt, and solve the regenerator pinch problem. Taking overall efficiency and cabin space constraints into consideration, the SCRBC layout was selected.

### *2.1. Flue Gas Waste Heat Recovery Layout*

Figure 1 shows the SCRBC system and its working process. The working medium absorbs heat from the flue gas in the flue gas heat exchanger of the engine, and becomes a supercritical working medium with high temperature and pressure through the process of isobaric heat absorption (1-2). After the turbine, the working medium performs isentropic expansion (2-3), and then enters HTR (High Temperature Regenerator) LTR (Low Temperature Regenerator) for heat recovery. The working medium after heating and compression increases the circulating efficiency (3-4, 4-5); After LTR, the working medium is shunt compressed, and isentropic compression (6-7, 5B-8B) is completed in the main compressor and the Re-compressor respectively. There is cooler in front of the main compressor to complete the intercooling process (5A-6) to reduce the power consumption of the compressor. The compressed working medium completes the heat recovery and confluence process (7-8A, 8-1) in HTR and LTR, and then enters the next cycle.

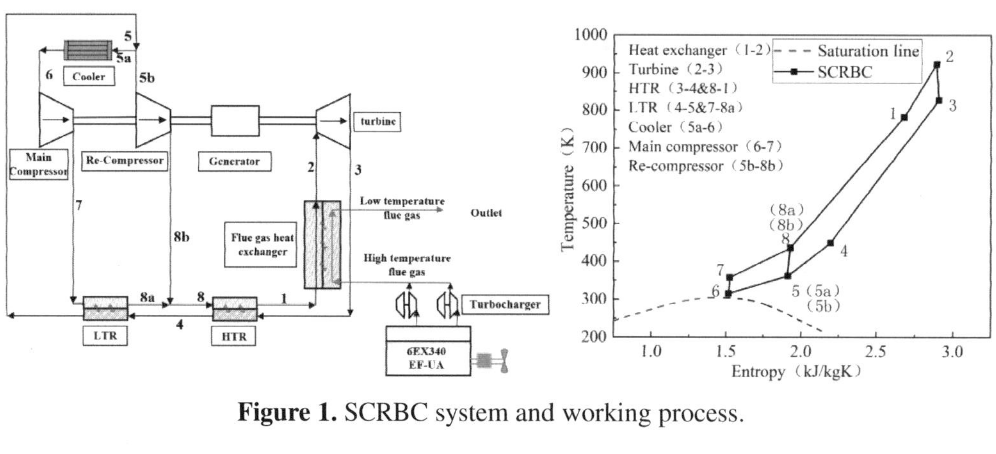


**Figure 1.** SCRBC system and working process.

6EX340EF-UA marine low-speed diesel engine (hereinafter referred to as low-speed engine) is chosen, its main parameters are shown in table 1.

**Table 1.** Main engine parameters.

| Parameter | Value | Parameter | Value |
|---|---|---|---|
| Number of cylinders | 6 | Fire order | 1-6-2-4-3-5 |
| Fuel | Diesel | Rated speed/r·min$^{-1}$ | 157 |
| Bore/mm | 340 | Rated power/kW | 4896 |
| Stroke/mm | 1600 | Fuel injection | High pressure common rail (HPCR) |
| Compression ratio | 20.5 | Form | turbocharging |

The thermophysical property parameters of flue gas measured on the low-speed engine bench are shown in table 2. The components of flue gas in different working conditions are different, and the corresponding thermophysical property of flue gas also changes, which directly affect the ability of flue gas waste heat recovery. Table 3 shows the measured flue gas composition and enthalpy entropy value of low speed engine. In this paper, the measured flue gas composition parameters are defined in the flue gas heat exchanger of the engine. The measured flue gas thermophysical parameters are used as the input of S-$CO_2$ Brayton cycle to analyze the recovery potential and influencing factors of flue gas waste heat recovery.

**Table 2.** Thermophysical properties of flue gas of low-speed engine.

| Load (%) | Speed (r/min) | Mass flow (kg/s) | Pressure (MPa) | Temperature (°C) | Waste heat (kJ) |
|---|---|---|---|---|---|
| 25 | 157 | 3.14 | 0.04 | 246 | 711.5 |
| 50 | 157 | 5.90 | 0.13 | 273 | 1512.2 |
| 75 | 157 | 8.58 | 0.24 | 295 | 2312.3 |
| 100 | 157 | 10.60 | 0.32 | 343 | 3532.9 |

**Table 3.** Composition of the low-speed engine flue gas.

| Load (%) | HC (ppm) | $NO_X$ (ppm) | $CO_2$ (%) | CO (ppm) | $O_2$ (%) | Enthalpy (kJ/kg) | Entropy (kJ/kgK) |
|---|---|---|---|---|---|---|---|
| 25 | 811.7 | 700.7 | 4.0 | 166.6 | 15.1 | 243.03 | 4.88 |
| 50 | 672.1 | 727.4 | 4.4 | 70.8 | 14.8 | 271.62 | 4.73 |
| 75 | 444.7 | 782.3 | 4.6 | 66.1 | 14.8 | 295.13 | 4.67 |
| 100 | 333.8 | 812.7 | 5.0 | 48.9 | 14.3 | 347.13 | 4.71 |

*2.2. Thermodynamic Model*

The SCBC model is established based on Ebsilon platform to simulate the process and calculate the related parameters. The calculation assumptions are as follows:

(1) The system is in a stable state, and the change of kinetic energy is ignored;

(2) Isentropic efficiency is adopted for turbines and compressors;

(3) Except cooler, heat transfer loss of the whole cycle and environment is ignored;

(4) Considering the effectiveness of HTR and LTR;

(5) Considering the end difference at the cooler, regenerator and heat exchanger;

(6) The parameters of each state point do not change with time. The thermodynamic model of each component of the cycle is shown in table 4.

**Table 4.** Thermodynamic model of the components.

| Components | Formula |
|---|---|
| Turbine | $W_T = h_2 - h_3 \quad \eta_{T,S} = \dfrac{h_2 - h_3}{h_2 - h_{3,S}}$ |
| Main compressor | $W_{MC} = (1-r)(h_7 - h_6) \quad \eta_{MC,S} = \dfrac{h_{7,S} - h_6}{h_7 - h_6}$ |
| Re-compressor | $W_{RC} = r(h_{8b} - h_{5b}) \quad \eta_{RC,S} = \dfrac{h_{8b,S} - h_{5b}}{h_{8b} - h_{5b}}$ |
| Heat exchanger | $q_{Heater} = h_2 - h_1$ |
| HTR | $h_3 - h_4 = h_1 - \left[ rh_{8b} + (1-r)h_{8a} \right]$ |
| LTR | $h_4 - h_5 = h_{8a} - h_7$ |
| Cooler | $q_{Cooler} = \Delta h_{Cooler} = (1-r)(h_6 - h_{5a})$ |
| Overall | $\eta_{Total} = \dfrac{(1-r)(\Delta h_T - \Delta h_C) + r(\Delta h_T - \Delta h_{AC})}{(1-r)(\Delta h_T - \Delta h_C + \Delta h_{Cooler}) + r(\Delta h_T - \Delta h_{AC})}$ |

In the table, $h_i$ is the enthalpy of each measurement point, $r$ is the Shunt coefficient, $\Delta h_C$ is the enthalpy difference between the outlet and the inlet of the main compressor, $\Delta h_{AC}$ is the enthalpy difference between the outlet and the inlet of the re-compressor, $\Delta h_{Heater}$ is the enthalpy difference between the outlet and the inlet of the heat exchanger, $\Delta h_T$ is the enthalpy difference between the inlet and the outlet of the turbine.

### *2.3. Model Validation*

The component parameters in the SCRBC were set and modelled [8]. Figure 2 shows the S-$CO_2$ recompression Brayton cycle built on EBsilon platform. Table 5 compares the simulation with the test of the recompression cycle equipment.

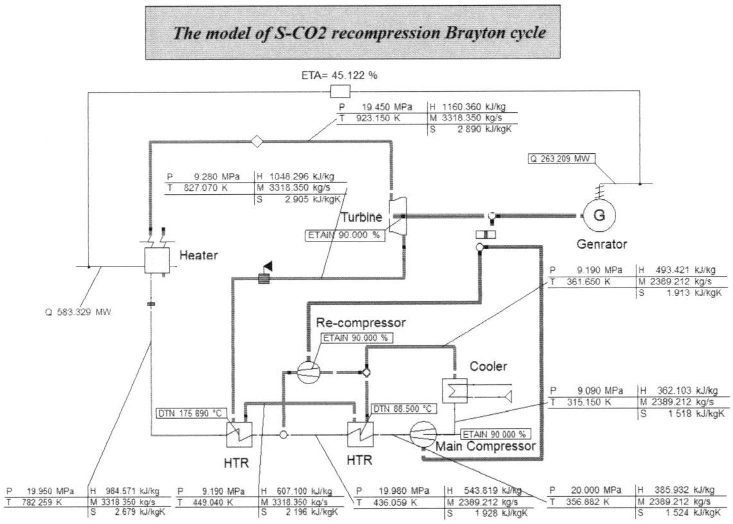


**Figure 2.** The model of S-$CO_2$ recompression Brayton cycle.

Table 5 shows a comparison between the simulated and the experimental values of the recompression cycle equipment. It can be seen from the table that the simulated value calculated by the model is close to the test value in Sandia laboratory, and the deviation is within 5%. Because the isentropic efficiency of the model turbine is set slightly lower, the turbine power is low, and the efficiency is 5.21% lower than the experimental value. It is considered that the established model is accurate and can be used for subsequent calculation.

**Table 5.** Comparison of simulated and experimental values of recompression equipment.

| Power/MW | Sandia | Simulation | Deviation/% |
|---|---|---|---|
| Cooler | 315.5 | 313.75 | 0.55 |
| Main compressor | 56.30 | 57.51 | 2.2 |
| Re-compressor | 45.89 | 45.81 | 0.17 |
| LTR | 375.00 | 377.23 | 0.59 |
| HTR | 1452.40 | 1466.11 | 0.94 |
| Heat exchanger | 600.00 | 583.33 | 2.78 |
| Turbine | 388.00 | 371.12 | 4.35 |
| Efficiency (%) | 47.60 | 45.12 | 5.21 |

## 3. Brayton Cycle Working Medium Optimization

### *3.1. Influence of Brayton Cycle Working Medium on SCRBC*

Supercritical carbon dioxide has a higher density. The carbon dioxide fluid in supercritical state has the properties of both gas and liquid. Its diffusion coefficient is much greater than that of common liquids. These characteristics make supercritical carbon dioxide superior to the unique heat transfer and flow characteristics of ordinary gases. The change of temperature and pressure in the supercritical zone will cause the drastic change of carbon dioxide physical properties. The change of temperature and pressure in the supercritical zone will only cause a slight change in physical properties. Such complex physical changes make the thermodynamic properties of carbon dioxide very complicated.

Figure 3 shows the diagram of temperature and enthalpy of $CO_2$. $CO_2$ is divided into four regions by critical pressure isobaric line, saturated liquid line and saturated steam line: supercritical zone, liquid zone, gas-liquid mixed zone and superheated steam zone. SCBC takes advantage of the low critical parameters of $CO_2$ [9] to make $CO_2$ in the supercritical zone, which reduces the power consumption of the compressor and thus improves the efficiency.

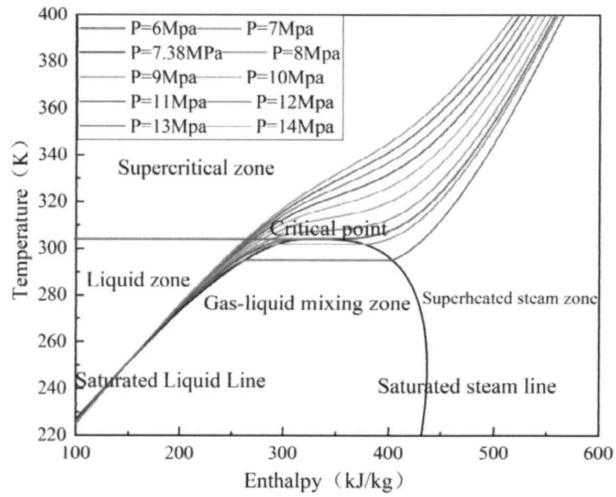


**Figure 3.** The temperature and enthalpy of $CO_2$.

The main compressor inlet temperature is the key factor affecting SCRBC performance. Figure 4 shows the influence of main compressor inlet pressure and inlet temperature on the efficiency, the influence of the figure shows that working medium at the entrance to the main compressor temperature near the critical temperature as far as possible, however, under certain conditions by environmental impact, such as temperature, power and cooling water temperature cooler, will lead to higher main compressor inlet temperature, away from the $CO_2$ critical point, The cycle will lose the advantage of low power consumption near the critical point, and the thermal efficiency of the cycle will decrease rapidly. In this paper, by adding other components in $CO_2$ to adjust the critical temperature of the working fluid, the compressor inlet temperature changes around the critical temperature of the working fluid, so as to improve the circulation efficiency.

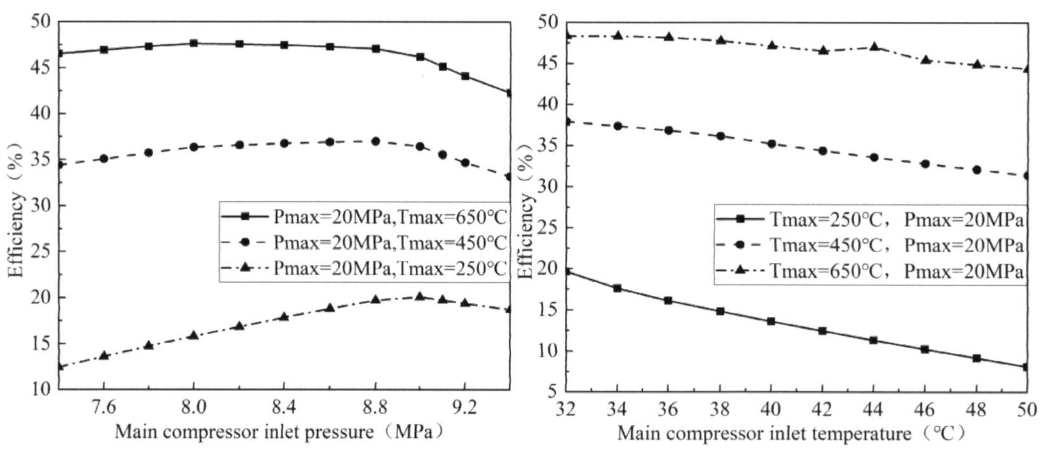


**Figure 4.** Influence of inlet pressure and inlet temperature of main compressor on efficiency.

Taking SCRBC as the research object, the above thermodynamic performance analysis shows that the critical parameters are the key parameters affecting the cycle performance. The critical pressure determines the maximum expansion ratio of the cycle and the critical temperature determines the minimum inlet temperature of the cycle compressor. Increasing the turbine expansion ratio and reducing the compressor inlet temperature can improve the cycle efficiency, that is, reducing the $CO_2$ critical point parameters can further improve the cycle efficiency.

Therefore, $CO_2$ critical parameters can be adjusted by mixing $CO_2$ with other gases [9]. The selection of added gases should be considered as follows: gas physical properties are known and available; Safe and non-toxic, no harm to the environment, no spontaneous combustion; After mixing with $CO_2$, no reaction occurs at any temperature or pressure; After mixing with $CO_2$, the critical temperature of the mixed gas should not be lower than the surface temperature, so as to facilitate the selection of appropriate cooling methods.

### *3.2. Working Medium Screening Requirements for Waste Heat Breton Cycle of Flue Gas of Low Speed Engine*

For circulating working medium, in addition to the limitation of circulating operation parameters, thermodynamic stability, appropriate critical pressure, corrosiveness and environmental protection should be taken into consideration when selecting $CO_2$-based mixed working medium [10].

Since the physical property parameters of supercritical fluid vary dramatically near the critical point, accurate physical property data are needed to simulate the recompression Brayton cycle of $CO_2$-based mixed working medium, which is also an important prerequisite to ensure the accuracy of this study. The physical properties in the calculation process in this paper were obtained from the ref-PROP9.1 software of NIST (National Institute of Standards and Technology). For pure working medium and mixed working medium, the software uses the improved Helmholtz free energy equation and the improved Helmholtz mixed rule to calculate. Table 6 shows the physical parameters and environmental data of various working medium.

**Table 6.** The physical parameters and environmental data of various working medium.

| Working medium | Molar mass(kg/kmol) | Critical temperature(K) | Critical pressure(MPa) | ODP | GWP |
|---|---|---|---|---|---|
| $CO_2$ | 44.01 | 304.1 | 7.38 | 0 | 1 |
| Propane | 44.10 | 370.0 | 4.25 | 0 | 0-20 |
| Butane | 58.12 | 425.1 | 3.79 | 0 | 0-20 |
| Isobutane | 58.12 | 408.2 | 3.65 | 0 | 0-20 |
| Hydrogen sulphide | 34.08 | 373.2 | 9.01 | - | - |
| Argon | 39.94 | 150.8 | 4.86 | - | - |

Note: ODP (ozone depletion potential); GWP (Global Warming Potential)

### *3.3. Fluid Physical Properties Model of $CO_2$-Based Mixed Working Medium*

Figure 5 shows the thermal properties of various working medium changing with temperature, from which $CO_2$, low carbon number alkanes (propane, butane, isobutane), inert gas argon and hydrogen sulfide are selected as additives of $CO_2$-based mixed working medium. The thermal physical parameters (thermal conductivity and specific heat capacity) of all the selected working medium on the isobars above the critical pressure have abrupt changes [11], and with the increase of pressure, the temperature of the abrupt change point moves to the right, but the abrupt change amplitude decreases. This characteristic makes the supercritical working medium absorb more waste heat energy of flue gas at a certain temperature change in the heat exchanger of waste heat recovery system of Marine low-speed engine. In the selected additives, except for argon, the critical temperature of the other working medium is greater than $CO_2$. Therefore, when argon is used as the additive of $CO_2$-based mixed working medium, the critical temperature will decrease accordingly, so as to meet the requirements of the inlet temperature of SCBC compressor. On the contrary, the critical temperature of $CO_2$-based mixed working medium can be increased by adding low carbon number alkanes and

Journal of Physics: Conference Series **2329** (2022) 012024 doi:10.1088/1742-6596/2329/1/012024

hydrogen sulfide.

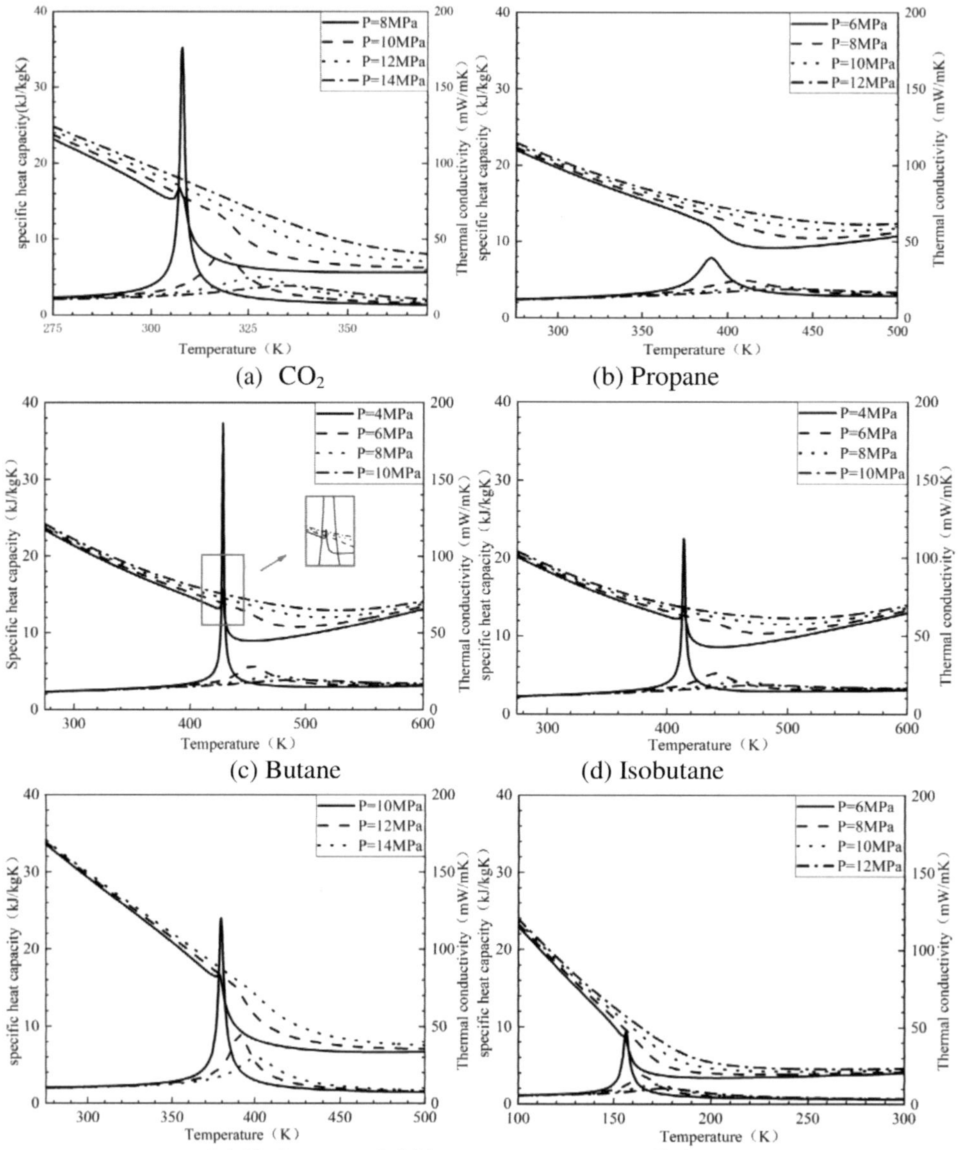


(a) $CO_2$ (b) Propane

(c) Butane (d) Isobutane

(e) Hydrogen sulphide (f) Argon

**Figure 5.** Thermophysical properties of various working fluids vary with temperature.

Figure 6 shows the density, thermal conductivity and viscosity decrease with the increase of temperature, and the decreasing speed increases first and then decreases, reaching the maximum near the critical point. The specific heat increases first and then decreases with the increase of temperature. The specific heat reaches the maximum near the critical point. When the specific heat reaches the maximum, the temperature at this time is the false critical temperature of $CO_2$.

$$T_{pc} = -31.40 + 12.15P - 0.6927P^2 + 0.03160P^3 - 0.0007521P^4 \quad (1)$$

$$7.5\text{MPa} \le P \le 14\text{MPa}$$

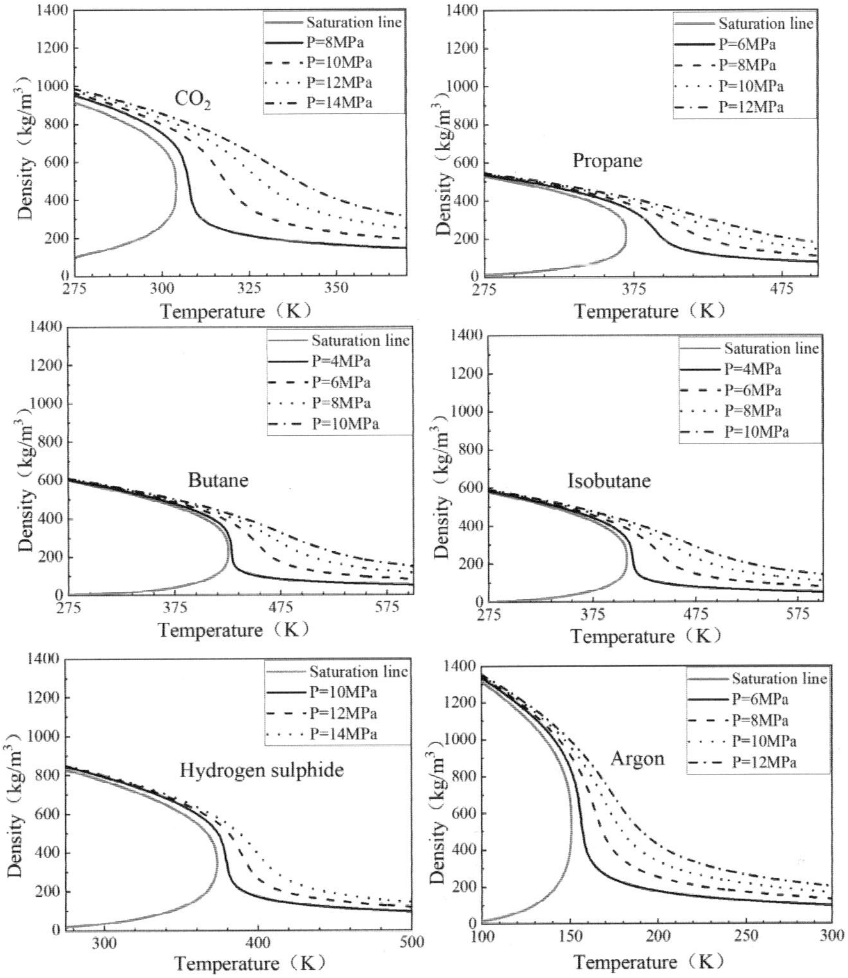


**Figure 6.** Density and temperature changes of various working fluids.

The isobars of the selected working medium above the critical pressure also have the phenomenon of density mutation near the critical temperature point, and with the increase of pressure, the temperature of the mutation point increases and the amplitude decreases. This characteristic makes the supercritical working medium consume less compression work when rising the same pressure in the compression process.

Using $CO_2$-based mixed working medium, the introduction of additives into $CO_2$ can change the critical point of the produced working medium. Assuming that the compressor inlet must operate above (or close to) the critical point, an optimized design of the working medium will enable the cycle to operate at a temperature suitable for the local climate, improving the overall efficiency of the cycle.

## 4. Optimization Analysis of Working Medium

Figure 7 shows the change of critical temperature and pressure of $CO_2$-based mixed working medium with the mass fraction of the additives. It can be seen that with the addition of low carbon alkane as ethane, propane and hydrogen sulfide, the critical temperature of working medium rises gradually, mixed working medium biggest rise in the naphthene, propane and hydrogen sulfide critical temperature is close to, However, with the addition of inert gas argon, the critical temperature of $CO_2$-based mixed working medium presents an opposite trend. The trend of critical pressure of

$CO_2$-based mixed working medium is as follows: with the addition of cycloalkanes, the critical pressure increases first and then decreases, and the variation range is the largest. With the addition of $H_2S$, the critical pressure rises slowly. With the addition of low-carbon alkanes such as butane, isobutane and propane, the critical pressure decreases slowly. The critical temperature and pressure of hydrogen sulfide are close to that of $CO_2$.

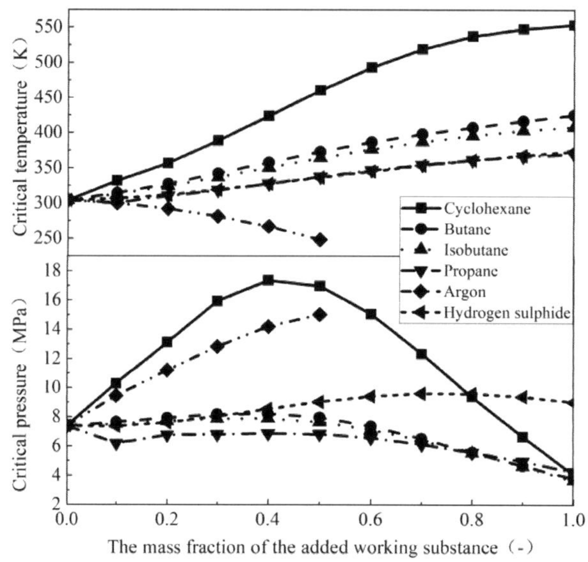


**Figure 7.** The mass fraction of additives in $CO_2$ based mixed working medium.

Table 7 shows the mass fraction of additives in $CO_2$ based mixed working medium. The mass fraction of other working medium added to $CO_2$ is set according to the critical temperature, the composition and proportion of working medium in SCRBC model are defined. In order to explore the influence on the design parameters of the Brayton cycle for recompression of $CO_2$-based mixed working medium, the cycle performance of the mixed working medium with the main compressor inlet temperature of 280K-330K is evaluated. In order to make full use of the characteristics near the critical point, the inlet temperature of the compressor is set as the critical temperature +1K [12], and the cycle operates in the supercritical zone.

**Table 7.** Mass fraction of additives in $CO_2$-based mixtures.

| Working medium | Proportions of mixture(%) | | | | |
|---|---|---|---|---|---|
| (Critical temperature) | 280 | 310 | 320 | 330 | 340 |
| Butane | - | 5.75 | 14.02 | 21.71 | 28.53 |
| Isobutane | - | 6.75 | 16.14 | 24.60 | 32.33 |
| Propane | - | 20.11 | 32.03 | 42.67 | 52.53 |
| Hydrogen sulphide | - | 16.48 | 31.10 | 42.75 | 54.56 |
| Argon | 30.71 | - | - | - | - |

The thermal physical parameters and component ratio of flue gas under 100% load are used as the input of SCRBC model. Figure 8 shows the relationship between the optimal thermal efficiency of the Brayton cycle for recompression of $CO_2$-based mixed working medium and the inlet temperature of the main compressor. When the working medium is $CO_2$, the total energy recovered from the flue gas does not change with the inlet temperature of the main compressor. As a working medium, $CO_2$ can recover more flue gas energy than $CO_2$-based mixed working medium, but the net recovery work of

$CO_2$-based mixed working medium is greater than that of pure $CO_2$ due to the low efficiency. With the rise of the false critical temperature of the working medium, the mixing proportion of the additive working medium rose, and the working medium is always in the critical state, but the total energy of flue gas recovery, efficiency and net recovery work all decrease.

Using the total energy recovered from the flue gas, the maximum total energy recovered is 750.55 kW with $CO_2$ as the medium. The net recovery work of $CO_2$-based mixed medium with 20.11% propane is 34.02kW, which is 17.02kW higher than $CO_2$. The $CO_2$-based mixed working medium with 32.03% propane has the highest efficiency and the efficiency is 8.31%, which is 5.42% higher than $CO_2$.

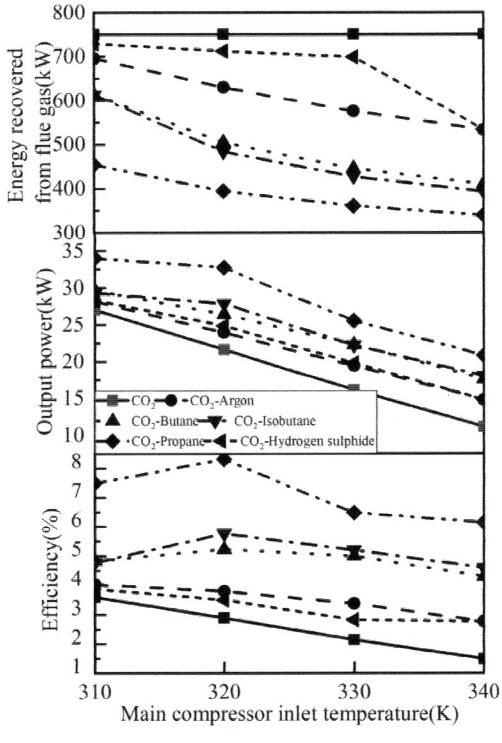


**Figure 8.** Optimum thermal efficiency of $CO_2$-based mixed medium SCRBC.

## 5. Conclusion

(1) In the Marine low-speed diesel engine flue gas waste heat recovery SCRBC system, the critical point of the working medium can be changed by additives to form a $CO_2$-based mixed working medium, and then the cycle can be optimized by the inlet temperature of the main compressor.

(2) In SCRBC, compared with pure $CO_2$ as the working medium, the net recovery of $CO_2$-based mixed working medium supplemented with 20.11% propane increased by 17.02kW. The efficiency of $CO_2$ based mixed medium with 32.03% propane was increased by 5.42%.

## Credit Author Statement

Liangtao Xie: Writing-original draft. Jianguo Yang: Conceptualization, Writing-review & editing.

## Declaration of Competing Interest

The author declare that they have no known competing financial interests or personal relationships that could have appeared to influence the work reported in this paper.

**Acknowledgements**
This work is sponsored by the program "Researches on the principle prototype of 340mm Marine low-speed diesel engine" from Ministry of Industry and Information Technology of the People's Republic of China.

**Reference**

[1] Zhang J Z 2014 *Research of Advanced Ship Type Selection in International Dry Bulk Shipping Market* (Tianjin: Tianjin university).
[2] Shi L F, Tian H and Shu G Q 2020 Multi-mode analysis of a $CO_2$-based combined refrigeration and power cycle for engine waste heat recovery *Applied Energy* **264** 114670.
[3] Weerasinghe W M, Stobart R K and Hounsham S M 2010 Thermal efficiency improvement in high output diesel engines a comparison of a Rankine cycle with turbo-compounding *Applied Thermal Engineering* **30**(14) 11.
[4] Tian H, Jing D Z, Shu G Q, Wang X and Liu P 2018 Comparative analysis on the off-design performance of organic rankine cycle systems with different working fluids *Journal of Xi'an Jiaotong University* **52**(03) 25-33.
[5] Guo J Q, Wang K, Zhu H H and He Y L 2017 Thermodynamic analysis of brayton cycles using supercritical carbon dioxide and its mixture as working fluid *Journal of Engineering Thermophysics* **38**(04) 695-702.
[6] Conboy T, Ames D, Lewis T and Wright S 2011 Operation of a Supercritical Fluid Compression Loop Using $CO_2$-Based Mixtures *Sandia National Laboratories* SAND2011-0220C.
[7] Cao C H 2018 *Optimization and Analysis of Supercritical Carbon Dioxide Brayton Cycle Based on Tower Solar Power Generation System* (Tianjin: Tianjin university).
[8] Wright S A, Radel R F, Vernon M E, et al. 2010 Operation and analysis of a supercritical $CO_2$ Brayton cycle *Sandia National Laboratories* SAND2010-0171.
[9] Xi H, Li M J, He Y L and Tao W Q 2015 A graphical criterion for working fluid selection and thermodynamic system comparison in waste heat recovery *Applied Thermal Engineering* **89** 772-782.
[10] Dai X Y, An Q S, Shi L and Zhai H X 2017 A study on working fluid selection for organic rankine cycle of engine waste heat *Automotive Engineering* **39**(02) 121-126.
[11] Wang J Y, Guan Z Q, Gurgenci H, Sun Y B, et al. 2020 A comprehensive review on numerical approaches to simulate heat transfer of turbulent supercritical $CO_2$ flows *Numerical Heat Transfer, Part B: Fundamentals* **77**(5) 349-400.
[12] Hu L, Chen D, Huang Y, et al. 2015 Investigation on the performance of the supercritical brayton cycle with $CO_2$-based binary mixture as working fluid for an energy transportation system: of a nuclear reactor *Energy* **89** 874-886.

# Hydraulic Structure Design of Changzhou Fishway Based on Optimization Calculation of Hydrodynamic Model

**Rui Ji[1*], Jing Li[2] and Bingsen Chen[1]**

[1] Guangxi Vocational College of Water Resources and Electric Power, Nanning, 530023, Guangxi, China,
[2] China Energy Construction Group Guangxi Electric Power Design and Research Institute Co., Ltd. Nanning, 530023, Guangxi, China.
Email: Hrry@foxmail.com

**Abstract.** The Changzhou Water Conservancy Project is the first large-scale water conservancy project above the Pearl River Estuary, and its location is extremely important. This section of the river has historically been the main channel for the migration and fattening of various fishes in the Pearl River system; in order to meet the requirements of fish in this section, protect these rare fishes and achieve sustainable development and ecological balance of biological diversity. Purpose, it was decided to build a fish building structure and fish channel in the Cheung Chau Water Conservancy Project. The Cheung Chau Fishway has a total length of 1423 m, and the water head difference between the upstream and downstream is 15 m. It is the first large-scale fishway designed in China. The water head difference of domestic fishways is generally below 5.0 m, and there is very little experience to learn from. Since there was not much successful experience in the domestic fishway at that time, and the domestic fishway construction has been out of gear for nearly 30 years, through the safety appraisal and evaluation of the Cheung Chau fishway during the operation period, valuable opinions were summarized, which will be an important reference for future projects. The experience can be used as a reference for similar projects and promotion.
**Keywords.** Fishway hydraulic structure design, hydrodynamic design and model test, safety evaluation.

## 1. Introduction

The fishway is located on the left side of the installation room of the Waijiang plant on Sihuazhou Island, with pile numbers ranging from 0-134.690 m to 0-112.665 m. The fishway is arranged from bottom to top with an inlet section, a fishway pool, a rest pool, an observation room, a flood gate section and an outlet section, with a total length of 1443.301 m. The downstream inlet is located about 100m downstream of the tail water of the powerhouse [1]. It is open and has a bottom width of 5 m. On the right side of the inlet gate, a guide wall of about 8m long is set at about 30° from the center line of the fishway to improve the flow pattern of the inlet and strengthen the guide. The upstream outlet is set on the outer river side of Sihuazhou Island, upstream of the dam. There is a flood gate at the intersection of the fishway and the axis of the dam, and an overhaul gate at the entrance and exit of the fishway. The designed water level of the inlet overhaul gate is 9.0 m, and the designed water level of the exit overhaul gate is 20.6m. In order to determine the basic parameters of fishway design and the rationality of fishway layout of Changzhou water control project, the flow conditions at the inlet and outlet of fishway are verified through the hydrodynamic calculation of two-dimensional mathematical simulation model, and the hydraulic characteristics of fishway are studied through physical model. Verify the rationality of the overall stability and base stress analysis of the overall design of the fishway, and give evaluation and suggestions through the detection of a third-party company. Layout refer to figure 1.

## 2. Overall Stability and Base Stress Analysis

### *2.1. Fishway Flood Sluice Dam Section*

The calculated load includes the dam body weight, water weight, water pressure, earth pressure, uplift pressure, etc., divided into normal operating conditions (upstream water level 20.6 m, downstream water level 5.05 m), check conditions and completion conditions (both inside and outside anhydrous) Carry out overall stability and base stress analysis.

According to JTJ307-2001 "Code for Design of Ship Lock Hydraulic Structures", calculate the stability of anti-sliding, anti-tilting and anti-floating stability; according to "Code for Design of Sluices", calculate the base stress of the lock chamber wall.

Basic parameters: the shear friction coefficient of the contact surface between the lock chamber and the ground weakly weathered granite f'= 0.9, and the shear cohesion c' = 0.7 MPa;

Control standard: anti-sliding stability safety factor $Kc \geq 3.0$ calculated by anti-shearing, anti-overturning stability safety factor $K_0 \geq 1.5$, anti-floating stability safety factor $K_f \geq 1.1$, and foundation allowable bearing capacity $[\sigma]$=3.0 MPa.

According to the results of geological exploration and excavation, the low-dip structural surfaces of the rock mass below the foundation surface are not developed, and the strength of the rock mass is relatively high. The anti-sliding stability of the foundation is controlled by the shear strength of the concrete and the foundation surface rock mass, so the deep and shallow anti-sliding stability of the dam foundation along the weak structural surface are not calculated, and only the anti-sliding stability of the foundation surface is calculated and analyzed. The calculation results of the overall stability of the fishway flood gate dam section and the base stress are shown in table 1.

**Table 1.** The overall stability and foundation stress calculation results of the fishway flood gate dam section.

| Item | Base stress (kPa) | | Base stress (kPa) | |
|---|---|---|---|---|
| | Base stress (kPa)<br>$K_c$ | Base stress (kPa)<br>$K_f$ | Base stress (kPa)<br>$\sigma_{max}$ | Base stress (kPa)<br>$\sigma_{min}$ |
| Normal operating conditions | 15.65 | 8.67 | 690 | 102 |
| Check condition | 32.31 | 2.19 | 421 | 67 |
| Completed conditions | 18.15 | - | 672 | 144 |
| Standard or geological requirement value | 3.0/2.5 | 1.1 | 3000 | 0 |

The control condition of the base stress at the edge of the flood gate parallel to the flow direction of dam section of fishway is the normal operating condition; the control condition of the base stress at the edge of the vertical water flow direction is the completion condition. The overall stability and stress of the fishway flood gate meets the requirements of the specification.

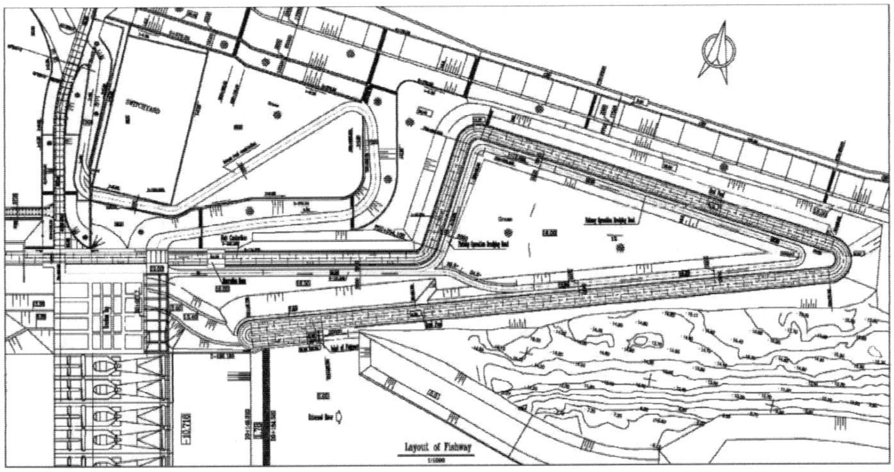


**Figure 1.** Layout of Fishway Changzhou.

### 2.2. *Imported Inspection Gate*

The calculated load includes dam body weight, water weight, water pressure, uplift pressure [2], etc., divided into normal operating conditions (the normal tail water level inside and outside the gate is 9.6 m), and overhaul conditions (outside the gate is the overhaul water level of 9.0 m, and there is no water in the gate. ), the completion condition (there is no water inside and outside), and the check condition (check water level is 30.75 m) for overall stability and base stress analysis [3]. Basic parameters: the internal friction angle between the structure of the contact surface of the lock chamber wall and the foundation and the soil foundation $\varphi_0$=14.4°, the cohesive force $C_0$=15 kPa; the compressive modulus Es of the upper and lower saturated soils in the settlement calculation are 5.0 MPa and respectively 16.0 MPa. Control standards: soil foundation anti-sliding stability safety factor $K_c \geq 1.2$, anti-overturning stability safety factor $K_0 \geq 1.4$, anti-floating stability safety factor $K_f \geq 1.1$, foundation allowable bearing capacity [σ]=260 kPa, and settlement less than 400 mm. The calculation results of the overall stability of the imported maintenance gate and the base stress are shown in table 2.

**Table 2.** Overall stability and base stress calculation results of imported inspection gates.

| Calculating item | | Operating conditions | Inspection and repair conditions | Completion conditions | Check conditions |
|---|---|---|---|---|---|
| Anti-tilt stability Safety factor ($K_0$) | BC edge | 2.2 | 5.4 | ∞ | ∞ |
| | CD edge | 2.2 | 5.7 | ∞ | ∞ |
| Anti-sliding stability safety factor ($K_c$) | | ∞ | 2.4 | ∞ | ∞ |
| Anti-sliding stability safety factor ($K_f$) | | 2.2 | 3.5 | ∞ | 1.3 |
| Base stress(kPa) | $P_A$ | 82.1 | 240.4 | 132.8 | 77.5 |
| | $P_B$ | 120.2 | 140.3 | 175.6 | 102.4 |
| | $P_C$ | 136.0 | -12.2 | 185.3 | 108.1 |
| | $P_D$ | 97.9 | 87.8 | 142.5 | 83.1 |
| Base stress(m) | | 0.239 | 0.250 | 0.348 | 0.262 |

Imported inspection gates have anti-sliding stability, anti-tilting stability, anti-floating stability, stress and base settlement all meet the requirements of the specification.

### *2.3. Exit Inspection Gate*

The calculated load includes the dam body weight, water weight, water pressure, lateral earth pressure, uplift pressure, etc., divided into normal operating conditions (the water level inside and outside the sluice is 20.6 m), maintenance conditions (upstream water level 20.6 m, downstream no water), the completion condition (there is no water inside and outside the inspection gate), and the check condition (the water level is 30.88 m) for overall stability and base stress analysis. Basic parameters: The internal friction angle between the structure of the contact surface of the lock chamber wall and the foundation and the soil foundation is $\varphi0=14.4^{\circ}$, and the cohesive force $C_0=$ 15 kPa; the compressive modulus Es of the upper and lower saturated soils in the settlement calculation are 5.0 MPa and respectively 16.0 MPa.

Control standards: soil foundation anti-sliding stability safety factor $K_c \geq 1.2$; anti-overturning stability safety factor $K_0 \geq 1.4$; anti-floating stability safety factor $K_f \geq 1.1$; fishway exit inspection gate foundation is located on Holocene river alluvial $Q_{al4}$, dark gray powder for high-quality clay, the allowable bearing capacity of the foundation is 130 kPa; the settlement is less than 400mm.

The calculation results of the overall stability [4] of the exit inspection gate and the base stress are shown in table 3.

**Table 3.** Overall stability of the exit inspection gate and calculation results of the base stress.

| Calculating item | | Operating conditions | Inspection and repair conditions | Completion conditions | Check conditions |
|---|---|---|---|---|---|
| Anti-tilt stability Safety factor ($K_0$) | BC Edge | 13.5 | 11.7 | 20.7 | 24.2 |
| | CD Edge | ∞ | 37.8 | ∞ | ∞ |
| Anti-sliding stability safety factor ($K_c$) | | | 1.7 | 2.6 | 3.8 |
| Anti-sliding stability safety factor ($K_f$) | | | 2.2 | ∞ | 1.4 |
| Base stress(kPa) | $P_A$ | 47.9 | 42.3 | 80.3 | 47.9 |
| | $P_B$ | 75.5 | 69.9 | 107.9 | 61.7 |
| | $P_C$ | 76.6 | 65.7 | 110.3 | 63.2 |
| | $P_D$ | 49.0 | 38.1 | 82.7 | 49.4 |
| Base stress(m) | | | 0.152 | 0.269 | 0.157 |

The anti-sliding stability, anti-tilting stability, anti-floating stability, stress and base settlement of the exit inspection gate all meet the requirements of the specification.

### *2.4. Fishway High Retaining Wall*

The fishway high retaining wall is located on the left side of the downstream of the flood gate. The elevation of the top of the wall is 34.4 m, the elevation of the bottom of the wall is 14.12 m, and the left side is the outer river rolling earth dam section. The calculated load includes the dam body weight, water weight, water pressure, earth pressure, bridge deck precast beam weight, 55 t heavy vehicle pressure, etc., divided into normal operating conditions (inside is 3 m design water depth, outside is 9.6 m water level), maintenance work Condition (no water on the inner side and 9.0 m water level on the outer side) for overall stability and base stress analysis. Basic parameters: the internal friction angle between the structure of the contact surface of the lock chamber wall and the foundation and the soil foundation $\varphi_0=14.4^{\circ}$, the cohesive force $C_0=15$ kPa; the compressive modulus Es of the upper and lower saturated soils in the settlement calculation are 5.0 MPa and respectively 16.0 MPa. Control standards: soil foundation anti-sliding stability safety factor $K_c \geq 1.2$; anti-overturning stability safety factor $K_0 \geq 1.4$; anti-floating stability safety factor $K_f \geq 1.1$; the bottom of the retaining wall is backfilled, and the allowable foundation bearing capacity is taken as [σ]=360 kPa ; The settlement is less than 40 cm. The calculation results of the overall stability of the fishway high retaining wall and the base stress are shown in table 4.

**Table 4.** The overall stability of the fishway high retaining wall and the calculation results of the base stress.

| Calculation item | Normal operating condition | Normal operating condition |
|---|---|---|
| Normal operating condition($K_0$) | 3.47 | 2.68 |
| Normal operating condition($K_c$) | 2.24 | 1.73 |
| Anti-floating stability safety factor($K_f$) | +∞ | +∞ |
| Anti-floating stability safety factor $P_{max}$(kPa) | 298.74 | 195.74 |
| Minimum base stress $P_{min}$(kPa) | 104.12 | 113.95 |
| The final settlement of the base S(mm) | 10.62 | 10.62 |

The anti-sliding stability, anti-tilting stability, anti-floating stability, stress and settlement of the basement of the fishway high retaining wall all meet the requirements of the specification [5].

## 3. Structural Design

The fishway flood gate is a monolithic structure, using three-dimensional finite element to calculate [6] the stress and strain. According to the calculation results, under normal operating conditions, the flood gate is biased to the right in the direction of the dam axis as a whole, and the maximum offset is 12.7 mm. This is mainly because the earth pressure on the left side of the flood barrier is greater than the earth pressure on the right side. In the height direction, the flood gate has a sinking trend as a whole, and the maximum settlement is 5.9mm. In the direction of the water flow, the flood gate is biased towards the downstream as a whole, with a maximum offset of 3.2 mm. The deformation value of the flood gate is small, which meets the requirements of safe operation of the fishway.

According to the stress results calculated by the three-dimensional finite element [7], the bottom of the fishway parapet wall is equipped with steel bar 6Φ28, the top surface of the bottom plate is equipped with steel bar 5Φ20, and the side wall of the gate observation room is equipped with steel bar 5Φ22. The entrance and exit inspection gates adopt the material mechanics method to calculate the structure. The bottom plate is simplified to a fixed beam with a single width of 1m and a height of 1.5 m for calculation. The structure is reinforced according to the structural requirements. The fishway high retaining wall adopts the PKPM program to calculate the internal force of the structure according to the eccentric compression member, and it is reinforced according to the structural requirements.

The reinforcement of the flood gate, the entrance inspection gate, the exit inspection gate and the fishway high retaining wall all meet the requirements of the "Design Code for Hydraulic Concrete Structures".

## 4. Hydraulic Design and Model Test

### *4.1. Hydraulic Calculation*

The upstream exit gate can be closed after the fish pass every year after the fish season to save water and for maintenance; when the upstream exit gate is opened for the first time every year when the fish season comes, a free flow state is formed under the gate, and high-speed outflow from the bottom of the gate is possible. Scouring the bottom plate and the first few partitions upstream [8], so hydraulic calculations are required and verified by overall model tests. The free outflow flow rate under the slab gate can be calculated as follows [9]:

$$Q = \mu b e \sqrt{2gH} \tag{1}$$

In the formula: b is the gate width, the fishway is 5 m,e is the gate opening, 0.3 m<e<1.95 m for this project; H is the water depth in front of the gate, and the fishway is 3m; μ is the flow coefficient,

$$\mu=0.60-0.18*\frac{e}{H} \tag{2}$$

this formula is suitable for

$$0.1<\frac{e}{H}<0.65 \tag{3}$$

Refer to table 5 for the flow rate and the average flow velocity of the section at each gate opening.

**Table 5.** Hydraulic calculation table for free outflow of upstream exit gate.

| e(m) | μ | Q($m^3$/s) | v(m/s) |
|---|---|---|---|
| 0.3 | 0.582 | 6.70 | 4.46 |
| 0.4 | 0.576 | 8.84 | 4.42 |
| 0.5 | 0.570 | 10.93 | 4.37 |
| 0.8 | 0.552 | 16.94 | 4.23 |
| 1.0 | 0.540 | 20.71 | 4.14 |
| 1.2 | 0.528 | 24.30 | 4.05 |
| 1.5 | 0.510 | 29.35 | 3.91 |
| 1.8 | 0.500 | 34.52 | 3.84 |

### *4.2. Hydraulic Model Test*

The hydraulic design is combined with hydraulic model tests. Through hydraulic model tests, compare and confirm structural design schemes such as the type of partitions in the fishway and the layout of the fish passages, and verify the flow rate and flow pattern during the water filling process of the fishway and the connection relationship between the water flow of the fishway and the upstream and downstream water levels. Recheck fishway flow and comprehensive velocity coefficient.

*4.2.1. Local Hydraulic Model Test Results.* (1) For fish passages that need to pass both large fish and small fishes at the same time, a single rectangular section cannot meet the flow velocity requirements, and a comprehensive “vertical hole-sloping hole” section with a small triangular slope angle must be adopted, and the “average flow velocity” must be designed according to the design The requirements of the applicant are allocated to different areas, forming a "large flow rate zone" and a "small flow rate zone" in the same section [10].

(2) To form a "small velocity zone", the adjacent two partitions must be arranged in a staggered manner, otherwise, a straight water flow from top to bottom will form, directly impacting the triangular slope, and the "small velocity zone" will no longer exist.

(3) Since the energy dissipation effect of the straight water flow from top to bottom is not good, it is necessary to form a curved main flow and a certain strength and range of return flow in the pool to meet the energy dissipation and flow rate requirements;

(4) Through the test and analysis, the partition type used in the model test can meet the requirements.

*4.2.2. Test Results of Overall Hydraulics Model.* (1) When the upstream exit gate has a large opening, the flow velocity of the gate orifice is very consistent with the hydraulic calculation results. At a small opening, that is, when the gate opening is 0.3 m, the flow velocity at the rear contraction section of the gate can reach 6.2 m/s, and the water flows directly into the upstream partitions of the first and second partitions, which need to be reinforced to ensure stability.

(2) The measured fishway flow rate of the triangular weir is 6.64 $m^3/s$. After calculation, the average flow velocity at the baffle passing through the fish hole is 0.948 m/s, and the fishway comprehensive velocity coefficient is 0.73. The baffle type adopted has a better energy dissipation effect. .

(3) When the upstream is at the design water level of 20.6 m and the downstream is at the average low water level of 8.0 m, that is, when the design water level difference is 12.6 m (generally in the initial water state), no water flows down when the upstream gate is closed, and the downstream inlet of the fishway has a design The water depth is 3 m, so the 210m long fishway (35 ponds) is in a watery state, and only the upstream part of the pond is in an anhydrous state. When the gate opening is gradually increased, the water levels in the first few upstream pools will quickly become high, and the flow velocity behind the gate will gradually decrease. The water level in the downstream is high, and it has been partially filled with water. At this time, the fishway is at the upper and lower ends of the pool where there is water, and most of the pools in the middle are in a state of water passing at the bottom. Continue to raise the gate to a fully open state (opening of 3 m), and the water level of the pool will rise gradually from bottom to top until the fishway reaches the designed water depth throughout the entire process and is connected to the upstream and downstream water levels. When the upstream water level is 20.6 m and the downstream is the average high water level 11.95 m (usually in the flood season) At this time, the water depth at the downstream inlet has reached 6.95 m, that is, 81 pools have been submerged with water, and only the upstream part of the pool has no water. At this time, because it is not the main fishing season, the upstream gate has been closed. If the gate needs to be opened, there is still a maximum flow rate of 6.2 m/s behind the upstream gate, the entire fishway filling process remains the same as described above, and the filling time will be greatly shortened.

(4) Except for the slope angle measuring points, the flow velocity of each measuring point in the vertical hole-the slope hole and the bottom hole has no obvious increasing trend from top to bottom, indicating that the partition type is more reasonable, the water flow energy dissipation is sufficient, and the accumulation from top to bottom is not obvious. The model test results show that the final recommended partition type can basically meet the requirements of the design flow rate.

## 5. Basic Treatment Design

The foundation of the fishway flood sluice is located in the weakly weathered upper granite, and the rock mass cracks are relatively developed. 6 rows of consolidation grouting holes are arranged in the foundation of the fishway flood gate, the hole spacing is 3m, and the hole depth is 8m. Fishway flood gate dam section with stake number 0-103.915 m～0-134.690 m, 3Lu line is higher, foundation excavation has been excavated below or close to 3Lu line, and set at 0+003.55 m below the stake fish A row of consolidation grouting also serves as curtain grouting, with a hole spacing of 2 m and a depth of 8m. The grouting pressure is 0.4 MPa～0.8 MPa, and it is actually adjusted according to the site conditions based on the principle of not lifting the rock mass.

## 6. Conclusion

### *6.1. Evaluation Opinion*

(1) The layout of the fishway is reasonable. After calculation and review, the flood gate, the entrance inspection gate, the exit inspection gate and the fishway high retaining wall are stable against sliding, tilting, and anti-floating, and the base stress and settlement meet the requirements of the specification [11].

(2) The main structure of the fishway is calculated using finite element and material mechanics methods, and the structural reinforcement meets the requirements of the current specifications.

(3) The structural layout and hydraulic design of the fishway have been verified by hydraulic model tests, which are basically in line with the design expectations and basically meet the requirements for fishing.

(4) The basic treatment design of the fish pass is appropriate.

### *6.2. Recommendation*

Considering that there is a great contradiction between the setting of fish trap in Neijiang River and navigation, and the setting of fish trap in Waijiang river is greatly affected by the operation of Waijiang sluice gate, the downstream section of fish trap in Neijiang River and fish trap in Waijiang River and sluice gate is temporarily cancelled. It is suggested to decide whether to build these temporarily cancelled fish blocking grids according to the fish blocking effect of the fishway during the operation period.

## References

[1] Nong J 2008 Fishway design of Changzhou Water Conservancy Project *Hong Shui He* (05) 50-54.

[2] Yang X R, Zhu C D and Fan S X 2020 Discussion on key technical issues in fishway design *Water Conservancy Planning and Design* (12) 114-120.

[3] Cai Y P, Liu H J and Liu Z X 2021 Study on hydrodynamic characteristics of vertical slotted fishway trapezoidal cross-section pool chamber *People's Yangtze River* **52**(07) 204-209+230 DOI: 10.16232/j.cnki.1001-4179.2021.07.034.

[4] Zheng J X, Han D J, Hu W B, Wang X and Zhang X M 2010 Research on fish swimming behaviour related to fishway design *Journal of Aquatic Ecology* **31**(05) 104-110.DOI:10.15928/j .1674-3075.2010.05.023.

[5] Yan P D, Tan J J, Gao Z, Dai H C, Shi X T and Huang T A 2018 Analysis of fish movement behaviour in vertical slit fishway based on video tracking *Acta Aquatic Biology* **42**(02) 250-254.

[6] Zhang G Q and Sun S K 2012 Influence of vertical slit width on water flow structure of vertical slit fishway *Journal of Hydroelectric Power Generation* **31**(01) 151-156.

[7] Bian Y H and Sun S K 2013 Research on hydraulic characteristics of vertical slot fish pass *Journal of Hydraulic Engineering* **44**(12) 1462-1467.

[8] Zhang D Y 2011 Influence of water conservancy and hydropower projects on fish and protection measures *Water Resources Protection* **27**(05) 75-77.

[9] Nanjing Institute of Water Conservancy Science 1982 *Fish Road* (Beijing: Electric Power Industry Press).

[10] Cao N, Zhong Z G and Cao X H 2016 Current status of fishway construction in my country and analysis of typical cases *Water Resources Protection* **32**(06) 156-162.

[11] China Hydropower Engineering Consulting Group Corporation 2010 *Guangxi Changzhou Water Conservancy Project Completion Safety Appraisal Report*.

# Numerical Simulation Research on the Effect of the Rotor Passage Cross-Sectional Dimension on the Performance of Gas Wave Ejector

**Yiming Zhao, Qing Feng and Dapeng Hu***

Department of Chemical Machinery, Dalian University of Technology, Dalian, 116012, China.
Email: blockbusterzym@163.com

**Abstract.** The gas wave ejector (GWE) that utilizes pressure waves to realize the efficient energy transfer between gases has broad application prospects in chemical industry. In this study, the numerical simulation method was used to investigate the influence mechanism and law of the rotor passage height (*h*), width (*w*) and aspect ratio (*h*/*w*) on the equipment performance. The main conclusions are as follows: Increasing the passage height or width alone within a certain range may help improve the equipment performance, but if the passage width is too large, the sealing time of passage may be too short and the gas expansion may be incomplete, resulting in a significant reduction in performance. When the passage aspect ratio is fixed, the effects of passage width and height on the equipment performance are independent of one another, and the performance is determined by the superposition effect of these two. Within the range of passage cross-sectional area ($40mm^2$~$360mm^2$) investigated in this study, the passage aspect ratio for achieving the best performance is in the range of 0.71~1.6, and the efficiency and ejection rate under the same *h*/*w* increase with the increment of the cross-sectional area when $h/w \geq 0.71$.
**Keywords.** Ejector, Wave rotor, Cross-Sectional dimension, Fluid inside passage, Simulation.

## 1. Introduction

The employment of pressure exchange technology to achieve efficient energy transfer between high and low-pressure fluids is critical for the recovery and utilization of pressure energy in chemical industrial production. For example, in the field of natural gas extraction, the gas from high-pressure wells with pressure higher than the gathering and transportation requirements can be utilized to eject and supercharge gas from the low-pressure wells with pressure lower than the requirements. It may avoid the energy waste caused by the throttling and depressurization of high-pressure gas and reduce the external power consumption required to pressurize the low-pressure well with the compressor, which is significant for improving the overall production efficiency. The static ejector and expander-booster compressor are the commonly used equipment for the pressure energy transfer at present. The expander-booster compressor is efficient, but it has the disadvantages of high rotation speed, complex structure, and high purity requirements of the gas medium [1]; in contrast, the static ejector has a simple structure and a strong ability to carry impurities, but its working mechanism of transferring energy by direct mixing of gases makes its working efficiency relatively low [2].

The gas wave ejector (GWE) combines the benefits of both technologies mentioned above to some extent. GWE employs the efficient gas pressure waves as the energy transmission medium, and it has a simple structure, low rotation speed and high strength of rotor passages, which enhance its liquid carrying capacity [3]. The concept of GWE was proposed by Spalding in 1958 [4]. Kentfield et al. [5] from the Imperial College conducted its first application test, which preliminarily confirmed that GWE had an isentropic efficiency similar to turbomachinery under certain working conditions. In the 2000s, Kharazi et al. [6] from the University of Michigan proved that GWE might replace the condenser and the first-stage compressor in the R718 refrigeration cycle, resulting in a 50% reduction in the system investment. In the 2010s, Hu et al. [7] of Dalian University of Technology (DUT) started to study on GWE and obtained the influence of several structure and operating parameters on its performance [8].

The wave rotor and pressure ports are the core components of GWE. There are several identical passages uniformly distributed on the rotor, and the pressure waves utilized to transfer energy between gases are created when passages connect or disconnect with the pressure ports during the rotor rotation. According to this basic working principle, analyzing the influence mechanism of the passage cross-sectional dimension on equipment performance is significant for improving the mechanism and design theory of GWE, as well as promoting its development and application.

## 2. Research Methodology

Manufacturing several different wave rotors for comparative experiments is expensive, and measuring the distribution of state parameters in the rotor passages for related mechanism analysis is complex. As a result, the CFD commercial software Fluent was employed to carry out numerical simulation and comparative analysis of multiple models with various passage cross-sectional dimensions in this study.

### *2.1. Model Foundation of Numerical Simulation*

As shown in figure 1, the computational domain of model primarily consists of three fixed ports with different pressures, several moving rotor passages (a new type of curved ejection-rotor passage was employed in this study [9]), and the gaps between these two. The end faces of each pressure port are defined as pressure boundaries as illustrated in figure 1, and the overlapping end faces of each domain are set as interface. Because the motion process of each rotor passage is exactly identical, this model only includes a portion of the passages to conserve computing resources, and the periodic boundary is set to make passages reciprocate between the pressure ports. The gas does not have enough time to exchange heat with walls due to the fast propagation speed of pressure waves and the tiny temperature difference in passages. Therefore, the walls in model are set as adiabatic walls. Because the shape of the geometric model is rather regular, hexahedral mesh is chosen for the mesh division. The mesh refinement is carried out near each interface, and the boundary layer meshing is performed on the walls to simulate the unsteady flow process in domains more accurately. The cell number of a single passage in the model is about $4.2\times10^4$, and the total cell number is around $6.5\times10^5$.

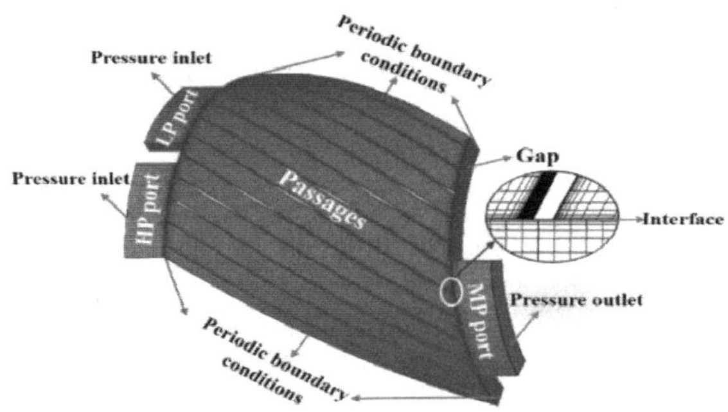


**Figure 1.** Schematic diagram of numerical calculation physical model and meshing.

The fluid medium used in this study is air with pressure within 0.3 MPa and temperature above 273K, which can be assumed to be ideal gas, and its flow process satisfies the ideal gas state equation and the governing equation of fluid mechanics. A new type of scale adaptive model, SST-SAS turbulence model, is chosen for the turbulence calculation. This model is very suitable for calculating the primary flow phenomena in GWE such as the strong swirling and strong interaction flow, and it may produce results similar to the LES model in the unsteady flow domain, allowing for the capture of more turbulent pulsations [10]. Through the PIV monitoring experiments, Kurec et al. [11] proved that this model could be ideal for the research and design of wave rotor equipment. In order to accurately calculate the changes of state parameters before and after discontinuities such as shock wave and contact surface, the AUSM method is used in this study for spatial discretization, and the second-order implicit double time step method is used for time discretization.

### *2.2. Validation of the Numerical Model*

*2.2.1. Performance Evaluation.* The main parameters for evaluating the performance of GWE are the ejection rate $\xi$ and the isentropic efficiency $\eta$. The ejection rate $\xi$:

$$\xi = \frac{m_{lp}}{m_{hp}} \tag{1}$$

represents the equipment capacity for sucking the low-pressure gas, and the isentropic efficiency $\eta$:

$$\eta = \frac{m_{lp} T_{lt} \left( \left( \frac{p_{mt}}{p_{lt}} \right)^{\frac{\gamma-1}{\gamma}} - 1 \right)}{m_{hp} T_{ht} \left( 1 - \left( \frac{p_{mt}}{p_{ht}} \right)^{\frac{\gamma-1}{\gamma}} \right)} \tag{2}$$

is used for denoting the energy transfer efficiency, where $m_{hp}$ and $m_{lp}$ represent the mass flow rate of high and low-pressure gas, respectively; $p_{ht}$, $p_{mt}$ and $p_{lt}$ represent the total pressure of the high, medium and low-pressure gas; $T_{ht}$ and $T_{lt}$ represent the total temperature of the high and low-pressure gas; and $\gamma$ is the adiabatic exponent of gas.

*2.2.2. Model Verification.* This paper cites the data of the pressure fluctuation measurement experiment completed by Okamoto et al. [12] from the University of Tokyo (figure 2(a) is the schematic diagram of the relevant experimental device) to verify the calculation accuracy of the model for simulating the flow inside passage.

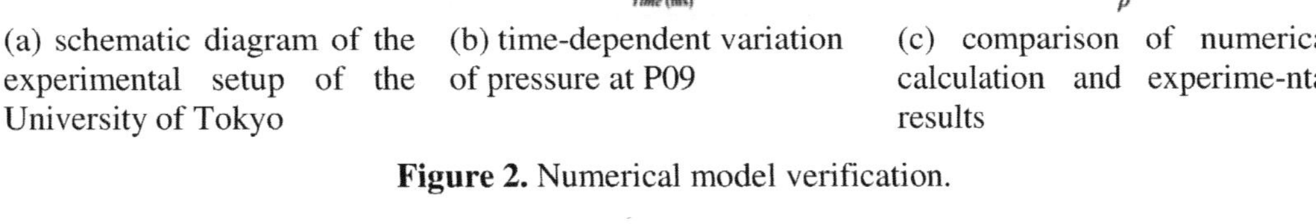


(a) schematic diagram of the experimental setup of the University of Tokyo
(b) time-dependent variation of pressure at P09
(c) comparison of numerical calculation and experime-ntal results

**Figure 2.** Numerical model verification.

Figure 3(b) shows the comparison results between the imulation computation and the experimental measurement values of pressure at P09 point. Due to the smooth wall used for the numerical simulation, the calculation pressure values of the P09 point after the shock wave under different gaps are slightly different from the experimental values, while the shape and peak position of the changing curves are basically identical. It demonstrates that the calculation model is suitable for the simulation of the flow field such as the pressure distribution in the rotor passages. Figure 3(c) shows the comparison curves of the equipment ejection rate $\xi$ and isentropic efficiency $\eta$ between the numerical simulation and experiment when the expansion ratio $\alpha$=1.5 ($\alpha=p_{ht}/p_{lt}$). The experimental results were obtained from the GWE experiments conducted by Li et al. [9] from DUT, and the structure as well as operating parameters of the simulation models are exactly identical to Li's experiments. The variation laws of the computational performance parameter values with the compression ratio $\beta$ ($\beta=p_{mt}/p_{lt}$) under each working condition are completely consistent with the experimental results, and the difference between these two is within 7%, confirming the accuracy of the GWE model used in this study to some extent.

## 3. Result and Analysis

As shown in figure 3, the passage width $w$ and the passage height $h$ are the key cross-sectional dimension parameters of the rotor passage. Therefore, each model in this study only varies in the values of $w$ and $h$, while other parameters such as the passage length, passage bending angles, and the rotor mean diameter remain constant and consistent with the rotor in Reference 9.

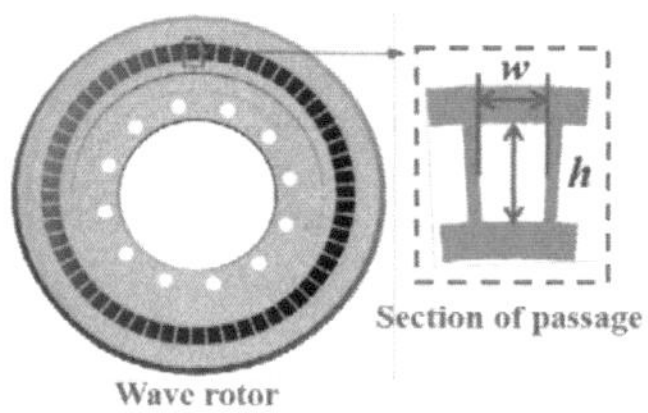


**Figure 3.** Schematic diagram of the passage cross-section.

Figure 4 shows the variation of performance parameters under the conditions of $p_{ht}$=152 kPa, $p_{lt}$=101 kPa, and $p_{ms}$=115 kPa ($p_{ms}$ represents the static pressure of medium-pressure gas) with the passage height $h$, when the passage width is fixed at $w$=10 mm. As shown in figure 4(a), when $h \leq 16$mm, the isentropic efficiency $\eta$ rises with the increase of $h$, but the growth rate gradually decreases; when $h>16$mm, $\eta$ decreases slightly with the increase of $h$. As shown in figure 4(b), the total pressure of medium-pressure gas $p_{mt}$ at each $h$ is almost identical, and the changing law of ejection rate $\xi$ is nearly consistent with the changing law of $\eta$. When 5 mm<$h$<16 mm, the cross-sectional area of gas flow in the passage rises with the increase of $h$, reducing the viscous friction loss of gas flow, which is the reason for the rapid improvement of the performance parameter values as $h$ increases.

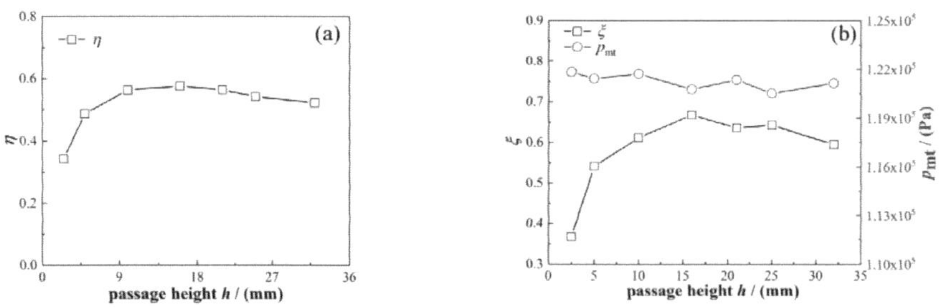


(a) isentropic efficiency $\eta$

(b) ejection rate $\xi$ and the total pressure of medium-pressure gas $p_{mt}$

**Figure 4.** The variation of performance parameters with passage height $h$.

Figure 5 shows the velocity distribution at the passage inlet end of different models during the process of high-pressure gas incidence. When the passage height reaches to 21 mm, the three-dimensional flow complexity of the gas incidence increases, significantly enhancing the inhomogeneity and gradient of velocity distribution along the height direction at the passage inlet end. Therefore, the loss of the gas incidence process may be significantly increased when $h$>16mm, reducing the performance improvement effect of enlarging the passage cross-sectional area, and as a result of which, the performance parameters may decrease slightly.

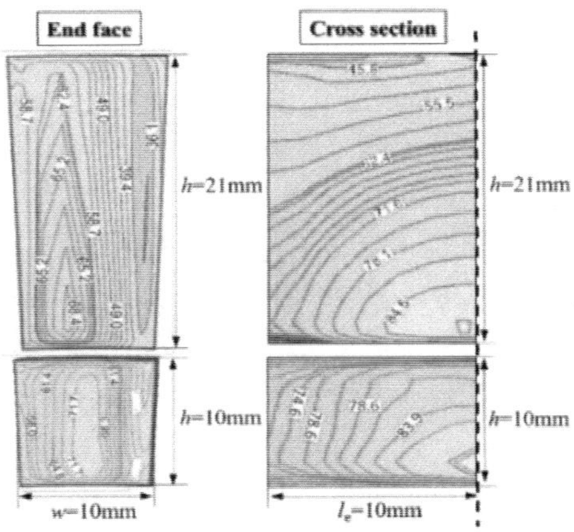


**Figure 5.** Velocity distribution at the inlet end of the passages with different heights.

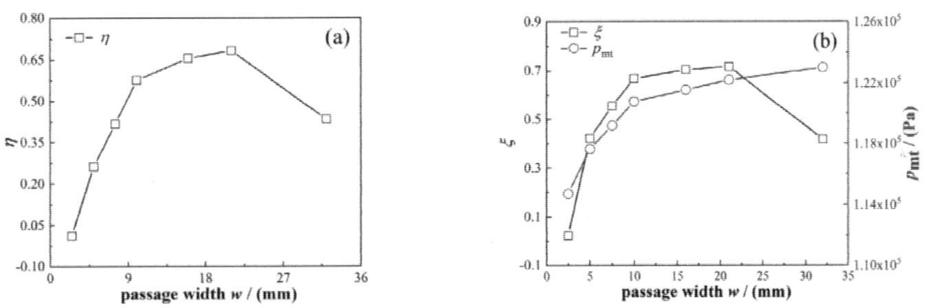


(a) isentropic efficiency $\eta$

(b) ejection rate $\xi$ and the total pressure of medium-pressure gas $p_{\mathrm{mt}}$

**Figure 6.** The variation of performance parameters with passage width $w$.

Figure 6 shows the variation law of the performance parameters with the passage width $w$ when the passage height $h$ is fixed. As shown in figure 6(a), when $w$≤21 mm, the isentropic efficiency $\eta$ increases continuously with the increase of $w$, and the growth rate progressively declines; however, $\eta$ of the model with $w$=32 mm is lower than that of the model with $w$=21 mm. As shown in figure 6(b), different from the variation laws of the passage height changing alone, the total pressure of medium-pressure gas $p_{\mathrm{mt}}$ rises with the increase of $w$, and the growth rate decreases. The variation law of the ejection rate $\xi$ is basically consistent with $\eta$, and the passage width required for attain the maximum $\xi$ is the same as that for achieving the highest efficiency within the calculation range.

The isentropic efficiency $\eta$ of the $w$=5 mm, $h$=16 mm model shown in figure 6(a) is lower than that of the $w$=10 mm, $h$=5 mm model with a smaller passage cross-sectional area as shown in figure 4(a). It indicates that the gas incidence loss enhanced by reducing passage width is more than the loss increment caused by lessening passage height. As shown in figure 7(a), when the passage width is too large, the substantial velocity unevenness at the inlet end of the passage may increase the gas incidence loss of the high-pressure gas to some extent. However, the main reasons for the obvious performance decline are the incomplete expansion and gas backflow, which occur as a result of the lengthy connection and disconnection processes between the passages and the pressure ports. As shown in figure 7(b), because the structure size of high-pressure port, the passage length, and the passage rotation speed are all the same in different models, the open and closed positions of the medium and low-pressure ports determined according to the ideal wave diagram remain nearly unchanged. As a result, when the passage with $w$=32 mm moves between the high and low-pressure ports, a portion of the passage end face may always connect to the pressure port. The passage may connect to the low-pressure port almost without going through the complete closing process, resulting in incomplete gas expansion and a chaotic wave system in the passage. Backflow occurs in the low-pressure port as a result of these negative phenomena, reducing the ejection ability and overall working efficiency of the rotor passages.

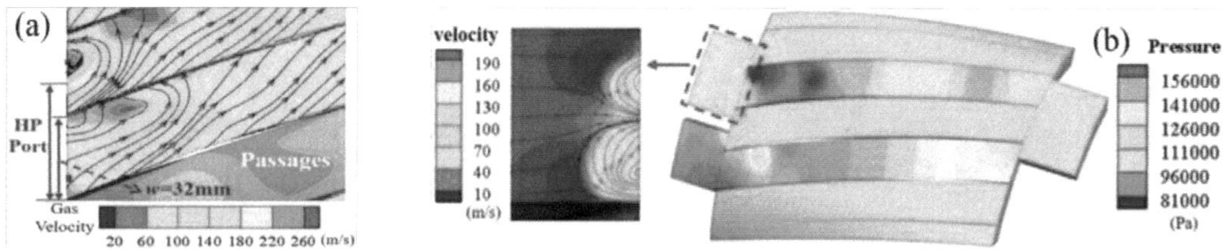


(a) velocity distribution at the passage inlet end

(b) pressure distribution

**Figure 7.** The velocity and pressure distribution in the passages with $w$=32mm, $h$=16mm.

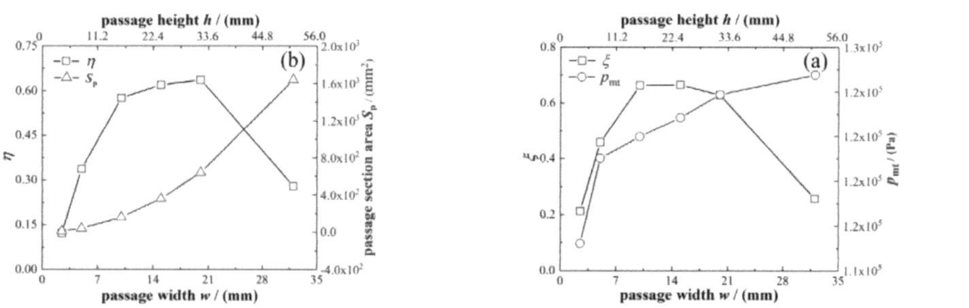


(a) isentropic efficiency $\eta$ and the passage cross-sectional area

(b) ejection rate $\xi$ and the total pressure of medium-pressure gas $p_{mt}$

**Figure 8.** The variation of performance parameters with the size of passages with constant height-to-width ratio.

A single size modification may affect the aspect ratio of passage in these two groups of models mentioned above because of the usage of the control variable method. In order to study the influence of the cross-sectional area of passage $S_p$ on the equipment performance under a fixed aspect ratio, models with different passage sizes were built at a fixed height-to-width ratio (aspect ratio) of 1.6, and the ejection performance was computed under the same working conditions as before. As shown in figure 8, the effects of passage width and height on performance do not conflict with each other under a fixed aspect ratio, and the performance of the model is determined by the superimposed effect of these two. When the passage width $w \leq 10$ mm (passage height $h \leq 16$ mm), the isentropic efficiency $\eta$ increases gradually with the increment of passage size, and the growth rate is higher than that of computational models where $w$ or $h$ is varied alone, since increasing $w$ and $h$ within this range are both beneficial to the efficiency improvement. When 10 mm<w≤21 mm (16 mm<h≤33.6 mm), $\eta$ continues to rise with the increase of passage size. However, the growth rate decreases compared to figure 6(a), owing to the fact that the gas incidence loss increases with the increasing $h$ when $h$>16 mm. For the same reason, when the passage size continues to increase from $w$=21 mm, $h$=33.6 mm to $w$=32 mm, $h$=51.2 mm, the decline degree of $\eta$ is also larger than that in figure 6(a). Because increasing $w$ alone may improve the total pressure of the medium-pressure gas $p_{mt}$, while changing $h$ alone almost has no effect on $p_{mt}$, $p_{mt}$ in figure 8(b) increases as the passage size rises under the combined influence of these two. The ejection rate $\xi$ increases first and then decreases with the increase of passage size, and the changing rule is the same as that of passage width changing alone. Because of the negative effect of excessive passage height, $\xi$ reaches to its maximum value when $w$=16 mm, which is less than the passage width for achieving the maximal efficiency when the passage width changes alone.

Although the cross-sectional area of passage $S_p$ may affect the gas flow loss, it is always set according to the capacity requirement of the equipment. As a result, studying the influence of the aspect ratio $h/w$ of passage on the equipment performance when $S_p$ is fixed is significant for the rational design of the wave rotor. In view of this issue, 4 different cross-sectional area values were chosen for comparing the performance of models with different aspect ratios as shown in figure 9. Under each fixed $S_p$, the isentropic efficiency $\eta$ increases first, then decreases as $h/w$ increases, and may maintain a high level when $0.71 \leq h/w \leq 1.6$, as shown in figure 9(a). When the $h/w$ deviates from this range, the width or height of passage may have a negative impact on the equipment performance. When $h/w \leq 0.5$ and $S_p \geq 250$ mm$^2$, $w$ increases to more than 22.4 mm, causing the issues of short sealing time and gas incidence loss mentioned above and leads to a great decline in performance as $h/w$ decreases. When $S_p$ of passage is fixed, $w$ decreases continuously with the increase of $h/w$. Therefore, as shown in figure 9(b), the total pressure of medium-pressure gas $p_{mt}$ decreases monotonically with the increasing $h/w$ under each cross-sectional area. As shown in figure 9(c), the changing trend of $\xi$ with $h/w$ is the same as that of $\eta$ under each $S_p$, and the $h/w$ values for achieving the maximal $\xi$ are also in the range of $0.71 \leq h/w \leq 1.6$. Because $p_{mt}$ decreases with the increase of $h/w$ (namely the decrease of $w$), $h/w$ values of maximal $\xi$ are slightly larger than the corresponding $h/w$ values when $\eta$ reaches to its maximum, and the decline degree of $\xi$ with $h/w$ is also significantly smaller. When comparing the performance curves under different cross-sectional areas, it can be found that increasing $S_p$ helps to reduce the gas flow loss, which makes $\eta$ and $\xi$ increase with the increment of $S_p$ under the fixed $h/w$ when $h/w \geq 0.71$.

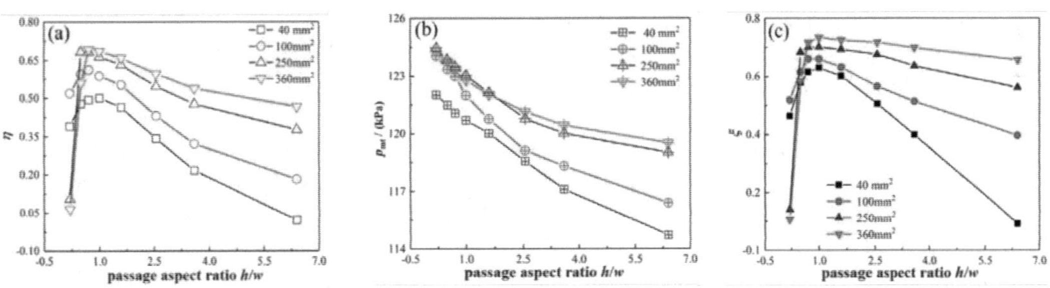


(a) isentropic efficiency $\eta$ and the passage cross-sectional area

(b) total pressure of medium-pressure gas $p_{mt}$

(c) ejection rate $\xi$

**Figure 9.** The variation of performance parameters with height-to-width ratio of passages with constant sectional area.

## 4. Conclusion

As an efficient pressure energy transmission device, the gas wave ejector (GWE) has significant application value in the fields of pressure energy recovery and utilization in the chemical industry. Wave rotor is the core component of GWE, and its passage cross-sectional dimension has an important impact on the equipment performance. In this study, the numerical simulation method was utilized to investigate the influence mechanism and law of the passage height, width and aspect ratio on the equipment performance. The main conclusions are as follows:

1. Increasing the height and width of passage within a certain range can help improve the equipment performance by increasing the gas flow area. However, if the passage height is too large, the unevenness of the gas incident velocity along the height direction may increase the gas incidence loss, resulting in a slight reduction in equipment performance; while if the passage width is too large, the sealing time of passage may be too short, and the gas expansion may not be sufficient, resulting in a significant decrease in the equipment performance.

2. When the passage aspect ratio is fixed (1.6 in this study), the impacts of passage width and height on the performance of the device are independent of one another, and the performance of device is determined by their combined effect.

3. Within the range of passage cross-sectional area investigated in this study ($40mm^2$~$360mm^2$), the equipment performance is optimal when $0.71 \leq h/w \leq 1.6$. When $h/w \geq 0.71$, the efficiency and ejection rate under the same $h/w$ increase with the increment of cross-sectional area.

4. Within the working conditions of this study, the isentropic efficiency and ejection rate of GWE can reach up to 68.5% and 68.4%, respectively, which proves that GWE theoretically has excellent ejection and pressurization performance.

**Acknowledgment**
This work was funded by the National Key Research and Development Program of China (grant no. 2018YFA0704600).

**References**

[1] Li J 2015 Application status and reliability analysis of natural gas compressor in China *Chemical Management* **2**(20) 15-16.

[2] Bulusu K V, Gould D M and Jr. Garris C A 2008 Evaluation of efficiency in compressible flow ejectors *Asme International Mechanical Engineering Congress & Exposition* **8** 531-554.

[3] Hu D, Zhao Y, Wu T, et al. 2019 The complete performance map of gas wave ejector and analysis on the variation laws and limitation of performance *Journal of Engineering for Gas Turbines and Power-Transactions of The Asme* **142**(2) 021012.

[4] Akbari P, Nalim R and Norbert M 2006 A review of wave rotor technology and its applications *Journal of Engineering for Gas Turbines and Power* **128**(4) 717-735.

[5] Kentfield J A C 1969 The performance of pressure-exchange dividers and equalizers *Journal of Basic Engineering* **91**(3) 361-370.

[6] Kharazi Amir A, Akbari Pezhman and Müller N 2006 Implementation of 3-Port condensing wave rotors in R718 cycles *Journal of Energy Resources Technology* **128**(4) 325-334.

[7] Zhao W, Hu D, Liu P, et al. 2012 The port width and position determination for gas wave ejector *Journal of Engineering for Gas Turbines & Power* **134**(6) 064502.

[8] Hu D, Zhao Y, Teng W, et al. 2019 The experimental research and mechanism analysis on the influence of wave rotor rotational speed on the wave system and flow losses of gas wave ejector *Chemical Engineering & Processing: Process Intensification* **144** 107638.

[9] Li Z, Zhao Y, Ji Y, et al. 2020 Design and performance study of forward-curved channel wave rotor for gas wave ejection *Journal of Dalian University of Technology* **60**(06):577-583.

[10] Bai J, Wang C and Zhang Y 2014 Application of a turbulence model based on von karmen length scale in steady and unsteady flow simulation *Engineering Mechanics* **31**(11) 39-45.

[11] Kurec K, Piechna J, Gumowski K, et al. 2017 Investigations on unsteady flow within a stationary passage of a pressure wave exchanger, by means of PIV measurements and CFD calculations *Applied Thermal Engineering* **112** 610-620.

[12] Okamoto K and Nagashima T 2007 Visualization of wave rotor inner flow dynamics *Journal of Propulsion and Power* **23**(2) 292-300.

Journal of Physics: Conference Series **2329** (2022) 012027 doi:10.1088/1742-6596/2329/1/012027

# Singularly Perturbed Solution of Non-Fourier Temperature Field with Mixed Boundary Value Dual-Phase Delay with Thermal Conductivity Jumps

**Liping Bao*, Yanan Jiang and Liqun Wu**

Hangzhou Dianzi University, Hangzhou, China
Email: baolp@hdu.edu.cn

**Abstract.** Based on the properties of laminates, a class of nonlinear singularly perturbed mixture equations with discontinuous coefficients over bounded domains is constructed by using a dual-phase delayed heat conduction model. First, the singular perturbation expansion method is used, combined with the corresponding boundary conditions, the partial differential equation method and the Laplace transform method are used to obtain the external solution, the boundary layer, and the corner layer. Secondly, the time-varying temperature field at the discontinuity is obtained, which leads to the asymptotic expansion of the solution. Finally, the consistent validity of the asymptotic solution is obtained through residual estimation.
**Keywords.** Singular perturbations, discontinuity coefficients, dual-phase delay, mixed equations.

## 1. Introduction

Although the heat conduction model based on Fourier's law has achieved many achievements in dealing with most engineering application problems, it cannot predict the temperature distribution in special cases such as ultra-rapid, micro-scale, and low temperature. Therefore, more and more attention has been paid to the study of the law of supernormal heat conduction. Cattaneo [1] and Vernotte [2] coupled the hyperbolic heat conduction equation with the local heat balance to obtain a non-Fourier heat conduction equation model and called it the Cattaneo-Vernotte (CV) model. However, this model still has limitations and is not suitable for the microstructure heat transfer process. Therefore, Tzou [3-4] attributed the entire effect of the microstructure to the delay of the macroscopic equation, the dual-phase-lagging (DPL) model is proposed. Maes et al. [5] investigated an isotropic thermoelastic system with the DPL heat conduction. They consider a homogeneous Dirichlet boundary condition for displacement and temperature and formulate the variational formulation for the associated fully hyperbolic coupled system. By using a time discretization method, they show the existence of a unique weak solution under lower regularity assumptions on the data than established before in literature. Mao et al. [6] based on the DPL heat conduction model, the CV model, and the improved CV model investigate the one-dimensional heat conduction in gold films with nano-scale thickness exposed to ultrafast laser heating. The influence of system parameters on the temperature field is explored. They find that a larger Knudsen number normally leads to a higher temperature for all the non-Fourier heat conduction models considered in this work. References [7-10] are studied on the application of the DPL model to biological tissues, characterizing non-Fourier heat conduction through thermal hysteresis during the interaction of pulsed lasers with biological tissues. Tan et al. [11] established a hyperbolic heat conduction model based on the DPL model. The results show that the ratio of the two delay times significantly affects the temperature characteristics and the ablation of the target during the femtosecond laser ablation process. The femtosecond laser ablation model based on the DPL model can effectively simulate the femtosecond laser ablation process. Ghasemi et al. [12] studied DPL transient non-Fourier heat conduction in a functional graded cylindrical material that is analytically solved under axial heat flux conditions.

With the improvement of coating technology, the materials of the laminated structure have been widely used. Due to the characteristics of their structures, the situation of coefficient discontinuity is caused. How to solve them accurately and efficiently is a big challenge. Li et al. [13] established an analytical model for the non-Fourier heat conduction of a layered material plate subjected to thermal shock on its surfaces. Numerical solutions are obtained by using the backward difference method. The stability of the proposed method is discussed. Variation of temperature field with time, thermal diffusivity, the ratio of space step size to time step size, and relaxation time are discussed in detail. In addition, the thermal stress field associated with the calculated temperature field is given. Prabha et al. [14] studied a singularly perturbed second-order ordinary differential equation having two small parameters with a discontinuous source term. The presence of two parameters gives rise to boundary layers of different widths and the discontinuous source term generates interior layers on both sides of the discontinuous point. Maurya et al. [15] studied the treatment of infected tumorous cells through mathematical modeling and simulation of heat transfer in tissues by using a nonlinear DPL bioheat transfer model with Dirichlet boundary condition. Li et al. [16] proposed a new analytical method for solving heat transfer problems of laminate materials based on the differential theory.

In this paper, the DPL model is used to describe the temperature field problem caused by the action of a short-pulse laser on the surface of the laminated material, and the nonlinear singular perturbation mixing equation is obtained by considering the jump in thermal conductivity due to the sharp temperature change. The asymptotic expansion equation of the equation is constructed by the singular perturbation expansion method, the external solution is solved, the boundary layer and the corners are corrected, the temperature field distribution at the interval is determined, and the existence of the asymptotic solution is proved, to obtain a complete temperature field distribution, which provides a reference for the application of non-Fourier heat conduction in the field of laminated materials.

## 2. Model Establishment

Now make the following hypotheses:

[$H_1$] $\phi_1(x), \phi_2(x), \phi_3(t), \phi_4(t)$ are known continuously differentiable functions of arbitrary order

[$H_2$] $T$ is finite value, $a$, $b$, $c$ are all constants, and $a<0$, $b>0$

[$H_3$] Thermal conductivity

$$k(u) = \begin{cases} k_1, 0 \le x \le x_0 \\ k_2, x_0 < x \le 1 \end{cases}$$

The dual-phase delayed heat conduction model of non-Fourier heat conduction can be obtained from the literatures [3-4]. In the case of no internal heat source, the phase delay is considered to be introduced into both the heat flow $q$ and the temperature gradient $u_x$. The delay characteristics can be described by the following equation

$$q(x, t+\sigma) = -k(u)u_x(x, t+\mu) \tag{1}$$

Equation (1) states that the material is at position $x$, and the temperature gradient is formed at the time $t+\mu$, causing heat flow at different time $t+\sigma$. Here $k(u)$ is the thermal conductivity coefficient, and the phase delay times $\sigma$ and $\varepsilon$ are both positive values. Do the Taylor series expansion of equation (1) concerning time $t$, we can get

$$q(x,t)+\sigma\frac{\partial q}{\partial t}(x,t)\approx -k(u)\{u_x(x,t)+\mu\frac{\partial}{\partial t}[u_x(x,t)]\} \tag{2}$$

Substitute equation (2) into the energy conservation equation $-q_x(x,t)=cu_t$ we can obtain

$$(k(u)u_x)_x+\mu(k(u)u_{xt})_x=c(u_t+\sigma u_{tt}) \tag{3}$$

where $c=\rho\cdot c_p$, $\rho$ and $c_p$ represent density and specific heat capacity, respectively.

This article considers the case of $\sigma=\mu$ and for the convenience of subsequent comparisons of the same power, let $\mu=\varepsilon^2$, that is

$$(k(u)u_x)_x+\varepsilon^2(k(u)u_{tx})_x=c(u_t+\varepsilon^2 u_{tt}) \tag{4}$$

The initial and boundary conditions are

$$\begin{cases} u(x,0)=\phi_1(x), & 0\le x\le 1 \\ u_t(x,0)=\phi_2(x), & 0\le x\le 1 \\ u_x(0,t)+au(0,t)=\phi_3(t), & 0<t<T \\ u_x(1,t)+bu(1,t)=\phi_4(t), & 0<t<T \end{cases} \tag{5}$$

## 3. Form Expansion

Given hypothesis [H3], $k(u)$ is discontinuous. Let $\mathbf{\Omega}_1=[0,x_0]\times(0,T)$, $\mathbf{\Omega}_2=[x_0,1]\times(0,T)$, then there is

$$\begin{cases} u(x,t)=u^{(-)}(x,t), & (x,t)\in\mathbf{\Omega}_1 \\ u(x,t)=u^{(+)}(x,t), & (x,t)\in\mathbf{\Omega}_2 \end{cases}$$

Therefore, equation (4) and equation (5) can be rewritten as

$$(-)\begin{cases} k_1(u_{xx}^{(-)}+\varepsilon^2 u_{xxt}^{(-)})=c(u_t^{(-)}+\varepsilon^2 u_{tt}^{(-)}) \\ u^{(-)}(x,0)=\phi_1(x), & 0\le x<x_0 \\ u_t^{(-)}(x,0)=\phi_2(x), & 0\le x<x_0 \\ u_x^{(-)}(0,t)+au^{(-)}(0,t)=\phi_3(t), & 0<t<T \\ u^{(-)}(x_0,t)=\psi(t,\varepsilon), & 0<t<T \end{cases} \tag{6}$$

$$(+)\begin{cases} k_2(u_{xx}^{(+)}+\varepsilon^2 u_{xxt}^{(+)})=c(u_t^{(+)}+\varepsilon^2 u_{tt}^{(+)}) \\ u^{(+)}(x,0)=\phi_1(x), & x_0<x\le 1 \\ u_t^{(+)}(x,0)=\phi_2(x), & x_0<x\le 1 \\ u^{(+)}(x_0,t)=\psi(t,\varepsilon), & 0<t<T \\ u_x^{(+)}(1,t)+bu^{(+)}(1,t)=\phi_4(t), & 0<t<T \end{cases} \tag{7}$$

Here the temperature $\psi(t,\varepsilon) \sim \sum_{n=0}^{\infty} \psi_n(t)\varepsilon^n$ at the discontinuity in the thermal conductivity coefficient is introduced, which is to be determined. Let stretch variables

$$\tau = \frac{t}{\varepsilon^2}, \xi = \frac{x}{\varepsilon}, \eta = \frac{1-x}{\varepsilon}, \varsigma = \frac{x - x_0}{\varepsilon}$$

*3.1. Left Part*

First, construct the formal asymptotic solution of (6)

$$u^{(-)}(x,t) \sim \sum_{n}^{\infty} \bar{u}_n^{(-)}(x,t)\varepsilon^n + \sum_{n}^{\infty} U_n^{(-)}(x,\tau)\varepsilon^n + \sum_{n}^{\infty} V_n^{(-)}(\xi,\tau)\varepsilon^n + \sum_{n}^{\infty} I_n^{(-)}(\zeta,\tau)\varepsilon^n \tag{8}$$

*3.1.1. External Solution.* Do regular expansion, substitute $\sum_{n}^{\infty} \bar{u}_n^{(-)}(x,t)\varepsilon^n$ into equation (6), comparing the coefficients of the same power of $\varepsilon$, we can get

$$(-)\begin{cases} k_1 \bar{u}_{0xx}^{(-)} = c\bar{u}_{0t}^{(-)} \\ \bar{u}_0^{(-)}(x,0) = \phi_1^{(-)}(x), & 0 \le x < x_0 \\ \bar{u}_{0x}^{(-)}(0,t) + a\bar{u}_0^{(-)}(0,t) = \phi_3(t), 0 < t < T \\ \bar{u}_0^{(-)}(x_0,t) = \psi_0(t), & 0 < t < T \end{cases} \tag{9}$$

$$(-)\begin{cases} k_1 \bar{u}_{1xx}^{(-)} = c\bar{u}_{1t}^{(-)} \\ \bar{u}_1^{(-)}(x,0) = 0, & 0 \le x < x_0 \\ \bar{u}_{1x}^{(-)}(0,t) + a\bar{u}_1^{(-)}(0,t) = 0, 0 < t < T \\ \bar{u}_1^{(-)}(x_0,t) = \psi_1(t), & 0 < t < T \end{cases} \tag{10}$$

$$(-)\begin{cases} k_1 \bar{u}_{nxx}^{(-)} - c\bar{u}_{nt}^{(-)} = c\bar{u}_{n-2,tt}^{(-)} - k_1 \bar{u}_{n-2,xxt}^{(-)} = f_n^{(-)} \\ \bar{u}_n^{(-)}(x,0) = -U_n^{(-)}(x,0), & 0 \le x < x_0 \\ \bar{u}_{nx}^{(-)}(0,t) + a\bar{u}_n^{(-)}(0,t) = 0, 0 < t < T \\ \bar{u}_n^{(-)}(x_0,t) = \psi_n(t), & 0 < t < T \end{cases} \tag{11}$$

By boundary homogenization and separation of variables method to solve the equations (9)-(11), we get

$$\begin{aligned} \bar{u}_n^{(-)}(x,t) &= w_n(x,t) + \sum_{m=1}^{\infty} A_{nm} \frac{\sin\sqrt{\gamma_m}(x - x_0)}{\cos\sqrt{\gamma_m}x_0} \exp(-\frac{k_1\gamma_m t}{c}) \\ &+ \int_0^t \sum_{m=1}^{\infty} B_{nm} \frac{\sin\sqrt{\gamma_m}(x - x_0)}{\cos\sqrt{\gamma_m}x_0} \exp(-\frac{k_1\gamma_m(t-\theta)}{c}) d\theta \qquad n = 0,1,... \end{aligned} \tag{12}$$

Where

$$w_0(x,t) = \frac{\phi_3(t) - a\psi_0(t)}{1 - ax_0}(x - x_0) + \psi_0(t) \qquad w_n(x,t) = \frac{\psi_n(t)}{1 - ax_0}(1 - ax)$$

$$A_{0m}=\frac{\cos\sqrt{\gamma_m}x_0}{L_m}\int_0^{x_0}(\phi_1^{(-)}(x)-w_0(x,0))\sin\sqrt{\gamma_m}(x-x_0)\mathrm{d}x$$

$$A_{1m}=\frac{\cos\sqrt{\gamma_m}x_0}{L_m}\int_0^{x_0}-w_1(x,0)\sin\sqrt{\gamma_m}(x-x_0)\mathrm{d}x$$

$$A_{nm}=\frac{\cos\sqrt{\gamma_m}x_0}{L_m}\int_0^{x_0}-(U_n(x,0)+w_n(x,0))\sin\sqrt{\gamma_m}(x-x_0)\mathrm{d}x$$

$$B_{0m}=\frac{c\cos\sqrt{\gamma_m}x_0}{L_m}\int_0^{x_0}w_{0t}(x,\theta)\sin\sqrt{\gamma_m}(x-x_0)\mathrm{d}x\quad B_{1m}=\frac{c\cos\sqrt{\gamma_m}x_0}{L_m}\int_0^{x_0}w_{1t}(x,\theta)\sin\sqrt{\gamma_m}(x-x_0)\mathrm{d}x$$

$$B_{nm}=\frac{\cos\sqrt{\gamma_m}x_0}{L_m}\int_0^{x_0}(f_n^{(-)}+cw_{nt}(x,\theta))\sin\sqrt{\gamma_m}(x-x_0)\mathrm{d}x$$

$L_m=\int_0^{x_0}\sin^2\sqrt{\gamma_m}(x-x_0)\mathrm{d}x$, $\gamma_m$ satisfies $\tan\sqrt{\gamma_m}x_0=\frac{\sqrt{\gamma_m}}{a}, m=1,2,\ldots$.

*3.1.2. Initial Layer.* Do initial layer correction at *t=0*, substitute $\sum_n^{\infty}\bar{u}_n^{(-)}(x,t)\varepsilon^n+\sum_{n=0}^{\infty}U_n^{(-)}(x,\tau)\varepsilon^n$ into equation (6), comparing the coefficients of the same power of $\varepsilon$, we can get

$$(-)\begin{cases}U_{l\tau}^{(-)}+U_{l\tau\tau}^{(-)}=0\\ U_l^{(-)}(x,\infty)=0,\ 0\le x<x_0 \qquad l=0,1\\ U_{l\tau}^{(-)}(x,0)=0,\ \ 0\le x<x_0\end{cases}\tag{13}$$

$$(-)\begin{cases}U_{2\tau}^{(-)}+U_{2\tau\tau}^{(-)}=0\\ U_2^{(-)}(x,\infty)=0, \qquad 0\le x<x_0\\ U_{2\tau}^{(-)}(x,0)=\phi_2^{(-)}(x)-\bar{u}_{0t}^{(-)}(x,0),0\le x<x_0\end{cases}\tag{14}$$

$$(-)\begin{cases}U_{n\tau}^{(-)}+U_{n\tau\tau}^{(-)}=\frac{k_1}{c}(U_{n-2,xx}^{(-)}+U_{n-2,xx\tau}^{(-)})\\ U_n^{(-)}(x,\infty)=0, \qquad 0\le x<x_0\\ U_{n\tau}^{(-)}(x,0)=-\bar{u}_{n-2,t}^{(-)}(x,0),0\le x<x_0\end{cases}\tag{15}$$

Solve equations (13)-(15) to get

$$U_l^{(-)}(x,\tau)=0\quad U_2^{(-)}(x,\tau)=(\bar{u}_{0t}^{(-)}(x,0)-\phi_2^{(-)}(x))\mathrm{e}^{-\tau}\quad U_n^{(-)}(x,\tau)=\bar{u}_{\mathrm{n}-2,t}(x,0)\mathrm{e}^{-\tau}$$

*3.1.3. (0,0) Near the Corner.* To ensure the continuity of the solution and the derivative of the solution around the point (0,0), a correction needs to be done around this point, substitute

$$\sum_n^{\infty}\bar{u}_n^{(-)}(x,t)\varepsilon^n+\sum_{n=0}^{\infty}U_n^{(-)}(x,\tau)\varepsilon^n+\sum_{n=0}^{\infty}V_n^{(-)}(\xi,\tau)\varepsilon^n$$

into equation (6), we get

$$(-)\begin{cases} V_{0\xi\xi}^{(-)}+V_{0\xi\xi\tau}^{(-)}=\dfrac{c}{k_1}(V_{0\tau}^{(-)}+V_{0\tau\tau}^{(-)}) \\ V_0^{(-)}(\xi,0)=0, \quad 0\le x<x_0 \\ V_{0\tau}^{(-)}(\xi,0)=0, \quad 0\le x<x_0 \\ V_{0\xi}^{(-)}(0,\tau)=0, \quad 0<t<T \\ V_0^{(-)}(+\infty,\tau)=0, 0<t<T \end{cases} \tag{16}$$

$$(-)\begin{cases} V_{n\xi\xi}^{(-)}+V_{n\xi\xi\tau}^{(-)}=\dfrac{c}{k_1}(V_{n\tau}^{(-)}+V_{n\tau\tau}^{(-)}) \\ V_n^{(-)}(\xi,0)=0, \quad 0\le x<x_0 \\ V_{n\tau}^{(-)}(\xi,0)=0, \quad 0\le x<x_0 \\ V_{n\xi}^{(-)}(0,\tau)=H_n, \quad 0<t<T \\ V_n^{(-)}(+\infty,\tau)=0, 0<t<T \end{cases} \quad n=1,2,... \tag{17}$$

where $H_n$ is composed of known functions related to $\overline{u}_0^{(-)},\overline{u}_0^{(-)},...,\overline{u}_{n-1}^{(-)},U_2^{(-)},...,U_{n-1}^{(-)},V_{n-1}^{(-)}$.

Taking the Laplace transform of equations (16)-(18) and solving it by the constant variation method, we get

$$V_0^{(-)}(\xi,\tau)=0 \qquad V_n^{(-)}(\xi,\tau)=\frac{1}{2\pi \mathrm{i}}\int_{\beta-\mathrm{i}\infty}^{\beta+\mathrm{i}\infty} \mathrm{e}^{s\tau}\hat{V}_n(\xi,s)\mathrm{d}s$$

$\hat{V}_n^{(-)}(\xi,s)=L[V_n^{(-)}(\xi,\tau)]=-\dfrac{\hat{H}_n}{\sqrt{as/k_1}}\mathrm{e}^{-\sqrt{as/k_1}\xi}$, where $L$ is the Laplace transform symbol.

*3.1.4.* $(x_0,0)$ *Near the Corner.* To ensure the continuity of the solution and the derivative of the solution around the point ( $x_0$,0), a correction needs to be done around this point, substitute

$$\sum_{n}^{\infty}\overline{u}_n^{(-)}(x,t)\varepsilon^n+\sum_{n=0}^{\infty}U_n^{(-)}(x,\tau)\varepsilon^n+\sum_{n=0}^{\infty}V_n^{(-)}(\xi,\tau)\varepsilon^n+\sum_{n=0}^{\infty}I_n^{(-)}(\varsigma,\tau)\varepsilon^n$$

into equation (6), we get

$$(-)\begin{cases} I_{l\varsigma\varsigma}^{(-)}+I_{l\varsigma\varsigma\tau}^{(-)}=\dfrac{c}{k_1}(I_{l\tau}^{(-)}+I_{l\tau\tau}^{(-)}) \\ I_l^{(-)}(\varsigma,0)=0, \quad 0\le x<x_0 \\ I_{l\tau}^{(-)}(\varsigma,0)=0, \quad 0\le x<x_0 \\ I_l^{(-)}(0,\tau)=0, \quad 0<t<T \\ I_{l\varsigma}^{(-)}(-\infty,\tau)=0, \ 0<t<T \end{cases} \quad l=0,1 \tag{18}$$

$$(-)\begin{cases} I_{n\varsigma\varsigma}^{(-)}+I_{n\varsigma\varsigma\tau}^{(-)}=\dfrac{c}{k_1}(I_{n\tau}^{(-)}+I_{n\tau\tau}^{(-)}) \\ I_n^{(-)}(\varsigma,0)=0, & 0\le x<x_0 \\ I_{n\tau}^{(-)}(\varsigma,0)=0, & 0\le x<x_0 \\ I_n^{(-)}(0,\tau)=-U_n^{(-)}(x_0,\tau)\mathrm{e}^{-\tau}, & 0<t<T \\ I_{n\varsigma}^{(-)}(-\infty,\tau)=0, & 0<t<T \end{cases} \quad n=3,4,... \tag{19}$$

Solving equations (19)-(20) in the same way as equations (16)-(18), we get

$$I_l^{(-)}(\varsigma,\tau)=0 \quad I_n^{(-)}(\varsigma,\tau)=\frac{1}{2\pi\mathrm{i}}\int_{\beta\text{-i}\infty}^{\beta\text{+i}\infty}\mathrm{e}^{s\tau}\hat{I}_n(\varsigma,s)\mathrm{d}s \quad \hat{I}_n^{(-)}(\varsigma,s)=L[I_n^{(-)}(\varsigma,\tau)]=-\frac{\hat{I}_n^{(-)}(0,s)}{\sqrt{as/k_1}}\mathrm{e}^{-\sqrt{as/k_1}\varsigma}$$

*3.2. Right Part*
Construct asymptotic solutions of equation (7)

$$u^{(+)}(x,t)=\sum_n^{\infty}\overline{u}_n^{(+)}(x,t)\varepsilon^n+\sum_n^{\infty}U_n^{(+)}(x,\tau)\varepsilon^n+\sum_n^{\infty}S_n^{(+)}(\eta,\tau)\varepsilon^n+\sum_n^{\infty}I_n^{(+)}(\zeta,\tau)\varepsilon^n \tag{20}$$

*3.2.1. External Solution.* The same method as the left part can be obtained

$$(+)\begin{cases} k_2\overline{u}_{0xx}^{(+)}=c\overline{u}_{0t}^{(+)} \\ \overline{u}_0^{(+)}(x,0)=\phi_1^{(+)}(x), & x_0<x\le 1 \\ \overline{u}_0^{(+)}(x_0,t)=\psi_0(t), & 0<t<T \\ \overline{u}_{0x}^{(+)}(1,t)+b\overline{u}_0^{(+)}(1,t)=\phi_4(t), 0<t<T \end{cases} \tag{21}$$

$$(+)\begin{cases} k_2\overline{u}_{1xx}^{(+)}=c\overline{u}_{1t}^{(+)} \\ \overline{u}_1^{(+)}(x,0)=0, & x_0<x\le 1 \\ \overline{u}_1^{(+)}(x_0,t)=\psi_1(t), & 0<t<T \\ \overline{u}_{1x}^{(+)}(1,t)+b\overline{u}_1^{(+)}(1,t)=0, 0<t<T \end{cases} \tag{22}$$

$$(+)\begin{cases} k_2\overline{u}_{nxx}^{(+)}-c\overline{u}_{nt}^{(+)}=c\overline{u}_{n-2,tt}^{(+)}-k_2\overline{u}_{n-2,xxt}^{(+)}=f_n^{(+)} \\ \overline{u}_n^{(+)}(x,0)=-U_n^{(+)}(x,0), & x_0<x\le 1 \\ \overline{u}_n^{(+)}(x_0,t)=\psi_n(t), & 0<t<T \\ \overline{u}_{nx}^{(+)}(1,t)+b\overline{u}_n^{(+)}(1,t)=0, & 0<t<T \end{cases} \tag{23}$$

Similar to the solution method of equation (9), equations (22)-(24) is solved

$$\begin{aligned} \overline{u}_n^{(+)}(x,t)&=\omega_n(x,t)+\sum_{m=1}^{\infty}C_{nm}\frac{\sin\sqrt{\lambda_m}(x-x_0)}{\cos\sqrt{\lambda_m}x_0}\exp(-\frac{k_2\lambda_m t}{c}) \\ &+\int_0^t\sum_{m=1}^{\infty}D_{nm}\frac{\sin\sqrt{\lambda_m}(x-x_0)}{\cos\sqrt{\lambda_m}x_0}\exp(-\frac{k_2\lambda_m(t-\theta)}{c})\mathrm{d}\theta \qquad n=0,1,... \end{aligned} \tag{24}$$

Where

$$\omega_0(x,t)=\frac{\phi_4(t)-b\psi_0(t)}{1+b(1-x_0)}(x-x_0)+\psi_0(t) \qquad \omega_n(x,t)=\frac{\psi_n(t)}{1+b(1-x_0)}[1+b(1-x)]$$

$$C_{0m}=\frac{\cos\sqrt{\lambda_m}x_0}{E_m}\int_{x_0}^{1}(\phi_1^{(+)}(x)-\omega_0(x,0)\sin\sqrt{\lambda_m}(x-x_0)\mathrm{d}x \quad C_{1m}=\frac{\cos\sqrt{\lambda_m}x_0}{E_m}\int_{x_0}^{1}-\omega_1(x,0)\sin\sqrt{\lambda_m}(x-x_0)\mathrm{d}x$$

$$C_{nm}=\frac{\cos\sqrt{\lambda_m}x_0}{E_m}\int_{x_0}^{1}-(U_n^{(+)}(x,0)+\omega_n(x,0))\sin\sqrt{\lambda_m}(x-x_0)\mathrm{d}x$$

$$D_{0m}=\frac{\cos\sqrt{\lambda_m}x_0}{E_m}\int_{x_0}^{1}(c\omega_{0t}(x,\theta))\sin\sqrt{\lambda_m}(x-x_0)\mathrm{d}x \quad D_{1m}=\frac{\cos\sqrt{\lambda_m}x_0}{E_m}\int_{x_0}^{1}(c\omega_{1t}(x,\theta))\sin\sqrt{\lambda_m}(x-x_0)\mathrm{d}x$$

$$D_{nm}=\frac{\cos\sqrt{\lambda_m}x_0}{E_m}\int_{x_0}^{1}(f_n^{(+)}+c\omega_{nt}(x,\theta))\sin\sqrt{\lambda_m}(x-x_0)\mathrm{d}x$$

$E_m=\int_{x_0}^{1}\sin^2\sqrt{\lambda_m}(x-x_0)\mathrm{d}x$, $\lambda_m$ satisfies $\tan\sqrt{\lambda_m}(x_0-1)=\frac{\sqrt{\lambda_m}}{b}$, $m=1,2,\ldots$.

*3.2.2 Initial Layer.* In the same way as *3.1.2*, we can get

$$(+)\begin{cases}U_{l\tau}^{(+)}+U_{l\tau\tau}^{(+)}=0 & \\ U_l^{(+)}(x,\infty)=0, \quad x_0<x\le 1 & l=0,1 \\ U_{l\tau}^{(+)}(x,0)=0, \quad x_0<x\le 1 & \end{cases} \tag{25}$$

$$(+)\begin{cases}U_{2\tau}^{(+)}+U_{2\tau\tau}^{(+)}=0 \\ U_2^{(+)}(x,\infty)=0, \qquad x_0<x\le 1 \\ U_{2\tau}^{(+)}(x,0)=\phi_2^{(+)}(x)-\bar{u}_{0t}^{(+)}(x,0), x_0<x\le 1\end{cases} \tag{26}$$

$$(+)\begin{cases}U_{n\tau}^{(+)}+U_{n\tau\tau}^{(+)}=\frac{k_2}{c}(U_{n-2,xx}^{(+)}+U_{n-2,xx\tau}^{(+)}) & \\ U_n^{(+)}(x,\infty)=0, \qquad x_0<x\le 1 & n=3,4,\ldots \\ U_{n\tau}^{(+)}(x,0)=-\bar{u}_{n-2,t}^{(+)}(x,0), x_0<x\le 1 & \end{cases} \tag{27}$$

Solving equations (25)-(27) to get

$$U_l^{(+)}(x,\tau)=0 \quad U_2^{(+)}(x,\tau)=(\bar{u}_{0t}^{(+)}(x,0)-\phi_2^{(+)}(x))e^{-\tau} \quad U_n^{(+)}(x,\tau)=\bar{u}_{n-2,t}^{(+)}(x,0)e^{-\tau}$$

*3.2.3.* (1,0) *Near the Corner.* To ensure the consistency of the solution and the derivative of the solution at (1,0), make corner correction around (1,0), substitute

$$\sum_{n}^{\infty}\bar{u}_n^{(+)}(x,t)\varepsilon^n+\sum_{n=0}^{\infty}U_n^{(+)}(x,\tau)\varepsilon^n+\sum_{n=0}^{\infty}S_n^{(+)}(\eta,\tau)\varepsilon^n$$

into equation (7), we obtain

$$(+)\begin{cases} S_{0\eta\eta}^{(+)}+S_{0\eta\eta\tau}^{(+)}=\dfrac{c}{k_2}(S_{0\tau}^{(+)}+S_{0\tau\tau}^{(+)}) \\ S_0^{(+)}(\eta,0)=0, \quad x_0<x\le 1 \\ S_{0\tau}^{(+)}(\eta,0)=0, \quad x_0<x\le 1 \\ S_{0\eta}^{(+)}(0,\tau)=0, \quad 0<t<T \\ S_0^{(+)}(+\infty,\tau)=0, 0<t<T \end{cases} \tag{28}$$

$$(+)\begin{cases} S_{n\eta\eta}^{(+)}+S_{n\eta\eta\tau}^{(+)}=\dfrac{c}{k_2}(S_{n\tau}^{(+)}+S_{n\tau\tau}^{(+)}) \\ S_n^{(+)}(\eta,0)=0, \quad x_0<x\le 1 \\ S_{n\tau}^{(+)}(\eta,0)=0, \quad x_0<x\le 1 \\ S_{n\eta}^{(+)}(0,\tau)=P_n, \ 0<t<T \\ S_n^{(+)}(+\infty,\tau)=0, 0<t<T \end{cases} \quad n=1,2,... \tag{29}$$

where $P_n$ is composed of known functions related to $\overline{u}_0^{(+)},\overline{u}_1^{(+)},...,\overline{u}_{n-1}^{(+)},U_2^{(+)},...,U_{n-1}^{(+)},S_{n-1}^{(+)}$.

Solving equations (28)-(29) to get

$$S_0^{(+)}(\eta,\tau)=0 \quad S_n^{(+)}(\eta,\tau)=\frac{1}{2\pi i}\int_{\beta-i\infty}^{\beta+i\infty} e^{s\tau}\hat{S}_n^{(+)}(\eta,s)ds \quad \hat{S}_n^{(+)}(\eta,s)=L[S_n^{(+)}(\eta,\tau)]=-\frac{\hat{S}_n^{(+)}(0,s)}{\sqrt{as/k_2}}e^{-\sqrt{as/k_2}\eta}$$

*3.2.4.* $(x_0, 0)$ *Near the Corner.* Same method as 3.1.4, we get

$$(+)\begin{cases} I_{l\varsigma\varsigma}^{(+)}+I_{l\varsigma\varsigma\tau}^{(+)}=\dfrac{c}{k_2}(I_{l\tau}^{(+)}+I_{l\tau\tau}^{(+)}) \\ I_l^{(+)}(\varsigma,0)=0, \ x_0<x\le 1 \\ I_{l\tau}^{(+)}(\varsigma,0)=0, \ x_0<x\le 1 \\ I_l^{(+)}(0,\tau)=0, \ 0<t<T \\ I_{l\varsigma}^{(+)}(\infty,\tau)=0, 0<t<T \end{cases} \quad l=0,1 \tag{30}$$

$$(+)\begin{cases} I_{n\varsigma\varsigma}^{(+)}+I_{n\varsigma\varsigma\tau}^{(+)}=\dfrac{c}{k_2}(I_{n\tau}^{(+)}+I_{n\tau\tau}^{(+)}) \\ I_n^{(+)}(\varsigma,0)=0, \quad x_0<x\le 1 \\ I_{n\tau}^{(+)}(\varsigma,0)=0, \quad x_0<x\le 1 \\ I_n^{(+)}(0,\tau)=-U_n^{(+)}(x_0,\tau)e^{-\tau}, 0<t<T \\ I_{n\varsigma}^{(+)}(\infty,\tau)=0, \quad 0<t<T \end{cases} \quad n=2,3,... \tag{31}$$

Solving equations (30)-(31) to get

$$I_l^{(+)}(\varsigma,\tau)=0 \quad I_n^{(+)}(\varsigma,\tau)=\frac{1}{2\pi i}\int_{\beta-i\infty}^{\beta+i\infty} e^{s\tau}\hat{I}_n(\varsigma,s)ds \quad \hat{I}_n^{(+)}(\varsigma,s)=L[I_n(\varsigma,\tau)]=-\frac{\hat{I}_n^{(+)}(0,s)}{\sqrt{as/k_2}}e^{-\sqrt{as/k_2}\varsigma}$$

## 4. Determination of the Junction

Next we consider the existence of connection condition $\psi(t,\varepsilon)$, there is the following theorem:

**Theorem 1.** The temperature at which the thermal conductivity $k(u)$ in equation (1) jumps at $x = x_0$ is

$$\psi(t,\varepsilon) \sim \sum_{n=0}^{\infty} \psi_n(t)\varepsilon^n \tag{32}$$

then $\psi(t,\varepsilon)$ exists.

**Proof.** Since temperature is smooth and continuously changing, let $u_x^{(-)}(x_0,t) = u_x^{(+)}(x_0,t)$, compare the coefficients of the same power of $\varepsilon$, we can get

$$\begin{cases} \bar{u}_{0x}^{(-)}(x_0,t) - \bar{u}_{0x}^{(+)}(x_0,t) = 0 \\ \bar{u}_{nx}^{(-)}(x_0,t) - \bar{u}_{nx}^{(+)}(x_0,t) = g_n(t) \end{cases}$$

where $g_n(t)$ is a known function, so

$$\begin{aligned}
&\bar{u}_{0x}^{(-)}(x_0,t) - \bar{u}_{0x}^{(+)}(x_0,t) = \Big(\frac{b}{1+b(1-x_0)} - \frac{a}{1-ax_0} + \sum_{m=1}^{\infty} B_{0m,2}\sqrt{\gamma_m}\sec\sqrt{\gamma_m}x_0 - \sum_{m=1}^{\infty} D_{0m,2}\sqrt{\lambda_m}\sec\sqrt{\lambda_m}x_0\Big)\psi_0(t) \\
&+\Big[\sum_{m=0}^{\infty}(A_{0m,2}-B_{0m,2})\sqrt{\gamma_m}\exp\Big(-\frac{k_1\gamma_m t}{c}\Big)\sec\sqrt{\gamma_m}x_0 - \sum_{m=0}^{\infty}(C_{0m,2}-D_{0m,2})\sqrt{\lambda_m}\exp\Big(-\frac{k_2\lambda_m t}{c}\Big)\sec\sqrt{\lambda_m}x_0\Big]\psi_0(0) \\
&-\int_0^t\Big[\sum_{m=1}^{\infty}\frac{k_1B_{0m,2}}{c}\gamma_m^{3/2}\sec\sqrt{\gamma_m}x_0\exp\Big(-\frac{k_1\gamma_m(t-\theta)}{c}\Big) + \sum_{m=1}^{\infty}\frac{k_2D_{0m,2}}{c}\lambda_m^{3/2}\sec\sqrt{\lambda_m}x_0\exp\Big(-\frac{k_2\lambda_m(t-\theta)}{c}\Big)\Big]\psi_0(\theta)\mathrm{d}\theta \\
&+\sum_{m=1}^{\infty}A_{0m,1}\sqrt{\gamma_m}\sec\sqrt{\gamma_m}x_0\exp\Big(-\frac{k_1\gamma_m t}{c}\Big) + \sum_{m=1}^{\infty}\int_0^t B_{0m,1}(\theta)\sqrt{\gamma_m}\sec\sqrt{\gamma_m}x_0\exp\Big(-\frac{k_1\gamma_m(t-\theta)}{c}\Big)\mathrm{d}\theta + \frac{\phi_3(t)}{1-ax_0} \\
&-\frac{\phi_4(t)}{1+b(1-x_0)} - \sum_{m=1}^{\infty}C_{0m,1}\sqrt{\lambda_m}\sec\sqrt{\lambda_m}x_0\exp\Big(-\frac{k_1\lambda_m t}{c}\Big) - \sum_{m=1}^{\infty}\int_0^t D_{0m,1}(\theta)\sqrt{\lambda_m}\sec\sqrt{\lambda_m}x_0\exp\Big(-\frac{k_1\lambda_m(t-\theta)}{c}\Big)\mathrm{d}\theta \\
&= 0
\end{aligned} \tag{33}$$

$$\begin{aligned}
&\bar{u}_{nx}^{(-)}(x_0,t) - \bar{u}_{nx}^{(+)}(x_0,t) = \Big(\frac{b}{1+b(1-x_0)} - \frac{a}{1-ax_0} + \sum_{m=1}^{\infty} B_{nm,2}\sqrt{\gamma_m}\sec\sqrt{\gamma_m}x_0 - \sum_{m=1}^{\infty} D_{nm,2}\sqrt{\lambda_m}\sec\sqrt{\lambda_m}x_0\Big)\psi_n(t) \\
&+\Big[\sum_{m=0}^{\infty}(A_{nm,2}-B_{nm,2})\sqrt{\gamma_m}\exp\Big(-\frac{k_1\gamma_m t}{c}\Big)\sec\sqrt{\gamma_m}x_0 - \sum_{m=0}^{\infty}(C_{nm,2}-D_{nm,2})\sec\sqrt{\lambda_m}x_0\sqrt{\lambda_m}\exp\Big(-\frac{k_2\lambda_m t}{c}\Big)\Big]\psi_n(0) \\
&-\int_0^t\Big[\sum_{m=1}^{\infty}\frac{k_1B_{nm,2}}{c}\gamma_m^{3/2}\sec\sqrt{\gamma_m}x_0\exp\Big(-\frac{k_1\gamma_m(t-\theta)}{c}\Big) + \sum_{m=1}^{\infty}\frac{k_2D_{nm,2}}{c}\lambda_m^{3/2}\sec\sqrt{\lambda_m}x_0\exp\Big(-\frac{k_2\lambda_m(t-\theta)}{c}\Big)\Big]\psi_n(\theta)\mathrm{d}\theta \\
&+\sum_{m=1}^{\infty}A_{nm,1}\sqrt{\gamma_m}\sec\sqrt{\gamma_m}x_0\exp\Big(-\frac{k_1\gamma_m t}{c}\Big) + \sum_{m=1}^{\infty}\int_0^t B_{nm,1}(\theta)\sqrt{\gamma_m}\sec\sqrt{\gamma_m}x_0\exp\Big(-\frac{k_1\gamma_m(t-\theta)}{c}\Big)\mathrm{d}\theta \\
&-\sum_{m=1}^{\infty}C_{nm,1}\sqrt{\lambda_m}\sec\sqrt{\lambda_m}x_0\exp\Big(-\frac{k_1\lambda_m t}{c}\Big) - \sum_{m=1}^{\infty}\int_0^t D_{nm,1}(\theta)\sqrt{\lambda_m}\sec\sqrt{\lambda_m}x_0\exp\Big(-\frac{k_1\lambda_m(t-\theta)}{c}\Big)\mathrm{d}\theta = g_n(t)
\end{aligned} \tag{34}$$

where

$$\begin{aligned}
A_{0m} &= \frac{\cos\sqrt{\gamma_m}x_0}{L_m}\int_0^{x_0}\Big(\phi_1^{(-)}(x) - \frac{\phi_3(0)(x-x_0)}{1-ax_0}\Big)\sin\sqrt{\gamma_m}(x-x_0)\mathrm{d}x + \psi_0(0)\frac{\cos\sqrt{\gamma_m}x_0}{L_m}\int_0^{x_0}\frac{ax-1}{1-ax_0}\sin\sqrt{\gamma_m}(x-x_0)\mathrm{d}x \\
&= A_{0m,1} + A_{0m,2}\psi_0(0)
\end{aligned}$$

$$A_{1m}=\psi_1(0)\frac{\cos\sqrt{\gamma_m}x_0}{L_m}\int_0^{x_0}\frac{ax-1}{1-ax_0}\sin\sqrt{\gamma_m}(x-x_0)\mathrm{d}x=A_{1m,2}\psi_1(0)\qquad (A_{1m,1}=0)$$

$$A_{nm}=\frac{\cos\sqrt{\gamma_m}x_0}{-L_m}\int_0^{x_0}U_n(x,0)\sin\sqrt{\gamma_m}(x-x_0)\mathrm{d}x+\psi_n(0)\frac{\cos\sqrt{\gamma_m}x_0}{L_m}\int_0^{x_0}\frac{ax-1}{1-ax_0}\sin\sqrt{\gamma_m}(x-x_0)\mathrm{d}x$$
$$=A_{nm,1}+A_{nm,2}\psi_n(0)$$

$$B_{0m}=\frac{c\cos\sqrt{\gamma_m}x_0}{L_m}\int_0^{x_0}\frac{\phi_3'(\theta)(x-x_0)}{1-ax_0}\sin\sqrt{\gamma_m}(x-x_0)\mathrm{d}x+\frac{c\cos\sqrt{\gamma_m}x_0}{L_m}\psi_0'(\theta)\int_0^{x_0}\frac{1-ax}{1-ax_0}\sin\sqrt{\gamma_m}(x-x_0)\mathrm{d}x$$
$$=B_{0m,1}(\theta)+B_{0m,2}\psi_0'(\theta)$$

$$B_{1m}=\psi_1'(\theta)\frac{c\cos\sqrt{\gamma_m}x_0}{L_m}\int_0^{x_0}\frac{1-ax}{1-ax_0}\sin\sqrt{\gamma_m}(x-x_0)\mathrm{d}x=B_{1m,2}\psi_n'(\theta)\qquad (B_{1m,1}(\theta)=0)$$

$$B_{nm}=\frac{\cos\sqrt{\gamma_m}x_0}{L_m}\int_0^{x_0}f_n^{(-)}(\theta)\sin\sqrt{\gamma_m}(x-x_0)\mathrm{d}x+\psi_n'(\theta)\frac{c\cos\sqrt{\gamma_m}x_0}{L_m}\int_0^{x_0}\frac{1-ax}{1-ax_0}\sin\sqrt{\gamma_m}(x-x_0)\mathrm{d}x$$
$$=B_{nm,1}(\theta)+B_{nm,2}\psi_n'(\theta)$$

$$C_{0m}=\frac{\cos\sqrt{\lambda_m}x_0}{E_m}\int_{x_0}^1(\phi_1^{(-)}(x)-\frac{\phi_4(0)(x-x_0)}{1+b(1-x_0)})\sin\sqrt{\lambda_m}(x-x_0)\mathrm{d}x+\frac{\cos\sqrt{\lambda_m}x_0}{E_m}\psi_0(0)\int_{x_0}^1\frac{bx-1-b}{1+b(1-x_0)}\sin\sqrt{\lambda_m}(x-x_0)\mathrm{d}x$$
$$=C_{0m,1}+C_{0m,2}\psi_0(0)$$

$$C_{1m}=\psi_1(0)\frac{\cos\sqrt{\lambda_m}x_0}{E_m}\int_{x_0}^1\frac{bx-1-b}{1+b(1-x_0)}\sin\sqrt{\lambda_m}(x-x_0)\mathrm{d}x=C_{1m,2}\psi_1(0)\qquad (C_{1m,1}=0)$$

$$C_{nm}=\frac{\cos\sqrt{\lambda_m}x_0}{-E_m}\int_{x_0}^1U_n^{(+)}(x,0)\sin\sqrt{\lambda_m}(x-x_0)\mathrm{d}x+\frac{\cos\sqrt{\lambda_m}x_0}{E_m}\psi_n(0)\int_{x_0}^1\frac{bx-1-b}{1+b(1-x_0)}\sin\sqrt{\lambda_m}(x-x_0)\mathrm{d}x$$
$$=C_{nm,1}+C_{nm,2}\psi_n(0)$$

$$D_{0m}=\frac{c\cos\sqrt{\lambda_m}x_0}{E_m}\int_{x_0}^1\frac{\phi_4'(\theta)(x-x_0)}{1+b(1-x_0)}\sin\sqrt{\lambda_m}(x-x_0)\mathrm{d}x+\psi_0'(\theta)\frac{c\cos\sqrt{\lambda_m}x_0}{E_m}\int_{x_0}^1\frac{1+b(1-x)}{1+b(1-x_0)}\sin\sqrt{\lambda_m}(x-x_0)\mathrm{d}x$$
$$=D_{0m,1}(\theta)+D_{0m,2}\psi_0'(\theta)$$

$$D_{1m}=\psi_1'(\theta)\frac{c\cos\sqrt{\lambda_m}x_0}{E_m}\int_{x_0}^1\frac{1+b(1-x)}{1+b(1-x_0)}\sin\sqrt{\lambda_m}(x-x_0)\mathrm{d}x=D_{1m,2}\psi_1'(\theta)\qquad (D_{1m,1}(\theta)=0)$$

$$D_{nm}=\frac{\cos\sqrt{\lambda_m}x_0}{E_m}\int_{x_0}^1f_n^{(+)}(\theta)\sin\sqrt{\lambda_m}(x-x_0)\mathrm{d}x+\psi_n'(\theta)\frac{c\cos\sqrt{\lambda_m}x_0}{E_m}\int_{x_0}^1\frac{1+b(1-x)}{1+b(1-x_0)}\sin\sqrt{\lambda_m}(x-x_0)\mathrm{d}x$$
$$=D_{nm,1}(\theta)+D_{nm,2}\psi_n'(\theta)$$

Equation (33) and equation (34) can be transformed into the following form

$$\psi_n(t)=\beta_n(t)+\frac{1}{r_n}\int_0^t\kappa_n(t,\theta)\psi_n(\theta)\mathrm{d}\theta+\alpha_n(t)\psi_n(0)\qquad n=0,1,\ldots \tag{35}$$

where

$$\kappa_n(t,\theta)=\sum_{m=1}^{\infty}\frac{k_1B_{nm,2}}{c}\gamma_m^{3/2}\sec\sqrt{\gamma_m}x_0\exp(-\frac{k_1\gamma_m(t-\theta)}{c})+\sum_{m=1}^{\infty}\frac{k_2D_{nm,2}}{c}\lambda_m^{3/2}\sec\sqrt{\lambda_m}x_0\exp(-\frac{k_2\lambda_m(t-\theta)}{c})$$

$$r_n=\frac{b}{1+b(1-x_0)}-\frac{a}{1-ax_0}+\sum_{m=1}^{\infty}B_{nm,2}\sqrt{\gamma_m}\sec\sqrt{\gamma_m}x_0+\sum_{m=1}^{\infty}D_{nm,2}\sqrt{\lambda_m}\sec\sqrt{\lambda_m}x_0$$

$$\alpha_n(t)=\frac{\sum_{m=1}^{\infty}(B_{nm,2}-A_{nm,2})\sqrt{\gamma_m}\sec\sqrt{\gamma_m}x_0\mathrm{e}^{-\frac{k_1\gamma_m t}{c}}+\sum_{m=1}^{\infty}(C_{nm,2}+D_{nm,2})\sqrt{\lambda_m}\sec\sqrt{\lambda_m}x_0\mathrm{e}^{-\frac{k_2\lambda_m t}{c}}}{r_n} \quad (36)$$

Assuming the equation (36) is not 1 at $t$=0.

$$\beta_0(t)=\frac{1}{r_0}[\frac{\phi_4(t)}{1+b(1-x_0)}-\frac{\phi_3(t)}{1-ax_0}-\sum_{m=1}^{\infty}A_{0m,1}\sqrt{\gamma_m}\sec\sqrt{\gamma_m}x_0\exp(-\frac{k_1\gamma_m t}{c})+\sum_{m=1}^{\infty}C_{0m,1}\sqrt{\lambda_m}\sec\sqrt{\lambda_m}x_0\exp(-\frac{k_1\lambda_m t}{c})$$
$$-\sum_{m=1}^{\infty}\int_0^t B_{0m,1}(\theta)\sqrt{\gamma_m}\sec\sqrt{\gamma_m}x_0\exp(-\frac{k_1\gamma_m(t-\theta)}{c})\mathrm{d}\theta+\sum_{m=1}^{\infty}\int_0^t D_{0m,1}(\theta)\sqrt{\lambda_m}\sec\sqrt{\lambda_m}x_0\exp(-\frac{k_1\lambda_m(t-\theta)}{c})\mathrm{d}\theta]$$

$$\beta_n(t)=\frac{1}{r_n}[g_n(t)-\sum_{m=1}^{\infty}A_{nm,1}\sqrt{\gamma_m}\sec\sqrt{\gamma_m}x_0\exp(-\frac{k_1\gamma_m t}{c})-\sum_{m=1}^{\infty}\int_0^t B_{nm,1}(\theta)\sqrt{\gamma_m}\sec\sqrt{\gamma_m}x_0\exp(-\frac{k_1\gamma_m(t-\theta)}{c})\mathrm{d}\theta$$
$$+\sum_{m=1}^{\infty}C_{nm,1}\sqrt{\lambda_m}\sec\sqrt{\lambda_m}x_0\exp(-\frac{k_1\lambda_m t}{c})+\sum_{m=1}^{\infty}\int_0^t D_{nm,1}(\theta)\sqrt{\lambda_m}\sec\sqrt{\lambda_m}x_0\exp(-\frac{k_1\lambda_m(t-\theta)}{c})\mathrm{d}\theta]$$

Available from equation (35), $\psi_n(0)=\beta_n(0)+\alpha_n(0)\psi_n(0)$, therefore $\psi_n(0)=\frac{\beta_n(0)}{1-\alpha_n(0)}$. Equation (35) can be reduced to

$$\psi_n(t)=\chi_n(t)+\frac{1}{r_n}\int_0^t\kappa_n(t,\theta)\psi_n(\theta)\mathrm{d}\theta$$

where

$$\chi_n(t)=\beta_n(t)+\alpha_n(t)\psi_n(0).$$

This is the Fredholm integral equation of the second kind, it is integral kernel satisfies $\kappa_n(t,\theta)=\kappa_n(t-\theta)$ and the integral interval is $(0,t)$. It can be obtained by Laplace transform

$$\hat{\psi}_n(s)-\frac{1}{r_n}\hat{\kappa}_n(s)\hat{\psi}_n(s)=\hat{\chi}_n(s) \qquad \hat{\psi}_n(s)=\frac{\hat{\chi}_n(s)}{1-\hat{\kappa}_n(s)/r_n}$$

Thus

$$\psi_n(t)=L^{-1}[\frac{\hat{\chi}_n(s)}{1-\hat{\kappa}_n(s)/r_n}]$$

That is $\psi(t,\varepsilon)$ exists.

## 5. Remainder Estimation

**Theorem 2.** The solution of equation (4)

$$u(x,t)=\sum_{i=0}^{N}\overline{u}_i(x,t)\varepsilon^i+\sum_{i=0}^{N}U_i(x,\tau)\varepsilon^i+\sum_{i=0}^{N}V_i(\xi,\tau)\varepsilon^i+\sum_{i=0}^{N}S_i(\eta,\tau)\varepsilon^i+\sum_{i=0}^{N}I_i(\varsigma,\tau)\varepsilon^i+R\varepsilon^{N+1} \quad (37)$$

the remainder $R$ satisfies

$$\|R\|_{L^2(\Omega)}+\|R_x\|_{L^2(\Omega)}+\|R_t\|_{L^2(\Omega)}\le \mathrm{M}$$

**Proof.** Set $\Omega=[0,1]\times(0,T)$, $\tilde{k}=k(u)$, substituting (37) into (4) and (5), we can get

$$\begin{cases} g(x,t)+\tilde{k}(R_{xx}+\varepsilon^2R_{xxt})=c(R_t+\varepsilon^2R_{tt}) \\ R|_{t=0}=0 \quad R_t|_{t=0}=0 \quad R|_{x=x_0}=0 \\ R_x|_{x=0}+aR|_{x=0}=0 \quad R_x|_{x=1}+bR|_{x=1}=0 \end{cases} \quad (38)$$

where $g(x,t)=\tilde{k}\overline{u}_{N-1,xxt}-c\overline{u}_{N-1,tt}$ is a known bounded function, then $\int_\Omega g^2(x,t)\mathrm{d}\Omega$ is also bounded.

Multiply both sides of equation (38) by $2R_t$, and integrate on $\Omega$ to getting

$$\int_\Omega 2aR_t^2\mathrm{d}\Omega+\int_\Omega 2a\varepsilon^2R_{tt}R_t\mathrm{d}\Omega-\int_\Omega 2\tilde{k}R_{xx}R_t\mathrm{d}\Omega-\int_\Omega 2\tilde{k}\varepsilon^2R_{xxt}R_t\mathrm{d}\Omega=\int_\Omega 2R_tg\mathrm{d}\Omega \quad (39)$$

$$R_x(x,0)=\lim_{\Delta x\to 0}\frac{R(x+\Delta x,0)-R(x,0)}{\Delta x}=0 \qquad R_t(x_0,t)=\lim_{\Delta t\to 0}\frac{R(x_0,t+\Delta t)-R(x_0,t)}{\Delta t}=0 \quad (40)$$

where

$$\int_\Omega 2c\varepsilon^2R_{tt}R_t\mathrm{d}\Omega=c\varepsilon^2\int_0^1[R_t^2(x,t)-R_t^2(x,0)]\mathrm{d}x=c\varepsilon^2\int_0^1 R_t^2(x,t)\mathrm{d}x \quad (41)$$

$$\int_\Omega 2\tilde{k}R_{xx}R_t\mathrm{d}\Omega=\int_\Omega \tilde{k}[2(R_tR_x)_x-(R_x^2)_t]\mathrm{d}\Omega \quad (42)$$

$$\int_\Omega \tilde{k}(R_x^2)_t\mathrm{d}\Omega=\int_0^1\tilde{k}R_x^2(x,t)|_0^t\mathrm{d}x=\int_0^1\tilde{k}[R_x^2(x,t)-R_x^2(x,0)]\mathrm{d}x=\int_0^1\tilde{k}R_x^2(x,t)\mathrm{d}x \quad (43)$$

$$\begin{aligned} &2\int_\Omega \tilde{k}(R_tR_x)_x\mathrm{d}\Omega=2k_1\int_0^{x_0}\int_0^t(R_tR_x)_x\mathrm{d}t\mathrm{d}x+2k_2\int_{x_0}^1\int_0^t(R_tR_x)_x\mathrm{d}t\mathrm{d}x \\ &=2ak_1\int_0^t R_t(0,t)R(0,t)\mathrm{d}t-2bk_2\int_0^t R_t(1,t)R(1,t)\mathrm{d}t=ak_1R^2(0,t)-bk_2R^2(1,t) \end{aligned} \quad (44)$$

$$-\int_\Omega 2\tilde{k}\varepsilon^2R_{xxt}R_t\mathrm{d}\Omega=\int_\Omega 2\tilde{k}\varepsilon^2R_{tx}^2\mathrm{d}\Omega-\int_\Omega \tilde{k}\varepsilon^2(R_t^2)_{xx}\mathrm{d}\Omega \quad (45)$$

$$\begin{aligned} &-\int_\Omega \tilde{k}\varepsilon^2(R_t^2)_{xx}\mathrm{d}\Omega=-(\int_0^t k_1\varepsilon^2(R_t^2)_x|_0^{x_0}\,\mathrm{d}t+\int_0^t k_2\varepsilon^2(R_t^2)_x|_{x_0}^1\,\mathrm{d}t) \\ &=2bk_2\varepsilon^2\int_0^t R_t^2(1,t)\mathrm{d}t-2ak_1\varepsilon^2\int_0^t R_t^2(0,t)\mathrm{d}t \end{aligned} \quad (46)$$

With simultaneous equations (42)-(46), we can get

$$\begin{aligned} &\int_0^1\tilde{k}R_x^2(x,t)\mathrm{d}x+bk_2R^2(1,t)-ak_1R^2(0,t)+2\varepsilon^2(\int_\Omega\tilde{k}R_{tx}^2\mathrm{d}\Omega+bk_2\int_0^t R_t^2(1,t)\mathrm{d}t-ak_1\int_0^t R_t^2(0,t)\mathrm{d}t) \\ &+2c\int_\Omega R_t^2\mathrm{d}\Omega+c\varepsilon^2\int_0^1 R_t^2(x,t)\mathrm{d}x=\frac{1}{r}\int_\Omega 2rg(x,t)R_t\mathrm{d}\Omega\le\frac{1}{r}(\int_\Omega g^2(x,t)\mathrm{d}\Omega+r^2\int_\Omega R_t^2\mathrm{d}\Omega) \end{aligned} \quad (47)$$

Let $r=c$, inequality (47) becomes

$$\begin{aligned}&c\int_{\Omega}R_t^{\ 2}\mathrm{d}\Omega+c\varepsilon^2\int_0^1 R_t^{\ 2}(x,t)\mathrm{d}x+\int_0^1\tilde{k}R_x^{\ 2}(x,t)\mathrm{d}x+bk_2R^2(1,t)-ak_1R^2(0,t)\\&+2\varepsilon^2\big(\int_{\Omega}\tilde{k}R_{tx}^{\ 2}\mathrm{d}\Omega+bk_2\int_0^t R_t^2(1,\mathrm{t})\mathrm{d}t-ak_1\int_0^t R_t^2(0,\mathrm{t})\mathrm{d}t\big)\le\frac{1}{c}\int_{\Omega}g^2\mathrm{d}\Omega\end{aligned}\tag{48}$$

According to hypothesis [$H_2$], we get $\int_{\Omega}R_t^{\ 2}\mathrm{d}\Omega\le\frac{\int_{\Omega}g^2\mathrm{d}\Omega}{c^2}$.

Let $q=\int_0^1 R^2\mathrm{d}x$, we get

$$\int_{\Omega}(R^2)_t\mathrm{d}\Omega=2\int_{\Omega}RR_t\mathrm{d}\Omega\le\int_{\Omega}R^2\mathrm{d}\Omega+\int_{\Omega}R_t^{\ 2}\mathrm{d}\Omega$$

$$\int_{\Omega}(R^2)_t\mathrm{d}\Omega=\int_0^1[R^2(x,t)-R^2(x,0)]\mathrm{d}x=\int_0^1 R^2(x,t)\mathrm{d}x=q$$

$$q\le\int_0^t q\mathrm{d}t+\int_{\Omega}R_t^{\ 2}\mathrm{d}\Omega\le\int_0^t q\mathrm{d}t+\frac{\int_{\Omega}g^2\mathrm{d}\Omega}{c^2}$$

Order again $p=\int_0^t q\mathrm{d}t$, according to Gronwall inequality, $\int_{\Omega}R^2\mathrm{d}\Omega\le\frac{\|g\|^2(\mathrm{e}^T-1)}{c^2}$. Because of $\int_0^1\tilde{k}R_x^{\ 2}(x,t)\mathrm{d}x\le\frac{1}{c}\int_{\Omega}g^2\mathrm{d}\Omega$, $\int_0^{\mathrm{T}}\int_0^1\tilde{k}R_x^{\ 2}(x,t)\mathrm{d}x\mathrm{d}t=\int_{\Omega}\tilde{k}R_x^{\ 2}(x,t)\mathrm{d}\Omega\le\frac{1}{c}\int_0^{\mathrm{T}}\int_{\Omega}g^2\mathrm{d}\Omega\mathrm{d}t$, we get

$$\int_{\Omega}R_x^{\ 2}(x,t)\mathrm{d}\Omega\le\frac{\mathrm{T}}{c\min\{k_1,k_2\}}\|g\|^2$$

Because of $\int_{\Omega}g^2\mathrm{d}\Omega$ is bounded, $\int_{\Omega}|R_x|^2\,\mathrm{d}\Omega$ is also bounded. Therefore, there is a positive number M, such that

$$\|R\|_{L^2(\Omega)}+\|R_x\|_{L^2(\Omega)}+\|R_t\|_{L^2(\Omega)}\le\mathrm{M}$$

## 6. Conclusions

In this paper, the non-Fourier heat conduction law is used to establish the temperature field model of the laminate under the action of short-pulse laser heating. The model is a boundary value problem of nonlinear mixed equations with discontinuous coefficients, and it is difficult to solve the exact solution. We apply the singular perturbation method to transform the nonlinear discontinuous equation model into a series of linear parabolic equation problems, so formal external solution can be obtained by well-known methods such as the separation of variables method. To eliminate the error caused by perturbation expansion, we perform initial layer and boundary layer corrections at $t=0$ and the angular domain of the region, respectively, and obtain models satisfied by the correction terms, which are a series of linear mixed equations Therefore, the correction term can be obtained by applying methods such as Laplace transform, so as to obtain the internal solution.

Since the thermal conductivity is discontinuous in the laminate material problem, and the temperature is continuous and smooth, we need to determine the corresponding connection condition $\psi(t,\varepsilon)\sim\sum_{n=0}^{\infty}\psi_n(t)\varepsilon^n$, and we obtain the Fredholm integral equation of the second kind satisfied by

$\psi_n(t)$. Applying tools such as Laplace transform, we obtain the solution of the integral equation, which leads to the formal asymptotic expansion of the temperature field.

To ensure the accuracy of the asymptotic expansion, we estimate the remainder. A series of tools such as Gronwall inequality and integral estimation are comprehensively applied, and consistent validity in the sense of $L^2$ is obtained, thereby obtaining the distribution of the temperature field.

**Acknowledgment**
The research was supported by National Natural Science Foundation of China under Grant No.52175460 and 51775154.

**References**

[1] Cattaneo C 1958 A form of heat-conduction equations which eliminates the paradox of instantaneous propagation *Comptes Rendus* **247** 431
[2] Vernotte P 1958 Les paradoxes de la theorie continue de l'equation de la chaleur *Comptes Rendus* **246** 3154-3155
[3] Tzou D Y 1995 A unified field approach for heat conduction from macro-to micro-scales *ASME Journal of Heat* **117**(1) 8-16
[4] Tzou D Y 1995 The generalized lagging response in small-scale and high-rate heating *International Journal of Heat and Mass Transfer* **38**(17) 3231-3240
[5] Maes F and Van B K 2021 Thermoelastic problem in the setting of dual-phase-lag heat conduction: Existence and uniqueness of a weak solution *Journal of Mathematical Analysis and Applications* **503**(1) 125304
[6] Mao Y D and Xu M T 2015 Non-Fourier heat conduction in a thin gold film heated by an ultra-fast-laser Science China *Technological Sciences* **58**(4) 638-649
[7] ZhangY, Chen B, Li D and Ying Z X 2018 Investigation of non-Fourier heat transfer in biological tissues irradiated by pulsed laser *Journal of Engineering Thermophysics* **39**(3) 598-602
[8] Askarizadeh H and Ahmadikia H 2014 Analytical analysis of the dual-phase-lag model of bioheat transfer equation during transient heating of skin tissue *Heat and Mass Transfer* **50**(12) 1673-1684
[9] Zhang Y, Chen B and Li D 2017 Non-Fourier effect of laser-mediated thermal behaviors in bio-tissues: A numerical study by the dual-phase-lag model *International Journal of Heat and Mass Transfer* **108** 1428-1438
[10] Li X Y and Tian X G 2021 Fractional order thermo-viscoelastic theory of biological tissue with dual phase lag heat conduction model *Applied Mathematical Modelling* **95** 612-622
[11] Tan S, Wu J J, Huang Q, Zhang Y, Du X R 2019 A model of femtosecond laser ablation of metalbased on dual-phase-lag model *Acta Physica Sinica* **68**(5) 233-244
[12] Ghasemi M H, Hoseinzadeh S and Memon S 2022 A dual-phase-lag (DPL) transient non-Fourier heat transfer analysis of functional graded cylindrical material under axial heat flux *International Communications in Heat and Mass Transfer* **131** 105858
[13] Li J E, Wang B L and Chang D M 2011 Non-fourier heat conduction and thermal stress of layeredmaterials *Acta Mechanica Solida Sinica* **32**(S1) 248-253
[14] Prabha T, Chandru M and Shanthi V 2017 Hybrid difference scheme for singularly perturbed reaction-convection-diffusion problem with boundary and interior layers *Applied Mathematics and Computation* **314** 237-256
[15] Maurya P, Kumar D, Sharma S K and Upadhyay V 2022 A numerical study on nonlinear DPL model for analyzing heat transfer in tissue during thermal therapy *Heat Transfer* **51**(2) 2020-2033
[16] Li X X, Dai H Y and Li C Y 2018 A new method for solving heat transfer problems of laminate materials based on the differential theory *Applied Mathematics and Mechanics* **39**(6) 728-73 6

# Numerical Study on Gas Blown in Tundish Side Walls

**Feng Zhou[1], Yonglin Huang[2], Hexuan Niu[2], Mengjiao Geng[2], Tianyang Wang[2], Yuqian Li[2], Binshan Zhang[2], and Chao Chen[2*]**

1. No.2 Steelmaking Plant, Shanxi Taigang Stainless Steel Co., Ltd., 030030, China.
2. College of Materials Science and Engineering, Taiyuan University of Technology, Taiyuan 030024, China.
Email: chenchao@tyut.edu.cn

**Abstract.** The performance of gas blown in side walls of a single strand tundish is numerically simulated. Three cases are studied, i.e. tundish without gas blown, tundish with gas blown from the right side wall, and from the front wall. The vortex circulating flow is the main flow structure in tundish without gas blown. The gas blown from the right side wall of tundish will significantly change the flow field. The short-circuit flow that the tracer flow along the bottom of tundish to the outlet is becoming the main flow pattern. The RTD curve shows a rapid increase tendency. An anticlockwise flow near the top surface is formed. The mixing of tracer in the tundish is delayed and the dead volume fraction is increased when compared with the tundish without gas blown case. In contrast, the gas blown from the front wall of tundish will divide the tundish into two parts. The strong circulation flow in two parts is formed. Besides, the mixing of tracer is faster and the dead volume fraction is lower than that of the case without gas blown. This case may be an optimized gas blown technology in tundish.
**Keywords.** Tundish, gas blown, side blown, front blown, numerical study.

## 1. Introduction

Tundish is a vessel located between a teeming ladle and casting mold. Tundish now plays a significant role in inclusion removal. Many new technologies have been developed to improve the inclusion floatation and removal in tundishes. Among them, gas bubbling or gas curtain was a promising technology since developed in 1980's [1-2]. A number of studies focused on industrial application [3-4] and numerical and physical model studies [5-11]. However, most of the design of gas bubbling is located at the bottom of tundish wall. The cause of slag entrapment and temperature drop made it difficult to apply in steelmaking industry. So the motivation of this study is to test the performance of gas blown in tundish side walls. The gas blown from the side wall and front wall are studied and compared with the tundish without gas blown. The fluid flow, dead volume and RTD (residence time distribution) curves will be studied to evaluate the performance of three tundish designs.

## 2. Tundish Designs and Numerical Model

### *2.1. Tundish Designs*

A single strand tundish without flow control devices is studied. And the size is shown in figure 1. The size in figure 1 is the fluid part. The left and right sides of the walls are inclined by angle of 15°. The height of the liquid level is 260mm. The gas blown from the right side wall and the front wall are studied and the locations of gas blown orifice is shown in figure 2. Specifically, in the first case, the blowing orifice is located at one third of liquid level height and at the vertical centre line. In the second case, the blowing orifice is located at 50 mm from bottom height and at the half of the tundish length. The gas blown velocity is 0.5 m/s in both cases.

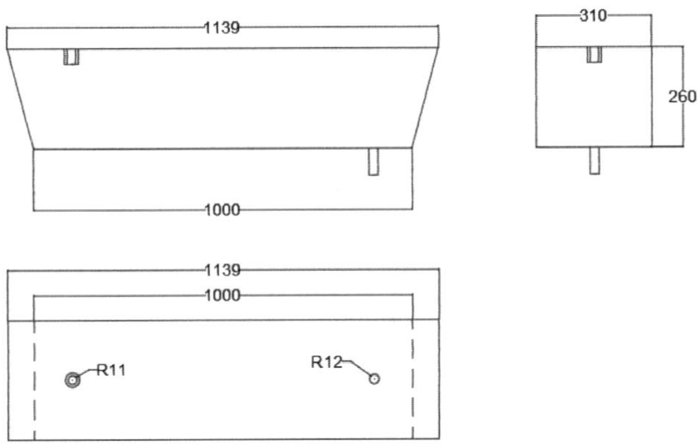


**Figure 1.** Size of the single strand tundish (all size are in mm).

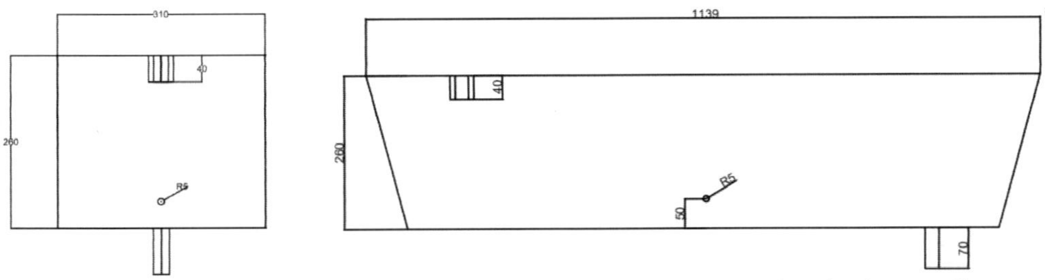


(a) Gas blown located at right side wall (b) Gas blown located at the front wall

**Figure 2.** Gas blown locations in side wall and front wall (all size are in mm).

*2.2. Numerical Models*

The flow in tundish without gas blown is a single phase flow and can be simulated by solving the mass and momentum conservation equations. The turbulence is modelled by the Realizable Two-Layer (RKE-2L) model, which combines the two-layer model with turbulence model to obtain the effective viscosity field by solving for the turbulent kinetic energy $k$ and the turbulent dissipation rate $\varepsilon$. For gas blown cases, the argon gas is modelled by a discrete phase model: discrete multiphase (DMP) model. The interphase force is dominated by the drag force, and the drag coefficient is set as 0.44.

*2.2.1. Governing Equation.* The equations for continuity and momentum can be written as follows:

$$\rho\frac{\partial\phi}{\partial t}+\rho\mathbf{u}\frac{\partial\phi}{\partial x}=\frac{\partial}{\partial x}\left[\Gamma_{\phi,eff}\frac{\partial\phi}{\partial x}\right]+S_{\phi} \quad (1)$$

where $\varphi$ represents the solved variable, such as concentration, turbulent kinetic energy, turbulent dissipation rate, etc. $\rho$ is the density, kg/m$^3$. $u$ is the velocity vector, m/s. $t$ is the time, s. $\Gamma_{\varphi,\mathrm{eff}}$ is the effective diffusion coefficient, m$^2$/s and $S_\varphi$ is the source term.

*2.2.2. Tracer Transport Model.* The passive scalar transport model is used to predict the transport process of tracer in the fluid. The model can be described as follows:

$$\frac{\partial}{\partial t}(\rho\omega)+\nabla\cdot(\rho\mathbf{u}\omega)=\nabla\cdot(\rho D_{eff}\nabla\omega) \quad (2)$$

where $\omega$ is the volume fraction of the tracer in the computational domain, that is the proportion of the tracer in the grid to the grid volume, $\omega$=1 means the grid is filled with tracer, $\omega$=0 means the grid is filled with fluid. $D_{eff}$ is Effective diffusion coefficient of passive scalar, $m^2/s$.

The tracer injection is a pulsating process within a time interval, when the flow field calculation converges, a passive scalar for a certain period of time is injected to calculate its dispersion process [12-13]. The cross-section average of volume fraction of tracer at the outlet is used to plot RTD curve.

*2.2.3. Solutions Procedure.* The models were solved by using Simcenter Star-CCM+, a software based on the finite volume method. The steady-state flow field was calculated firstly. The initialization of the tracer transport transient simulation case was based on the converged steady-state simulation result. The pressure-velocity coupling is solved by the semi-implicit (SIMPLE) method. The convergence criterion is that the residual values of all variables are less than $1\times10^{-3}$. For the transient simulation, the initial time step is 0.01 s, the maximum time is 1400 s, and each time step includes 30 iterations.

## 3. Results

### *3.1. Streamlines of Tundish under Different Gas Blown Conditions*

*3.1.1. Numerical Simulation Results of Tundish without Gas Blown.* Figure 3 is the streamline evolution in the tundish. The tracer flow from the inlet hits the bottom of the tundish, and then moves upward along the sidewall to the free surface. The tracer disperses from left part to the outlet by a vortex circulating flow. The tracer gradually mixed in the left area at about 22 s. While, a major portion of tracer flow upward to the free surface on the right side, then flows downwardly to the outlet at about 62.6 s. A small portion of tracer flow along the bottom of tundish to the outlet at 54.6 s. The above mentioned flow structure could be grouped as main vortex circulating flow, upward circulating flow, and short-circuit flow, respectively.

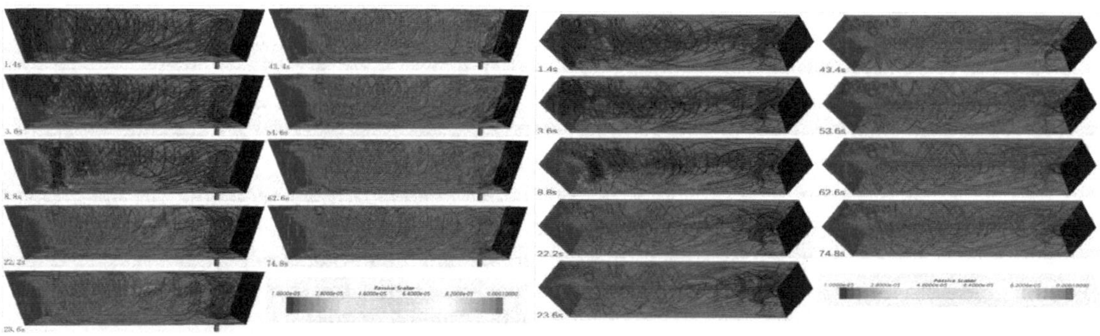

(a) front view (b) top and front view

**Figure 3.** Streamline evolution of tundish without gas blown.

*3.1.2. Numerical Simulation Results of Gas Blown from the Right Side Wall in Tundish.* Figure 4 is the streamline evolution of this case. The inlet flow area is not influenced by the gas blown. However, the vortex flow and upward circulating flow (clockwise) are destroyed by the gas blown. A strong circulating flow along the top surface is formed (anti-clockwise). In addition, the short-circuit flow that the tracer flow along the bottom of tundish to the outlet is becoming the main flow pattern. For the tracer dispersion, a major portion of tracer flow downwardly to the outlet at about 52.8s. Thereafter, a small portion tracer flows by the anti-clockwise circulation and disperses along the free surface to the inlet. Followed by the circulation, the tracer flow to the outlet by the short-circuit flow. At about 118.8s, a stable flow field was formed in the tundish. The tracer in tundish is gradually mixed at about 164.4s.

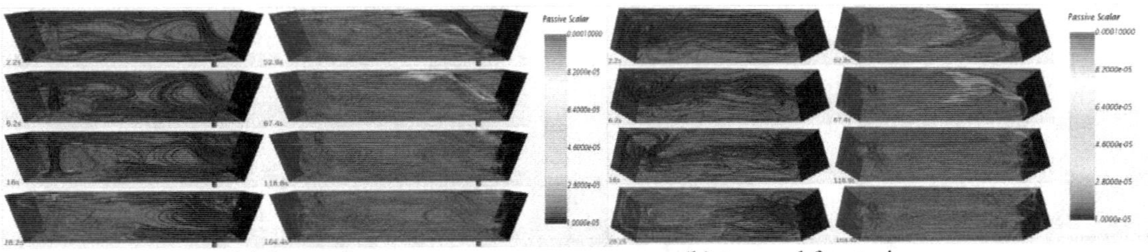

(a) front view (b) top and front view

**Figure 4.** Streamline evolution of tundish with gas blown from the right side wall.

*3.1.3. Numerical Simulation Results of Gas Blown from the Front Wall in Tundish.* Figure 5 is the streamline evolution of this case. The flow structure is reshaped by the gas blown. The flow structure in tundish is divided into two distinguishable parts. The first part is located between the inlet and gas blown area and the second part is between the gas blown area and outlet. The anti-clockwise circulations (both from front and top views) are formed in the first part. The tracer is firstly mixed in the inlet area and flow downwardly to the gas orifice location at about 9.4s. A portion of tracer disperses to the inlet again by the anti-clockwise circulation from top view. In addition, a portion of tracer disperses to the second part followed by the near surface flow. The clockwise circulations (both from front and top views) are the main flow structure in the second part. Thereafter, the tracer flows to the outlet at about 45.2s. The tracer in tundish is gradually mixed at about 78.8s.

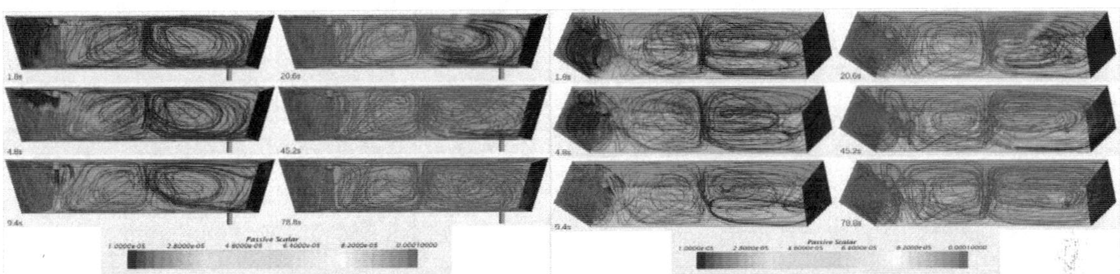

(a) front view (b) top and front view

**Figure 5.** Streamline evolution of tundish with gas blown from the front wall.

*3.2. Dead Volume Distribution and Its Fraction*

He et al. [14] defined a critical velocity and used to evaluate the dead volume in simulations. In this study, the critical velocity of 0.002 m/s is used. The grid that has a velocity value lower than this critical velocity is marked as dead volume. The dead volume distribution of three cases is shown in figure 6 to figure 8, respectively. The ratio of dead volume to the fluid volume is further analysed and defined as dead volume fraction. The results of the fraction are summarized in table 1. The dead volume is located in the upper right part of the tundish for the case without gas blown. The dead volume fraction is 5.38%, which is acceptable.

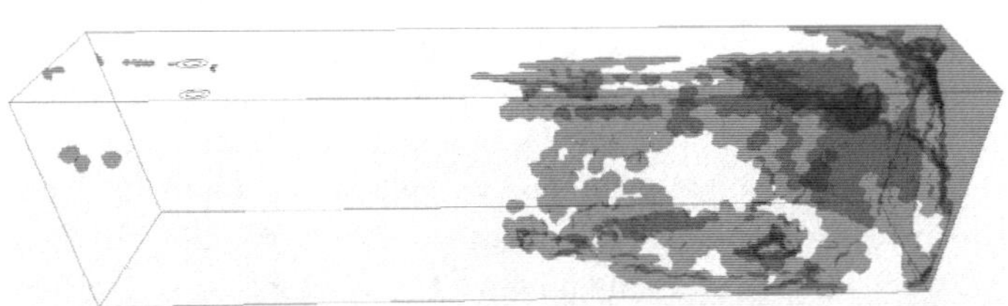

**Figure 6.** Dead volume of tundish without gas blown.

For the case that the gas blown from the right wall, the anticlockwise circulating flow and the short-circuit flow are the dominate flow pattern. Thus, the dead volume is located near the walls, in the inlet part, and other parts. The dead volume fraction is increased to 12.17% when compared with the tundish without gas blown.

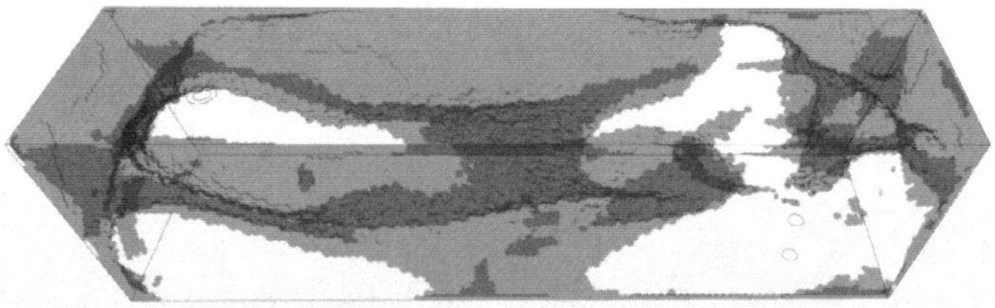

**Figure 7.** Dead volume of tundish with gas blown from the right side wall.

For the case that the gas blown from the front wall, the flow and mixing in the tundish is improved. The dead volume is reduced significantly and the fraction is only 0.95%.

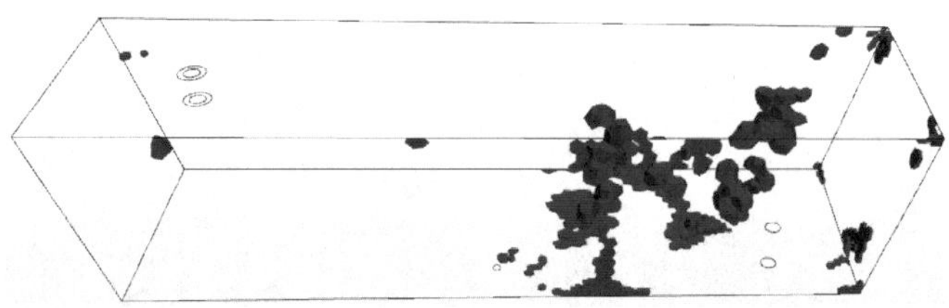

**Figure 8.** Dead volume of tundish with gas blown from the front wall.

**Table 1.** Dead volume fraction of three designs.

| Cases | Total volume/m$^3$ | Dead volume/m$^3$ | Dead volume fraction |
|---|---|---|---|
| Tundish without gas blown | $8.62\times10^{-2}$ | $4.63\times10^{-3}$ | 5.38% |
| Tundish with gas blown from the right side wall | $8.62\times10^{-2}$ | $1.05\times10^{-2}$ | 12.17% |
| Tundish with gas blown from the front wall | $8.62\times10^{-2}$ | $8.14\times10^{-4}$ | 0.95% |

### *3.3. RTD Curve of Three Designs*

The RTD curve of three cases are shown in figure 9, and the characteristic parameters are summarized in table 2. The response time of the case of tundish without gas blown is 35.28s, the peak concentration time is 164.09s, and the dimensionless peak concentration is 0.95. The RTD curve of the case that gas blown from the front wall is similar to the RTD curve of the tundish without gas blown. The response time is 33.47s, and the peak concentration time is 198.04s. The dimensionless peak concentration is 1.01. The slightly delayed peak time is beneficial to the longer residence time of fluid in tundish, which is thought to be a key factor for inclusions removal. However, the RTD curve of the case that gas blown from the right side wall shows a sharp increase tendency. The dimensionless peak concentration is 5.79 and the peak concentration time is only 65.05s. This is due to the flow out of the major portion of tracer by the short-circuit flow downward to the outlet.

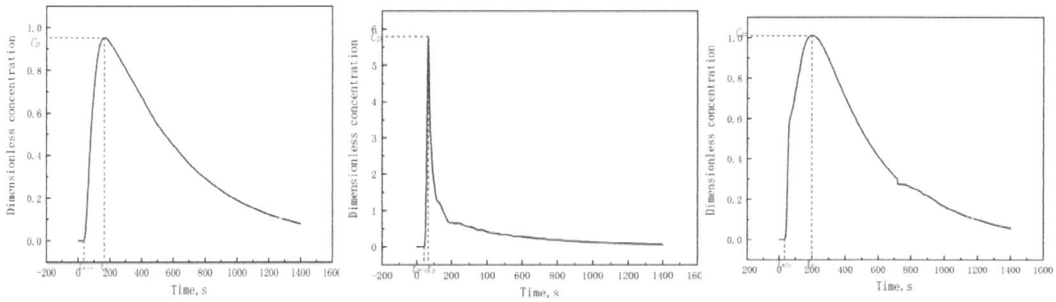


(a) tundish without gas blown (b) gas blown from the right side wall (c) gas blown from the front wall

**Figure 9.** RTD curve comparison of three cases.

**Table 2.** RTD curve analysis of three designs.

| Cases | Response time/s | Peak concentration time/s | Dimensionless Concentrations |
|---|---|---|---|
| Tundish without gas blown | 35.28 | 164.09 | 0.95 |
| Tundish with gas blown from the right side wall | 42.66 | 65.05 | 5.79 |
| Tundish with gas blown from the front wall | 33.47 | 198.04 | 1.01 |

## 4. Conclusions

The following conclusions could be summarized:

1) The gas blown from the right side wall of tundish will significantly change the flow field in the tundish. The short-circuit flow that the tracer flow along the bottom of tundish to the outlet is becoming the main flow pattern. A second backward anticlockwise flow near the top surface is formed. The mixing of tracer in the tundish is delayed somehow.

2) The gas blown from the front wall of tundish will divided the tundish into two parts. The strong circulation flow in two parts is formed. Besides, the mixing of tracer is faster than that of the case with gas blown from the right side.

3) When use the critical velocity of 0.002 m/s, the dead volume of three cases are compared. The dead volume fraction of tundish without gas blown, tundish with gas blown from the right side wall, and gas blown from the front wall are 5.38%, 12.17%, and 0.95%, respectively.

4) The RTD curves of the three cases are compared. The RTD result of tundish without gas blown and the tundish with gas blown from the front wall are similar. However, the RTD curve of the tundish with gas blown from the right side wall shows a rapid increase tendency, which indicate a strong short-circuit flow in the design.

5) Compared with the three studied cases from the view of tracer dispersion, flow structure, dead volume fraction and RTD curves, the tundish with gas blown from the front wall may be an optimized design.

## Acknowledgments

The authors are grateful to the financial support of the Shanxi Provincial Science and Technology Maj or Special Project (20191102004), and undergraduate innovation and entrepreneurship training progra m of TYUT (22100).

## References

[1] Dressel L, Shrader R and Dukelow A 1983 *Proc. Steelmaking Conf.* AIME p 205-10
[2] Nakajima K and Kawasaki M 1987 *Tetsu to Hagané* **73** 860-67
[3] Marique C, Dony A and Nyssen P 1990 *Proc. Steelmaking Conf.* AIME p 461-67
[4] Riley M F 1996 *Proc. Steelmaking Conf.* AIME p 513-18
[5] Moravec R, Stancik M and Adolf Z 2001 *Proc. Steelmaking Conf.* AIME p 139-46
[6] Ramos-Banderas A, Morales R D, García-Demedices L and Díaz-cruz M 2003 *ISIJ Int.* **43** 653-62
[7] Zhong L, Li L, Wang B, Jiang M, Zhu L, Zhang L and Chen R 2006 *Steel Res. Int.* **77** 103-06
[8] Wang J, Zhu M, Zhou H and Wang Y 2008 *J. Iron Steel Res. Int.* **15** 26-31
[9] Kumar D S, Rajendra T, Prasad R, Sarkar A and Ranjan M 2009 *Ironmak. Steelmak.* **36** 470-75
[10] Chen D, Xie X, Long M, Zhang M, Zhang L and Liao Q 2014 *Metall. Mater. Trans. B* **45** 392-98
[11] Cwudzinski A 2018 *J. South Africa Inst. Min. Metall.* **118** 545-54
[12] Chen C, Jonsson L, Tilliander A, Cheng G and Jönsson P 2015 *Metall. Mater. Trans. B* **46** 169-90
[13] Chen C, Jonsson L, Tilliander A, Cheng G and Jönsson P 2015 *Chem. Eng. Sci.* **137** 914-37
[14] He F, Wang H and Zhu Z 2019 *ISIJ Int.* **59** 1250-58

# Research on the Performance on Frost-Free Air Source Heat Pump with Waste Heat Desorption

**Zhaorui Zhao[1*], Hao Yuan[2], Sen Gao[1] and Yafen Tian[1]**

[1] School of Energy and Power Engineering, University of Shanghai for Science and Technology, 516 Jungong Road, Shanghai, China.
[2] Shanghai Marine Diesel Engine Research Institute (SMDERI), 3111 Huaning Rd., Shanghai, China
Email: zrzhao@usst.edu.cn

**Abstract.** For the novel frost-free air source heat pump system (FASHP), the compressor waste heat is used to assist desorption for the desiccant on the surface of the desiccant coated heat exchange (DCHE). It can promote the desorption efficiency of the DCHE in the FASHP and reduce the energy loss, which means the improvement of the stability and performance. In this paper, a DCHE model is established, and its accuracy is verified through experimental data. The model is then used to analyze the effect of working conditions and structural parameters on the adsorption rate, desorption rate, heat exchange and proportion of latent heat in the adsorption and desorption processes. The results show that the latent heat ratio is higher, the heat exchange of the DCHE is lower and the system performance is better when the temperature of the analytical circulating air increases and the velocity decreases to 25 °C/0.5 m/s. In addition, the performance is greatly affected by the fin length and width. With the increase of parameters, the heat exchange and proportion of latent heat can be increased to 115.9 J and 69.0%.
**Keywords.** Frost-free air source heat pump, desiccant coated heat exchanger, numerical simulation, waste heat desorption.

## 1. Introduction

The traditional defrosting method of the air source heat pump has phenomena such as uneven defrosting and increased system energy consumption. Therefore, the air is pretreated to reduce the moisture content of the air, so that the frost-free air source heat pump of the evaporator can be fully researched and developed [1]. In the frost-free air source heat pump system, the heat of condensation is usually used to desorb the desiccant on the surface of the dehumidification heat exchanger, which reduces the performance of the heat pump system [2]. Since the desorption temperature of the desiccant on the surface of the dehumidification heat exchanger is low, it can be desorbed by using low-grade thermal energy such as solar energy [3] and waste heat [4] to avoid the waste of high-grade thermal energy. In the current large-scale air source heat pump water heater, the oil cooler of the oil-injected screw compressor has a large amount of waste heat, which usually needs to be cooled by the oil cooler, and the waste heat is often not utilized. And most of the waste heat below 100°C has the disadvantages of high recovery cost and difficulty in recovery, and will directly discharge the waste heat into the environment [5]. Therefore, using the waste heat to desorb the DCHE in the frost-free air source heat pump is an effective method for waste heat recovery and utilization.

Due to the low thermal conductivity of the desiccant on the surface of the DCHE, the desorption efficiency is affected. Many scholars have studied the influence of factors such as desiccant [6], regeneration temperature [7] and size [8] on the performance of DCHE, for improving DCHE performance. However, due to the high cost and time-consuming experiment of dehumidification heat exchangers, various theoretical models of dehumidification heat exchangers have been proposed. Ge et al. [9] introduced the Lewis number to calculate the heat and mass transfer coefficients of air and desiccant. By assuming the parallel flow of water and air, the model is simplified to a one-dimensional system, and the error between simulation and experiment results is less than 15%.

Therefore, based on the frost-free air source heat pump system, the lubricating oil in the screw compressor can be used to desorb the desiccant on the surface of the DCHE. The 3D model was established and simulated by COMSOL software. The influence of the structure and operating parameters of the DCHE on its heat and mass transfer performance was investigated to provide a reference for the design and optimization of the size and operating parameters of the DCHE of the frost-free air source heat pump.

## 2. System Description

Figure 1 shows the system diagram of a novel frost-free air source heat pump with waste heat desorption. During the adsorption process, the air A1 passes through the DCHE for dehumidification, so that the evaporator operates without frost. In the desorption process, the DCHE is desorbed by air A5 and the waste heat of the compressor. A closed loop is used to prevent energy dissipation. The specific process is shown in Reference [10].

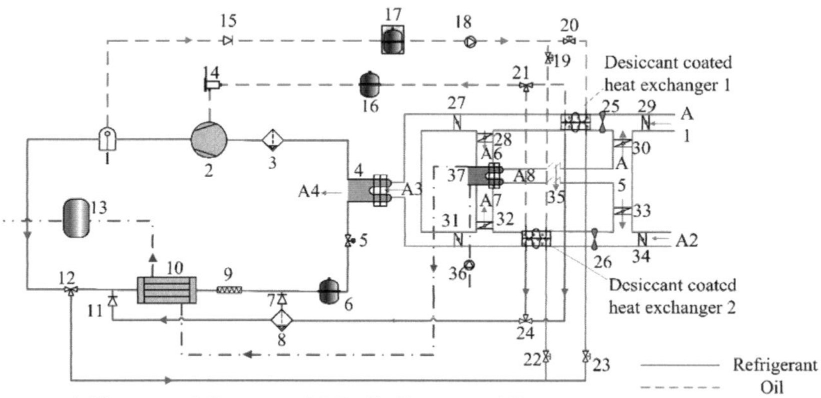


1-Oil separator, 2-Compressor, 3,8-Gas-liquid separator, 4-Evaporator, 5-Electronic expansion valve, 6,16-Liquid storage tank, 7,11,15-Check valve, 9-Dryer, 10-Condenser, 12-Three-way solenoid valve, 13-Water tank, 14-Pressure reducing device, 17-High-pressure thermal insulation storage tank, 18,36-Pump, 19,20,22,23-Electromagnetic valve, 21,24-Three-way valve, 25,26-Fan, 27~34-Damper, 35-Heat exchanger, 37-Preheating device

**Figure 1.** Schematic diagram of frost-free air source heat pump system with waste heat desorption.

## 3. Simulation Model

### *3.1. Physical Model*

The desorption performance is improved by introducing the internal heat source and the higher heat transfer coefficient of the heat exchanger. It will simulate the state of the adsorption and desorption process for designing a reasonable size. In order to reduce the amount of calculation, the heat and mass transfer performance of the single-finned tube is studied by the symmetrical structure, as shown in figure 2. The DCHE uses eight rows of tubes. The model includes copper tubes, aluminum fins, air flow area, waste heat flow area, and the Taiyo Kagaku Meso-porous Silica-1.5 (TMPS-1.5) which is selected for desiccant coated on the surface of the DCHE [11]. The physical parameters of lubricating oil refer to literature [12].

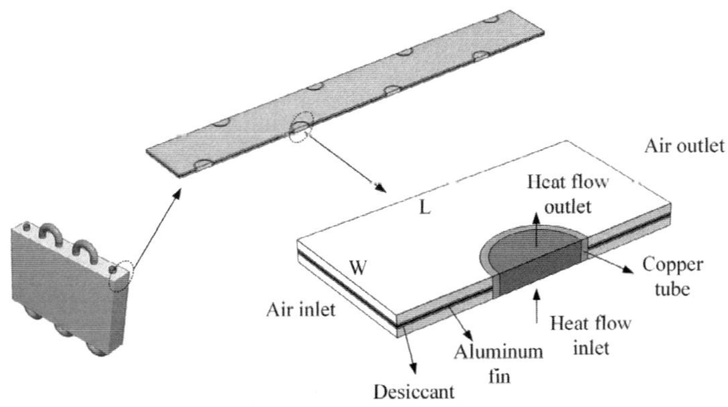


**Figure 2.** Simplified diagram of DCHE.

### *3.2. Mathematical Mode*

In this paper, a mathematical model of the local DCHE is established, and the lubricating oil in the pipeline is used to analyse the desiccant on the surface of the DCHE. Make the following assumptions:

- The thickness of the desiccant on the fin is uniform.
- Tubes and fins are isotropic, and their thermophysical properties are constant within the selected temperature range.
- The fluid temperature in the pipe changes linearly.
- The gravity of the fluid and the heat loss on the boundary are neglected.

### *3.3. Auxiliary Equation Setting*

The dehumidification and regeneration velocity of the desiccant is related to the equilibrium adsorption capacity and mass transfer coefficient of the desiccant. The calculation formula [11] is:

$$\phi_m = KH_\mathrm{d}\left(c_\mathrm{d,max} - c_\mathrm{d}\right) \tag{1}$$

$$\phi_h = \phi_m \Delta H \tag{2}$$

where $\Phi_\mathrm{m}$ refers to the adsorption or desorption flux of desiccant, $\varphi$ refers to the relative humidity of air, $M_\mathrm{w}$ refers to the molar mass of water, $K$ refers to kinetic constant, $q_\mathrm{e}$ refers to the desiccant moisture content, $\Phi_\mathrm{h}$ refers to the adsorption or desorption heat flux of desiccant, $q_\mathrm{st}$ refers to the latent heat of phase change.

Conversion of moisture content and concentration of desiccant:

$$c_d = \frac{\rho_\mathrm{d} c_w}{M_\mathrm{w}} \tag{3}$$

Conversion of air humidity and concentration:

$$c_\mathrm{a} = \frac{\omega_\mathrm{a,i} P_0 M_\mathrm{a}}{\left(M_\mathrm{a}\omega_\mathrm{a,i} + M_\mathrm{w}\right) RT_\mathrm{a,i}} \tag{4}$$

where $c_\mathrm{w}$ refers to moisture concentration in desiccant, $T_\mathrm{a}$ refers to air temperature, $R$ refers to gas molar constant.

### 3.4. Simulation Settings

COMSOL Multiphysics 5.4 is used to calculate the mathematical model. The simulation is made using a direct solver, and the backward finite difference method was employed as the time stepping method with a maximum time step of 1s. The residual error of the continuity equation was set to be less than $10^{-3}$, and the residual error of the momentum equation, energy equation and mass transfer equation was set to be less than $10^{-6}$ for transient calculation.

### 3.5. Simulation Settings

The total energy required for desorption is:

$$Q_o = m_o C_{p,o} \left(T_{o,in} - T_{o,out}\right)_{sim} N_f N_t \tag{5}$$

where $(T_{o,in}-T_{o,out})_{sim}$ refers to the temperature difference of simulation between the inlet and outlet in the tube, $N_f$ refers to the number of fins, $N_t$ refers to the number of tubes.

The ratio of latent heat(LHF) is defined to indicates the proportion of energy required for desorption in the desorption process. Adsorption rate and desorption rate($\Delta M_{ad/de}$) represent the performance of the DCHE. They can be expressed as:

$$LHF = 1 - \frac{Q_a}{Q_o} \tag{6}$$

$$\Delta M_{ad/de} = \frac{m_a}{t_{cyc}} \int_0^{t_{cyc}} \left|\omega_{a,in} - \omega_{a,out}\right| dt \tag{7}$$

where $Q_a$ refers to the sensible heat of the air temperature increase during the desorption process, $\Delta\omega_{ad/de}$ refers to the instantaneous adsorption or desorption amount, $t_{cyc}$ refers to the time required for adsorption or desorption process.

## 4. Results and Analysis

### 4.1. Model Validations

The DCHE is the core component of the frost-free ASHP system, and its calculation model accuracy directly affects the accuracy of the heat pump system model. A DCHE model was established with the same parameters as in the literature [11]. Comparison is made of the simulated and tested result on the moisture content of the air outlet and the temperature of the heat flow outlet to verify the accuracy of the established model. The result is shown in figure 3. The relative error range between the simulated value and the experimental value of the air outlet humidity is 0.6%~10.3%. The relative error of the outlet temperature in the pipe is less than 1.6%. It indicates that the model has high accuracy and can be predicted.

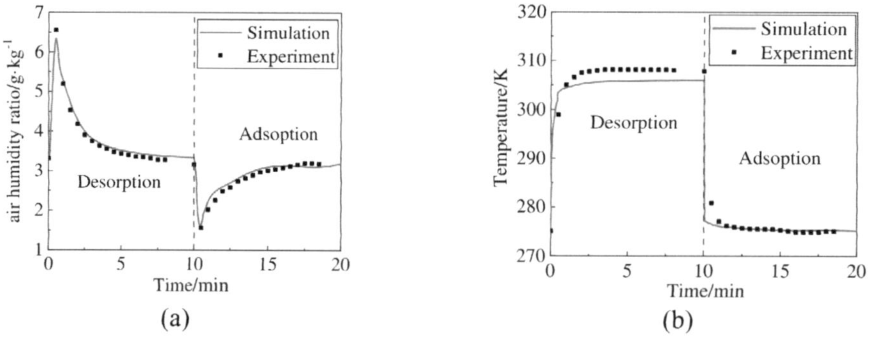


**Figure 3.** Accuracy of steam production capacity.

### *4.2. The Effect of Working Condition Parameters*

Figure 4 shows the effect of desorption circulating air temperature ($T_{a,i}$) on the performance of the DCHE. In this paper, the performance of the DCHE is described using the average value of the parameters in the adsorption or desorption process. In the desorption process, the desorption circulating air velocity is 0.15m/s. During the desorption process, due to the increase in $T_{a,i}$, the equilibrium moisture content at the interface between the air and the desiccant increases, resulting in a decrease in the desorption rate. Similarly, the increase in air temperature causes the temperature difference between the fins and the air to decrease, so the heat exchange is reduced, and the reduction of the heat exchange is greater than that of the desorption process, so the LHF increases from 64.8% to 66.5%.

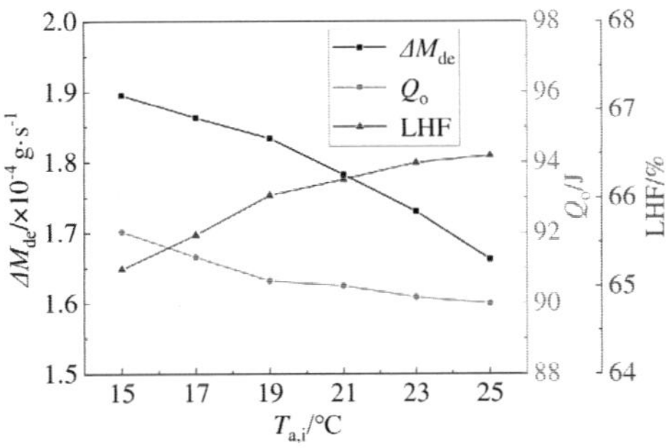


**Figure 4.** The effect of $T_{a,i}$ on the performance of the DCHE.

Figure 5 show the effect of desorption circulating air velocity ($v_{a,i}$) on the performance of the DCHE, when the temperature of the desorption circulating air is 15°C. With the increase of $v_{a,i}$, the desorption rate reaches the maximum value of $2.62\times10^{-4}$g/s at the velocity of 0.9m/s. The reason is that with the increase of desorption circulating air velocity, the air mass flow increases, the difference between the current moisture content and the maximum moisture content decreases, and the velocity decreases, resulting in the increase of desorption rate. However, it can be seen from the figure that the desorption rate decreases with the increase of velocity. This is due to the change of air temperature decreases, and the decline rate of air relative humidity slows down, resulting in the decrease of the difference between the maximum moisture content of desiccant and the current moisture content, the mass transfer decreases. Relative to the heat exchange, the air mass flow increases, and the outlet temperature is basically unchanged, resulting in the heat exchange rising from 92.0J to 196.7J. However, the increase of heat exchange is mainly caused by sensible heat, and the change of desorption heat required in the analysis process is small, so the proportion of latent heat is reduced to 30.4%.

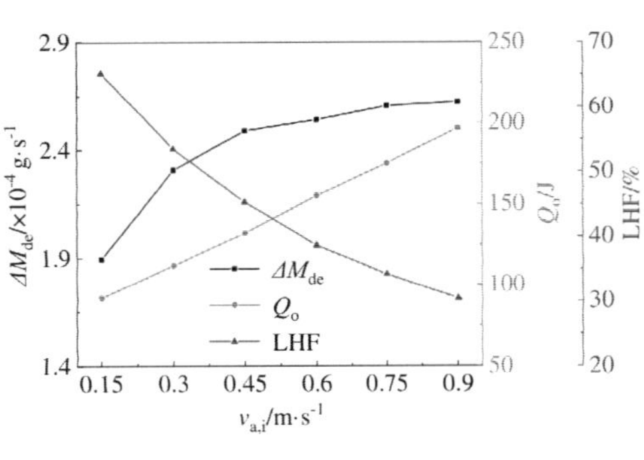


**Figure 5.** The effect of $v_{a,i}$ on the performance of the DCHE.

### *4.3. The Effect of Desiccant Coated Heat Exchanger Size*

Figure 6(a) illustrates the effect of fin width on DCHE, average moisture content of air outlet and temperature. During the adsorption process, with the increase of the fin width, the contact area between the air and the desiccant increases, resulting in an increase in the adsorption rate from $7.56\times10^{-5}$g/s to $1.11\times10^{-5}$g/s. The mass flow rate remains the same, so the air outlet moisture content decreases with increasing fin width. The desiccant will release the adsorption heat during the adsorption process. The larger the adsorption rate, the more the adsorption heat will be released, and the air outlet temperature will increase with the increase of the fin width.

During the desorption process, with the increase of the fin width, the contact area between the air and the desiccant increases. At the same time the mass of the desiccant increases, resulting in an increase in the difference between the equilibrium moisture content of the air and the desiccant, and the mass transfer rate increases. resulting in an increase in the desorption rate. The air quality remains unchanged, so the moisture content at the outlet is proportional to the desorption rate, and the moisture content increases from 48.9g/(kg dry air) to 63.5g/(kg dry air). As the width increases, the air velocity decreases, the heat transfer coefficient decreases, and the required heat of desorption increases, so the outlet temperature decreases, as shown in figure 6(b).

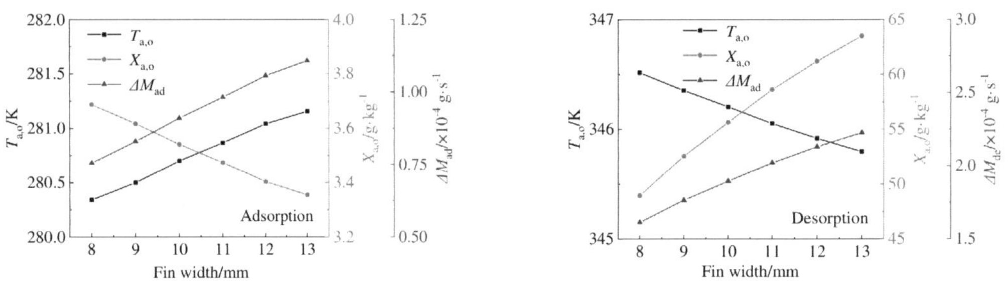


(a) (b)

**Figure 6.** The effect of fin width.

During the adsorption process, as the length of the fins increases, the area between the air and the desiccant increases, while the mass flow of the air remain unchanged. The time passing through the DCHE increases, resulting in a longer adsorption time. The adsorption rate increases, resulting the moisture content of the air outlet decreases, and the outlet temperature increases, as shown in figure 7(a).

During the desorption process, with the increase of the fin length, the heat absorbed by the desiccant increases due to the increase of thermal resistance and the increase of the desorption rate, so the air temperature slowly decreases from 346.4K to 345.9K with the increase of the fin length. As the fin length increases, the mass transfer area increases, the desorption rate increases, and the moisture content of the air outlet increases, as shown in figure 7(b).

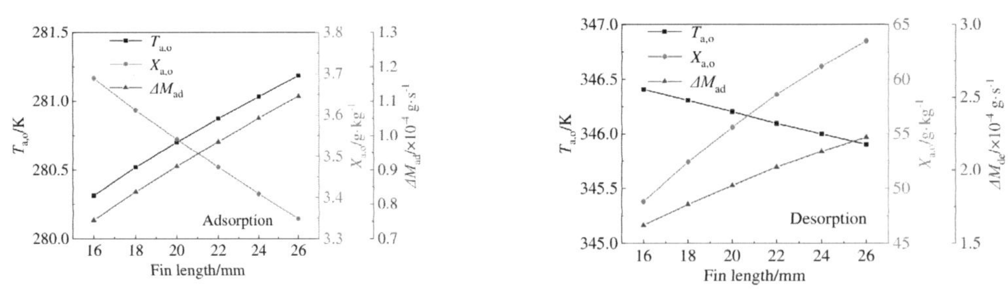


(a) (b)

**Figure 7.** The effect of fin length.

## 5. Conclusion

In this paper, a DCHE calculation model is established based on the frost-free air source heat pump system. Through experimental verification, the moisture content of the air outlet and the load error of the DCHE are both less than 10.3%, and the simulation results can accurately predict the performance of the heat exchanger. As the desorption circulating air velocity decreases and the temperature increases, the heat exchange between the air and the heat exchanger decreases to 90.0J, resulting in an increase in the proportion of latent heat to 66.5%. The LHF increases by 12.6% and 12.4% with the increase of the fin length and width. Therefore, it is necessary to focus on the influence of the fin length and width, when designing the DCHE. The fin pitch has a certain influence on the performance of the DCHE.

## Acknowledgments

This work is supported and sponsored by National Natural Science Foundation of China (Grant No. 51206022), and Shanghai Sailing Program under Grant No. 20YF1431700.

## References

[1] Song M J, Deng S M, Dang C B, Mao M and Wang Z H 2018 Review on improvement for air source heat pump units during frosting and defrosting *Applied Energy* **211** 1150–1170

[2] Chai S W, Sun X Y, Zhao Y and Dai Y J 2019 Experimental investigation on a fresh air dehumidification system using heat pump with desiccant coated heat exchanger *Energy* **171** 306–314.

[3] Xu H, Sun X Y and Dai Y J 2019 Thermodynamic study on an enhanced humidification-dehumidification solar desalination system with weakly compressed air and internal heat recovery *Energy Conversion & Management* **181** 68–79

[4] Sun X Y, Chen J L, Zhao Y, Li X, Ge T S, Wang C H and Dai Y J 2021 Experimental investigation on a dehumidification unit with heat recovery using desiccant coated heat exchanger in waste to energy system *Applied Thermal Engineering* **185** 116342

[5] Hussam J, Navid K, Sulaiman A, Bertrand D, Amisha C and Savvas A 2018 Waste heat recovery technologies and applications *Thermal Science and Engineering Progress* **6** 268–289

[6] Zheng X, Wang R Z, Ge T S and Hu L M 2015 Performance study of SAPO-34 and FAPO-34 desiccants for desiccant coated heat exchanger systems *Energy* **93** 88–94

[7] Seung J, Kim C, Chun W and Chu K 2017 Evaluation of a dehumidifier with adsorbent coated heat exchangers for tropical climate operations *Energy* **137** 441–448

[8] Sun X Y, Dai Y J, Ge T S, Zhao Y and Wang R Z 2018 Experimental and comparison study on heat and moisture transfer characteristics of desiccant coated heat exchanger with variable structure sizes *Applied Thermal Engineering* **137** 32–46

[9] Ge T S, Dai Y J and Wang R Z 2011 Performance study of silica gel coated fin-tube heat exchanger cooling system based on a developed mathematical model *Energy Conversion & Management* **52** 2329–2338

[10] GAO S, ZHAO Z R, Chen X and Zhang H 2022 Study on Characteristics of Large Capacity Screw Frost-free Air Source Heat Pump System *Journal of Refrigeration* (Beijing: Journal of Refrigeration) **43** 46–52

[11] Zhang Li, Michiyuki S and Takeshi F 2015 Experimental study on heat and mass transfer characteristics for a desiccant coated fin-tube heat exchanger *International Journal of Heat and Mass Transfer* **89** 641–651

[12] Shi H Y and Wu J H 2017 Physical Property Models of R32 and POE and PVE Oil *Journal of Refrigeration* (Beijing: Journal of Refrigeration) **38** 13–28

# Study on the Mechanism and Kinetics of $O_3/H_2O_2$ Advanced Oxidation Reaction

**Leilei Feng**

Yangling Vocational and Technical College, Yangling 712100, Shaanxi, China.
Email: 330940540@qq.com

**Abstract.** With the advent of the era of Industry 4.0, the composition of industrial wastewater is complex and toxic, and at the same time, the country and people have higher and higher requirements for the environment, making water pollution prevention and control a lot of attention. The article systematically expounds the reaction mechanism of $O_3$ single oxidation and $O_3/H_2O_2$ advanced oxidation process, analyzes the superiority of $O_3/H_2O_2$ process compared with $O_3$ single oxidation, and experimentally explores the effect of $O_3/H_2O_2$ process on DDNP production wastewater. The orthogonal experiment results show that the primary and secondary relationship of the influencing factors of $O_3/H_2O_2$ process is pH>reaction time>$O_3$ concentration>$H_2O_2$ dosage, the optimal process is pH=9, reaction time 30min, O3 concentration 8.64mg/L, $H_2O_2$ dosage 2ml/L, the effluent after treatment reaches the industrial discharge standard. By increasing the acidity and inhibiting the reaction of OH radicals under the condition of pH 3, the reaction order of the characteristic pollutants (nitrophenols) in the direct oxidation of O3 was determined to be first-order, and the rate constant was 0.148 L/moL·min.
**Keywords.** Ozone oxidation, $O_3/H_2O_2$ oxidation, orthogonal experimental design, reaction mechanism, kinetics.

## 1. Introduction

With the continuous improvement of my country's industrial level and the advent of the era of Industry 4.0, we have entered the transition from "made in China" to "created in China". The wastewater produced by industrial production has increased biorefractory groups, more complex components, and more and more toxicity. Larger, the chroma is getting higher and higher. On the other hand, the great strategic blueprint for the development of the national green economy and environmental protection is gradually being implemented, and people's requirements for the environment are getting higher and higher. Therefore, under the dual effect, the pollution problem of industrial production wastewater has become more and more prominent, and the prevention and control of water pollution has received extensive attention. At present, the main methods of industrial wastewater treatment are: flocculation method, adsorption method, micro-electrolysis method, wet oxidation method, Fenton oxidation method and ozone oxidation method [1]. Among them, ozone oxidation treatment of wastewater has the following advantages: strong oxidizing ability (2 times that of chlorine, and sterilizing ability is hundreds of times that of chlorine), fast reaction rate, simple process (ozone is generally produced by air and electricity, and there is no raw material transportation and storage. problems), easy to use in conjunction with other technologies, etc. However, industrial wastewater currently contains a variety of complex organic compounds, including nitrobenzene, phenols, polycyclic aromatic hydrocarbons, etc. It is difficult to oxidize these organic compounds by ozone oxidation alone, and some intermediate products will be generated during the oxidation process, which is difficult to degrade.causing secondary pollution.

$O_3/H_2O_2$ oxidation is an advanced oxidation technology developed on the basis of single oxidation. Compared with the single oxidation process, the final degradation products of organic matter are $CO_2$, $H_2O_2$ and other minerals, and there is no secondary pollution. At the same time. the degradation rate of organic pollutants can be significantly increased by using this treatment process, and the reaction conditions are mild. Thus, it is a promising wastewater oxidation treatment process [2]. Through literature research, the author has shown that $O_3/H_2O_2$ advanced oxidation technology has obvious effects in the treatment of phenol-containing wastewater, dye wastewater, cyanide-containing wastewater, pharmaceutical factory wastewater, pesticide wastewater, and ammonia nitrogen-containing wastewater. However, most of these literature reports focus on experimental exploration and process optimization, and there are few reports on the systematic reaction mechanism and kinetics of $O_3/H_2O_2$ advanced oxidation technology. In the process of ozonation of wastewater, it can be observed that various oxidation products are generated, and these products require a longer oxidation time to be further oxidized or even cannot be further oxidized. Therefore, in order to evaluate the applicable organic range of the ozonation reaction and the product distribution after the ozonation of organics, the pathway and kinetics of the reaction need to be studied.

## 2. Mechanism of $O_3/H_2O_2$ Oxidation Reaction

### *2.1. Limitations of $O_3$ Oxidation Alone*

The oxidation reaction of $O_3$ with organic matter in aqueous solution can be divided into two reaction pathways: direct oxidation of ozone molecules and indirect oxidation of free radicals [3], as shown in figure 1.

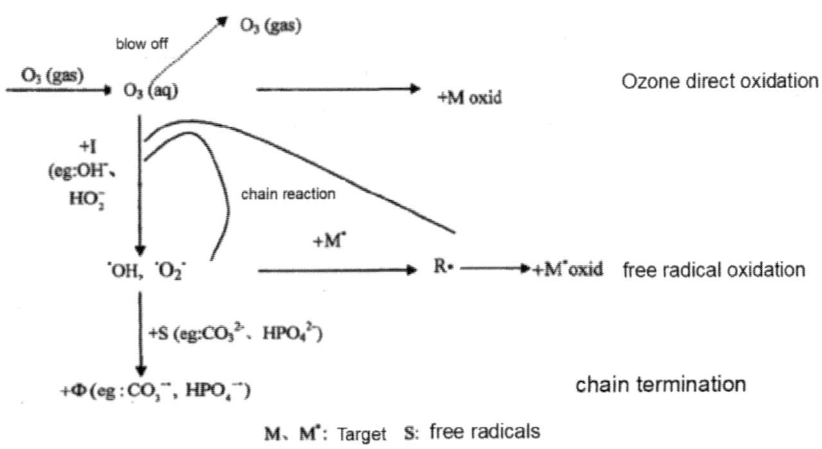


**Figure 1.** Schematic diagram of the reaction between ozone and target.

The amount of free radicals generated by $O_3$ alone in water is very small, and most of the reactions are direct reactions between ozone and organic matter. There are two theories for the direct oxidation reaction between $O_3$ molecules and organic substances [4]: one is that ozone molecules use their own dipole structure to have an addition reaction with organic substances containing unsaturated bonds; the other is that ozone molecules use electrophilic properties and active Aromatic compounds as well as deprotonated amines react. It shows that the direct oxidation reaction of $O_3$ molecule and organic matter has high selectivity, and the products are mostly aldehydes and carboxylic acids, which cannot be further reacted with ozone and cannot be completely mineralized, which makes the direct oxidation of organic matter in wastewater by ozone has certain limitations.

### 2.2. $O_3/H_2O_2$ Advanced Oxidation Mechanism

Aiming at the limitation of $O_3$ oxidation alone, $O_3/H_2O_2$ advanced oxidation technology was applied, which is a widely used and efficient oxidation technology in wastewater treatment. In the $O_3/H_2O_2$ oxidation system, the conjugated group of $H_2O_2$ can accelerate the ozonolysis to generate more hydroxyl radicals (·OH), so the oxidation reaction of ·OH plays a leading role. The specific mechanism is as follows:

$H_2O_2$ is a weak acid that partially dissociates into $HO_2^-$ ions in aqueous solution:

$$H_2O_2 + H_2O \leftrightarrow HO_2^- + H_3O^+ \qquad \kappa_a = 10^{-11.8} \tag{1}$$

Under alkaline conditions, it will promote the decomposition of ozone and accelerate the generation of $HO_2^-$ ions:

$$OH^- + O_3 \rightarrow O_2^- + HO_2 \bullet \qquad \kappa = 70L \bullet mol^- \bullet s^{-1} \tag{2}$$

The $HO_2^-$ ions generated by the above reactions are the inducers of ·OH production:

$$\begin{aligned} & HO_2^- + O_3 \rightarrow O_3^- + HO_2 \qquad \kappa = 2.8\times10^6 L\cdot mol^{-1}s^{-1} \\ & HO_2 \leftrightarrow H^+ + O_2^- \qquad p\kappa_a = 4.8 \\ & O_2^- + O_3 \rightarrow O_3^- + O_2 \qquad \kappa = 1.6\times10^9 L\cdot mol^{-1}s^{-1} \\ & H^+ + O_3^- \rightarrow HO_3 \qquad \kappa = 5.2\times10^{10} L\cdot mol^{-1}s^{-1} \\ & HO_3 \rightarrow HO\bullet + O_2 \qquad \kappa = 4.96 \end{aligned} \tag{3}$$

The standard redox potential of ·OH is as high as 2.80V, which is similar to the oxidizing ability of fluorine. It has strong oxidizing property and negative charge electrophilicity. The rate constant of the oxidation reaction with organic matter is generally $10^6 M^{-1}\cdot s^{-1}$~$10^{10} M^{-1}\cdot s^{-1}$, compared with the direct oxidation reaction of ozone molecules, the reaction rate constant of ·OH and organic matter is nearly 7 orders of magnitude higher, the selectivity is low, and it can almost completely oxidize most organic matter, which has obvious advantages [5].

Yan Yichao et al. [6] used $O_3$ and $O_3/H_2O_2$ oxidation technology to treat printing and dyeing wastewater. The experiment found that the $O_3/H_2O_2$ method was more effective than the single $O_3$ treatment, and the COD removal rates of the two were 49.4% and 67.7% respectively; BOD /COD increased from the initial 0.1 to 0.165 and 0.223, respectively. The results of ESR experiments to determine the presence of ·OH radicals show that no obvious characteristic splitting peaks are detected under ozone alone, which confirms that the $O_3/H_2O_2$ oxidation system can produce more ·OH with stronger oxidation efficiency, thereby improving its oxidative degradation. Thereby, the $O_3/H_2O_2$ oxidation system improves the efficiency of oxidation degradation of pollutants in wastewater.

The lack of selectivity of ·OH also has disadvantages. ·OH can not only remove the target pollutants, but also be easily captured by other dissolved organic matter and inorganic carbon in the aqueous solution to induce side reactions. Yang Demin et al. [7] studied the effect of various hydroxyl radical inhibitors on the ozonation degradation of phenol, and confirmed that the addition of free radical inhibitors can significantly reduce the degradation rate of phenol under neutral and alkaline conditions (tertiary The butanol concentration increased from 0 to 50 mg/L, and the phenol degradation rate decreased from 99% to 69%). Therefore, it is worth noting that when the concentration of $H_2O_2$ is high, it will be an indirect inhibitor of ozone decomposition, thereby reducing the effect of $O_3/H_2O_2$ oxidation treatment of wastewater. $H_2O_2$ reacts with ·OH as follows:

$$\begin{aligned} & H_2O_2 + HO\bullet \rightarrow H_2O + HO_2 \\ & HO^- + HO\bullet \rightarrow OH^- + HO_2 \end{aligned} \tag{4}$$

### 2.3. The Mechanism of ·OH Radical Oxidation of Pollutants

The essence of $O_3/H_2O_2$ oxidation process is to generate more hydroxyl radicals with stronger oxidation and lower selectivity by promoting ozone, which can degrade organic compounds with stable structure and low biodegradability in various wastewaters. There are three main reaction mechanisms for the oxidation of pollutants by ·OH radicals [8]:

(1) Hydroxyl substitution reaction

The hydroxyl radical attacks the hydrogen on the aromatic ring, and a displacement reaction occurs, and the reaction rate constant is about $6\times10^9$mol/s. Due to the presence of hydroxyl groups, it is easy to generate dihydroxyl substitutions of aromatic rings, resulting in ortho or meta cleavage of the aromatic rings.

(2) Dehydrogenation reaction

Hydroxyl radicals can directly react with hydrogen on alkane molecules to generate water and organic radicals R·, and the reaction rate constant is about $10^7$~$10^9$mol/s. The generated R· radicals can react with each other or with dissolved oxygen in water:

$$RA\bullet + O_2 \rightarrow ROO\bullet \tag{5}$$

The formed peroxy radical (ROO·) acts as a strong oxidant to remove hydrogen atoms from organic matter:

$$ROO\bullet + R'H \rightarrow ROOH\bullet + R'\bullet \tag{6}$$

The generated R'· radical can add an oxygen molecule to the molecule, and the chain reaction of auto-oxidation can continue until the organic matter is completely oxidized.

(3) Electron transfer reaction

The production of hydroxyl radicals and their reactions with organic molecules are accomplished by a series of complex chain reactions. Many reaction products, such as $CO_3^{2-}$, $HCO_3^-$, $HPO_4^{2-}$, etc. also react with ·OH:

$$\begin{aligned} &\bullet OH + CO_3^{2-} \rightarrow OH^- + CO_3^-\bullet \\ &\bullet OH + HCO_3^- \rightarrow OH^- + HCO_3\bullet \\ &\bullet OH + HPO_4^{2-} \rightarrow OH^- + HPO_4^-\bullet \end{aligned} \tag{7}$$

Since the reaction product will no longer induce the oxidative group to generate hydroxyl radical, it will mask the hydroxyl radical, and the chain reaction will be terminated.

## 3. Experimental Research on $O_3/H_2O_2$ Oxidation Process

The experimental wastewater was selected from the DDNP production wastewater of a certain factory. The wastewater contained biorefractory groups such as diazo and nitro groups, with complex components, high chroma and high toxicity. DDNP production wastewater pretreatment: adding polymeric ferric sulfate coagulant, fine suspended particles and colloidal particles aggregate into relatively coarse particles, and the wastewater is clarified from turbidity to sedimentation and separation for 1 hour [9]. After pretreatment, the DDNP wastewater was passed through a self-made microporous diffusion reactor (the air flow was transformed into fine bubbles through the sand core air distribution device to increase the mass transfer contact area with the wastewater), and the $O_3/H_2O_2$ oxidation treatment experiment was carried out, as shown in figure 2.

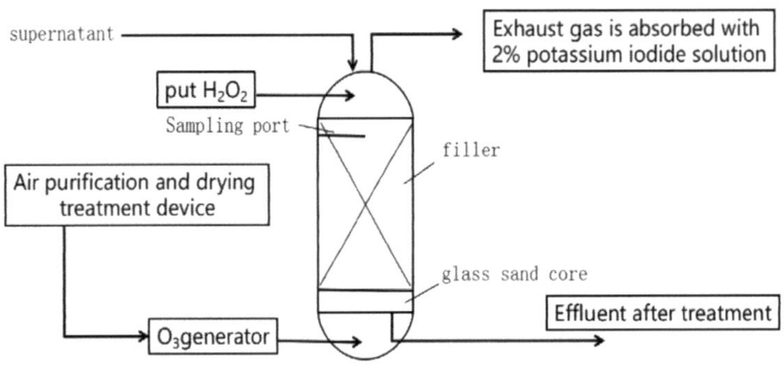


**Figure 2.** Homemade Microporous Diffusion Reactor.

The main factors affecting the effect of $O_3/H_2O_2$ treatment were pH, reaction time, $O_3$ concentration and $H_2O_2$ dosage. Therefore, the $L_9(3^4)$ orthogonal experiment was determined, and the CODcr removal rate was used as the evaluation index. The results of the orthogonal experiment are shown in table 1.

**Table 1.** Orthogonal test design.

| Test number | Factor A | Factor B | Factor C | Factor D | COD removal rate/% |
|---|---|---|---|---|---|
| | pH | Reaction time/min | $O_3$ concentration/mg/L | $H_2O_2$ Dosage/ml/L | |
| 1 | 5 | 10 | 4.32 | 2 | 22.34 |
| 2 | 5 | 20 | 8.64 | 4 | 44.15 |
| 3 | 5 | 30 | 17.28 | 6 | 42.63 |
| 4 | 7 | 10 | 8.64 | 6 | 32.36 |
| 5 | 7 | 20 | 17.28 | 2 | 43.28 |
| 6 | 7 | 30 | 4.32 | 4 | 51.44 |
| 7 | 9 | 10 | 17.28 | 4 | 49.25 |
| 8 | 9 | 20 | 4.32 | 6 | 55.83 |
| 9 | 9 | 30 | 8.64 | 2 | 68.12 |
| K1 | 36.37 | 34.65 | 43.20 | 44.58 | |
| K2 | 42.36 | 47.75 | 48.21 | 48.28 | |
| K3 | 57.73 | 54.06 | 45.05 | 43.61 | |
| R | 21.36 | 19.41 | 5.01 | 4.67 | |

It can be seen from table 1 that the primary and secondary relationship of each factor is: pH > reaction time > $O_3$ concentration > $H_2O_2$ dosage, the optimal oxidation process conditions are pH=9, reaction time 30min, $O_3$ concentration 8.64mg/L, $H_2O_2$ dosage 2ml/L. In order to further determine the optimal reaction conditions, it is necessary to carry out single-factor optimization experiments for each factor. After testing, the $O_3/H_2O_2$ process treats DDNP wastewater. Under the best process conditions, the CODcr of the treated effluent is 90.51mg/L, colorless, and the nitrophenol content is 0.95mg/L, which meets the GB 14470.2-2002 industrial discharge standard.

## 4. Kinetics of $O_3$ Direct Oxidation

From a macroscopic point of view, the initial concentration method was used to conduct kinetic experiments on characteristic pollutants (nitrophenols) in DDNP wastewater. $O_3$ reacts with nitrophenols as follows: nitrophenols + z $O_3$ = reaction product, and its kinetic rate equation is [10]:

$$r = -d[N]/dt = \kappa[O_3]^{\alpha}[N]^{\beta}[OH]^{\gamma} \tag{8}$$

Since the ozone concentration in the liquid phase does not change much during the reaction, the kinetic rate equation can be simplified to:

$$r = -d[N]/dt = \kappa_1[N]^{\beta}[OH]^{\gamma} \quad (9)$$

During the experiments with different initial concentrations of wastewater, the pH value of the solution remained basically unchanged. For each experiment, $[OH]^{\gamma}$ was a constant, so the above formula can be simplified as:

$$r = d[N]/dt = \kappa_2[N]^{\beta} \quad (10)$$

In equation (9) and (10), $k_1 = k[O_3]^{\alpha}$; $k_2 = k_1[OH]^{\gamma}$

logarithm on both sides: $\ln(r) = \ln(-d[N]/dt) = \ln(\kappa_2) + \beta\ln(N)$

To accurately determine the rate constant of the direct oxidation reaction initiated by $O_3$ molecule, it needs to be studied separately from the secondary reaction in which $O_3$ decomposes to generate ·OH radicals. The usual practice is to add ·OH inhibitors or increase acidity (add $H^+$) to the solution, the oxidation of ·OH was masked, and only the direct oxidation of $O_3$ molecules was measured. By increasing the acidity of the wastewater, the relationship between the reaction time and the logarithmic concentration of a series of different initial concentration systems was determined under the conditions of pH 3, reaction time 30 min, and $O_3$ concentration 8.64 mg/L. The results are shown in figure 3.

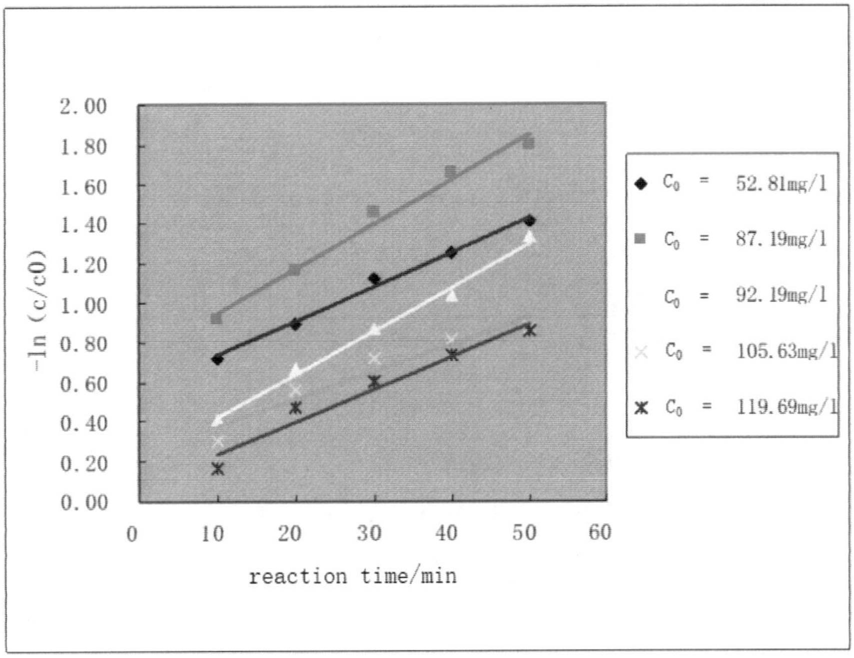


**Figure 3.** The relation curve of reaction time and concentration.

Since the slope of the curve at the initial concentration is difficult to determine, the method of numerical integration is used for processing. According to the integral median theorem, there is always a derivative $dC_A/dt$ at a certain moment in [t+△t] equal to $\triangle C_A/\triangle t$, so $\triangle C_A/\triangle t$ is approximately used to replace $dC_A/dt$ at [t+0.5t]. Then, according to the relationship curve of concentration and time, the nitrophenol concentration at [t+0.5t] was obtained from the curve interpolation. The calculation results are listed in table 2.

**Table 2.** The computational data of concentration interpolate.

| The initial concentration/mg /L | At the initial concentration-Ln ($c_0$/moL/L) | Regression curve with -Ln ($c/c_0$) and time | 5min reaction-Ln ($c/c_0$) | Concentration at 5min of reaction/moL/L | At the initial concentration-Ln (r) |
|---|---|---|---|---|---|
| 52.81 | 5.986 | y=0.0175x +0.5586 | 0.524 | 0.000132 | 7.649 |
| 87.19 | 5.559 | y=0.0226x +0.5586 | 0.466 | 0.000180 | 7.216 |
| 92.19 | 5.429 | y = 0.022x +0.204 | 0.731 | 0.000321 | 7.114 |
| 105.63 | 5.293 | y=0.0139x +0.2371 | 0.736 | 0.000370 | 6.979 |
| 119.69 | 5.168 | y=0.0165x +0.0699 | 0.859 | 0.000489 | 6.867 |

According to the chemical reaction rate at different initial concentrations, the relationship curve between the corresponding rate constant and the initial concentration is made, and the rate constant and reaction order of the reaction are obtained. From the data in table 2, the relationship curve between -Ln($c_0$) and -Ln(r) is drawn, as shown in figure 4. From the measurement results of the initial concentration method in figure 4, it can be obtained that the slope of the fitted straight line is 0.9611, and the intercept of the straight line is -1.9116, that is, the reaction order is first-order, and the rate constant Ln(k) = -1.9116, k = 0.148 L/ mol·min.

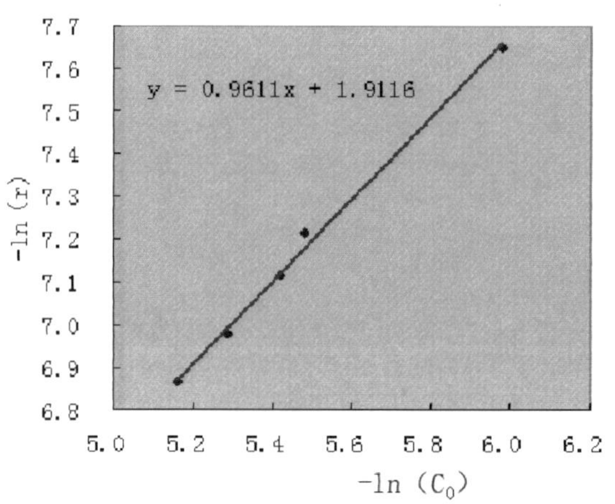


**Figure 4.** The relation curve of -Ln$c_0$ and –Lnr.

## 5. Results and Discussion

- $O_3$ alone is mainly oxidized by ozone, which has high selectivity and cannot mineralize pollutants with complex structures, which has certain limitations. $O_3/H_2O_2$ advanced oxidation application, $H_2O_2$ can promote the initiator to produce ·OH, mainly ·OH radical reaction, the oxidation reaction rate is high, the selectivity is low, and it can oxidatively degrade various types of organic matter in wastewater into $CO_2$ and $H_2O$, no secondary pollution. ·OH radical oxidation pollutants mainly include three reaction mechanisms: hydroxyl substitution reaction, dehydrogenation reaction and electron transfer reaction. It is worth noting that when the concentration of $H_2O_2$ is high, it will be an indirect inhibitor of ·OH radicals.

- Through the orthogonal experimental design, the primary and secondary relationship of the factors affecting the treatment of DDNP production wastewater by $O_3/H_2O_2$ process is: pH>reaction time>$O_3$ concentration>$H_2O_2$ dosage, the optimal oxidation process conditions are pH=9, reaction time 30min, The concentration of $O_3$ is 8.64mg/L, the dosage of $H_2O_2$ is 2ml/L, the CODcr of the treated effluent is 90.51mg/L, colorless, and the nitrophenol content is 0.95mg/L, meeting the industrial emission standard. Using the initial concentration method, under the conditions of pH 3, reaction time 30 min, and $O_3$ concentration 8.64 mg/L, the reaction order of the characteristic pollutants (nitrophenols) in the direct ozone oxidation of DDNP wastewater is first-order, and the rate constant k = 0.148 L/moL min.
- Ozone advanced oxidation process, in addition to $O_3/H_2O_2$, there are also developed $O_3$/UV, $O_3$ catalytic oxidation and other combined processes, the essence of which is to promote the strengthening of $O_3$ to produce more oxidizing and lowering selectivity. ·OH free radicals can degrade organic compounds with stable structure and low biodegradability in various wastewaters without causing secondary pollution. The price of $H_2O_2$ is not high, and it can be put into use on a large scale. Therefore, the $O_3/H_2O_2$ oxidation process has high practical value in industrial wastewater treatment.

## References

[1] Verma S, Prasad B and Mishra I M 2011 Thermochemical treatment (thermolysis) of petrochemical wastewater: COD removal mechanism and floc formation *Industrial and Engineering Chemistry Research* **50**(9) 5352-5359.

[2] Qian Z Y, Wang T, An Y M, etc. 2015 Research progress of advanced oxidation technology *Journal of Chongqing Technology and Business University: Natural Science Edition* **32**(3) 83-87.

[3] Hoigne J and Bader H 2014 Rate constants of reactions of ozone with organic and inorganic compounds in water-I: Non-dissociating organic compounds *Water Research* **17**(2) 173-183.

[4] Balciogia A 2016 Advanced oxidation of synthetic dye house effluent by $O_3$, $H_2O_2$ and $H_2O_2$/UV processes *Environmental Technology* **20**(9) 921-931.

[5] Jiang G G, Zhao Y, Ma C J and Wang X Q 2018 Research progress on kinetics of ozone oxidation reaction *Contemporary Chemical Industry* **47**(05) 1071-1074.

[6] Yan Y C 2015 *O3/H2O2 Advanced Treatment of Printing and Dyeing Wastewater* (Dalian: Dalian University of Technology).

[7] Yang D M, Yuan J M and Xia H 2014 Effects of hydroxyl free radical inhibitors on ozonated degradation of phenol *Chemical Environmental Protection* **34**(01) 24-27.

[8] Wang F, Smith D W and Gamal E M 2006 Aged raw landfill leachate: membrane fractionation, $O_3$ only and $O_3/H_2O_2$ oxidation, and molecular size distribution analysis *Water Research* **40**(3) 463-474.

[9] He F 2019 *Two-Stage AO-Coagulation-Adsorption Process for Waste Water Treatment of Waste Transfer Station* (Heilongjiang: Northeast Forestry University).

[10] Xu X H and Zhao W R 2003 *Ozone Treatment of Water and Wastewater* (Beijing: Chemical Industry Press) 10.

# Analysis of Pavement Performance for Nano-$CaCO_3$/SBS Modified Asphalt Mixture

**Pei Sun[1*], Ke Zhang[2], Yun Xiao[1] and Zijun Liang[1]**

[1] School of Urban Construction and Transportation, Hefei University, Jinxiu Avenue, Hefei, 230601, China.
[2] College of Information Engineering, Fuyang Normal University, Qinghe west road, Fuyang, 236041, China.
Email: sunpeixiaoming@163.com

**Abstract.** In order to improve the technical properties of SBS modified asphalt, firstly, nano-$CaCO_3$/SBS modified asphalt with nano-$CaCO_3$ dosage of 1%, 2%, 3%, 4%, 5% and 6% were prepared, while the optimal nano-$CaCO_3$ dosage was optimized by conventional laboratory experiments. Secondly, the pavement performance of SBS modified asphalt mixture and nano-$CaCO_3$/SBS modified asphalt mixture was analyzed by relevant laboratory tests. The experimental results show that nano-$CaCO_3$/SBS modified asphalt achieves the best performance when nano-$CaCO_3$ dosage is 4%. Compared with SBS modified asphalt mixture, the high-temperature performance, water stability, indirect tensile strength and fatigue resistance of nano-$CaCO_3$/SBS modified asphalt mixture are improved obviously, and the crack resistance at low-temperature is slightly decreased, but it still far meets the specification requirement. This indicates that it is possible to enhance the property of SBS modified asphalt and the performance of mixture by adding nano-$CaCO_3$.
**Keywords.** Nano-$CaCO_3$, SBS modified asphalt, pavement performance, fatigue resistance.

## 1. Introduction

Due to the excellent technical properties, SBS modified asphalt currently occupies most of the asphalt market in China [1]. However, SBS modified asphalt also has some shortcomings, among which the most significant one is poor thermal storage stability and easily segregation. This is because the properties of SBS modifier are quite different from those of base asphalt, which makes it difficult for SBS modifier to be evenly distributed in base asphalt, thus affecting the properties of SBS modified asphalt prepared. Furthermore, SBS modifier and base asphalt only coexist and melt in physical sense without chemical reaction, and the system formed by the two is very unstable, resulting in poor compatibility between the two [2]. Improving dispersion uniformity of SBS modifier in matrix asphalt and improving compatibility between SBS modifier and matrix asphalt have become an urgent problem to be solved.

Considering the strong surface effect of nanometer materials, many scholars tried to use nanometer materials to improve the performance of polymer modified asphalt [3-6]. Among the nanometer materials being studied, nano-$CaCO_3$ has been widely used due to its low-cost and mature production technology. Liu and Ma et al. explored the modification mechanism of nano-$CaCO_3$ modified asphalt, and found that a series of reactions occur between nanometer materials and SBS modified asphalt and they formed a unified system, then achieved the purpose of improving performance [7, 8]. Sun et al proposed that nano-$CaCO_3$ improved the compatibility of SBR modifier and matrix asphalt significantly, thus playing an active role in enhancing the high-temperature performance of the prepared asphalt mixture [9].

In conclusion, it is an effective means to use nano-$CaCO_3$ to improve the storage stability of polymer modified asphalt. Although extensive researches have focused on the application of nano-$CaCO_3$ in asphalt, few of these achievements involve the influence of nano-$CaCO_3$ on the performance of mixture. In this study, nano-$CaCO_3$ and SBS modifier were selected, and nano-$CaCO_3$ /SBS modified asphalt with different nano-$CaCO_3$ dosage were produced using the high-speed shear method. Firstly, the dosage of nano-$CaCO_3$ was optimized through the conventional performance test of asphalt. Secondly, through laboratory tests, the detailed effect of nano-$CaCO_3$ on SBS modified asphalt mixture was further characterized.

## 2. Materials and Experiments

### *2.1. Materials*

The nano-$CaCO_3$ used in the paper was white powder with an average particle size of 15-40nm. SBS modifier was linear structure with a block ratio (i.e. S/B) of 30:70. Karamay 70# asphalt binder was adopted due to its excellent properties, while the technical properties are listed in table 1.

**Table 1.** Properties of base asphalt.

| Penetration (0.1mm) | Softening point(°C) | 10°C Ductility (cm) | RTFOT | | |
|---|---|---|---|---|---|
| | | | Mass loss (%) | Penetration ratio(%) | 10°C Residual ductility(cm) |
| 74.2 | 47.4 | 26.7 | 0.56 | 65 | 8.9 |

Basalt gravel and the machine-made sand prepared by limestone were selected as coarse aggregate and fine aggregate respectively, and the mineral powder was limestone. All the properties have reached the specification requirements.

### *2.2. Development of Nano-polymer Composite Modified Asphalt*

*2.2.1. Preparation Method.* At present, the preparation methods of modified asphalt mainly include solvent method, mechanical agitation method and high-speed shear method. Considering the advantages and easy operation of the two mentioned methods, this study adopted the method of "mechanical stirring first and high-speed shear later" to prepare modified asphalt, while the nano-$CaCO_3$ and SBS modifier were evenly dispersed in matrix asphalt by high-speed shear force, see figure 1.

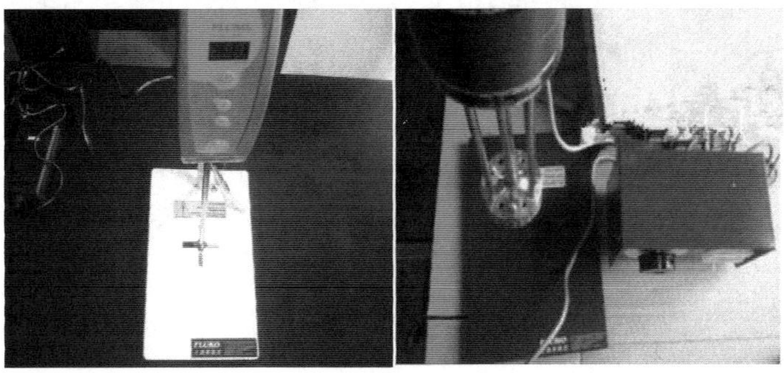

(a) (b)

**Figure 1.** Preparation method: (a) mechanical agitation device, (b) high-speed shearing machine.

Considering the large specific surface area and strong surface effect of nano-$CaCO_3$, if nano-$CaCO_3$ is directly mixed with base asphalt and SBS modifier, it is easy to appear agglomeration phenomenon, which leads to the difficult dispersion of nano-$CaCO_3$. Therefore, it is necessary to treat the surface of nano-$CaCO_3$ before adding to the base asphalt. Based on the previous research results, the silane coupling agent was finally selected to activate the surface of nano-$CaCO_3$, so as to reduce the surface energy of nano-$CaCO_3$ [10].

The specific process of preparing nano-$CaCO_3$/SBS modified asphalt mainly includes the following steps:

(1) The matrix asphalt selected was heated to molten condition, and 4% SBS modifier was added for stirring at a speed of 2000r/min for 10min.

(2) Nano-$CaCO_3$ with different dosages were added into asphalt and stirred for 10-15min.

(3) A shearing machine with high speed was adopted to shear for 45min at a speed of 4000-5000r/min. The test temperature was 165-175°C during the whole process. After the preceding operations were complete, the prepared modified asphalt was placed in an oven at 120°C for 2h.

*2.2.2. Determination of Modifier Dosage.* To determine the optimal dosage of nano-$CaCO_3$, seven kinds of dosages of 0%, 1%, 2%, 3%, 4%, 5% and 6% were selected. The experimental results are showed in table 2 below.

**Table 2.** Technology indexes of asphalt binders.

| Test item | Dosage of nano-$CaCO_3$(%) | | | | | | |
|---|---|---|---|---|---|---|---|
| | 0 | 1 | 2 | 3 | 4 | 5 | 6 |
| Penetration(0.1mm) | 56 | 57 | 55 | 55 | 54 | 52 | 52 |
| Softening point(°C) | 64.2 | 65.3 | 66.5 | 67.9 | 69.3 | 69.5 | 70.1 |
| 5°CDuctility(cm) | 28.3 | 27.2 | 26.5 | 25.4 | 25.0 | 24.3 | 23.6 |
| 135°C kinematic viscosity(Pa•s) | 1.42 | 1.46 | 1.53 | 1.62 | 1.69 | 1.84 | 2.01 |
| Penetration index | 1.232 | 1.278 | 1.336 | 1.415 | 1.493 | 1.558 | 1.587 |

The following information can be obtained from table 2, along with the nano-$CaCO_3$ dosage increase, the range of penetration of modified asphalt had little change, but the penetration index(i.e. PI) and 135°C kinematic viscosity gradually increased, indicating that nano-$CaCO_3$ is contribute to the enhancement of the temperature sensitivity of the SBS modified asphalt. This is mainly based on the on the characteristics of nano-$CaCO_3$ itself, which can absorb the light components of matrix asphalt, so as to improve the temperature sensitivity of asphalt. The value of softening point also presented an increasing trend with the increase of the dosage of nano-$CaCO_3$, and the increase rate decreased when the dosage of nano-$CaCO_3$ exceeded 4%, indicating that addition of nano-$CaCO_3$ is beneficial to the high-temperature property. Moreover, the value of ductility showed a decreasing trend, but the decreasing range was small, still far enough to meet the specification requirements.

The properties of nano-$CaCO_3$/SBS modified asphalt with different nano-$CaCO_3$ dosages were comprehensively compared. Considering economy, 4% was finally selected as the optimal nano-$CaCO_3$ dosage. In order to further analyze the influence of nano-$CaCO_3$ on SBS modified asphalt, the pavement performance of the mixtures prepared by SBS modified asphalt and nano-$CaCO_3$/SBS modified asphalt were tested. Here, SBS modified ($1^{\#}$ asphalt) refers to the modified asphalt mixed with the 4% SBS modifier, while nano-$CaCO_3$/SBS modified asphalt ($2^{\#}$ asphalt) refers to the modified asphalt mixed with the 4% SBS modifier and 4% nano-$CaCO_3$.

### *2.3. Aggregate Gradation*

AC-13 target was selected, as shown in table 3. The optimum asphalt-aggregate ratio of the two asphalt mixtures was determined to be 4.9% according to Marshall test.

**Table 3.** Aggregate gradation.

| Sieve size(mm) | 16 | 13.2 | 9.5 | 4.75 | 2.36 | 1.18 | 0.6 | 0.3 | 0.15 | 0.075 |
|---|---|---|---|---|---|---|---|---|---|---|
| Upper limit | 100 | 100 | 85 | 68 | 50 | 38 | 28 | 20 | 15 | 8 |
| Lower limit | 100 | 90 | 68 | 38 | 24 | 15 | 10 | 7 | 5 | 4 |
| Target | 100 | 95 | 79 | 56.5 | 43 | 29 | 23 | 15 | 12 | 6 |

## 3. Pavement Performance of Two Asphalt Mixtures

### *3.1. Stability at High-Temperature*

The high-temperature performance of the two asphalt mixtures was characterized by wheel tracking test, and the detailed test results are listed in table 4.

**Table 4.** Wheel tracking test results.

| Asphalt binder type | 45min rut depth(mm) | 60min rut depth (mm) | Dynamic stability (times·mm$^{-1}$) |
|---|---|---|---|
| 1# asphalt | 1.737 | 1.893 | 4038 |
| 2# asphalt | 1.732 | 1.866 | 4701 |

The following information can be obtained from table 4 that the addition of nano-$CaCO_3$ is beneficial to the enhancement of high-temperature performance, and the dynamic stability of nano-$CaCO_3$/SBS modified asphalt mixture is increased by 16.4% compared with SBS modified asphalt mixture. This is because the addition of nano-$CaCO_3$ makes the viscosity of modified asphalt increase, thus further improving the high-temperature rutting resistance of mixture.

### *3.2. Stability at Low-Temperature*

In this paper, the beam bending test was adopted to characterize the low-temperature cracking resistance of the two asphalt mixtures, and the experimental temperature was set at -10°C. Table 5 shows the specific test results. The larger of the flexural tensile strength, the better of the crack resistance of the tested asphalt mixture. The larger the flexural strain, the better the deformation resistance and stress relaxation capacity of the mixture.

**Table 5.** Results of beam bending test.

| Asphalt binder type | Flexural tensile strength (MPa) | Flexural strain ($10^{-6}$) |
|---|---|---|
| 1# asphalt | 12.32 | 3546 |
| 2# asphalt | 11.47 | 3348 |

It can be obtained from table 5 that after adding nanometer material, the flexural strain and flexural tensile strength of asphalt mixture decreased by 5.6% and 6.9% respectively, indicating that the low-temperature crack resistance of nano-$CaCO_3$/SBS modified asphalt mixture became worse, but the value is still far enough to meet the requirements of the specification.

### *3.3. Water Stability*

The water stability test results of asphalt mixture are displayed in table 6.

**Table 6.** Results of water stability.

| Asphalt binder type | Retained Marshall stability(%) | TSR(%) |
|---|---|---|
| $1^{\#}$ asphalt | 88.6 | 83.8 |
| $2^{\#}$ asphalt | 93.1 | 87.5 |

It can be obtained from table 6, the retained Marshall stability and TSR of nano-$CaCO_3$/SBS modified asphalt mixture were 5.08% and 4.42% higher than those of SBS modified asphalt mixture. Although the addition of nano-$CaCO_3$ had no obvious improvement effect on water stability of SBS modified asphalt mixture, it could still improve the water damage resistance slightly. This is because nano-$CaCO_3$/SBS modified asphalt has a strong surface effect and strong cohesion. During the mixing process with aggregate particles, the proportion of structural asphalt in the mixture increases, thus improving the water stability.

### *3.4. Indirect Tensile Performance*

The splitting test was adopted to characterize the mechanical properties of the two asphalt mixtures at the elastic stage. The experiment was carried out at 15°C and the poisson ratio was 0.3. The relevant test results are displayed in table 7 below.

**Table 7.** Results of indirect tensile test.

| Asphalt binder type | Splitting tensile strength/MPa | Stiffness modulus/MPa |
|---|---|---|
| $1^{\#}$ asphalt | 2.36 | 519.1 |
| $2^{\#}$ asphalt | 2.47 | 531.2 |

It was observed from table 7 that the splitting tensile strength of mixture increased by 4.66% after the addition nano-$CaCO_3$, which has a certain improvement effect on the tensile performance of SBS modified asphalt mixture.

It is well-known that the tensile strength of asphalt mixture mainly depends on the asphalt mastic. When choosing the same aggregate gradation and asphalt-aggregate ratio, the tensile strength mainly depends on the characteristics of the asphalt binder. The addition of nano-$CaCO_3$ makes the internal structure of asphalt binder denser, thus constrains the displacement and deformation of asphalt mastic under tension condition, and improves the tensile strength of the mixture.

### *3.5. Anti-fatigue Performance*

Four-point bending fatigue test was used to characterize the fatigue performance of the two asphalt mixtures. Three stress ratios of 0.5, 0.6 and 0.7 were selected in the experiment, and the selected stress level was the average failure load of the specimens at the loading rate of 10mm/min. The results are displayed in table 8.

**Table 8.** Results of fatigue test.

| Asphalt binder type | Stress ratio | $\sigma_t$/MPa | $\lg(\sigma_t)$ | $N_f$/time | $\lg(N_f)$ | $\lg N_f=B-n\lg\sigma_t$ |
|---|---|---|---|---|---|---|
| | 0.5 | 0.1042 | -0.98 | 31035 | 4.49 | |
| $1^{\#}$ asphalt | 0.6 | 0.1250 | -0.90 | 25843 | 4.41 | $\lg N_f=3.03-1.499\lg\sigma_t$ $R^2=0.9568$ |
| | 0.7 | 0.1459 | -0.84 | 18645 | 4.27 | |
| | 0.5 | 0.1165 | -0.93 | 34890 | 4.54 | |
| $2^{\#}$ asphalt | 0.6 | 0.1389 | -0.85 | 28864 | 4.46 | $\lg N_f=3.65-0.955\lg\sigma_t$ $R^2=0.9967$ |
| | 0.7 | 0.1631 | -0.79 | 25324 | 4.40 | |

The larger value of B in the regression equation refers to the better fatigue resistance of the asphalt mixture. The larger of the n value, the higher the sensitivity of the corresponding asphalt mixture to stress changes.

It was observed from table 8 that compared with SBS modified asphalt mixture, the fatigue resistance of nano-$CaCO_3$/SBS modified asphalt mixture was improved, and the sensitivity to stress changes was reduced, indicating that the addition of nano-$CaCO_3$ was beneficial to the anti-fatigue performance of SBS modified asphalt mixture.

## 4. Conclusions

(1) With the increase of nano-$CaCO_3$ dosage, the softening point, penetration index and viscosity of nano-$CaCO_3$/SBS modified asphalt increase, while the penetration basically does not change, the ductility decreases.

(2) The dynamic stability, residual stability and freeze-thaw splitting strength of SBS modified asphalt mixture increased by 16.4%, 5.08% and 4.42% after adding nanometer material. It shows that it is possible to improve the high-temperature stability and water stability of SBS modified asphalt mixture by using the advantage of nano-$CaCO_3$.

(3) The addition of nano-$CaCO_3$ reduces the flexural strain of SBS modified asphalt mixture by 6.9%, but it is still far enough to meet the specification requirement.

(4) The indirect tensile strength and fatigue resistance of SBS modified asphalt mixture can be significantly affected by adding 4% nano-$CaCO_3$, and the effects are positive.

## Acknowledgments

This work is supported by the University Natural Science Research Project of Anhui Province (Project Number: KJ2021A1003) and the Scientific Research Fund Project of Hefei University (Project Number: 18-19RC05). The authors gratefully acknowledge their financial support.

## References

[1] Song L, Wang C H, Shu C and Liu L Q 2021 Research progress and performance evaluation of SBS/CR-modified asphalt *China J. Highw. Transp.* **34** 17-33.
[2] Yildirim Y 2007 Polymer modified asphalt binders *Constr. Build. Mater.* **21** 66-72.
[3] Sun L, Xin X T and Ren J L 2013 Pavement performance of nanomaterial modified asphalt mixture *J. Southeast University* **43** 873-876.
[4] Ye C, Chen H X and Wang C 2010 Research on the pavement performance of nano-$TiO_2$ modified asphalt mixture *J. China & Foreign Highw.***30** 315-318.
[5] Sun L, Xin X T and Yu P 2013 Pavement performance of nano-$SiO_2$ modified asphalt mixture *J. Highw. Trans. Res. Develop* **30** 1-5.
[6] Sun P, Zhang H L, Guo G H and Niu D Y 2016 Research on pavement performance of nano-composite modified asphalt mixture *J. Build. Mater.***19** 672-677.
[7] Liu D L, Yao H B and Bao S Y 2007 Performance of nano-calcium carbonate and SBS compound modified asphalt *J. Central South University* **38** 579-582.
[8] Ma F, Zhang C and Fu Z 2007 Study on the properties and mechanism of nano-CaCO3 modified asphalt *J. Wuhan University of Technology* **31** 88-91.
[9] Sun P, Han S, Zhang H L, Wang Z Y and Xu O M 2016 High-temperature performance of nano-$CaCO_3$/SBR modified asphalt and mixture *Mater. Review* **30** 122-126.
[10] Zhang Y H, Zhai L L, Wang Y, Liu R W, Yuan J X and Lan Y J 2012 Surface modification of nano-$SiO_2$ by silane coupling agent KH-570 *J. Mater. Science & Eng.* **30** 752-756.

# New Paper-based Microfluidic Analytical Devices for Kojic Acid Assay Based on Double-sided Adhesive Tape

**Jian Wang, Jing Hu, Bo Xiao, Tong Yang, Jingye He and Peng Yu***

School of Materials Science and Engineering, Xiangtan University, Xiangtan, Hunan, 411105, China.

Email: yupeng@xtu.edu.cn

**Abstract.** Kojic acid is an antioxidant used by the cosmetics industry but the excessive usage of kojic acid may cause harm to human health. Therefore, it is important to detect the content of kojic acid in cosmetic products. In this work, the low-cost and simple paper-based analytical devices (μPADs) based on double-sided adhesive tape were developed for the colorimetric detection of kojic acid. The μPADs were fabricated by dissolving double-sided adhesive tape in toluene and then making the patterned adhesive tape penetrate into the filer paper. Ferric nitrate was used as the reagent to react with kojic acid, and the product of their chelating reaction caused the color change from colorless to orange-red. The ImageJ software was used to analyze the color intensity and the corresponding correction curve was determined. The μPADs for the detection of kojic acid exhibited a wide linear range (0.05 - 4 mM) and the detectable limit was 0.05 mM. The good selectivity and reproducibility were also demonstrated.

**Keywords.** μPADs, colorimetric detection, kojic acid, double-sided adhesive tape.

## 1. Introduction

Kojic acid is a natural metabolite of fungi, which has been widely used in the fields of food processing, cosmetics and medicine due to its functions of preventing oxidation, scavenging free radicals, chelating metal ions and inhibiting tyrosinase activity [1-3]. However, the 4-pyrone structure in the kojic acid molecule may cause harm to human health, such as carcinogenicity, teratogenicity, and embryotoxicity. Therefore, it is of great significance to detect the content of kojic acid [4-6].

The main methods commonly used to identify and measure kojic acid are gravimetric analysis [7], colorimetric analysis [8], electrochemical methods [9] and chromatography [10]. Among these methods, gravimetric analysis is cumbersome and difficult to determine trace components; chromatographic analysis usually requires several hours of analysis time and high daily maintenance costs; although electrochemical method is highly sensitive, it is susceptible to the coexistence of electroactive substances. Therefore, it is extremely necessary to develop a simple, rapid and effective method for the detection of kojic acid.

In recent years, microfluidic paper-based analytical devices (μPADs) have attracted widespread interest due to their low cost and simple operation. Many researchers combine μPADs with some simple colorimetric reaction systems to quantitatively detect the analyte. With the development and popularization of smartphones and tablet computers in recent years, the color of the reaction systems could be easily photographed and measured. Nuria et al. developed the μPADs for pH and nitrite detection using a smartphone and Android system analysis software [11]. Sechi et al. used three-dimensional μPADs to detect protein and glucose in urine [12]. The developed μPADs were generally disposable, portable, and user friendly, which were very suitable for on-site testing in some underdeveloped countries and regions. Therefore, the colorimetric μPADs would provide a good help for the effective detection of kojic acid.

In this work, the new μPADs based on double-sided adhesive tape was developed to quantitatively detect the kojic acid. A smartphone was used as the signal output terminal to construct a paper-based colorimetric analysis platform. The principle of color rendering is based on the complexing reaction between ferric nitrate and kojic acid [13], resulting the color change from colorless to orange-red. The imageJ software was employed to measure the color intensities in the RGB mode and the corresponding correction curve between the color intensity and the concentration of kojic acid was determined. The developed μPADs have the advantages of low cost, wide linear range, short assay time, good selectivity and good reproducibility.

## 2. Experimental

### *2.1. Experimental Reagents and Materials*

Kojic acid, ferric nitrate, anhydrous calcium chloride, magnesium chloride, potassium chloride, anhydrous sodium chloride, anhydrous sodium sulfate, trisodium citrate, glucose (monohydrate), sodium bicarbonate, anhydrous ethanol, toluene were of analytical grade and bought from Sinopharm Chemical Reagent Co., Ltd. (China). Whatman filter paper No.1 was purchased from Whatman International Ltd. (Maidstone, England). Double-sided adhesive tape (brand Youbisheng, width = 40 mm) was purchased from Hangzhou Shengneng Packaging Co., Ltd. (Hangzhou, China). The tape consists of a double-sided pressure-sensitive adhesive layer and a protective film.

### *2.2. The Fabrication of* μPADs

The fabrication process of the μPADs is shown in figure 1. First, the bottom of the A4 copy paper was pasted with double-sided adhesive tape to form a "sandwich" assembled structure of "copy paper-adhesive layer-protective film". Then, a craft punch was used to cut chrysanthemum holes from the "sandwich" assembled structure. The width of the chrysanthemum pattern is 2.4 cm. Next, toluene was sprayed on the filter paper with a sprayer. After the toluene was diffused on the filter paper, the protective film of the "sandwich" assembled structure was teared off. The "sandwich" assembled structure with chrysanthemum holes was pasted on the filter paper and dried in a fume hood for 10 min. During the drying process, the adhesive layer was gradually dissolved by toluene and then penetrated into the filter paper to form a hydrophobic barrier. After the tape was fully dissolved, the copy paper stuck on the filter paper was peeled off to obtain the μPADs. In the μPADs, the white region with the chrysanthemum shape was hydrophilic, and the gray region outside the chrysanthemum was hydrophobic. The center of the chrysanthemum flower was used as the sample introduction zone and the petals of the chrysanthemum were used as the detection zones. Using the craft punch with different designs can easily modify the hydrophilic channels' pattern of μPADs. The eight detection zones were labeled as 1, 2, 3, 4, 5, 6, 7 and 8 in clockwise order, respectively.

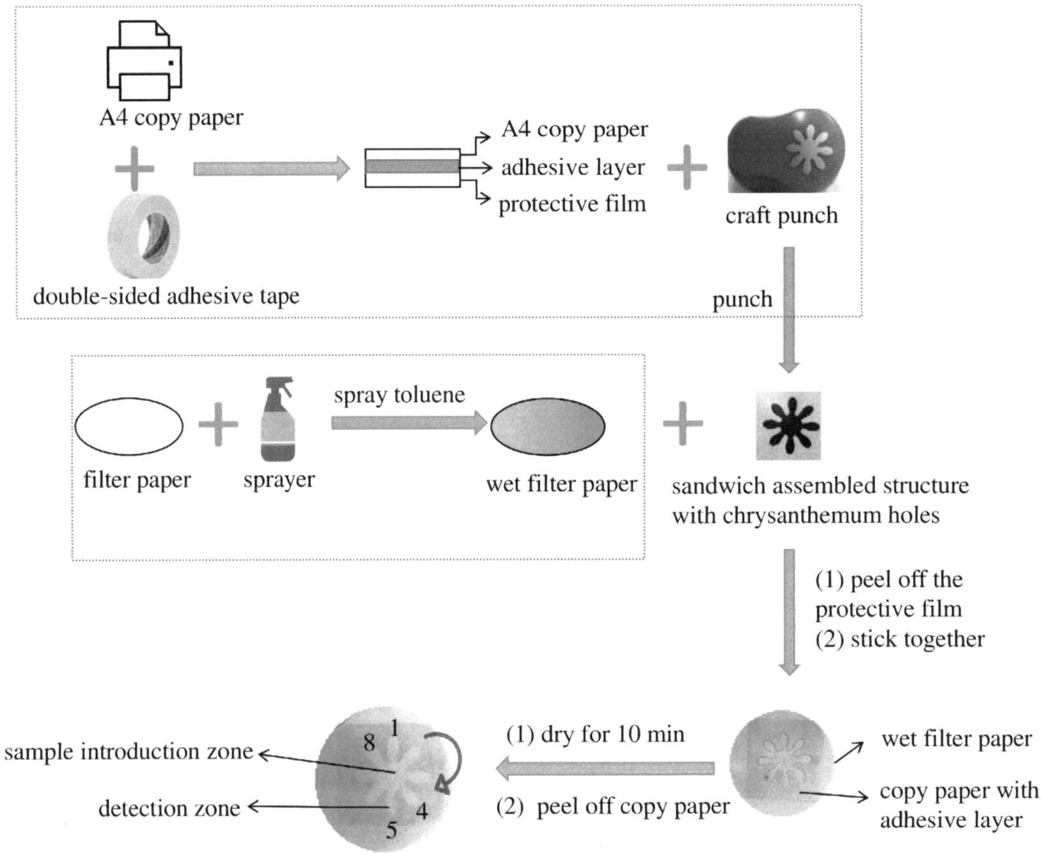


**Figure 1.** The fabrication process of μPADs based on double-sided adhesive tape.

*2.3. Colorimetric Detection of kojic Acid*

First, 1.2 μL of 15 mM ferric nitrate solution (pH = 4) was added to each detection zone of the μPADs, and dried under a fume hood for 10 min. Then 23 μL of kojic acid solution was added to the sample introduction zone. After 5 min for the reaction, a red-orange complex was formed in the detection zones. A photo was taken by a smart phone in JPEG format and analyzed using the ImageJ software by selecting the entire detection zone to calculate the average color intensity.

## 3. Results and Discussion

*3.1. The Feasibility of μPADs for kojic Acid Analysis*

To demonstrate the feasibility of the developed μPADs for kojic acid analysis, the feasibility experiments were performed with the blank (0 mM) and the kojic acid solution (5 mM) by measuring the color intensity of RGB channels. Figure 2 shows the intensity values for the different color channels. The mean color intensities of the red, green, and blue channels corresponding to the blank solution were 182, 180, and 174, respectively, while the mean color intensities of the red, green, and blue channels corresponding to 5 mM kojic acid were 195, 178, and 142, respectively. The absolute values of the difference in color intensity of the two solutions were 13, 2, and 32, respectively. These results indicated that the blue channel intensity exhibited the highest sensitivity for the detection of kojic acid. Therefore, this work selected the blue channel intensity to analyze kojic acid.

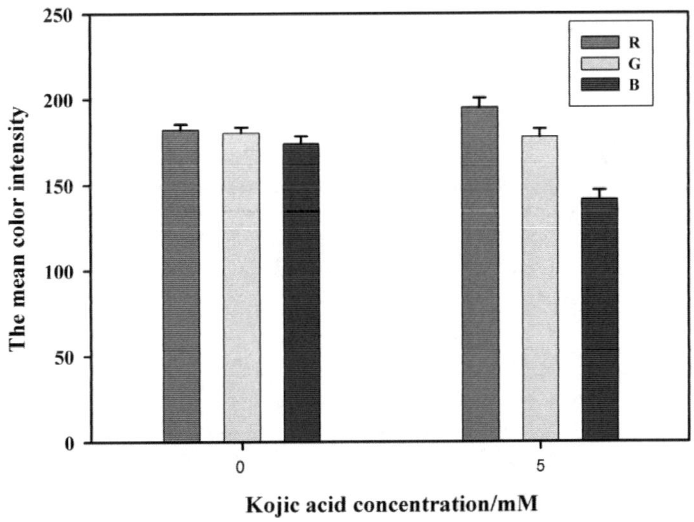


**Figure 2.** The mean color intensities for the developed μPADs after reacting with the blank and the kojic acid (5 mM). The red, green, and blue vertical bars represent the mean colour intensities for the Red, Green, and Blue channels, respectively.

### *3.2. The Analytical Performance of the Developed μPADs*

In order to investigate the analytical performance of the developed μPADs, a series of concentrations of kojic acid solution were detected, and the corresponding photos were taken using a smart phone. As shown in figure 3, when the concentration of kojic acid increased from 0 to 10 mM, the blue channel intensity continuously decreased. The kojic acid concentration and the corresponding color value showed a good linear range between 0.05 mM - 4 mM, and the linear regression equation was blue channel intensity = -6.96 $C_{\text{kojic acid}}$ + 166.06 ($R^2 = 0.9922$) with a detectable limit (LoD) of 0.05 mM.

Table 1 lists the comparison of different sensors for the detection of kojic acid. Although the methods such as ion-pair liquid chromatography (Ion-pair LC), high-performance liquid chromatography (HPLC), fluorescence spetrometry and electrochemistry exhibited lower detection limits for kojic acid detection, they required longer reaction times. The μPADs we developed needed obtained the good results within 5 minutes of analysis time. More importantly, the μPADs developed in this work did not use the expensive instruments (such as the laser cutter, the screen printer and the wax printer), the cost was as low as 0.08 $ and the fabrication process was simple and efficient.

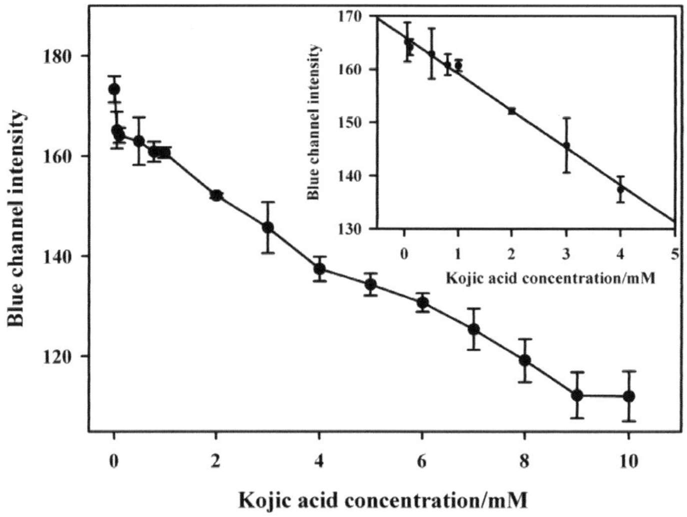


**Figure 3.** The dependence of the color intensity of the blue channel on the concentration of kojic acid. Inset: Linear fit plot of the blue channel intensity versus kojic acid concentration.

**Table 1.** Comparison of the analytical performances of the sensors for kojic acid assay.

| Method | Linear Range (mM) | LoD (mM) | Volume (μL) | Response Time (min) | Refs |
|---|---|---|---|---|---|
| Ion-pair LC | $1.7 \times 10^{-9} \sim 3.5\times 10^{-6}$ | $1.4 \times 10^{-10}$ | 5 | 5 | [14] |
| HPLC | $1.0 \times 10^{-4} \sim 7.0 \times 10^{-4}$ | $1.4 \times 10^{-5}$ | 20 | 8 | [15] |
| Fluorescence | $2.0 \times 10^{-4} \sim 5.0 \times 10^{-2}$ | $7.0 \times 10^{-5}$ | 1000 | 3 | [16] |
| Electrochemistry | $2.0 \times 10^{-4} \sim 6.47 \times 10^{-2}$ | $8.0 \times 10^{-5}$ | - | 30 | [17] |
| Colorimetric | 0.05 ~ 4 | 0.05 | 23 | 5 | This work |

### *3.3. The Selectivity and Reproducibility of the Developed μPADs*

To demonstrate the selectivity of the μPADs for the detection of kojic acid, the common interferents such as $SO_4^{2-}$, $Ca^{2+}$, $K^+$, $Mg^{2+}$, $Na^+$, citrate, glucose, $HCO_3^-$ and ethanol were reacted with ferric nitrate solution, respectively. The selectivity of the μPADs was investigated by comparing the color intensities of the solution containing only the interferent or kojic acid. The results are shown in figure 4. It is noted that kojic acid caused the obvious color intensity change while the changes caused by the other nine interfering substances were relatively small. The results showed that the μPADs performed good selectivity for the detection of kojic acid.

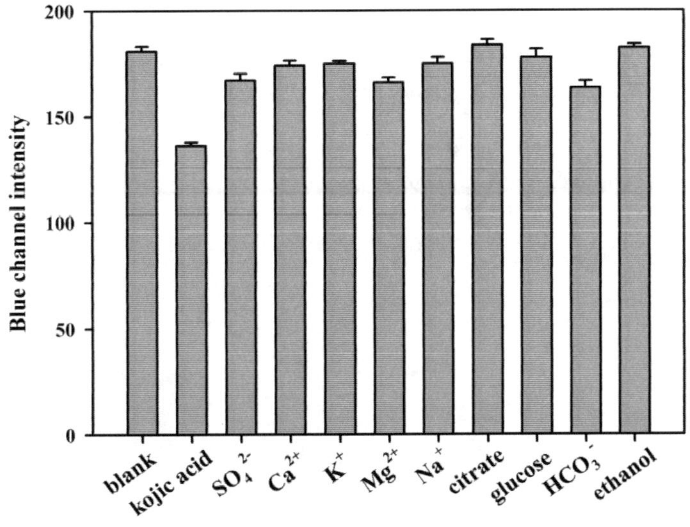


**Figure 4.** Selectivity of the developed μPADs with 5 mM kojic acid or 50 mM other interfering substances.

Next, we further examined the reproducibility of the μPADs. We used 3 different detection zone pairs of a μPADs for detection, and use the coefficient of variation to evaluate the reproducibility of the μPADs. For the low, medium and high concentrations of kojic acid, the results of the coefficient of variation of the μPADs were 6.83%, 11.53% and 2.43%, reflecting the good reproducibility of the μPADs.

## 4. Conclusions

In this work, the new μPADs based on the double-sided adhesive tape were developed for the sensitive detection of kojic acid. For the fabrication of the μPADs, the adhesive tape was dissolved by toluene and penetrated into the filter paper to form a hydrophobic barrier, thereby confining the hydrophilic area and hydrophobic area of the μPADs. The fabrication process is simple and efficient, without the usage of the expensive instruments (such as laser cutter, screen printer and wax printer). The analytical performance of the μPADs for kojic acid assay was demonstrated. The detection limit for kojic acid was 0.05 mM and the linear range was 0.05 - 4 mM, with good selectivity and reproducibility.

## Acknowledgments

The authors are grateful to the financial support from National Innovation Training Program for college students (grant number 202110530018) and the National Natural Science Foundation of China (grant number 21804114).

## References

[1] Buleandra M, Rabinca A A and Buleandra M 2017 *Sensor Actuat. B-Chem.* **241** 406–12
[2] Ma X and Chao M 2014 *Food Anal. Method* **7** 1458–64
[3] Sheikhshoaie M, Sheikhshoaie I and Ranjbar M 2017 *J. Mol. Liq.* **231** 597–01
[4] Gao Z, Su R and Qi W 2014 *Sensor Actuat. B-Chem.* **195** 359–64
[5] Chusiri Y, Wongpoomchai R and Kakehashi A 2011 *Food Chem. Toxicol* **49** 471–76
[6] Blumenthal C Z 2004 *Regul. Toxicol. Pharm.* **39** 214–28
[7] Barham H N 1939 *Ind. Eng. Chem. Anal. Edit.* **11** 31–3

[8] Moss M and Mellon M 1941 *Ind. Eng. Chem. Anal. Edit* **13** 612–14
[9] Wang L, Qi W, Su R X and He Z M 2014 *Food Anal. Method* **7** 109–15
[10] Lin C H, Wu H L and Huang Y L 2007 *Anal. Chim. Acta* **581** 102–07
[11] Lopez N, Curto V F and Erenas M M 2014 *Anal. Chim.* **86** 9554–62
[12] Sechi D, Greer B and Johnson 2013 *Anal. Chim.* **85** 10733–37
[13] Murakami Y 1962 *J. Inorg. Chem.* 24 679–88
[14] Shih Y 2001 *J. Aoac. Int.* **84** 1045–49
[15] Huang S C, Lin C C and Huang M 2004 *J. Food Drug Anal.* **12** 13–8
[16] Zhao G 2014 *Sensor Actuat. B-Chem.* **195** 359–64
[17] Yang Z, Yin Z and Chen F 2011 *Electrochimica Acta* **56** 1089–93

# Hydrogen-Oxygen Balance Method is used to Measure Hydrogen Consumption in Fuel Cell Vehicles

**Ting Guo[1,2*], Jun Li[1], Shiyu Wu[1,2], Fang Wang[2] and Rongliang Liang[1,2]**

[1] China Automotive Technology and Research Center Co., Ltd, Tianjin, China,
[2] CATARC Automotive Test Center (Tianjin) Co., Ltd, Tianjin, China.
Email: guoting@catarc.ac.cn

**Abstract.** At present, there are four test methods to measure hydrogen consumption of fuel cell vehicles, namely weight method, pressure method, flow method and current method. During the test process, hydrogen supply outside the vehicle is required, which requires high requirements for operation and testing equipment. This paper presents a method to measure the hydrogen and oxygen concentrations in the tail gas of fuel cell vehicles without changing the vehicle structure, and uses the mass conservation method to evaluate hydrogen consumption. This paper compares and analyzes the data of traditional hydrogen consumption measurement method, introduces the principle and testing method of the new method, and the test results show that the hydrogen and oxygen balance method can accurately evaluate the hydrogen consumption of fuel cell vehicles.
**Keywords.** Fuel cell vehicle; Hydrogen and oxygen equilibrium method; Hydrogen consumption

## 1. Introduction

Fuel cell vehicles are considered to be an important measure to solve "carbon neutrality", and hydrogen consumption is an important indicator to measure the economy of fuel cell vehicles. At present, there are three methods for measuring the hydrogen consumption of fuel cell vehicles in ISO23828 [1] and SAEJ2572 [2], namely the gravimetric method, the temperature-pressure method, and the flow method. These three methods all measure the hydrogen consumption by changing the vehicle structure and using the technical means of external hydrogen supply. Although these three measurement methods have high accuracy, they have certain testing difficulties.

Many researchers have studied the accuracy and technology of three methods for measuring hydrogen consumption [3-10]. The mass method is currently considered to be the most accurate method of measurement. Kuroda E et al. [11] measured the mass of the hydrogen supply cylinder outside the vehicle. The measurement accuracy mainly depends on the mass of the weighing scale, although the method does not require accuracy correction. However, the hydrogen supply cylinder is a high-pressure container which brings certain risks for testers. Yano M et al. [12] obtained the data of hydrogen consumption by measuring the temperature and pressure before and after the high-pressure hydrogen storage container. This method needs to measure the temperature inside the hydrogen storage container, which adds a certain degree of difficulty to the test. Hirata H et al. [13] measured hydrogen consumption by adding a flowmeter between the hydrogen supply system and the vehicle. This method is simpler than the temperature pressure method and the gravimetric method, but its test accuracy mainly depends on the flowmeter, and the level of accuracy of flowmeter varies greatly, especially at low flow rates, it is difficult to ensure the accuracy of the measurement. Above all, these three methods are all based on external hydrogen supply, which increases the difficulty and complexity of the test. In addition, due to the testing requirements of the hydrogen supply pipeline and high pressure, these three methods all need to put forward higher requirements on the testing equipment and testing environment, which greatly increases the testing cost. And different manufacturers and enterprises will choose the differences measurement method according to the conditions of their own laboratories, which cause differences in the test results

This paper analyses the accuracy and the difficulty of testing by comparing the three methods of measurements, the temperature pressure method and the flow method. Secondly, a method for measuring the concentration of hydrogen and oxygen is proposed. This alternative method is mainly to obtain the value of hydrogen consumption by calculating the oxygen concentration, the flow rate of the exhaust gas, and the hydrogen concentration. The difficulty of measurement will be greatly reduced without changing the original car in the actual test process.

## 2. Comparative Analysis of Measurement Methods of Hydrogen Consumption

### *2.1. Comparative Analysis of Measurement Methods*

In *GB/T 35178 Measurement Method for Hydrogen Consumption of Fuel Cell Electric Vehicles*, the test methods for hydrogen consumption of fuel cell vehicles are mainly three recommended methods: gravimetric method, temperature pressure method and flow method. Each measurement method has its own advantages and disadvantages, and the comparison is shown in the following table:

**Table 1.** Comparison of hydrogen consumption measurement methods.

| The test method | Advantage | Disadvantage | Precision |
|---|---|---|---|
| Pressure temperature method | Capable of measuring hydrogen consumption, including penetration, leakage and purge | Measure the temperature of the gas in the cylinder | High |
| Quality method | Simple, almost no need to fix | Cannot be measured under operation | Higher |
| Flow method | Be able to run real measurements | Flowmeters require a certain degree of accuracy | General |

The mass method is to measure the mass of the external hydrogen cylinder before and after the test, which is shown in figure 1. The error of this method mainly depends on the accuracy of the weighing scale and the error caused by moving the gas cylinder back and forth. The formula is as follows:

$$W = g1 - g2 \tag{1}$$

g1 The mass of the hydrogen tank before the test;
g2 The mass of the hydrogen tank after testing;

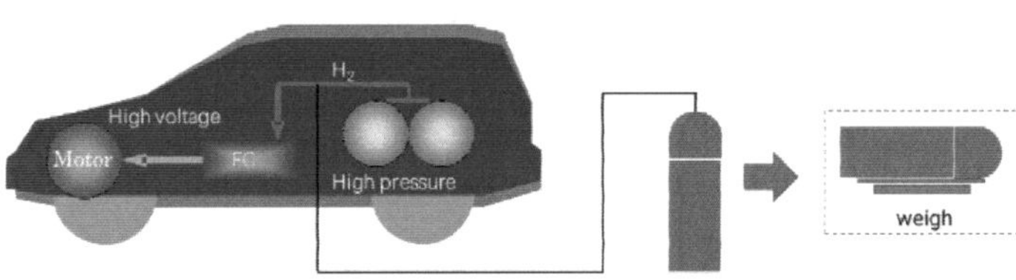


**Figure 1**. Schematic diagram of gravimetric test.

The principle of the temperature and pressure method test is mainly obtained by calculating the temperature and pressure changes before and after the test conditions. The disadvantage is that it is difficult to measure the temperature inside the hydrogen bottle. The formula is as follows:

$$W = \mathrm{m} \times \frac{\mathrm{V}}{\mathrm{R}} \times (\frac{\mathrm{P_1}}{\mathrm{z_1 \times T_1}} - \frac{\mathrm{P_2}}{\mathrm{z_2 \times T_{21}}}) \tag{2}$$

W is hydrogen consumption (g);
M is the molar mass of hydrogen;

Generally, it is calculated by the temperature of the gas cylinder or the ambient temperature after the gas cylinder is stabilized, which is shown in figure 2. V is the volume of the hydrogen tank (m3); R is the gas constant 8.314J/mol.K;

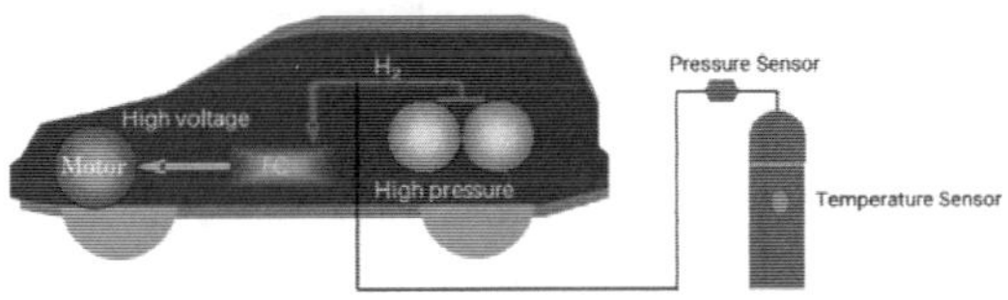


**Figure 2**. Schematic diagram of temperature and pressure measurement.

The flow method mainly records the consumed hydrogen during the entire conversion process by adding a flow meter between the external hydrogen supply system and the vehicle, which is shown in figure 3. This measurement method can measure all the hydrogen consumption, including the hydrogen lost by purging and leakage, and also achieve real-time measurement. The accuracy of this measurement method mainly depends on the accuracy of the gas flowmeter. When the hydrogen consumption varies greatly under different working conditions, there are certain requirements for the selected range of the flow meter, otherwise the test results will be greatly affected.

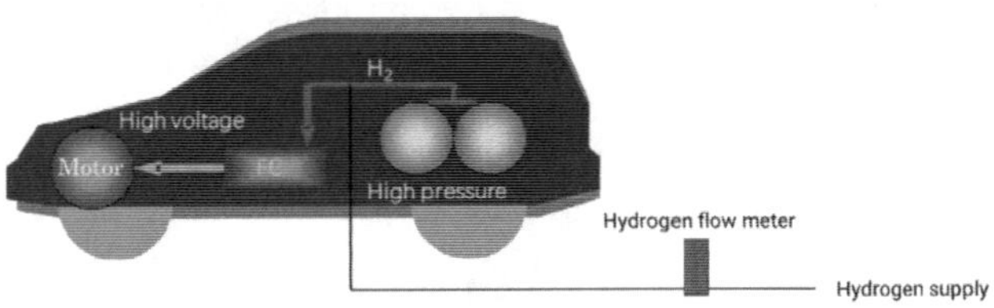


**Figure 3.** Schematic diagram of flow method test.

*2.2. Hydrogen and Oxygen Balance Test Principle*

*2.2.1. Total Hydrogen Consumption of Vehicles.* The schematic diagram of the hydrogen-oxygen balance method is shown in the figure below. The total amount of hydrogen consumed by fuel cell vehicles ($ML_{H2\_FC}$) can be decomposed into hydrogen consumed by electrochemical reactions ($ML_{H2\_\mathrm{Rec}}$) and hydrogen in vehicle exhaust ($m_{H2\_out}$). Hydrogen consumption is small and negligible. The hydrogen consumption of the electrochemical reaction ($ML_{H2\_\mathrm{Rec}}$) has a certain stoichiometric relationship with the amount of oxygen ($ML_{O2\_\mathrm{Rec}}$) that occurs in the electrochemical reaction.

$$ML_{H2\_FC} = ML_{H2\_Rec} + ML_{H2\_\mathrm{out}} \tag{3}$$

$$ML_{H2\_Rec} = 2 \times ML_{O2\_Rec} \tag{4}$$

*2.2.2. The Content of Hydrogen in the Tail Exhaust.* The content of hydrogen in vehicle exhaust can be directly measured by volume flowmeter Measure and calculate the mass consumption of hydrogen in the exhaust gas:

$$m_{H2\_out} = \rho_{H2} \times \frac{P}{P_0} \times \frac{T_0}{T} \times V_{H2\_out} \tag{5}$$

$$ML_{H2\_\mathrm{out}} = m_{H2\_out}/M_{H2} \tag{6}$$

*2.2.3. Oxygen Content in Tail Exhaust.* Since the oxygen in the electrochemical reaction comes from air, it can be calculated by measuring the difference between the air content in the intake air and the oxygen content in the vehicle exhaust. At the same time, the oxygen content in the exhaust gas can also be directly measured by a mass flow meter.

$$m_{O2\_Rec} = m_{O2\_in} - m_{O2\_out} \tag{7}$$

$$m_{O2\_in} = \rho_{O2} \times \frac{P}{P_0} \times \frac{T_0}{T} \times V_{O2\_in} \tag{8}$$

$$m_{O2\_out} = \rho_{O2} \times \frac{P}{P_0} \times \frac{T_0}{T} \times V_{O2\_out} \tag{9}$$

$$ML_{O2\_\mathrm{Rec}} = \frac{m_{O2\_\mathrm{Rec}}}{M_{O2}} \tag{10}$$

*2.2.4. Calculation of Hydrogen Consumption.* The hydrogen consumption of the final fuel cell vehicle is composed of:

$$ML_{H2\_FC} = 2 \times (ML_{O2\_in}(dry) - ML_{O2\_out}(dry)) + \frac{m_{H2\_out}(dry)}{M_{H2}} \tag{11}$$

Above all, the hydrogen consumption of fuel cell vehicles can be obtained by testing the oxygen content and hydrogen content in the tail exhaust

## 3. Test Equipment

In the process of using the hydrogen-oxygen balance method to test the hydrogen consumption of fuel cell vehicles, the equipment needs both the chassis dynamometer and the exhaust gas equipment. In order to increase the measurement accuracy of the hydrogen-oxygen balance method, the CMS2000 flowmeter is added to the exhaust gas equipment, which can ensure the test accuracy under the condition of low flow. The pluggable temperature and humidity sensor and pressure sensor are used to measure the concentration of hydrogen and oxygen at the tail exhaust port and then extract part of the exhaust gas at the tail exhaust port. As shown below:

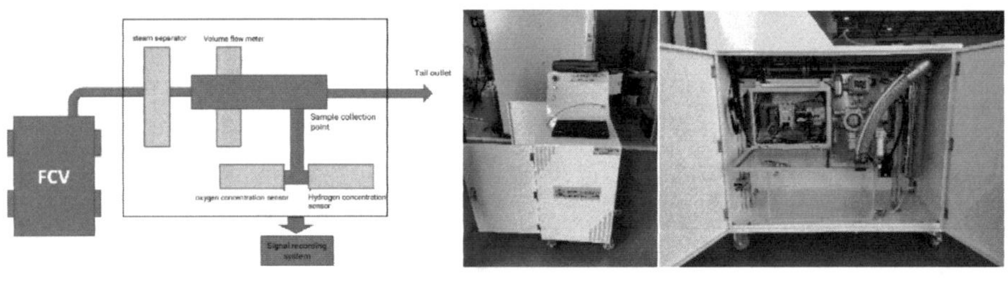


**Figure 4.** Hydrogen and oxygen balance measuring equipment.

In order to ensure the accuracy of the test results, it is necessary to install the steam-water separation device before the flowmeter, and then measure the total volume flow in the entire waste discharge process through the flowmeter. The temperature, humidity and pressure at the tailpipe are recorded by installing sensors inside the tailpipe. Further, the hydrogen concentration and oxygen concentration at the tail exhaust are recorded through the hydrogen concentration sensor and the oxygen concentration sensor.

## 4. Test Results

In order to compare the accuracy and difficulty of the test methods for hydrogen consumption of fuel cell vehicles, this paper selects an internationally advanced fuel cell vehicle for comparative test analysis. Since fuel cell vehicles have different energy supply methods under different driving conditions, it is necessary to construct vehicle driving conditions for comparative analysis. The current standard considers using the "Chinese working condition" to implement the working condition method to measure the hydrogen consumption.

### *4.1. Comparison of Traditional Measurement Methods*
Taking China comprehensive cycle condition as the target condition, the mass method, temperature pressure method and flow method were measured respectively. Details are as follows.

*4.1.1. Quality Method Data Analysis.* In order to ensure the accuracy of the test, the selected test vehicle was tested on the chassis dynamometer for 9 cycles of CLTC working conditions. Before and after the test, the mass of the off-board hydrogen supply equipment was weighed respectively. The results are shown in the following table:

**Table 2.** Quality test data.

| | 1 | 2 | 3 | 4 | 5 | 6 | 7 | 8 | 9 |
|---|---|---|---|---|---|---|---|---|---|
| Before | 67409.06 | 68916.08 | 68745.47 | 67289.37 | 68803.70 | 68633.69 | 69732.09 | 68691.06 | 68520.65 |
| After | 67288.99 | 68803.63 | 68633.30 | 67170.55 | 68691.06 | 68520.65 | 69615.72 | 68581.10 | 68409.85 |
| Consumed | 120.07 | 112.45 | 112.17 | 118.82 | 112.64 | 113.04 | 116.37 | 109.96 | 110.80 |

The data results show that the test results of each cycle condition are between 109g and 119g. The reason for this error may be due to the slight difference in the energy state of the vehicle, or it may be the error caused by the tester during the operation. In addition, the data results show that the energy consumption of the first cycle is the highest, which may be due to the fact that the temperature and state of the vehicle have not reached the optimum when the measurement is started, and the energy consumed is more used to balance the vehicle's own consumption. In order to reduce the influence of various errors, the average hydrogen consumption of 9 cycle conditions was selected as the final result of the hydrogen consumption of the mass method, and the hydrogen consumption of this passenger car was obtained as 113.28g.

*4.1.2. Data Analysis of Temperature and Pressure Method.* In order to accurately measure the temperature and pressure of the hydrogen supply cylinder outside the car, it is necessary to ensure that the environment of the gas cylinder is in a relatively stable state. During the test process, a relatively stable CLTC working condition is also selected for data comparison and analysis. The measured data and results. As shown in table 3 below:

**Table 3.** Test data of temperature and pressure method.

| **m** | **v** | **r** | **W1** | **P1** | **Z1** | **W2** | **P2** | **Z2** |
|---|---|---|---|---|---|---|---|---|
| 2.016 | **47** | 0.008314 | 501.23 | 14.143 | **1** | 388.51 | 10.681 | **1** |

The data results show that after the temperature correction, the hydrogen consumption is 130.40g under a Chinese working condition. It is quite different from the mass method because this measurement method is greatly affected by temperature and pressure and which are need to be corrected. When the correction coefficient is changed, Z1=1.082024703, Z2=1.0603058, the hydrogen consumption value is 112.72g.

During the test using the temperature-pressure method, it is possible to record the amount of hydrogen during use, including the amount of purging and leakage. The disadvantage is that it is difficult to measure the temperature inside the hydrogen bottle. In order to verify the influence of the cylinder temperature on the measurement results, the time correction of the temperature sensor was carried out, and the temperature and pressure changes after the test, 30 minutes, 60 minutes and 90 minutes, respectively. The results are shown in the following table:

**Table 4.** Changes of temperature and pressure method over time.

| Difference | Temperature | Pressure | Measuring time |
|---|---|---|---|
| 282.256 | 21.9 | 7.280 | 0min |
| 283.554 | 22.3 | 7.303 | 30min |
| 284.33 | 24.3 | 7.386 | 60min |
| 284.76 | 24.4 | 7.409 | 90min |

The results show that the temperature and pressure increase obviously with the change of time, but the influence on the results is not obvious, so the influence of time on the measurement results of hydrogen consumption can be ignored.

*4.1.3. Data Analysis by Flow Method.* The selection of hydrogen flow meter has a great influence on the accuracy of flow method. In order to ensure the accuracy of test results, the measurement results of six cycles in China are measured in the following table:

**Table 5.** Test data of flow method.

| | **1** | **2** | **3** | **4** | **5** | **6** |
|---|---|---|---|---|---|---|
| **Consumed** | 120.10 | 123.45 | 126.67 | 122.90 | 124.45 | 125.54 |

The data results show that the results of the flowmeter are all above 120g. Compared with the results of other measurement methods, the measured value is higher, which is caused by the accuracy of the flowmeter itself. At high power, the hydrogen consumption of fuel cell vehicles is high, and the measurement results are relatively accurate. When running at low power, the comprehensive measurement results are high due to the large measurement range of the flowmeter itself and low accuracy at low concentrations. The average value of 6 cycle conditions is taken in the test, and the hydrogen consumption value of the flow method is obtained as 123.85g:

*4.2. Analysis of Measurement Methods of Hydrogen and Oxygen Balance Method*

*4.2.1. Test Data of Hydrogen and Oxygen Balance Method.* In order to compare with the above measurement methods, the same fuel cell vehicles were also selected to analyse the hydrogen and oxygen concentrations under Chinese working conditions. The figure below shows the analysis of the measurement data of hydrogen and oxygen balance method in the driving cycle of China. Six cycle conditions are selected for test evaluation. The test data are shown in the following table.

**Table 6.** Hydrogen and oxygen balance test data.

| | **1** | **2** | **3** | **4** | **5** | **6** |
|---|---|---|---|---|---|---|
| **Consumed** | 125.60 | 127.90 | 130.54 | 137.70 | 129.56 | 125.45 |

The data results show that the results measured by the hydrogen-oxygen balance method are between 125g and 140g. The average value of the six cycle conditions is also taken. The calculated error is 14.28%. The error of several comparative hydrogen-oxygen balance methods is the largest, but also within a reasonable range.

*4.2.2. Error Analysis of Hydrogen-Oxygen Balance Method.* The reasons for the error are mainly analyzed in the following situations: the error of the flowmeter itself, the error of different types of flowmeters will be different, and the processing and collection of the rear exhaust will have errors; followed by the error of the vehicle structure itself, the amount of hydrogen leaking from the pipeline cannot be calculated, and the internal electrochemical reaction is not uniform, etc., which affect the overall error calculation; third, it is the error of the operator and processing. In the hydrogen-oxygen balance method, the exhaust gas needs to be collected. This error can be eliminated through experience. Although the overall measurement error is different from the flow method and the temperature pressure method, the results are also within a reasonable range. After the optimization of the equipment and the technical improvement of the test operators, the measurement accuracy will be greatly improved and the error will be reduced.

*4.2.3. Comparative Analysis of Five Test Methods.* Through the analysis of the results of the five measurement methods, if the mass method is used as the benchmark for measurement, the error of the temperature and pressure method is 4.94%, the error of the flow method is 9.33%, and the error of the current method is 8.29%. The measured values of hydrogen consumption under the driving cycle conditions in China are as follows in table 7, in which the measurement error of the hydrogen-oxygen balance method is the largest

**Table 7.** Hydrogen consumption comparison test data.

| | Quality standard | Temperature pressure method | Flow method | Hydrogen and oxygen equilibrium method |
|---|---|---|---|---|
| Hydrogen consumption（g） | 113.28 | 112.72 | 123.85 | 136.26 |
| Error | —— | 4.94% | 9.33% | 14.28% |

## 5. Summary

(1) This paper analyzes data from theoretical research and experiments, and proposes a method based on the conservation of exhaust gas quality of fuel cell vehicles. It is designed to calculate the hydrogen consumption of fuel cell vehicles by measuring the hydrogen and oxygen concentrations in the intake and exhaust gas. (2) By comparing the accuracy of several hydrogen consumption measurement methods, the test results show that, taking the mass method as the benchmark, the relative accuracy of the hydrogen-oxygen balance method is low and needs to be improved, but the measurement results are to a certain extent. It can be used to provide a better reference for the measurement of the hydrogen consumption of the subsequent fuel cell vehicle.

## References

[1] ISO 23828 2013 *Fuel Cell Road Vehicles – Energy Consumption Measurement - Vehicles Fuelled with Compressed Hydrogen.*

[2] SAE J 2572 : 2014 2014 recommended practice for measuring fuel consumption and range of fuel cell and hybrid fuel cell vehicles fuelled by compressed gaseous hydrogen *SAE International* 2014-08-2578,

[3] Aoyagi S, et al. 2004 development of fuel economy measurement method for fuel cell vehicle *SAE International* 2004-01-1305.

[4] Ding Y, et al. 2004 hydrogen consumption measurement for fuel cell vehicles *SAE International* 2004-01-1008.

[5] Ding Y, et al. 2004 hydrogen consumption measurement for fuel cell vehicles *SAE International* 2004-01-1008.

[6] Kuroda E, et al. 2002 Test methods to measure fuel economy in direct hydrogen fuel cell vehicles *JARI research Journal (in Japanese)* **24** 463-473.

[7] Kuroda E, et al. 2003 Test methods to measure fuel economy in direct hydrogen fuel cell vehicles *Engine Technology (in Japanese)* **25** 20-26.

[8] Kuroda E, Moriya K, Yano M, Watanabe S, Hirata H, et al. 2006 Development of fuel consumption measurement methods for hydrogen fuel cell vehicles *SAE Technical Paper* **115**(3) 144-154.

[9] Yano M, Kuroda E and Watanabe S 2004 Fuel economy measurements for fuel cell vehicles (weight method and pressure method) *JARI Research Journal* **26**(6) 257.

[10] Kuroda E, Moriya K, Tagami H and Yano M 2006 Hydrogen flowmeter for fuel consumption measurement of fuel cell vehicles *JARI Research Journal* **28**(7) 253.

[11] Kuroda E, Yano M, Hirata H and Watanabe S 2003 Test methods to measure fuel economy in direct hydrogen fuel cell vehicles *Society of Automotive Engineers of Japan Annual Congress Proceedings* 19: 03: 5.

[12] Yano M, Kuroda E and Watanabe S 2003 Fuel economy measurements for direct hydrogen fuel cell vehicles *The 23 rd Hydrogen Energy System Society of Japan Symposium Proceedings* p105.

[13] Hirata H, Yano M, Kuroda E and Watanabe S 2004 Test methods to measure fuel economy in direct hydrogen fuel cell vehicles (second report – The flow method) *Society of Automotive Engineers of Japan Annual Congress Proceedings* **76**(04) 13

# Discussion about the Relationship between Coating Life and Cathodic Protection Current Density of Offshore Pipeline

**Xue Li[1*], Dan Wang[1], Changsuo Dong[2], Weimin Jiang[1] and Jie Sun[1]**

[1] CNPC Project Management Company Tianjin Design Institute, Tianjin, China
[2] Pipe China Group Beijing Pipeline Co., Ltd., Beijing, China
Email: 270258134@qq.com

**Abstract.** This article studies characteristics of 3LPE external coating material and cathodic protection technology for offshore pipeline firstly, analysis the mechanism of cathodic disbondment and specified properties, carries out cathodic disbonding tests base on different cathodic protection current density applied on coated samples at given temperature, and then analyses the relationship between coating failure life and index of cathodic protection using Arrhenius model and inverse power law model, puts forward a recommended maximal cathodic protection current density for offshore pipeline at a certain design life of coating.
**Keywords.** Coating, 3LPE, Cathodic Protection, Cathodic Disbondment, Current Density.

## 1. Introduction

Because the offshore pipeline is always buried in sea water and mud, under the influence of the surrounding corrosive environment, pipe wall will be corroded inevitably. In order to avoid or reduce the corrosion damage and improve the service life of pipeline, effective anti-corrosion measures should be taken for offshore pipeline. In recent years, with increasing of offshore oil and gas exploration and production activities, more and more pipelines are used to transport submarine oil and gas, and the performance of anti-corrosion coating is required to be higher and higher.

According to the development trend of anticorrosion technology for offshore pipeline, a good pipeline anticorrosion system should have anti-moisture penetration performance, good field construction performance, anti-impacting, anti-cathode disbondment, anti-delamination performance, longer service life, lower cost during service, installation and maintenance.

Different from onshore pipelines, offshore pipelines will be affected by dissolved oxygen, chloride ions and other factors in the Marine environment, and the corrosion environment is more severe. Anticorrosive coating and cathodic protection are generally used in offshore pipeline projects to ensure the safety of pipeline operation together. The compatibility of anticorrosive coating and cathodic protection is the key factor to ensure the service life of pipeline.

## 2. Relationship between Pipeline Coating and Cathodic Protection

At present, most offshore pipelines often adopt 3PE structure as external anticorrosive coating, which is composed of fused bonding epoxy (FBE) as the inner layer, polymer adhesive as the middle layer, and polyethylene (PE) coating as the outer layer [1]. China first introduced 3PE coating technology and production line in 1994, and now 3PE anticorrosive technology has become more and more popular in China. The thermoset crosslinked molecular structure of 3PE coating and its chemical bond with steel determine its good performance as an anticorrosive coating. 3PE anticorrosion structure is used in offshore pipeline anticorrosion widely because of its superior anticorrosion performance and good production technology.

Generally, if 3PE anticorrosive coating has no defect, the influence of media factors in external environment such as ion content, dissolved oxygen and pH value on the coating is small, However, during production of 3PE raw materials, or 3PE coating fabrication on line, it is possible to cause 3PE anti-corrosion coating defects. Coating disbonding usually occurs in these defective areas. In view of the potential defects during coating preparation and later construction, the combined protection method of coating and cathodic protection is used to protect the exterior surface of the pipeline from corrosion. In Marine environments, as long as coating attached to the pipe continuously and cathodic protection system is set up properly, maintained and monitored as required, no corrosion will occur.

Cathodic protection methods of offshore pipelines include impressed current method and sacrificial anode method. According to the statistics, sacrificial anode cathodic protection is adopted as supplementary protection of anti-corrosion coating for offshore pipelines based on safety and reliability considerations mainly. During operation of cathodic protection system, sacrificial anode releases cathodic protection current that reaches to pipe surface and changes the pipe surface potential. When the cathodic protection current is too small, protection potential of pipeline is insufficient, and pipeline is under insufficient protection, and there is a risk of further corrosion. When the cathodic protection current is too large, pipeline is in the over-protection state, in this case, pipeline anti-corrosion coating may occur cathodic disbonding, bulging and other phenomena. Cathodic disbonding is one of the important indexes of the influence from cathodic protection system to anticorrosive coating.

At present, cathodic protection polarization potential is generally used as the evaluation index of whether the coating may occur cathode disbondment in the world. The Chinese national standard GB/T 21448 stipulates that the maximum cathodic protection polarization potential of pipeline is -1.2V, and the British BS EN 12954 stipulates that the maximum cathodic protection polarization potential of pipeline is -1.1V. The upper limit of polarization potential of cathodic protection is -1.2V according to NACE standard.

During offshore pipeline engineering design phase, cathodic protection current density is an important parameter to calculate the number and distribution of sacrificial anodes. According to the international standard ISO 15589-2 and DNV GL RP F103, cathodic protection current density of offshore pipeline is considered in three stages, namely, the initial, middle and final stages of construction. It is considered that the specified current density can cause the desired cathode polarization potential of pipeline without considering the influence of the design years of the pipeline. The design life of different offshore pipelines may be different, and the time to reach coating failure (when the cathode disbonding distance exceeds the specified value) may also be different due to the current density of cathodic protection, which has not been directly studied in the world. Therefore, it is very important to further clarify the relationship between cathodic protection current density and coating cathodic disbonding under different design life.

In this paper, cathode disbonding of the coating is taken as the accelerated experiment object. Under the simulated external service environment in the seabed, cathode disbonding experiment of 3LPE coating is carried out at different cathodic protection levels and at different test time. The change of cathode disbonding distance with cathodic protection current density is tested and fitted according to the coating life prediction model. Determine the upper limit of current density of cathodic protection for a given coating design life.

## 3. Cathode Disbonding Generation and Index Specification

When the anticorrosive coating is intact, metal is completely isolated from the corrosive medium, and no corrosion occurs. However, during the process of transportation, installation and operation, it is inevitable to produce leakage points due to damage of the anticorrosive layer. The purpose of applying cathodic protection to pipeline is to avoid corrosion when there are leakage points in the anticorrosive layer.

Under the action of cathodic protection, pipeline becomes cathode part of the cathodic protection system, and sacrificial anode is the anode part of the cathodic protection system. The following reactions may occur in the cathode part [2]:

$$O_2 + 2H_2O + 4e \rightarrow 4OH^- \quad (1)$$

$$H^+ + e \rightarrow H \quad (2)$$

Both of above cathodic reaction equation (1) and equation (2) lead to the formation of excess $OH^-$ in the cathode region, resulting in an alkaline environment. When pH value of the cathode area increases, a large number of $OH^-$ migrates to the metal/anticorrosive layer interface, so that the base metal oxide layer bonded to the anticorrosive layer dissolves, or erodes the polymer at the interface, or reduces the bonding force between the polymer and the metal, thus causing the anticorrosive layer to peel, that is cathode disbonding.

By referring to cathode disbonding indexes of domestic and foreign 3LPE coating specifications, it can be found that the cathode disbonding indexes are affected by peel time, test temperature and sample potential, as shown in table 1. The long-term performance index of cathodic disbonding can better reflect the anti-cathodic disbongding performance of the coating. Therefore, in this paper, the time when the cathodic disbonding distance of the coating reaches 15 mm is taken as the failure life or characteristic life of the coating as a conservative evaluation index.

**Table 1.** List of requirement for cathodic disbondment distance of different standards.

| Standard No. | Cathodic disbanding distance (mm) |
|---|---|
| GB/T 23257 | ≤6（65°C, 48h, -1.5V）;<br>≤15（max. operation temperature, 30d, -1.5V） |
| ISO 21809-1 | ≤7（23°C, 28d, -1.5V）;<br>≤7（65°C, 24h, -3.5V）;<br>≤ 15（max. operation temperature, 28 d, -1.5 V） |
| CSA_Z245.20/21 | ≤ 12（20°C, 28 d, -1.5 V）;<br>Purchase specification（max. operation temperature, 28 d, -1.5 V）;<br>≤7（65°C, 24h, -3.5V） |
| NFA-49-710 | ＜10（20mA, 40°C, 7d） |
| DIN 30670 | ＜7（23°C, 28d;60°C, 2d） |
| IPS-G-TP-335 | ＜5 |

## 4. Test of Cathode Disbonding with Current Density

Due to the action of electrolyte in the environment, pipeline anticorrosion layer may be corroded at damaged location, and application of cathodic protection potential will cause new leakage point or peel off at the leakage point. In order to judge the peeling resistance of the coating, the method of simulating the peeling occurrence and development of the coating at the damaged point was adopted. A simple schematic of this approach is shown in figure 1.

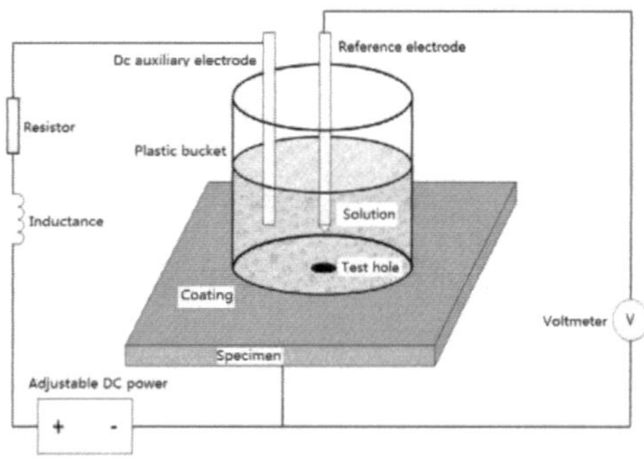


**Figure 1.** Schematic diagram of cathode disbonding test device.

**Figure 2.** Photo of cathode disbanding experimental facility.

This experiment was carried out at the maximum operating temperature (30°C in this paper) under atmospheric pressure, and cathodic protection was applied by using constant-current method. The corresponding current density was calculated according to defect hole area of the prefabricated coating, and corresponding cathodic protection potential was tested.

When a current density of 13.3 mA/cm$^2$ is applied to the 1$^{st}$ sample group, the potential of the sample was -1.25 V vs. CSE. The test results of cathode disbonding at different times are shown in table 2, and the macroscopic morphology after disbonding was shown in figure 3. It can be seen from the test results that the disbonding distance increases gradually with the increase of time.

**Table 2.** Test result of disbonding distance for 1$^{st}$ sample group.

| Time | Distance of cathodic disbonding (cm) | | | |
|---|---|---|---|---|
| | 1# | 2# | 3# | 4# |
| 3 days | 2.98 | 3.9 | | |
| 9 days | 4.47 | 4.08 | 4.85 | 4.65 |
| 16 days | 4.52 | 4.68 | 5.01 | 4.7 |
| 39 days | 5.43 | 5.36 | 5.68 | |

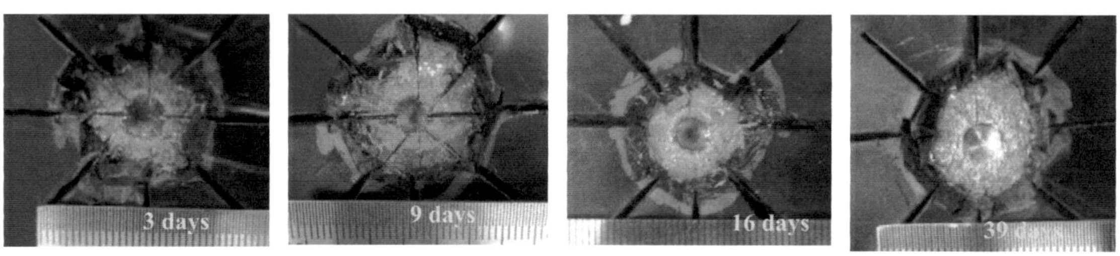


**Figure 3.** Morphology after disbandment of 1$^{st}$ sample group.

When a current density of 15.6 mA/cm$^2$ is applied to the 2$^{nd}$ sample group, the potential of the sample was measured to be -1.266Vvs.CSE. The test results of cathode disbonding after different times are shown in table 3. It can be seen from the test results that the current density will affect the cathode disbonding performance of the material.

**Table 3.** Test result of disbonding distance for 2nd sample group.

| Time | Distance of cathodic disbonding (cm) | | |
|---|---|---|---|
| | 1# | 2# | 3# |
| 3 days | 4.22 | 4.33 | |
| 9 days | 6 | 5.5 | 6.5 |
| 16 days | 6.78 | 5.98 | |
| 30 days | 7.23 | 6.53 | |
| 60 days | 8.04 | 6.64 | |

When a current density of 50.0 $mA/cm^2$ is applied to the 3rd sample group, the potential of the sample is -1.28V vs. CSE. The cathode disbonding test results at different times are shown in table 4.

**Table 4.** Test result of disbonding distance for 3rd sample group.

| Time | Distance of cathodic disbonding (cm) | | | |
|---|---|---|---|---|
| | 1# | 2# | 3# | 4# |
| 9 days | 3.45 | 3.65 | | |
| 16 days | 4.30 | 5.08 | 4.49 | 4 |
| 30 days | 4.57 | 5.58 | 5.16 | 4.17 |
| 39 days | 6.31 | 7.67 | | |

When a current density of 100.0mA $/cm^2$ is applied to the 4th sample group, the potential of the sample is -1.30 V vs. CSE. The test results of cathode disbonding at different times are shown in table 5.

**Table 5.** Test result of disbonding distance for 4th sample group.

| Time | Distance of cathodic disbonding (cm) | | | |
|---|---|---|---|---|
| | 1# | 2# | 3# | 4# |
| 9 days | 3.67 | 3.15 | | |
| 16 days | 5.09 | 5.11 | 5.29 | 6.1 |
| 30 days | 5.3 | 7.08 | 8.02 | 7.77 |
| 39 days | 6.18 | 7.63 | | |

When a current density of 156.2 $mA/cm^2$ is applied to the 5th sample group, the potential of the sample is -1.33Vvs.CSE. The test results of cathode disbonding after different times are shown in table 6.

**Table 6.** Test result of disbonding distance for 5th sample group.

| Time | Distance of cathodic disbonding (cm) | |
|---|---|---|
| | 1# | 2# |
| 9 days | 2.96 | 4.79 |
| 16 days | 8.46 | 8.23 |
| 30 days | 12.23 | 10.92 |
| 60 days | 12.53 | 10.48 |

When a current density of 312.5 mA/cm$^2$ is applied to the 6$^{th}$ sample group, the potential of the sample measured is -1.34 VvsCSE. The test results of cathode disbonding after different time are shown in table 7.

**Table 7.** Test result of disbonding distance for 6$^{th}$ sample group.

| Time | Distance of cathodic disbonding (cm) | |
|---|---|---|
| | 1# | 2# |
| 3 days | 7.89 | 7.62 |
| 9 days | 10.46 | 12.01 |
| 16 days | 11.95 | 11.66 |

When a current density of 1562.5 mA/cm$^2$ is applied to the 7$^{th}$ sample group, the potential of the sample measured is -1.36 VvsCSE. The test results of cathode disbonding after different times are shown in table 8.

**Table 8.** Test result of disbonding distance for 7$^{th}$ sample group.

| Time | Distance of cathodic disbonding (cm) | |
|---|---|---|
| | 1# | 2# |
| 3 days | 6.1 | 7.1 |
| 9 days | 9.12 | 12. 1 |
| 16 days | 10.35 | 11.77 |

## 5. Test of Cathode Disbonding with Current Density Change

In this paper, Arrhenius model and inverse power law model are used to calculate the accelerated coating failure experiment.

### *5.1. Arrhenius Model*

Arrhenius model equation is an acceleration model proposed by Swedish physical chemist Arrhenius in 1880 by summarizing a large number of experimental data when he studied the relationship between acceleration stress temperature and chemical reaction rates of electronic components and insulating materials [3].

$$\xi = A\exp\left(E / KT\right) \tag{3}$$

where:
ξis the life characteristic, such as median life and average life, etc;
A is A constant, and A > 0;
E is activation energy, related to the material, unit eV;
K is Boltzmann constant, with a value of 9.617×10-5 eV/°C;
T is the absolute temperature, which is Celsius plus 273.

According to the Arrhenius model shown in equation (3), the life characteristics of samples decrease exponentially with the increase of temperature. Temperature becomes a key factor affecting the life of materials. The increasing of ambient temperature will directly lead to the intensification of molecular chain movement in organic polymer materials, and when the molecular kinetic energy is greater than dissociation energy of chemical bonds, the thermal degradation behavior of molecular chain will occur. At the same time, under hot and humid conditions, when material is surrounded by water molecules, the force between the material molecules will change due to swelling and dissolution, which directly leads to material destruction, and then accelerates the aging of the material.

### *5.2. Inverse Power Law Model*

Many physical accelerated life experiments confirm that electrical stress (such as voltage, current, power, etc.) as the accelerated stress, can also promote product failure in advance. Through summary and calculation of a large number of relevant accelerated life experimental data, it is found that some life characteristics and stresses of products conform to the inverse power law model [4].

$$\xi = Av^{-c} \tag{4}$$

where:
ξ is the life characteristic, such as median life and average life, etc;
A is A constant, and A > 0;
C is the normal number related to activation energy;
v is an accelerating stress, such as voltage.

According to the inverse power law model shown in equation (4), the life characteristic of the sample is a negative power function of the accelerated stress v.

### *5.3. Arrhenius and Inverse Power Law Unified Model*

The logarithm of both sides of the Arrhenius model shown in equation (3) is obtained

$$\ln \xi = \ln A + E / (KT) \tag{5}$$

Similarly, take logarithms of both sides of the inverse power law model shown in equation (4), and get

$$\ln \xi = \ln A + (-c)\ln v \tag{6}$$

It can be seen that the life prediction formula can be unified as

$$\ln \xi = a + b\theta(s) \tag{7}$$

Where ξ is life characteristic, and θ(s) is a function of accelerating stress.

According to the life prediction model of equation (7), the relationship between lnξ and θ(s) is linear. After determining the values of constants a and b, the expected life of the coating at a given current density can be calculated according to the equation.

In this paper, the changes of cathode disbonding distance with time under seven different current densities were obtained. With the increase of experiment time, the disbonding distance gradually increased, and the increase rate is fast in the early stage of the experiment, but slow down in late stage of the experiment. According to the relationship between disbonding distance and experimental time, Oringin software can be used for mathematical analysis and calculation to obtain the characteristic life of coating when the disbonding distance reaches 15mm under the experimental current density (according to the requirements of the cathode disbonding distance in relevant standards, the time when disbonding distance reaches 15 mm is taken as the characteristic life of the coating). Then the accelerated life model is obtained according to the relationship between current density and characteristic life.

According to the test results, under the same current density, the disbonding distance and test time meet the function $L= a*t^b$, namely lnL= lna + b*lnt, and the logarithm of the distance is linear with the logarithm of the time.

The established model and fitting results are shown in figure 4 to figure 10. According to the curve fitting results, the time when the coating disbonding distance equal to 15mm under different experimental current densities can be obtained, as shown in table 9. It can be seen from the table that with increasing of current density, the life characteristic of coating decrease gradually.

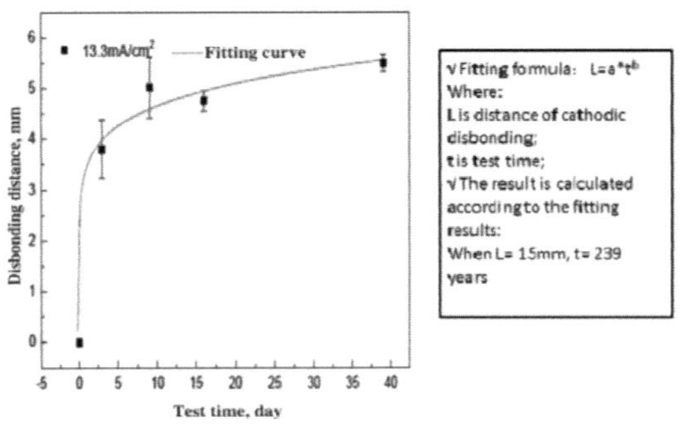


**Figure 4.** Analysis of life characteristics under 13.3mA/cm$^2$ current density.

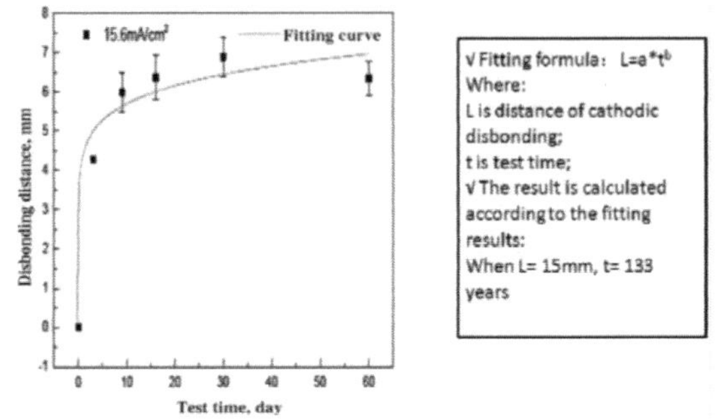


**Figure 5.** Analysis of life characteristics under 15.6mA/cm$^2$ current density.

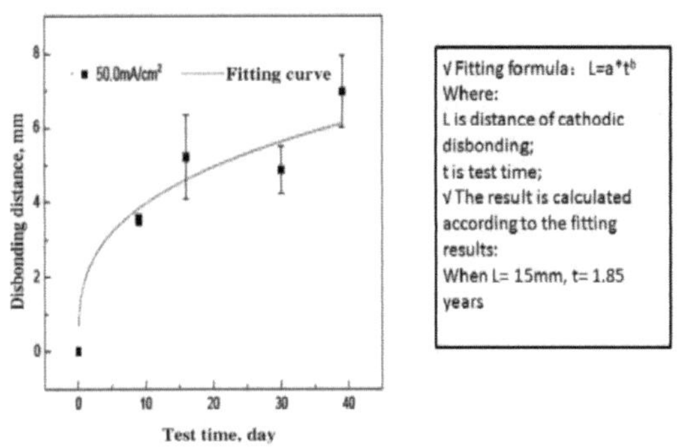


**Figure 6.** Analysis of life characteristics under 50mA/cm$^2$ current density.

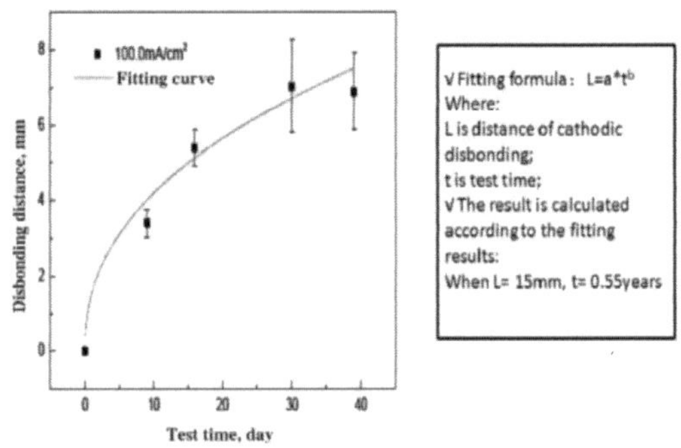


**Figure 7.** Analysis of life characteristics under 100mA/cm$^2$ current density.

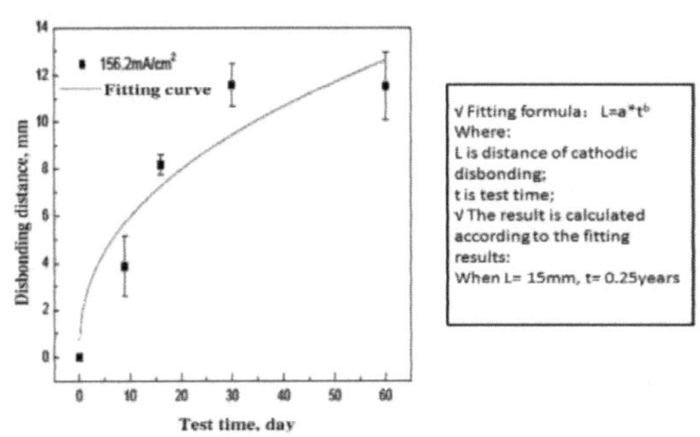


**Figure 8.** Analysis of life characteristics under 156.2mA/cm$^2$ current density.

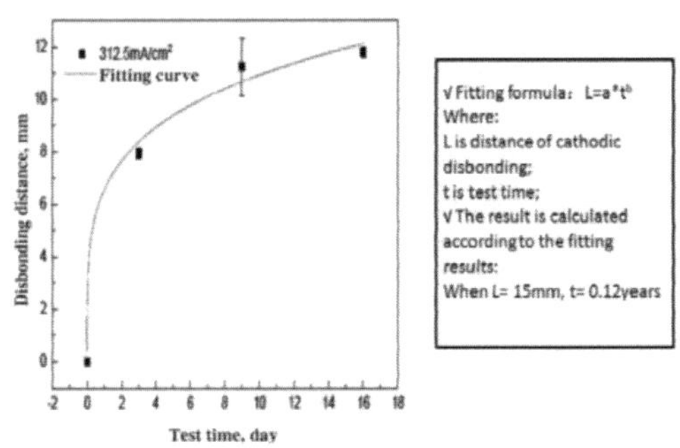


**Figure 9.** Analysis of life characteristics under 312.5mA/cm$^2$ current density.

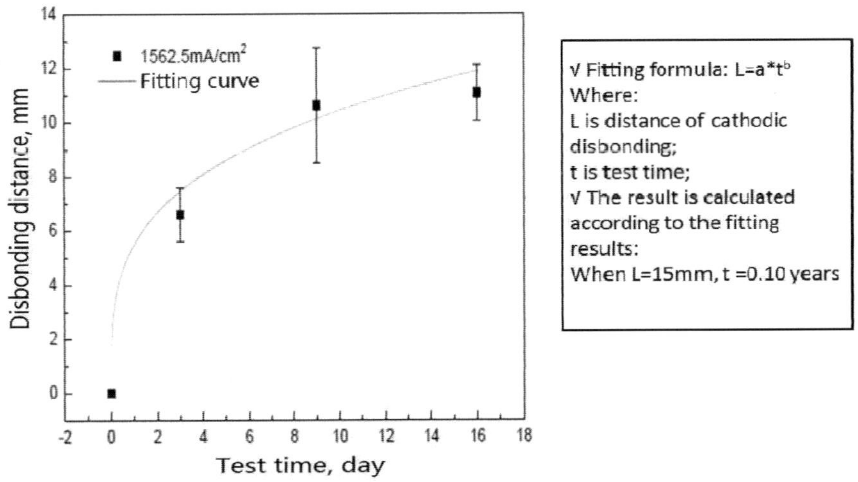


**Figure 10.** Analysis of life characteristics under 1562.5mA/cm$^2$ current density.

**Table 9.** Life characteristics of failure under different current densities.

| Item | Current densities ($mA \cdot cm^{-2}$) | Coating failure (year) |
|---|---|---|
| 1 | 13.3 | 239 |
| 2 | 15.6 | 133 |
| 3 | 50.0 | 1.85 |
| 4 | 100.0 | 0.55 |
| 5 | 156.2 | 0.25 |
| 6 | 312.5 | 0.12 |
| 7 | 1562.5 | 0.1 |

According to the failure life under the given experimental current density, a life expectancy model based on the results of cathode disbonding can be established according to equation (7). The life characteristic relationship between current density and coating is shown in figure 11. It can be seen from the figure that there is a linear relationship between the life characteristics and the reciprocal of current density. Origin software is used for linear fitting to obtain the life model as equation (8). Based on the experimental results of cathode disbonding, the lower the current density, the higher the coating life expectancy.

$$\xi = 105.9 \ln(1/I) - 2.06 \tag{8}$$

where:
$\xi$ is the expected life;
I is current density;

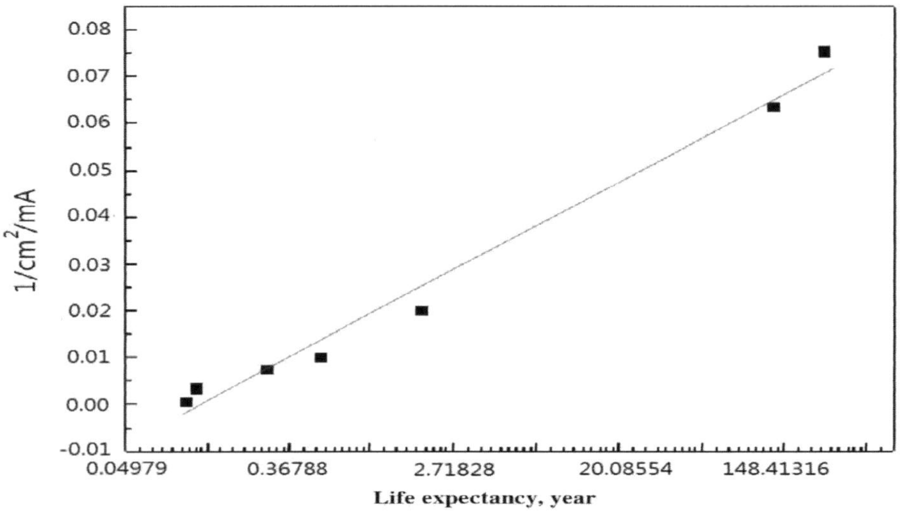


**Figure 11.** Life expectancy model based on the law of cathodic disbondment.

According to the specification, the maximum negative cathodic protection potential of steel structures should not exceed -1200 mV vs. CSE [5], and then cathodic protection current density is the maximum.

Assuming that the failure life of the pipeline coating is 50 years, the maximum current density of cathodic protection under the expected failure life is calculated according to the established model. As shown in figure 12, the current density of cathodic protection should not exceed 17.8 mA/cm$^2$, otherwise the cathode disbonding distance at the end of the coating life will exceed the specification value by 15 mm.

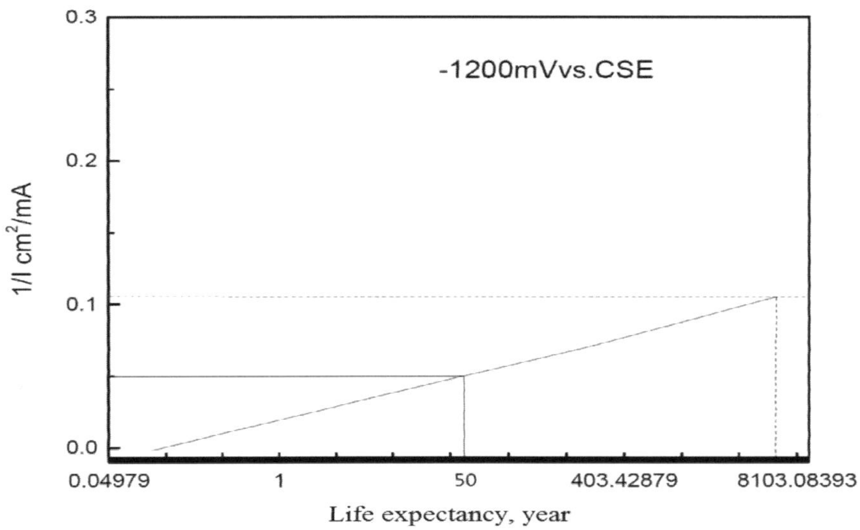


**Figure 12.** Life analysis of coating failure at -1200mVvs.CSE.

## 6. Summary

Corrosion control of offshore pipeline requires combination of external coating and cathodic protection to achieve desired protection target. The effect of cathodic protection current on coating can not be ignored, and too high cathodic protection current density may accelerate coating damage. In this paper, cathode disbonding is adopted as the coating failure acceleration experiment method. Under the simulated external service environment of offshore pipeline, cathode disbonding experiments are conducted on 3PE coating specimens at 7 different cathodic protection levels. The conclusions are as follows:

(1) With increasing of cathodic protection current density, the cathode disbonding distance of the sample coating increases.

(2) Under the same cathodic protection current density, the cathodic disbonding distance of coating increases with time increasing simultaneously.

(3) Different test conditions, such as sample source and test temperature, may affect the test results.

(4) The upper limit of cathodic protection current density for the sample coating can be calculated according to determined design life of coating and cathode disbonding index, which can be used in design of cathodic protection system.

## References

[1] International Organization for Standardization 2011 ISO 21809-1 Petroleum and natural gas industries - External coatings for buried or submerged pipelines used in pipeline transportation systems - Part 1: Polyolefin coatings (3-layer PE and 3-layer PP) chapter 8 p9

[2] Hu S X 1999 *Cathodic Protection Engineering Manual* Chemical Industry Press chapter 3 p45

[3] International Organization for Standardization 2014 ISO 11346 Rubber, vulcanized or thermoplastic — Estimation of life-time and maximum temperature of use chapter 11 p4

[4] Liu H C, Wu Y C and Zhao M 2008 Comparative analysis of accelerated life test and accelerated degradation test *Journal of Guiyang University* **4** 11-15.

[5] International Organization for Standardization 2012 ISO 15589-2 Petroleum, petrochemical and natural gas industries - Cathodic protection of pipeline transportation systems - Part 2: Offshore pipelines chapter 7 p8

# Research on Thermal Balance Test and Evaluation Technology of Fuel Cell Vehicle

**Ting Guo [1,2*], Xueyan Jin[1] Rongliang Liang [1,2], Shiyu Wu[1,2], Zhenyu Nie [1,2], Zhijun Wang [1,2], Guozhuo Wang [1,2] and Fang Wang[2]**

[1] China Automotive Technology and Research Center Co., Ltd, Tianjin, China
[2] CATARC Automotive Test Center (Tianjin) Co., Ltd, Tianjin, China
Email: guoting@catarc.ac.cn

**Abstract.** Thermal management system is importance to the efficiency and durability of fuel cell vehicles, so related researches are carried out on the thermal balance test and evaluation of fuel cells. Through theoretical and technical analysis, a test and evaluation method of thermal management system applied to various extreme conditions is proposed, and the test and evaluation conditions and evaluation indicators of each system of fuel cell vehicles are formed. In this paper, the thermal balance test results of a fuel cell vehicle are analyzed. The research results show that the thermal balance of fuel cell vehicles under extreme conditions reduces power and increases temperature difference to dissipate heat
**Keywords.** Fuel cell vehicle; thermal balance; test evaluation.

## 1. Introduction

During the commercial operation of fuel cell vehicles, thermal management is an important technical difficulty affecting its development [1-4], which has a decisive impact on the safety and lifetime of the entire vehicle. At present, the power source of fuel cell vehicles is mainly divided into two parts, which are battery and fuel cell. Due to the particularity of the fuel cell, its power generation efficiency is 40%-60%, and the rest power is dissipated in the form of heat. In the design process of the entire vehicle, 95% of the heat generated needs to be forced to dissipate through the radiator. The influence of the temperature on the fuel cell is crucial.

When the external temperature of the fuel cell is relatively high, the internal and external working temperature gradient will be small, which will cause problems of heat dissipation. On the other hand, due to the different heat sources on the whole vehicle and different heat dissipation requirements, generally three to five channels are required by vehicles for heat dissipation and cooling. It will cause the complexity of the structural design of the whole vehicle, especially for the passenger car, its complex structure is difficult to arrange. The fuel cell has a wide range of working conditions. When working in high-power conditions such as full acceleration or climbing, the output power of the fuel cell is relatively high. A large amount of heat will be generated in a short time, which will affect the heat dissipation performance of the vehicle. At present, the design of the most advanced fuel cell vehicles in the world try to avoid the operating time under extreme conditions as much as possible to meet the needs of heat dissipation.

At present, many researchers have carried out technical research on the thermal management system of fuel cell vehicles. Wang [4] analysed the thermal management system of typical international fuel cell vehicles, and introduced the detail of thermal management system of Mirai, NEXO and other fuel cell vehicles. Wang [5] established a thermal management model for fuel cell vehicles, and studied the influence of different fan selection and operating conditions. Kim et al. [6] carried out the influence of the driver's acceleration behavior and environmental conditions on the thermal management of electric vehicles. Pu [7] analysed the flow, pressure and heat related influence of the thermal management system, and put forward suggestions for system structure design and optimization. Qu [3] conducted a technical study on the thermal balance test of typical international fuel cell vehicles, and analyzed the thermal balance performance of NEXO. SAE [8] also published related introduction and analysis on the thermal management system of fuel cell vehicle.

Considering the structural differences between fuel cell vehicles and pure electric vehicles or traditional vehicles, the heat sources are diversity and the temperature control requirements are relatively large. It is necessary to comprehensively evaluate the thermal management system of the vehicle according to the control requirements of multiple heat sources and different temperatures. Existing standards for evaluating thermal balance systems "GB/T 12542-2020 Road Test Method for Thermal Balance Capability of Vehicles" [9], "T/CSAE 114-2019 Wind Tunnel Test Method for Cooling Environment of Automotive Powertrain" [10], etc. The structural characteristics of the fuel cell are not considered, so it is not suitable for the evaluation of fuel cell vehicles. This paper starts with the analysis of the test and evaluation of the thermal balance of the fuel cell vehicle, based on the working principle of the thermal management system, which also explores and verifies the thermal balance test and evaluation method of the fuel cell vehicle under the actual working conditions. The test and evaluation system have been complemented well.

## 2. The Working Principle of Fuel Cell Heat

The generation and discharge of heat in fuel cells run through the entire electrochemical reaction. The reasons for heat generation are mainly divided into four types, including chemical reaction heat generated by electrochemical reaction, Joule heat generated by ohmic polarization, and phase generated by condensation of water vapor, radiant heat from the environment. Among them, phase change heat and radiation heat account for less than 1% of the total heat, which can be ignored in the specific design process. Then the calculation formula of the heat generated by the fuel cell [11] is as follows:

$$Q_{gen} = (1.482 - V_{cell}) \times In_{cell} \tag{1}$$

In this formula, $Q_{gen}$ is the generated waste heat power, $In_{cell}$ is the transient output current of the stack, and $V_{cell}$ is the transient output voltage of the stack.

The fuel cell should ensure that all the heat generated is dissipated during operation, that is, the generated chemical reaction heat and ohmic scheme heat need to be taken away by the circulating cooling water. The heat taken away by the circulating cooling water is as follows:

$$\Delta Q_r = C \times N_1 \times A_w \times v \times \rho \times (T_{out} - T_{in}) \tag{2}$$

Among them: $C$ is the specific heat of water, $N_l$ is the number of flow channels, $A_w$ is the flow channel interception; $v$ is the flow velocity; $\rho$ is the density of water, $T$ is the stack outlet temperature. Therefore, according to the formula, the circulating cooling water is taken away as the inlet temperature of the stack. Therefore, according to the formula calculation, the heat taken away by the circulating cooling water is the main way of heat dissipation of the fuel cell. The heat produced is equal to the heat taken away when the fuel cell heat reaches equilibrium, as follows:

$$Q_{gen} = \Delta Q_r \tag{3}$$

## 3. Test Plan and Real Vehicle Test

### 3.1. Vehicle Information

This paper takes the Toyota fuel cell electric vehicle Mirai as the research object. The basic parameters of the vehicle are shown in table 1.

**Table 1.** Fuel cell vehicle parameters.

| Product name | Car | Product number | Toyota Mirai |
|---|---|---|---|
| Maximum allowable total mass (kg) | 2160 | Curb weight (kg) | 1850 |
| VIN | JTDBVRBDXGA000378 | Transmission ratio | 8.779 |
| Wheelbase (mm) | 2780 | number of seats | 5 |

*3.2. Test Conditions*

During the test preparation process, the vehicle must be able to meet the relevant safety technical requirements and the requirements in the "GB/T 24549-2020 Safety Requirements for Fuel Cell Electric Vehicles" [12]. It needs to check the main cooling components, such as a thermostat, cooling fan, radiator expansion valve.

The measurement location can be selected according to the situation, the test field that meets the temperature requirements, or the wind tunnel laboratory that meets the hydrogen-related safety requirements can be selected. This experiment is carried out in the wind tunnel laboratory of Tianjin China Automotive Technology Research Center Co., Ltd., which can control and record the temperature and energy sensor in the laboratory in real time. The measured temperature is generally recommended to be 40 °C, and which can also be determined according to the manufacturer. Generally, it is necessary to make test car stand in the test temperature environment for more than 12 hours, so that the parts on the car can reach the test ambient temperature. The requirements for the test equipment are detailed in table 2 below:

**Table 2.** Test equipment parameters.

| Device name | Equipment model | Manufacturer | The measured parameters | Measuring range | Measurement accuracy |
|---|---|---|---|---|---|
| Chassis Dynamometer | 48"MIM 4WD | AVL | Speed | (0～250)km/h | — |
| Environmental chamber | 763'/40-60/Ro,So | WEISS | Temperature<br>Humidity | (-40～60)°C<br>(10-90)%RH | ±0.5°C<br>±5%RH |
| Data collector | MW100 | YOKOGAWA | — | — | — |
| Temperature Sensor | Type K Thermocouple | Eurasian | temperature | (-200~+1400)°C | Class I |

*3.3. Test Method*

*3.3.1. Test Sensor Installation.* According to the structure and principle of the fuel cell vehicle, the measurement point that can reflect its thermal balance is selected to install the sensor, which is mainly divided into four types of sensors, namely temperature sensor, flow sensor, pressure sensor, and current and voltage sensor. The notification needs to record the state of the whole vehicle from time to time, especially the working state of the power system and the hydrogen supply system. The temperature sensor is the most important and most complex sensor installation for thermal balance testing. The recommended sensor locations are shown in table 3 below:

**Table 3.** Temperature sensor installation parameters.

| Serial number | | Measurement parameters | Sensor installation location |
|---|---|---|---|
| Important | 1 | Ambient temperature | Installed with the vehicle in a place with a height of 1.5m, away from heat sources, ventilation and no direct sunlight |
| | 2 | Stack surface temperature | Stack surface |
| | 3 | Stack inlet water temperature | At the center of the inlet pipe |

| Serial number | | Measurement parameters | Sensor installation location |
|---|---|---|---|
| | 4 | Stack outlet water temperature | At the center of the outlet pipe |
| | 5 | Radiator inlet water temperature | At the center of the radiator inlet pipe |
| | 6 | Radiator outlet water temperature | At the center of the radiator outlet pipe |
| | 7 | Power battery surface temperature | Power battery surface temperature |
| | 8 | Power battery intake temperature | At the center of the intake pipe |
| | 9 | Power battery outlet temperature | At the center of the air outlet |
| | 10 | Water pump temperature | At the center of the inlet pipe |
| | 11 | Water pump temperature | At the center of the outlet pipe |
| | 12 | Motor inlet water temperature | At the center of the inlet pipe |
| | 13 | Motor outlet water temperature | At the center of the outlet pipe |
| | 14 | Humidifier intake air temperature | At the center of the intake pipe |
| | 15 | Humidifier outlet temperature | At the center of the air outlet |
| | 16 | Intercooler intake air temperature | At the center of the intake pipe |
| | 17 | Intercooler outlet temperature | At the center of the air outlet |
| | 18 | Condenser temperature | At least 9 points, evenly distributed in front of the condenser, facing the air inlet direction |
| | 19 | Fan temperature | Air temperature at different locations |
| | 20 | Heater inlet water temperature | At the center of the inlet pipe |
| | 21 | Heater outlet water temperature | At the center of the outlet pipe |
| | 22 | Deionizer temperature | Mounting surface pipe wall, close to heat source |
| | 23 | Air compressor temperature | Mounting surface pipe wall, close to heat source |
| | 24 | Front cabin temperature | The middle position of front, rear, left, right, top and bottom and the highest temperature of the engine compartment (generally in the exhaust manifold position) |
| | 25 | Ambient relative humidity | Consistent with the installation position of the ambient temperature measurement sensor |
| | 26 | Sunlight intensity | On the roof of the vehicle, away from heat sources |
| Optional | 1 | Other measuring points | The measurement point selected by the enterprise itself |

The influence of flow and pressure on the mass and heat transfer of fuel cells is also the key to thermal management strategies. Generally, it is necessary to monitor the pressure and flow signals of the main heat-dissipating components, as shown in table 4 below

**Table 4.** Pressure flow measurement parameters.

| Serial number | Measurement parameters |
|---|---|
| 1 | Hydrogen bottle pressure |
| 2 | Hydrogen inlet pressure |
| 3 | Air inlet pressure |
| 4 | Cooling water inlet pressure |
| 5 | Cooling water flow |
| 6 | Hydrogen intake flow |
| 7 | Air intake flow |

The current and voltage measurement can fully test the heat loss and energy distribution. Generally, it is necessary to record the current and voltage electrical signals of the stack, key components, power batteries, etc. The details are shown in table 5 below:

**Table 5.** Current and voltage measurement parameters.

| Serial number | Measured parameters |
|---|---|
| 1 | Stack current |
| 2 | Stack voltage |
| 3 | Battery current |
| 4 | Battery voltage |
| 5 | Air compressor electrical flow |
| 6 | Air compressor mechanical pressure |
| 7 | Heater current |
| 8 | Heater voltage |

*3.3.2. Selection of Test Conditions.* Comparing with the traditional test methods for testing the thermal balance capability of automobiles and the analysis of the dynamic driving road spectrum conditions of the vehicle, it can be seen that the conditions varies from the single steady-state limit conditions of vehicle speed to dynamic driving conditions, such as high-speed conditions, high-speed climbing conditions, and low-speed climbing conditions. With the development requirements of automotive thermal management test capability and the actual road conditions in China, "China working condition" is taken as one of the assessment conditions for thermal balance capability. The detailed tests are shown in table 6 below.

**Table 6.** Test operating parameters of fuel cell vehicles.

| Working condition | Temperature | Preprocessing | Speed | Slope | post-processing | Travel time |
|---|---|---|---|---|---|---|
| High speed conditions | 40°C, humidity 40%, no light | Gas-liquid temperature reaches 40±1°C | 140km/h | 0% | Turn off the flame and immerse until the temperature of each measuring point begins to drop | Driving to thermal equilibrium |
| High-speed climbing conditions | | | 120km/h | 3% | Turn off the flame and immerse until the temperature of each measuring point begins to drop | Driving to thermal equilibrium |
| Low speed climbing condition | | | 40km/h | 10% | Turn off the flame and immerse until the temperature of each measuring point begins to drop | Driving to thermal equilibrium |
| Urban working conditions | | | China working condition | 0% | Turn off the flame and immerse until the temperature of each measuring point begins to drop | Driving to thermal equilibrium |

## 4. Data Analysis and Verification

According to the set test plan, Mirai, an advanced fuel cell vehicle in the world, is selected for thermal balance test. The rated power of the fuel cell stack is about 68kW, and the sensors are arranged in the engine compartment, passenger compartment, cargo compartment, hydrogen system compartment and chassis. After waiting for more than 100 test signals, move the fuel cell vehicle to the wind tunnel laboratory for thermal balance test at 43°C, as shown in the figure below.

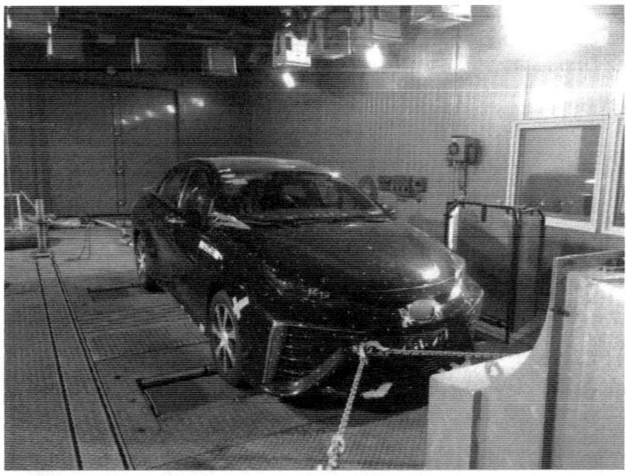

**Figure 1.** Thermal balance test of fuel cell vehicle.

First, the fuel cell vehicle is immersed in test temperature for more than 12 hours. At this time, the surface temperature of the vehicle has been maintained at about the test temperature of 43 °C, and the temperature of the coolant is mainly reached. At this time, the state of the vehicle and the SOC state of the battery are recorded. After completing the preparatory work, start the recording devices such as the data acquisition instrument and the tail exhaust concentration tester. The test process is mainly divided into three steps. First, start the vehicle according to the test method specified by the manufacturer; secondly, carry out the execution of the test conditions. According to the selected test conditions, achieve the specified test condition at the speed the vehicle can reach, and then carry out the test. finally, monitor the important sensors and signals. When the monitoring points are stable for 4 minutes and the temperature change does not exceed ±2°C, the vehicle is considered to have reached a thermal equilibrium state.

This test focuses on the thermal balance test of three test conditions, which are "China working condition", "0% slope, maximum speed" and "3% slope, the equilibrium temperature of the fuel cell stack under three working conditions, as shown in table 7 below.

**Table 7.** Fuel cell reactor inlet and outlet water temperature under test conditions of fuel cell vehicles.

| The test conditions | Reactor inlet water temperature(°C) | Reactor outlet water temperature (°C) | |
|---|---|---|---|
| | Equilibrium temperature | Equilibrium temperature | Maximum immersion temperature |
| China working condition | 73.7 | 78.3 | —— |
| Maximum speed, slope 3%+ immersion | 81.6 | 86.6 | 86.1 |
| Maximum speed, slope 0%+ immersion | 87.6 | 92.2 | 92.0 |

### *4.1. Working Condition of China*

The test adopts "China working condition" to test and evaluate the thermal balance, which can fully reflect the thermal balance ability under the comprehensive use of the working condition. The test focuses on analyzing the temperature change of different heat dissipation analysis under the fully immersed vehicle at 40°C. As shown in figures 2, 3 and 4, under the operation of "China working Condition", the temperature of various cooling components is relatively moderate and there will not be excessive temperature difference. The average temperature of condenser is about 40°C, and the average temperature of radiator of fuel cell stack is about 50°C. The temperature of the cooling system in each part is relatively stable. The temperature difference between inlet and outlet water temperature of the reactor is within ±5°C. It can be seen that under the operating conditions in China, the thermal management system of the vehicle can accurately control the reactor temperature, radiator temperature and other related components through good matching and calibration, so that the temperature is within a relatively comfortable range.

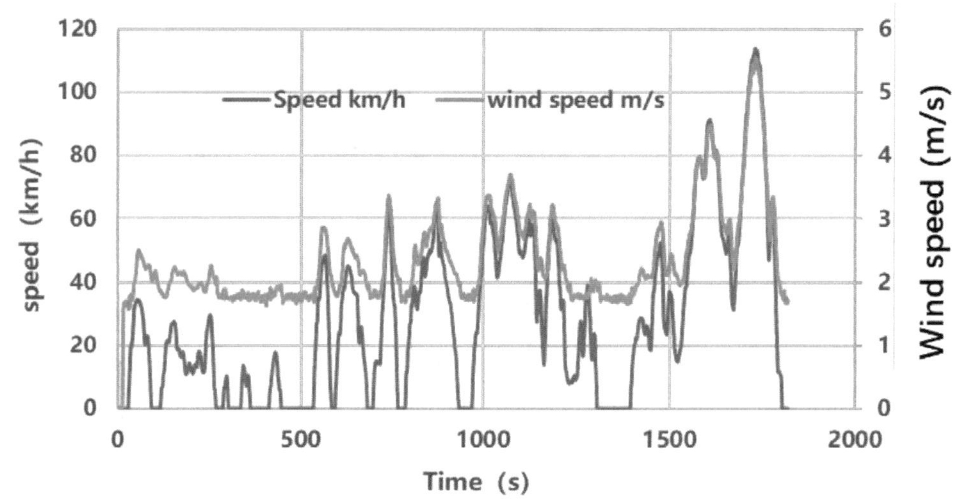


**Figure 2.** Average wind speed variation curve with speed and time in China working condition.

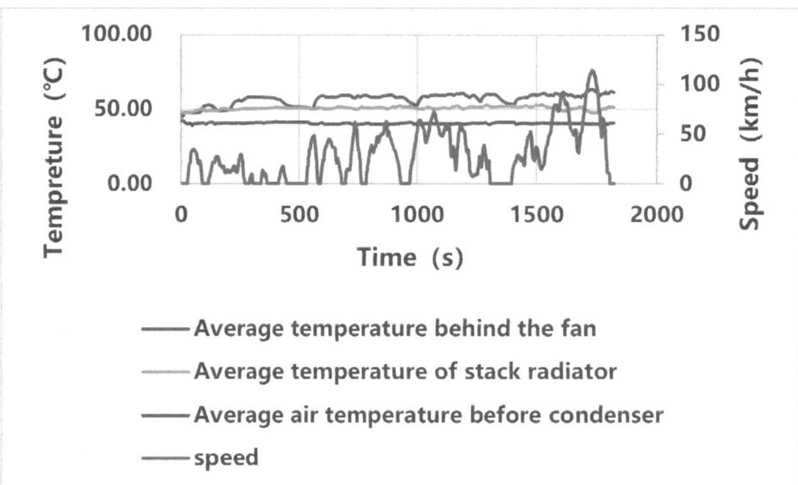


**Figure 3.** The variation curves of the average temperature behind the fan, the average temperature of the reactor radiator and the speed time under China working condition.

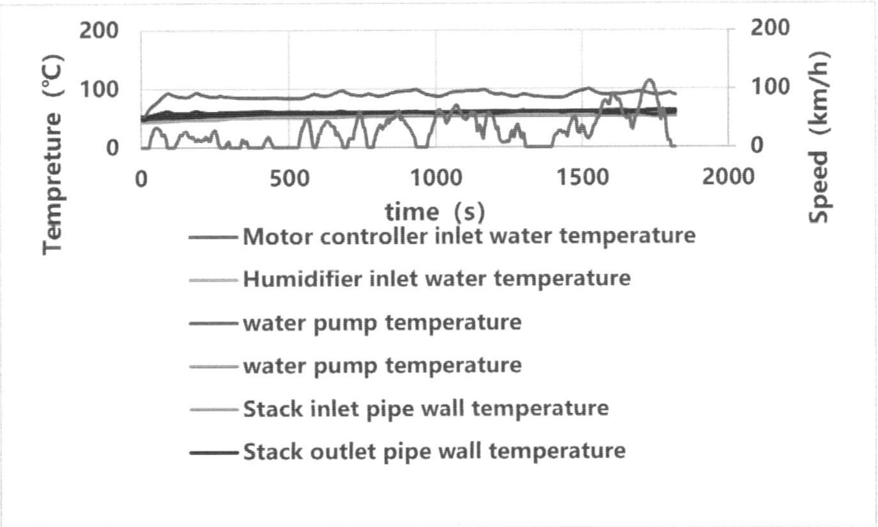


**Figure 4.** Variation curves of inlet temperature of motor controller, inlet temperature of humidifier, inlet temperature of water pump, outlet temperature of water pump, inlet temperature of reactor, outlet temperature of reactor, outlet pipe wall temperature of heater, inlet pipe wall temperature of condenser, back temperature of inlet pipe wall of condenser, and vehicle speed time under China working conditions.

### *4.2. Maximum Speed*

At the test temperature of 43°C, the vehicle was driven at the highest speed and kept running for 20 minutes. As can be seen from the following figure, although the maximum speed of the vehicle is set at 175km/h, in the actual operation, the maximum speed reaches 156km/h and remains stable at 150km/h in the follow-up. This is mainly due to the fuel cell stack at the highest speed, the operation power is constantly rising, when it reaches about 70KW, in order to balance the heat generated by the fuel cell stack, the vehicle reaches the target by regulating the operation power, and finally stabilizes at the rated power of the fuel cell stack at about 60kW. In this process, the SOC of the power battery rises briefly and then continues to decline and stabilize at about 38%, indicating that the driving power of the vehicle is jointly undertaken by the power battery and the fuel cell.

As shown in figure 5, during the process from starting to reaching the maximum speed of the vehicle, the maximum power of the fuel cell stack can reach about 80kW. At this time, its temperature is 86.6°C, that is, the power is limited when the temperature of the fuel cell stack reaches this temperature. When the fuel cell stack stability in 60 kw power, its temperature can reach 92 °C, the judge for the thermal equilibrium state, the temperature is basically is the limit of the fuel cell operating temperature, visible vehicle in the process of heat balance control, largely using the fuel cell stack heat capacity, but also use its temperature cooling capabilities such as the adjustment of the control strategy.

As shown in figure 5, in the process of rapid heating up of fuel cell stack, the initial temperature is 58.4°C, and the equilibrium temperature is 89.9°C, which takes a total of 59.1s. In the whole process of rapid heating up of the stack, the heating rate is 0.53°C/s. At this time, the fuel cell stack operating temperature has reached a certain height, belonging to the high temperature operation stage, in a short period of time, rapid warming activities cannot be carried out, the vehicle also needs to control the temperature according to the temperature characteristics.

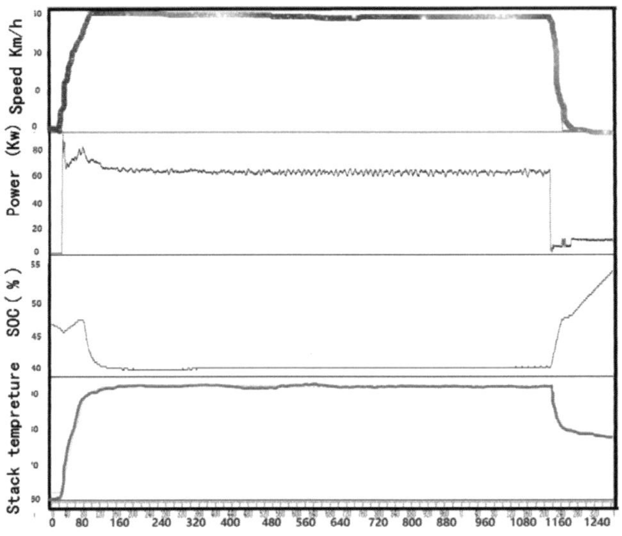


**Figure 5.** Variation Curves of fuel cell stack power, temperature and power cell SOC as a function of vehicle speed.

When the running speed of the vehicle is 140km/h and the immerging condition is 43°C, the temperature of the main cooling components is shown in figure 6 below. The upwind speed of the vehicle is 130km/h. From figure 7, it can be seen that the average temperature before the low-temperature radiator and the average temperature before the deputy radiator are relatively low, about 17 °C in control, after the radiator, the radiator of the post average temperature is higher, the post average temperature after the low-temperature radiator is about 45°C, the post average temperature of deputy FCS radiator is about 85 °C, both of them showed a large temperature rise.

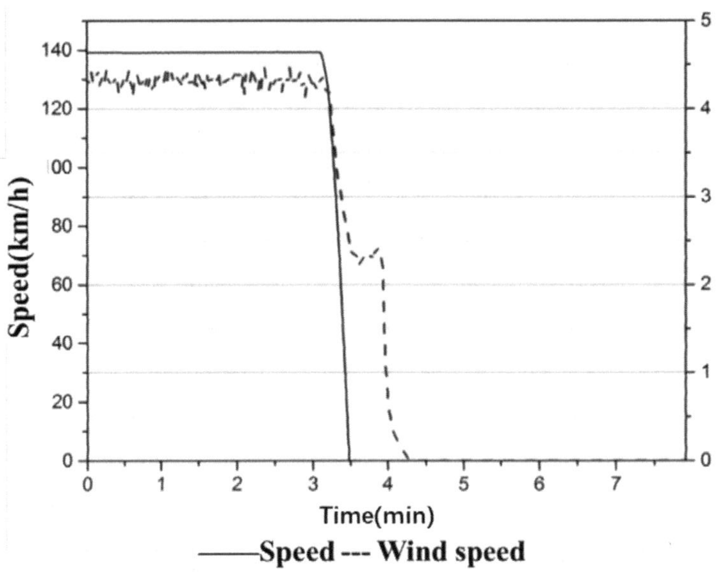


**Figure 6.** Variation curve of average wind speed with vehicle speed and time under 140km/h+ immersion condition (43°C).

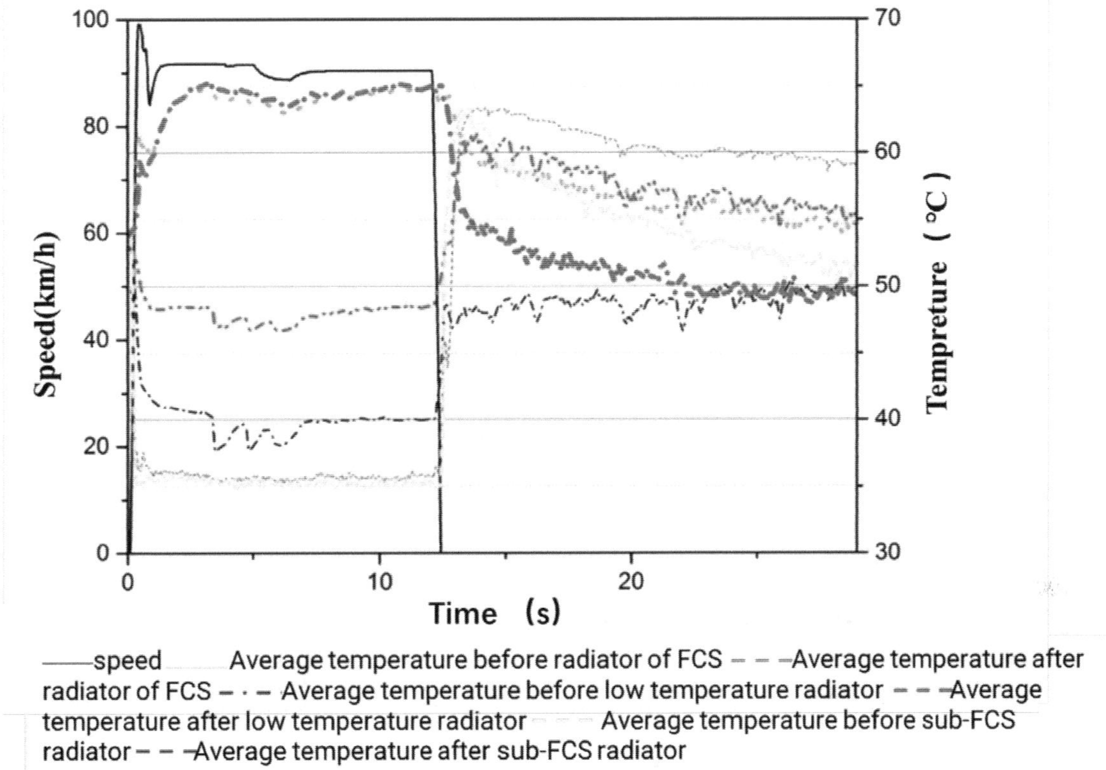


**Figure 7.** Variation curves of the average temperature before and after the heat dissipation module with vehicle speed and time at 140km/h+ immersion condition (43°C).

## 5. Conclusion

In this paper, based on the analysis of the principle of thermal management of fuel cell vehicles, a test and evaluation method for the thermal balance of fuel cell vehicles was developed, a test scheme and a signal acquisition scheme were developed, and test conditions were formulated based on the fuel cell operating conditions. The thermal balance of advanced fuel cell vehicles is tested and analyzed by this test scheme. The heat balance of fuel cell vehicles is dissipated by reducing power and increasing temperature difference under extreme conditions. This method can provide a good test basis for the thermal balance of fuel cell vehicles and help enterprises develop.

## Reference

[1] Zhao X Q 2015 *Research on Thermal Management System of Water-Cooled Proton Exchange Membrane Fuel Cell* (Chengdu: Southwest Jiaotong University).

[2] Chang G F, Zeng H J and Xu S C 2015 Research on fuel cell thermal management system *Automotive Engineering* (08) 103-107.

[3] Qu F F, Wang W and Wang Y 2020 Research on the thermal balance test of typical fuel cell vehicles *Annual Proceedings of the Society of Automotive Engineers of China.*

[4] Wang Y, Mu L S and Liu S X 2019 Analysis of water and thermal management systems of typical foreign fuel cell vehicles *Internal Combustion Engine Parts* (5) 118-121.

[5] Wang L F 2012 *Analysis and Optimization of Fuel Cell Vehicle Thermal Management System* (Shanghai: Shanghai Jiaotong University).

[6] Kim J, Rodriguez R, Kwak K H, Jung D and Kim Y 2020 The Effect of Driver's Behavior and Environmental Conditions on Thermal Management of Electric Vehicles *SAE Technical*

*Paper* 2020-01-1382.
[7] Pu J, Qin X J, Lu Y, Wang Y P, Ding T W and Zhao Z L 2019 Design and Research of Fuel Cell Thermal Management System *Automotive Engineering* (4) 24-27.
[8] Society of Automotive Engineers (SAE) 2021 Fuel Cell Vehicle Thermal Management (Standard J3193_202104) *SAE International*.
[9] GB/T 12542-2020 2020 *Road Test Method for Thermal Balance Capability of Automobiles* (Beijing: China Standard Press).
[10] T/CSAE 114-2019 2019 *Wind Tunnel Test Method for Automotive Powertrain Cooling Environment* (Beijing: China Standard Press).
[11] Huang Y C, Xiao X L, Kang H F, Lv J G, Zeng R and Shen J 2022 Thermal management of polymer electrolyte membrane fuel cell: A critical review of heat transfer mechanisms, cooling approaches, and advanced cooling techniques analysis *Energy Conversion and Management* (254) 115221-115256.
[12] GB/T 24549-2020 2020 *Safety Requirements for Fuel Cell Electric Vehicles* (Beijing: China Standard Press).

# The Use of Slag and Red Mud for the Synthesis of a Slag-Based Cementitious Material

**Hong Quan** [1,2] **, Huijuan Yu** [1,2] **, Dongpeng Lv** [1,2] **and Yuancheng Li** [1,2,3 *]

[1] College of Agriculture and Biological Sciences, Dali University, Dali 671003, Yunnan, China
[2] Key Laboratory of Ecological Microbial Remediation Technology of Yunnan Higher Education Institutes, Dali 671003, Yunnan, China
[3] Faculty of Metallurgical and Energy Engineering, Kunming University of Science and Technology, Kunming 650093, Yunnan, China
Email: LYC_DU@163.com

**Abstract.** A large amount of metallurgical slag storage occupies land and wastes a lot of resources. Therefore, this study uses slag and red mud (RM) as the main raw materials, sodium silicate, and sodium hydroxide (NaOH) as the composite alkaline activators, the slag-based cementitious materials were prepared by the secondary mixed grinding method. The mechanical properties of the cementitious materials were characterized and analyzed. The results show that when the ratio of slag to RM in the cementing material is 8:2, the strength increases initially and then decreases with the increase of the content of the alkaline activator. Meanwhile, the content of the alkaline activator is 10% (mass fraction), and the 28-day compressive strength of the sample is up to 67.8MPa, which meets the requirements of slag cement with a 42.5 strength grade.
**Keywords.** Solidification/stabilization, ettringite, cementitious material.

## 1. Introduction

Cement has become the most widely used building material worldwide because of its excellent mechanical properties and durability in different environments. The cement industry is one of the main consumption industries that take from natural resources. Cement faces three main problems as a basic part of modern social infrastructure. The first problem is carbon dioxide ($CO_2$) emissions in the production of Portland cement (PC), which accounts for approximately 7% of the world's $CO_2$ emissions. The second problem is the long-term durability of cement materials because PC is not as durable as many people think. The third problem is the high cost of treating solid waste with PC, faced by cement solidification/stabilization (s/s). To effectively reduce energy consumption and $CO_2$ emissions, the PC clinker consumption should be reduced. The desire to reduce $CO_2$ emissions and produce more durable concrete has driven further studies to create new adhesives.

In recent years, many studies have been carried out to find an alternative adhesive for PC. Some researchers have developed various types of adhesives, such as alkali-activated adhesives, calcium aluminate cement, calcium carbonate cement, belite sulfoaluminate cement, and partially hydrated calcium silicate (C-S-H) based adhesives. Some scholars predicted that alkali-activated adhesive will be the general trend of cement development in the future [1-4].

Slag is considered to be another industrial byproduct of the iron and steel industry, with an annual output of approximately 125 million tons [5]. Given the use of too much alkali during the extraction process of alumina, red mud (RM) is highly alkaline, and the pH is generally around 10.5–13, thus causing great harm to the environment [6]. At present, the utilization rate of slag, fly ash, red mud, and other waste residues in concrete accounts for a large part. It has been proven that using only silicon-rich and aluminum-rich materials without the PC, low-carbon concrete can be produced.

Slag-based cementitious material (SCM) is regarded as a low-carbon, environmentally friendly, and low-cost auxiliary cementitious material because it releases one-fifth of the $CO_2$ of PC (0.82 tons of $CO_2$ per ton of PC) [7-9]. Using multisource and multicomponent solid waste to realize effective solidification/stabilization underground filling technology has wide utilization prospects and may bring huge environmental and economic benefits given its low cost, low leaching rate, and high treatment efficiency.

There is a substantial amount of silicoaluminate in metallurgical waste residue. Previous studies have found that, sulfoaluminate has a good curing effect on heavy metals. Therefore, in this study, alkaline activators were used to activate the slag, and the activated slag reacted with RM to produce ettringite and C-S-H (hydrated calcium silicate). The properties of SCM synthesized by RM and slag under the alkaline activator were studied. First, the feasibility of synthesizing SCM using RM and slag is studied. Second, the effects of the alkaline activator on the BCM performance and hydration mechanism were discussed. Third, leaching toxicity tests were used in determining the long-term stability of SCM. This study provides a new idea for comprehensive utilization of the low-carbon industrial byproducts.

## 2. Materials and Methods

### *2.1. Materials*
The RM comes from Eastern China's alumina producers. Slag comes from a smelter in Southern China. The waste slag was dried up and ground to 0.0074mm. Sodium silicate solution is produced from Jinan Qianqi Chemical Co., LTD., its modulus is 3.35, solid content is 39%; NaOH was from Chengdu Puhe Chemical Co., LTD., its purity is 99% (mass fraction); The water is deionized. Table 1 shows the chemical composition of raw materials to be tested.

**Table 1.** Chemical composition of raw materials to be tested, w/%.

| Ingredient | $Na_2O$ | MgO | $Al_2O_3$ | $SiO_2$ | $SO_3$ | Cl | CaO | $Fe_2O_3$ | ZnO |
|---|---|---|---|---|---|---|---|---|---|
| RM | 11.34 | 0.74 | 37.56 | 17.71 | 1.23 | 0.22 | 12.79 | 10.68 | ND |
| BFS | 1.06 | 6.80 | 12.93 | 33.19 | 1.13 | ND | 38.52 | 0.598 | ND |

Note: ND-not detected (<0.01%)

### *2.2. Raw Material Ratio of Cementitious Material Sample*
The ratio of fixed slag to red mud is 8:2, and the ratio of the alkaline activator is shown in table 2.

**Table 2.** Alkali activated incorporation ratio.

| ID | Slag (%) | RM (%) | Activator $^b$ (%) |
|---|---|---|---|
| 1# | 80 | 20 | 0 |
| 2# | 80 | 20 | 6 |
| 3# | 80 | 20 | 8 |
| 4# | 80 | 20 | 10 |
| 5# | 80 | 20 | 12 |

### *2.3. Sample Preparation*
Slag, RM, and alkali activator were weighed according to the ratio in table 2. After adding deionized water, the mixture was stirred in a mechanical stirrer for three minutes. The mixture was removed after

mixing evenly and filled into a 20 mm cube mold for curing. The mixture is cured at room temperature and 95% relative humidity for 3, 7, and 28 days [10].

### *2.4. Performance Characterization of the Samples*

A cement mortar flexural and compressive testing machine (TYE-300D, Wuxi Jianyi) was used in testing the samples' compressive strength at various ages. The chemical composition of the raw materials was analyzed by x-ray fluorescence spectroscopy (ARL ADVANT 'x IntellipowerTM3600, Thermo Fisher, Germany), and the phase composition of the samples was analyzed by x-ray diffraction (D8ADVANCE, Brock, Germany). Terminated by ethanol hydration, a proper amount of sample fragments aged 7 days were taken and treated with vacuum ion sputtering for gold spraying. The microscopic morphology of the sample fragments was observed and analyzed by scanning electron microscope (JSM-6490LV, produced by Japan Electronics Company).

## 3. Results and Discussion

### *3.1. Phase Composition of Slag and Red Mud*

Figure 1 shows the x-ray diffraction (XRD) patterns of slag (a) and red mud (b) particles. It can be seen that the red mud contains bauxite, hydrated calcium silicate, hematite, quartz, and other substances, accompanied by some impurities. The main composition of slag is amorphous silicon-aluminum vitreous.

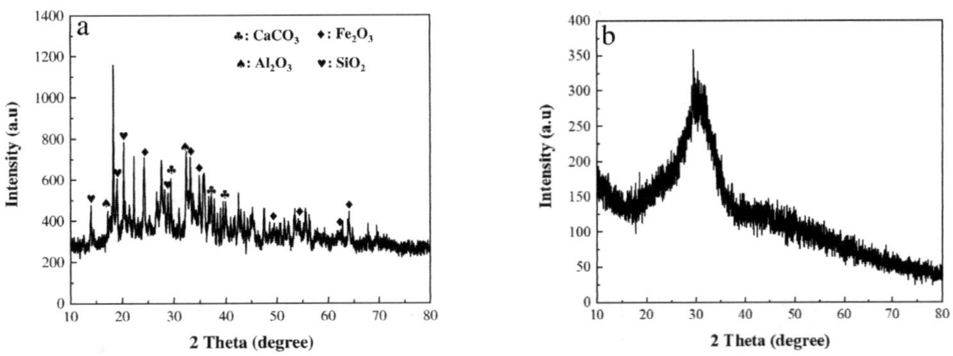


**Figure 1.** XRD patterns of slag and RM.

### *3.2. Effect of Alkaline Activator Content on Mechanical Properties of Samples*

Figure 2 shows the effect of alkaline activator on the compressive strength of samples at various ages. Figure 2 also shows that, with the increase of alkaline activator content of 10% (mass fraction), the compressive strength of samples at various ages increases initially and then decreases, and reaches the maximum value. The 28-day compressive strength of the sample is 67.8MPa, and the synthesized SCM meets the standard of MU20 fly ash brick. It is because the alkaline activator provides a high alkalinity environment for slag and red mud, accelerates the depolymerization and dissolution of aluminosilicate, and produces a large number of silicate ions, hence, accelerating the reaction speed. The hydration of $C_3S$, $C_3A$, and $C_2S$ in the system not only generates C-S-H and C-A-H gels but also produces a large amount of calcium hydroxide ($Ca(OH)_2$). Consequently, the content of $[OH^-]$ in the solution system increases, and the pH value increases, thus creating an excitation effect on the glass phase in the slag. However, if the amount of alkaline activator is too much, the polymerized product will be depolymerized again, hence, leading to the reduction of strength. Therefore, an alkaline activator plays a role in adjusting liquid basicity and stimulating slag activity and provides hydrated minerals for cementitious materials.

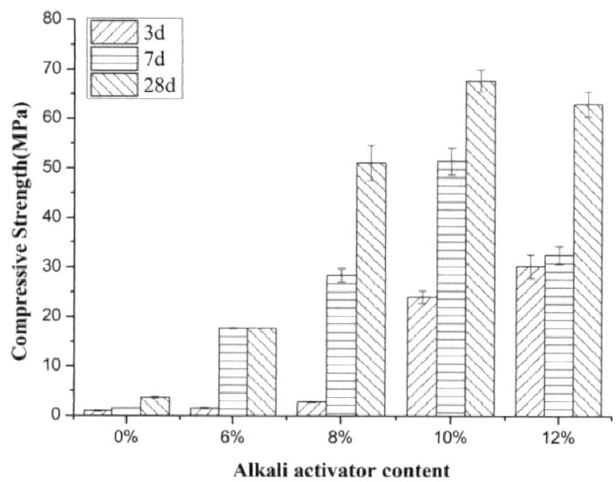


**Figure 2.** Influence of alkaline activator content on the compressive strengths at different curing ages of samples.

*3.3. Effect of Alkaline Activator Yield on Micromorphology of Samples*
Figure 3 shows the microscopic morphology photos of 7 d old colloidal samples with different dosages of alkaline activator. The amount of hydration products produced in the material with 0% alkaline activator content as shown in figure 3(a) is very small, and also only a small amount of columnar structure, ettringite, can be seen. However, after adding the alkaline activator, a large number of acicular ettringite and flake $Ca(OH)_2$ structures can be seen, as shown in figure 3(b). Therefore, the formation rate of ettringite and other hydration products is effectively increased by an alkaline activator. Hydration products increase SCM density and compressive strength after filling the pores.

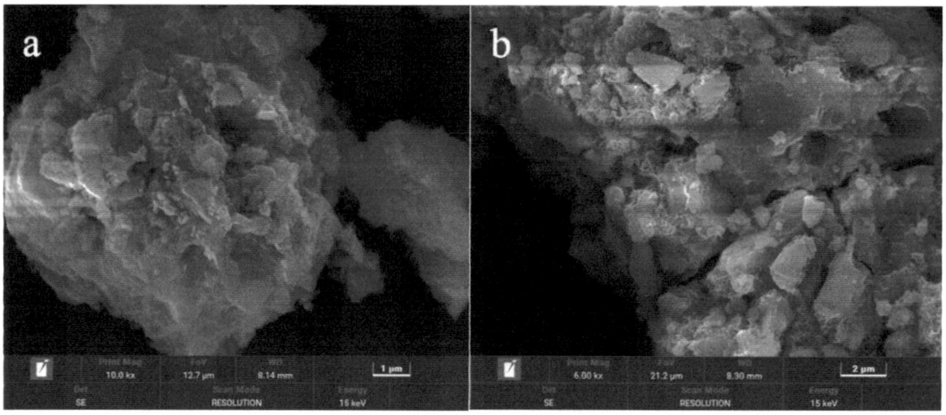


**Figure 3.** SEM photos of micromorphology of 7 d old specimens for binding materials with different alkaline activator cement contents.

*3.4. Leaching Test*
Table 3 shows the arsenic leaching concentration of samples before and after alkali activation measured by the CSLT method. The addition of an alkaline activator promotes the generation of

hydration products and reduces arsenic leaching through physical encapsulation, ion replacement, and adsorption among cement layers [11].

**Table 3.** A rsenic concentrations in the CSLT leachates and CSLT limits (≤ means under detection limit).

| | 3d (mg/L) | 7d (mg/L) | 28d (mg/L) | CSLT limits |
|---|---|---|---|---|
| Alkali activator 0 | 0.834 | 0.77 | 0.572 | 5 mg/L |
| Alkali activator 10 | ≤0.010 | ≤0.010 | ≤0.010 | |

*3.5. The Hydration Mechanism of the Slag*

In the cementitious system of sodium silicate as an alkaline activator of slag and red mud, the hydration reaction is as follows:

$$2(2CaO{\cdot}SiO_2) + 4H_2O \rightarrow Ca(OH)_2 + 3CaO{\cdot}2SiO_2{\cdot}3H_2O \quad (1)$$

$$2(3CaO{\cdot}Al_2O_3) + 6H_2O \rightarrow 3CaO + 3CaO{\cdot}Al_2O_3{\cdot}6H_2O \quad (2)$$

The vitreous body is the main composition of slag in a metastable state. During the event of a large number of alkaline activators, the vitreous structure of slag begins to decompose, and active silicon oxide and alumina enter the system. With the continuous steel slag hydration, the final particle surface is covered with a layer of $H_4SiO_4$. In the alkaline environment, a large amount of H4SiO4 dissociates to form $H_3SiO_4^-$, which reacts with $Ca(OH)_2$ in the system to form a hydrated calcium aluminate gel. $Al_2O_3$ in steel slag and slag structure will eventually dissociate into $H_3AlO_4^{2-}$ and $Al(OH)^{2+}$, thereby forming hydrated calcium aluminate gel under alkaline conditions.

$$H_3SiO_4^-{+}Ca^{2+}{+}OH^-{\rightarrow}C\text{-}S\text{-}H \quad (3)$$

$$H_3AlO_4^{2-}{+}Ca^{2+}{+}OH^-{\rightarrow}C\text{-}S\text{-}H \quad (4)$$

Reactions (4) and (5) accelerate the disintegration of the calcium-rich phase in slag vitreous. $Ca(OH)_2$ absorbs $CO_2$ in the air and carbonize $CaCO_3$ particles in it, thus increasing the density of the transition zone in the mortar, hence, avoiding the influence of disordered carbonization reaction on the structure of steel slag cement. The above reaction results in the production of a large number of gelled products and $CaCO_3$ and Aft ($3CaO{\cdot}Al_2O_3{\cdot}3CaSO_4{\cdot}32H_2O$) with an interstitial-filling effect in the colloidal sand system, thus developing high mechanical properties and densified structure of the samples at various ages.

## 4. Conclusions

With slag and RM as the main raw materials, and sodium silicate and NaOH as alkaline activators, the early cementing materials prepared by the secondary mixing method were characterized, and the main conclusions were as follows:

1) In the prepared gels, the mechanical strength of each age initially increases and then decreases with the increase of the content of the alkaline activator, and reaches the maximum when the content of the alkaline activator is at 10% (mass fraction). The 28-day compressive strength is 67.8MPa, meeting the national standard requirements of 42.5 strength grade slag PC.

2) With the content of alkaline activator increases; the amount of hydration product and the strength of slag cementitious materials gradually increase.

**Acknowledgements**

The authors are grateful for the financial supports from the Basic Research Project of Yunnan Provincial Science and Technology Department (202001AT070019), the Yunnan Province Postdoctoral Science Foundation (ynbh19029), the Grants from the project of Yunnan Province Postdoctoral (ynbh19029), and the Yunnan Province education department scientific research fund project (2022Y889).

**References**

[1] Armistead S J, Rawlings A E, Smith C C and Staniland S S 2020 Biopolymer stabilization/solidification of soils: a rapid, micro-macro, cross-disciplinary approach *Environmental Science & Technology* **54**(21) 13963-13972.
[2] Kiventerä J, Perumal P, Yliniemi J, and Illikainen M 2020 Mine tailings as a raw material in alkali activation: A review *International Journal of Minerals, Metallurgy and Materials* **27** 1009.
[3] Tan S T, Hashim H, Lim J S, Ho W S, Lee C T and Yan J 2014 Energy and emissions benefits of renewable energy derived from municipal solid waste: Analysis of a low carbon scenario in Malaysia *Applied Energy* **136** 797-804.
[4] Yin S, Shao Y, Wu A, Wang H, Liu X and Wang Y 2020 A systematic review of paste technology in metal mines for cleaner production in China *Journal of Cleaner Production* **247** 119590.
[5] Elibol C and Sengul O 2016 Effects of activator properties and ferrochrome slag aggregates on the properties of alkali-activated blast furnace slag mortars.*Arabian Journal for Science and Engineering* **41** 1561-1571.
[6] He J, Jie Y, Zhang J, Yu Y and Zhang G 2013 Synthesis and characterization of red mud and rice husk ash-based geopolymer composites *Cement and Concrete Composites* **37** 108-118.
[7] He Z, Zhu X, Wang J, Mu M and Wang Y 2019 Comparison of CO2 emissions from OPC and recycled cement production *Construction and Building Materials* **211** 965-973.
[8] Kavitha, Shanthi V, Arulraj G P and Sivakumar V 2016 Microstructural studies on eco-friendly and durable Self-compacting concrete blended with metakaolin *Applied Clay Science* **124** 143-149.
[9] Provis J L 2018 Alkali-activated materials *Cement and Concrete Research* **114** 40-48.
[10] Bayraktar O Y 2021 Possibilities of disposing silica fume and waste glass powder, which are environmental wastes, by using as a substitute for Portland cement *Environmental Science and Pollution Research* **28** 16843-16854.
[11] Quan H, Li Y C, Zhang M L, Lv D P and Zhu X 2022 Progress of synergistic arsenic fixation technology with metallurgical waste residue *Environmental Science and Management* **47**(04) 96-100.

# Parametric Analysis of the Static Behavior for Stainless-Steel Beam-Column Connections

**Xueliang Qiao[1*], Yuanqing Wang[2] , Lianguang Jia[3], Tianxiong Zhang[4] and Guangyu Qiao[1]**

[1] The Third Construction Engineering Company Ltd of China Construction Second Engineering Bureau, Beijing 100070, China;
[2] Tsinghua University, Department of Civil Engineering, Key Lab. of Civil Eng. Safety and Durability of China Education Ministry, Beijing, 100084, China;
[3] School of Civil Engineering, Shenyang Jianzhu University, Shenyang, 110168, China;
[4] School of Civil Engineering, Tianjin University, Tianjin 300072, China.
Email: qiaoxueliang1991@163.com

**Abstract.** To study the influence of geometric parameters on flexural capacity and stiffness of beam-column connections using stain-steel S31608 and get the key influence parameters, on the basis of test data, using the finite element analysis software ABAQUS, parameter analysis on three main geometrical parameters of flexural capacity and stiffness of stain-steel beam-column connections was conducted. The analysis results show that when flange thickness of I-shaped bars increases, bearing capacity and initial stiffness of the joints increase with large amplitude; when slipping coefficient increases, bearing capacity and initial stiffness of the joints increase with small amplitude; when weld ac-cess hole changes, bearing capacity increases and initial stiffness increase with small amplitude, but can decrease stress concentrate phenomenon; when "Dog-bone" is used, the position of plastic hinge appearance will out-shift, but bearing capacity and initial stiffness decrease with large amplitude. Using the weld access hole of FEME350 and increasing the flange thickness of I-shaped bars is a good way to improve the static behavior of stain-steel beam-column connections.
**Keywords.** Stain-steel Beam-Column Connections; flexural capacity; joint rigidity; FEM; parametric analysis

## 1. Introduction

The connection between the beam and the column in the steel frame plays the role of transmitting the bending moment and shear force between the two components, and it is the main component of the steel frame, and its performance is directly related to the overall response of the structure. The design of beam-column connection should satisfy sufficient strength and appropriate stiffness, and satisfy the principles of "strong joints, weak members" and "strong columns and weak beams" [1]. The beam-column connection of the steel frame structure is mostly designed according to the rigid connection [2], and the connection between the main beam and the column has sufficient rigidity. There are four main structural forms of beam-column rigid connection: all-welded joints; high-strength bolted joints; bolt-welded hybrid joints; three-sided surrounding welded joints. As shown in figure 1.

Journal of Physics: Conference Series **2329** (2022) 012037 doi:10.1088/1742-6596/2329/1/012037

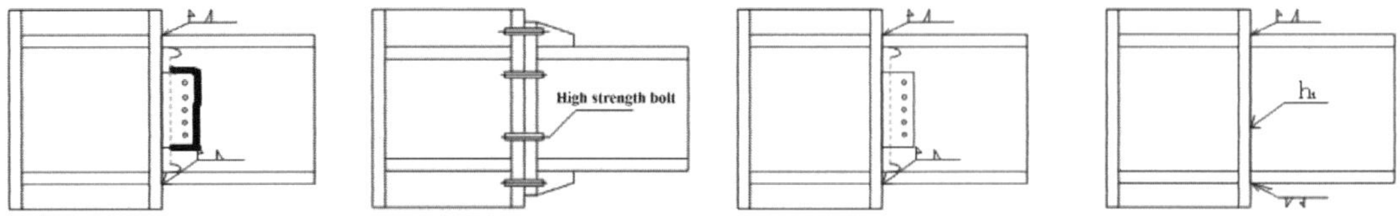


a. Full welding point

b. connections with welded flanges and bolted webs

c. high-strength bolt connection

d. Three-sided girth welding joint

**Figure 1.** The common joint form.

Beam-column joints are widely used in carbon steel structures, and the research on stainless steel is still in its beginning stage. Most of the existing researches focus on stainless steel materials and their components. There are few studies on the joint performance of stainless steel beam-column joints. For the end-plate connection[3] and bolt-welded connection[4] of a specific geometric size, there are few studies on the influence of geometric parameters on the bearing performance and stiffness of stainless steel beam-column joints. For the joint specimen, the large-scale general finite element analysis software ABAQUS is used to analyze the flexural bearing capacity, initial stiffness and flange stress concentration phenomenon of the joint with nonlinear finite element parameters, and the influence of each parameter on the flexural bearing capacity and initial stiffness is obtained. It provides reference for the joint design of stainless steel beam-column joints.

## 2. Finite Element Model and Verification Analysis

### *2.1. Main Parameters of Stainless Steel Beam - column Joints*

The geometric diagram of the beam-column bolt-welded joint of the stainless steel structure studied in this paper is shown in figure 2. The geometric parameters that affect the flexural bearing capacity, initial stiffness and stress concentration of the joints mainly include: flange plate thickness $t_f$, anti-slip coefficient $\mu$, structural hole type, flange plate structural type, parameter meanings and detailed dimensions are shown in figure 2. Take the specimen TS1 as the standard specimen and SP2-4 as the test-finite element verification specimen. The geometric parameters of the specimen are shown in table 1.

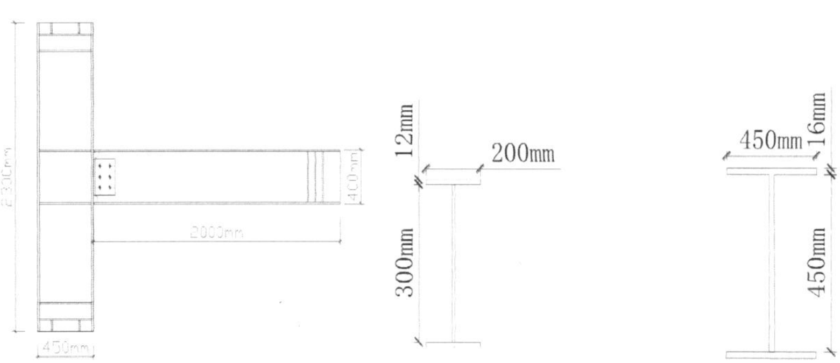


**Figure 2.** Schematic drawing stainless-steel beam-column connections.

**Table 1.** The specimens part list.

| Test specimen number | Flange thickness(mm) | Slip coefficient | Connection type | Bolt type | Variable |
|---|---|---|---|---|---|
| TS1-TS4 | 12 | 0.05-0.2 | Normalized form | A4-80 | Slip coefficient μ |
| TS5-TS7 | 10, 12, 14 | 0.2 | Normalized form | A4-80 | Flange plate thickness |
| TS8-TS11 | 12 | 0.05-0.2 | FEMA350 structural hole | A4-80 | Slip coefficient μ |
| TS12-TS14 | 10, 12, 14 | 0.2 | FEMA350 structural hole | A4-80 | Flange plate thickness |
| TS15-TS18 | 12 | 0.05-0.2 | Dog bone type | A4-80 | Slip coefficient μ |
| TS19-TS21 | 10, 12, 14 | 0.2 | Dog bone type | A4-80 | Flange plate thickness |

### *2.2. Finite Element Analysis Model*

*2.2.1. Material Model.* The constitutive model of the stainless steel material adopts the Ramberg-Osgood model. When verifying the finite element model and the test results, the mechanical property parameters of the stainless steel material take the material property test results[5]; when performing the finite element parameter analysis, the mechanical property parameters of the stainless steel material take the weighted average of the material property test results, and the parameter values are shown in the table 2, the Poisson's ratio of the stainless steel material is taken as 0.3.

**Table 2.** Mechanical property of stainless-steel.

| Test specimen number | Elastic modulus $E$/MPa | Yield strength $f_{0.2}$/MPa | Ultimate strength $f_u$/MPa | $n$ |
|---|---|---|---|---|
| TS1-TS21 | 193965 | 212 | 514 | 4.37 |

The fastening bolts are made of M24 stainless steel A4-80 bolts, and the stainless steel material is cold-worked austenitic stainless steel. Luo et al[6] analyzed the bearing performance of cast aluminum joints based on the ideal elastic-plastic constitutive model of austenitic stainless steel, which is basically consistent with the actual value. Therefore, the constitutive model of the material adopts the ideal elastic-plastic model, and the mechanical performance parameters are set according to the GB/T3098.6-2000 specification. As shown in table 3, the Poisson's ratio of the stainless steel material is set as 0.30. The schematic diagram of the stress-strain curve of the stainless steel material is shown in figure 3.

**Table 3.** Mechanical property of stainless steel used in bolts.

| Bolt type | Yield strength $f_y$/MPa | Tensile strength $f_{ub}$/MPa | Nut strength $f_{un}$/MPa | Final torque $M_{Bmin}$/N·m |
|---|---|---|---|---|
| M24 | 600 | 800 | 800 | 630 |

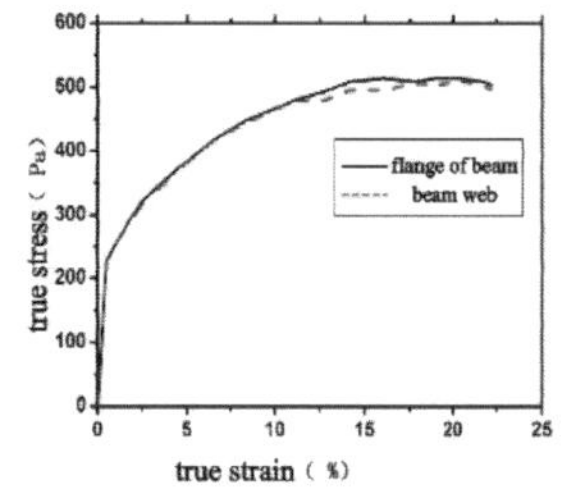

(a) Stress-strain of beam flange and web

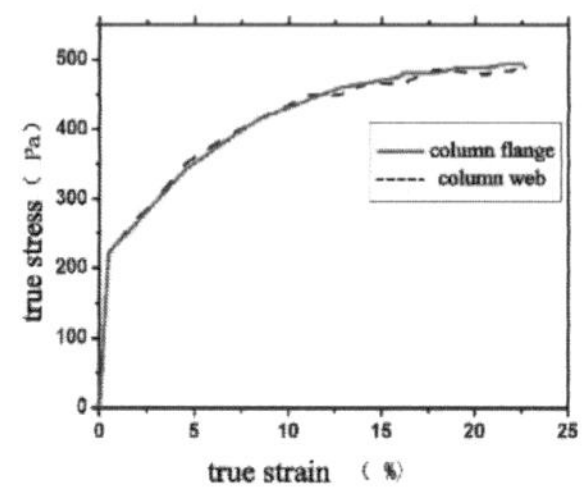

(b) Stress-strain of column flange and web

**Figure 3.** Material model of stainless steel.

*2.2.2. Element Type and Meshing.* Nodes select C3D8R solid elements to mesh the model. The geometric model and the mesh division of each part of the model are shown in figure 4. The elements and contact types of each part of the model are shown in table 4.

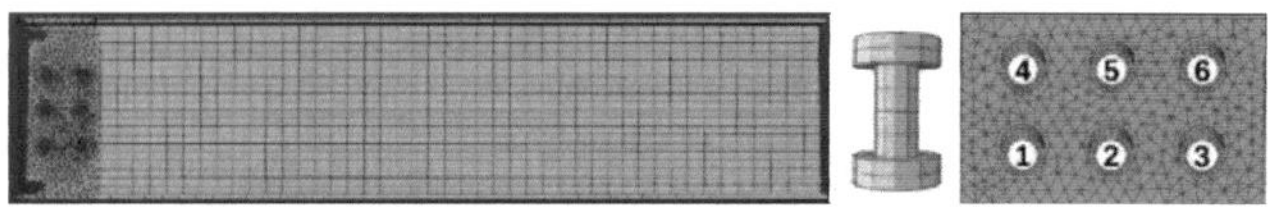

**Figure 4.** Finite element model and Meshing of each part.

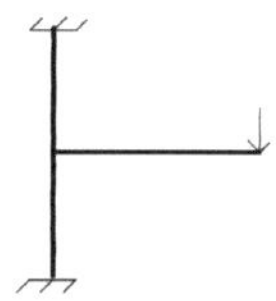

**Figure 5.** Schematic diagram of boundary conditions and loading method.

**Table 4.** Element and contact type of the finite element model.

| Units and contact types | Mock object |
|---|---|
| C3D8I | Shear plate, beam flange and web in joint area, ideal rigid connection model |
| C3D8R | External beam flange and web, column flange and web, bolt in joint area |
| Friction | Contacts between shear plates, bolts and beam webs |

*2.2.3. Boundary Conditions and Loading Methods.* According to the force and geometric symmetry of the specimen, the boundary conditions of the simplified model are shown in figure 5. The column ends are consolidated, and a vertical downward monotonic load is applied to the nodes by means of displacement loading at the two free ends.

*2.2.4. Stiffness Determination Criterion.* The joint stiffness is calculated according to the nodal bending moment and angle test and finite element data, using the provisions for the initial rotational stiffness Sj,ini in Eurocode 3[7-10], and the calculation formula is shown in equations (1)-(3).

$$S_{j,ini}=2M_{j,Rd}/3\varphi_0 \tag{1}$$

$$M_{j,Rd}=f_{0.2}\cdot W_{px} \tag{2}$$

$$W_{px}= h_w^2 t_f/4+bt_w^2+bht_w \tag{3}$$

where Sj,ini is the initial rotational stiffness of the node, Mj, Rd is the plastic flexural bearing capacity of the I-beam, f0.2 is the yield strength of the stainless steel material, and Wpx is the plastic cross-section resistance moment of I-shaped beam cross-section around x axis.

*2.2.5. Criteria for Determining Bearing Capacity.* The bearing capacity of the node is determined according to the load-strain curve of the node un-der the action of the above load, and the load at the extreme point of the load-strain curve is taken as the ultimate bearing capacity of the node.

### *2.3. Verification Analysis of Finite Element Model*

Using the above finite element models to establish the SP2-SP4 finite element verification model of the standard joint specimen , and comparing the finite element analysis results with the test data[9-10] to verify the accuracy of the finite element model. The comparison results are shown in figure 6 and table 5.

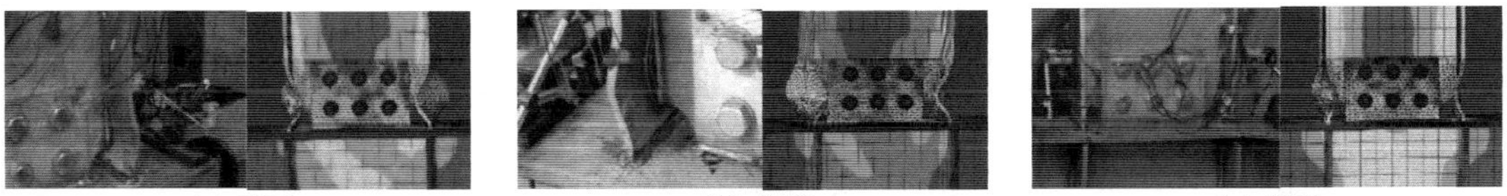

(a) SP2 (b) SP3 (c) SP4

**Figure 6.** Comparison between test and finite element of SP2-SP4.

**Table 5.** Comparison of bearing capacity between finite element analysis and experiment results of SP2-SP4.

| Name of specimen | Finite element calculation value /kN | Experimental value /kN | Inaccuracy /% |
|---|---|---|---|
| SP2 | 239.919 | 241.202 | 0.53 |
| SP3 | 240.153 | 235.406 | 2.02 |
| SP4 | 238.207 | 251.641 | 5.34 |

It can be seen from the chart data that the finite element data is in good agreement with the test data, and the average error is 2.63%. The reason for the error may be the influence of the weld of the specimen in the test. The finite element model is reasonable and effective for parametric analysis of beam-column joints of stainless steel structures.

## 3. Analysis on Influence Parameters of Hybrid Joint for Stainless Steel Beam - column Welding

### *3.1. Influence of Anti-Slip Coefficient*

Aiming at the analysis of the influence of the anti-slip coefficient μ on the joint performance, the finite element model is established for the standard type TS1-TS4, FEMA350 structural hole type TS8-TS11, and dog bone type TS15-TS18. The other geometric parameters of the model are selected according to table 1. The finite element analysis results of the influence of the anti-slip coefficient μ on the flexural bearing capacity, initial stiffness and the maximum bolt shear force of the shear plate and web slip of the hybrid joints of the stain-less steel structure are shown in the figures 7-9.

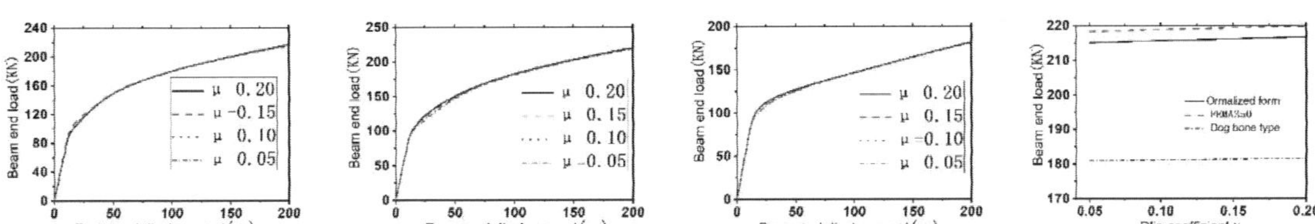


(a) TS1-TS4load-settlement curve (b) TS8-TS11 load-settlement curve TS15-TS18 load-settlement curve (e) variety of piles' ultimate vertiacal bearing capacity

**Figure 7.** The analysis of effect of slipping coefficient on bearing capacity.

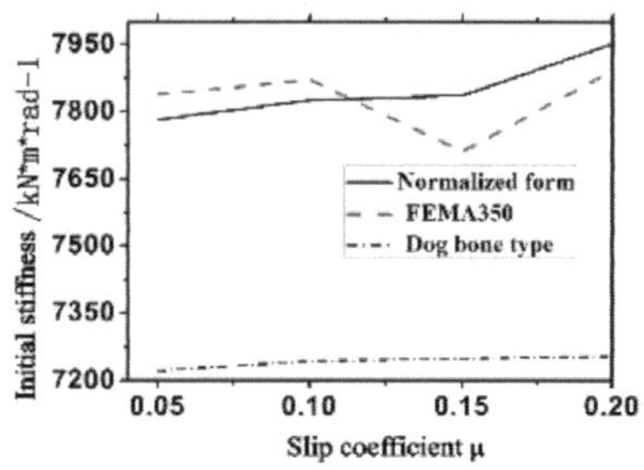


**Figure 8.** The analysis of effect of slipping coefficient on initial stiffness.

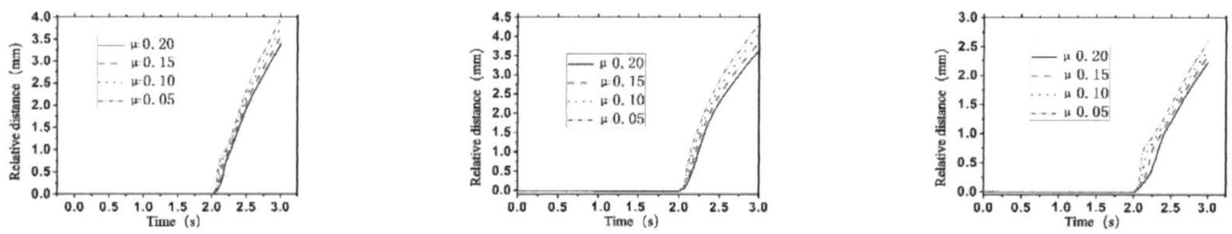


(a) TS1-TS4 slip amount (b) TS8-TS11 slip amount (c) TS15-TS18 slip amount

**Figure 9.** The analysis of effect of slipping coefficient on maximum axial force in bolt.

*3.1.1. Influence on Bearing Capacity Pu.* It can be seen from figure 7 and table 6 that for every 50% increase in the anti-slip coefficient, the bearing capacity of the standard joint increases: 0.26%, 0.46%, and 0.74%. The bearing capacity of FEMA350 structural hole type joints is improved: 0.25%, 0.47%, 0.68%. The bearing capacity of dog-bone joints is increased: 0.15%, 0.26%, 0.35%. With the increase of the anti-slip coefficient, the bearing capacity of the joints increases very little.

**Table 6.** Bearing capacity of specimens of different slipping coefficient.

| Slip coefficient μ | | 0.05 | 0.10 | 0.15 | 0.20 |
|---|---|---|---|---|---|
| Flexural bearing capacity/kN | Normalized form | 215.09 | 215.64 | 216.09 | 216.69 |
| | FEMA350 structural hole | 218.33 | 218.87 | 219.36 | 219.81 |
| | Dog bone type | 180.92 | 181.19 | 181.39 | 181.55 |

*3.1.2. Influence on Initial Stiffness $S_{j,ini}$.* It can be seen from figure 8 and table 7 that for every 50% increase in the anti-slip coefficient, the initial stiffness of the quasi-type joints increases: 0.55%, 0.69%, and 2.17%. The initial stiffness of FEMA350 structural pass joints is increased: 0.72%, -1.62%, 0.69%. The initial stiffness of dog-bone joints is increased: 0.31%, 0.37%, 0.46%. As the anti-slip coefficient increases, the initial stiffness of the joints differs very little.

**Table 7.** Initial stiffness of specimens of different slipping coefficient.

| Slip coefficient μ | | 0.05 | 0.10 | 0.15 | 0.20 |
|---|---|---|---|---|---|
| Initial stiffness /kN·m·rad$^{-1}$ | Normalized form | 7782 | 7825 | 7836 | 7951 |
| | FEMA350 structural hole | 7838 | 7871 | 7711 | 7892 |
| | Dog bone type | 7221 | 7243 | 7248 | 7254 |

*3.1.3. Influence on Slip between Shear Plate and Web.* It can be seen from figure 9 that:
(1) The influence of the anti-slip coefficient on the slip between the shear plate and the web is obvious. The specimens TS1-TS4 in figure 9(a) are standard nodes. It can be seen from the figure that with the increase of the loading step, the The displacement difference between the beam web and the shear

plate increases gradually. At the same loading step, the larger the anti-slip coefficient, the smaller the displacement difference between the beam web and the shear plate, that is, the smaller the slip.

(2) The test pieces TS8-TS11 in figure 9(b) are FEMA350 structural pass nodes. It can be seen from the figure that as the loading step increases, the displacement difference between the beam web and the shear plate also increases gradually. At the step time, the larger the anti-slip coefficient, the smaller the displacement difference between the beam web and the shear plate, that is, the smaller the slip.

(3) The specimens TS15-TS18 in figure 9(c) are dog-bone type nodes. It can be seen from the figure that, similar to the first two types of nodes, the displacement between the beam web and the shear plate is also increased with the increase of the loading step. The difference also increases gradually. At the same loading step, the larger the anti-slip coefficient, the smaller the displacement difference between the beam web and the shear plate, that is, the smaller the slip. However, because the web is weakened, the slip is not as big as the first two nodes.

### *3.2. Influence of Flange Thickness hf*

Aiming at the analysis of the influence of the flange thickness hf on the joint performance, the finite element models TS5-7, TS12-14 and TS19-TS21 are established, and hf is taken as 10 mm, 12 mm and 14 mm respectively. Other geometric pa-rameters of the model are selected according to table 1. The finite element analysis results of plate thickness tw on the flexural bearing capacity, initial stiffness and slippage of shear plates and webs of stainless steel structural bolt-welded joints are shown in figures 10-12.

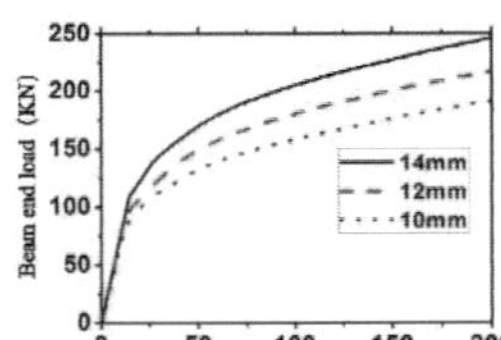

(a) TS5-TS7load-settlement curve

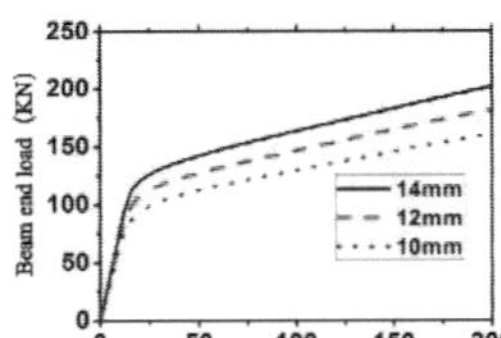

(b) TS12-TS14 load-settlement curve

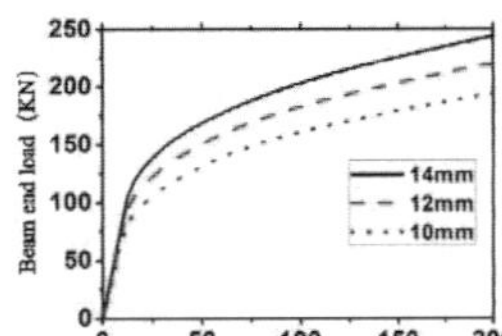

(c)TS19-TS21 load-settlement curve

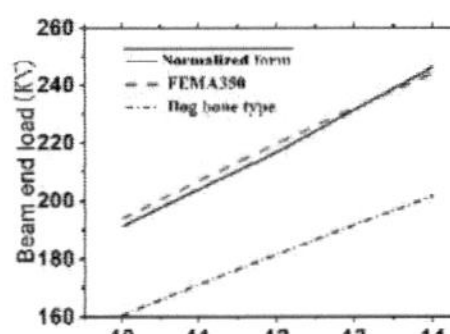

(e) variety of piles' ultimate vertiacal bearing capacity

**Figure 10.** The analysis of effect of flange thickness on bearing capacity.

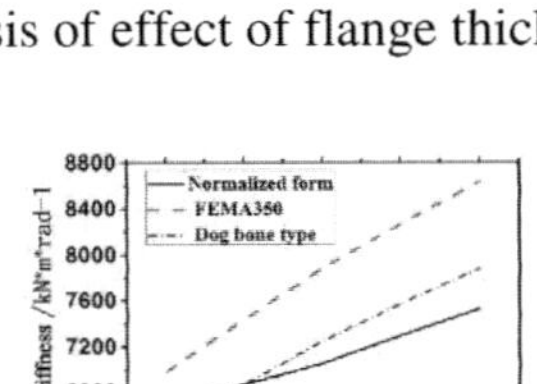


**Figure 11.** The analysis of effect of flange thickness on initial stiffness.

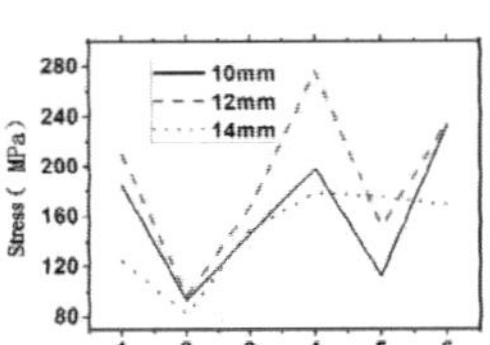

(a) TS5-TS7

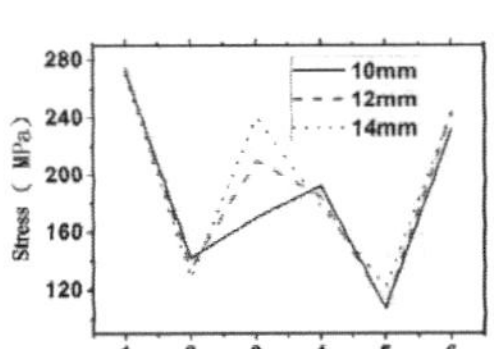

(b) TS12-TS14

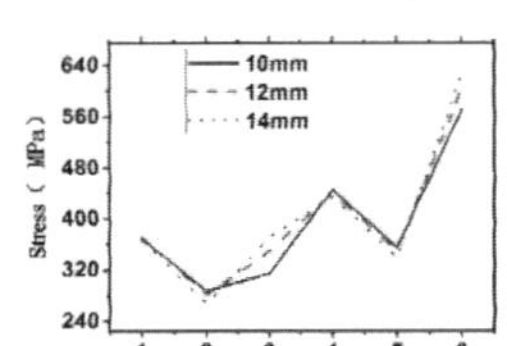

(c) TS19-TS21

**Figure 12.** The analysis of effect of flange thickness on maximum axial force in bolt.

*3.2.1. Influence on Bearing Capacity Pu.* It can be seen from figure 10 and table 8 that the bearing capacity of the standard joint increases by 13.35% and 13.53% for every 20% increase in the thickness of the flange plate. The bearing capacity of FEMA350 structural hole joints is improved: 13.38% and 11.06%. The bearing capacity of dog-bone joints is increased: 13.05%, 11.10%. With the increase of the thickness of the stainless steel beam web, the joint bearing capacity gradually increases, and it can also be seen from figure 11(d) that the bearing capacity increases linearly.

**Table 8.** Bearing capacity of specimens of different flange thickness.

| Flange plate thickness | | 10mm | 12mm | 14mm |
|---|---|---|---|---|
| Flexural bearing capacity/kN | Normalized form | 191.17 | 216.69 | 246.01 |
| | FEMA350 structural hole | 193.88 | 219.81 | 244.11 |
| | Dog bone type | 160.60 | 181.55 | 201.70 |

*3.2.2. Influence on Bearing Capacity $S_{j,ini}$.* It can be seen from figure 11 and table 9 that for every 20% increase in the thickness of the flange plate, the initial stiffness of the standard joint increases by 5.42% and 6.67%. The initial stiffness of FEMA350 structural pass joints is improved: 12.92%, 9.44%. The initial stiffness of dog-bone joints is increased: 11.56%, 8.74%. It can be seen that the initial stiffness increases gradually with the increase of the thickness of the flange plate, showing a linear growth trend.

**Table 9.** Initial stiffness of specimens of different flange thickness.

| Flange plate thickness | | 10mm | 12mm | 14mm |
|---|---|---|---|---|
| Initial stiffness /kN·m·rad$^{-1}$ | Normalized form | 6696 | 7059 | 7530 |
| | FEMA350 structural hole | 6990 | 7893 | 8638 |
| | Dog bone type | 6503 | 7255 | 7889 |

*3.2.3. Influence on Tangential Stress of Bolt.* It can be seen from figure 12 that: The change in the thickness of the flange plate has different effects on the tangential stress of the bolts in different parts. In the three different types of nodes, the data images all show that the bolts close to the flange, namely 1, 3, 4 and 6 is larger than that of bolts 2 and 5 in the middle, showing a trend of small in the middle and large on both sides, and the shear stress of the bolts close to the compression flange tends to decrease with the increase of the thickness of the flange plate. The shear stress of bolts in the middle position and near tensioned flange increases with the increase of flange thickness.

*3.3. The Influence of Structural Type*

The finite element models of standard TS1-TS4, FEMA350structure hole type TS8-TS11, dog bone type TS15-TS18were established according to the analysis of the influence of different structural types on the performance of the joint. The three types form four contrasts. The other geometric parameters of the model are selected according to table 1. The finite element analysis results of the influence of structural types on the bending capacity, initial stiffness, maximum bolt axial force and maximum bolt shear force of stainless steel structural bolted and welded joints are shown in figures 13-15.

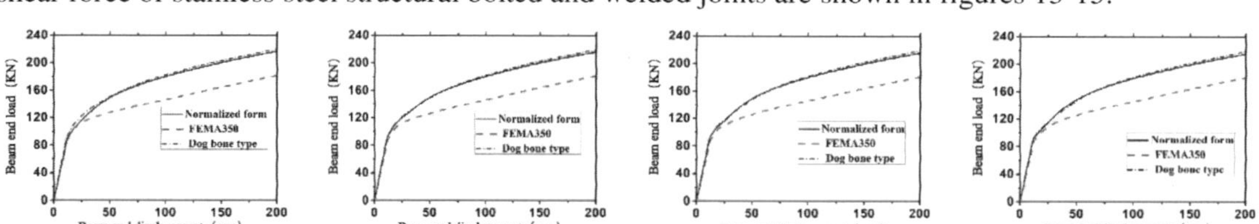


(a) μ=0.05(TS1, TS8, TS15) (b) μ=0.10(TS2, TS9, TS16) (c) μ=0.15(TS3, TS10, TS17) (e) μ=0.20(TS4, TS11, TS18)

**Figure 13.** The analysis of effect of structural pattern on bearing capacity.

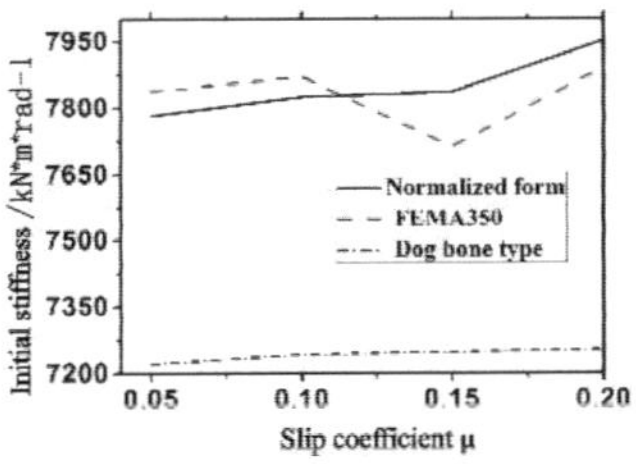


**Figure 14.** The analysis of effect of structural pattern on initial stiffness.

(a) top flange (b) lower cloud point flange

**Figure 15.** The analysis of effect of structural pattern on stress distribution.

*3.3.1. Influence on Bearing Capacity Pu.* It can be seen from figure 13 and table 10 that when μ=0.05, compared with the standard specimen, the bearing capacity distribution of the FEMA350 structural hole type and the dog-bone type specimen is improved: 1.51%, -15.89%. When μ=0.10, compared with the standard specimen, the bearing capacity distribution of the FEMA350 structural hole type and the dog-bone type specimen is improved: 1.50%, -15.98%. When μ=0.15, compared with the standard specimen, the bearing capacity distribution of the FEMA350 structural hole type and the dog-bone type specimen is improved: 1.51%, -16.06%. When μ=0.20, compared with the standard specimen, the bearing capacity distribution of the FEMA350 structural hole type and the dog-bone type specimen is improved: 1.44%, -16.22%. By improving the form of the structural holes, the bearing capacity of the nodes has been improved. After weakening the flange, the plastic hinge can move outward, but the bearing capacity is greatly reduced.

**Table 10.** Bearing capacity of specimens of different structural pattern.

| Structural style | | Normalized form | FEMA350 structural hole | Dog bone type |
|---|---|---|---|---|
| Flexural bearing capacity /kN | μ=0.05 | 215.09 | 218.33 | 180.92 |
| | μ=0.10 | 215.64 | 218.87 | 181.19 |
| | μ=0.15 | 216.09 | 219.36 | 181.39 |
| | μ=0.20 | 216.69 | 219.81 | 181.55 |

*3.3.2. Influence on Initial Stiffness* $S_{j,ini}$. It can be seen from figure 14 and table 11 that when μ=0.05, compared with the standard specimen, the initial stiffness of the FEMA350 structural hole type and the dog-bone type specimen is improved: 0.72%, -7.21%. When μ=0.10, compared with the standard specimen, the bearing capacity distribution of FEMA350 structural hole type and dog bone type specimen is improved: 0.59%, -7.44%. When μ=0.15, compared with the standard specimen, the bearing capacity distribution of the FEMA350 structural hole type and the dog-bone type specimen is improved: 0.57%, -7.50%. When μ=0.20, compared with the standard specimen, the bearing capacity distribution of the FEMA350 structural hole type and the dog-bone type specimen is improved: 0.52%, -7.61%. By improving the form of the construction holes, the increase in joint stiffness is not very obvious. The initial stiffness decreases significantly after weakening the flange.

**Table 11.** Initial stiffness of specimens of different structural pattern.

| Structural style | | Normalized form | FEMA350 structural hole | Dog bone type |
|---|---|---|---|---|
| Initial stiffness /kN·m·rad$^{-1}$ | μ=0.05 | 7782 | 7838 | 7221 |
| | μ=0.10 | 7825 | 7871 | 7243 |
| | μ=0.15 | 7836 | 7711 | 7248 |
| | μ=0.20 | 7951 | 7892 | 7254 |

*3.3.3. Effect of Structural Holes on Stress Distri-bution of Flange Plate.* It can be seen from figure 15 that:

(1) Figure 15(a) shows the stress distribution of the upper flange, that is the compression flange in three different structural types of specimens. It can be seen from the figure that by changing the type of the structural hole and the flange plate, the stress concentration of the flange plate at the structural hole has alleviated, but the minimum stress produced by weakening flange.

(2) Figure 15(b) shows the stress distribution of the lower flange, that is the tension flange in the structural type specimen. It can be seen from the figure that the stress concentration of the flange plate at the structural hole is reduced by changing the structural hole type but not very large. By changing the type of the flange plate, the stress concentration of the flange plate at the structural hole is alleviated more obviously. The stress distribution of the upper and lower flange plates shows a trend of being small in the middle and small at both ends.

## 4. Analysis on Influence Parameters of Hybrid Joint for Stainless Steel Beam - Column Welding

Based on the three main geometric parameters of the stainless steel structure beam-column hybrid joints, the nonlinear finite element parameter analysis results of the 21 joint specimens show that:

(1) The anti-slip coefficient increases: the bearing capacity of the joint increases approximately linearly, but the increase is small; when the anti-slip coefficient increases from 0.05 to 0.20, the initial stiffness of the joint increases linearly; the slip between the shear plate and the web plate decreases as the anti-slip coefficient increases.

(2) The thickness of the beam flange plate increases: the bearing capacity of the joint increases approximately linearly, the initial stiffness of the joint also increases approximately linearly, and the tangential stress of the bolt decreases.

(3) Improvement of the structural form: After using the FEMA350 structural hole, the bearing capacity of the node is improved but not obvious, but the stress concentration of the flange plate at the structural hole is greatly reduced. By selecting the weak flange plate and adopting the dog-bone type, although the stress concentration at the structural hole can be alleviated, the bearing capacity of the joint will be greatly reduced.

To sum up, it is suggested to adopt the structural hole type of FEMA350 specification and the method of increasing the thickness of the beam flange plate to improve the mechanical performance of the beam-column bolt-welded hybrid joint of stainless steel structure.

**Acknowledgements**

The financial support received from the National Natural Science Foundation of China (No. 51108007) and the Project of Tsinghua University Independent Scientific Research Program (Interdisciplinary Project) (No. 20131089288) is gratefully acknowledged.

**References**

[1] Ministry of Construction of the PRC 2010 *Code for Seismic Design of Buildings: GB50011—2010* (Beijing: China Architecture & Building Press) p371.

[2] Qiao X L, Wang Y Q and Jia L G 2017 The applications and research of the stainless steel form of connections with welded flanges and bolted webs *Industrial Construction* **17**(S) 154-162.

[3] Wang Y Q, Zhao Y P and Xu C Y 2017 Experimental study on seismic performance of stainless steel end-plate connections with different kinds of bolts *Journal of Tianjin University (Science and Technology)* **50**(B07) 140-146.
[4] Wang Y Q, Qiao X L and Jia L G 2018 Experimental research on seismic behavior of stainless-steel beam-column connections with different link method *Journal of Southeast University(Natural Science Edition)* **48**(2) 316-322.
[5] Wang Y Q, Qiao X L and Jia L G 2018 FEM Analysis on Seismic Behavior of Stainless-Steel Beam-Column Connections with Different Link Method *Journal of Southeast University (Natural Science Edition)* **34**(1) 1-10.
[6] Luo C, Wang Y Q, Shi Y J, Li G X and Luo Y 2010 Nonlinear analysis for the bearing properties of cast aluminum joint in the aluminum alloy lattice shell structures *Building Science* **26**(5) 57-61.
[7] Wang Y Q, Liu X C, Shi Y J, Yin J and Ouyang Y W 2014 Experimental study on mechanical performance of TEMCOR joints in aluminum alloy shell structures *Journal of Shenyang Jianzhu University (Natural Science)* **30**(5) 769-777.
[8] Hasan R, Kishi N and Chen W F 1998 A new nonlinear connection classification system *Journal of Constructional Steel Research* **47**(1-2) 119-140.
[9] Management Centre:rue de Stassart,36 B-1050 Brussels 2005 *Eurocode C E N. 3: Design of Steel Structures, Part 1-8: Design of Joints* Brussels: EN1993-1-8, European Committee for Standardization p89.
[10] Wang Y Q, Liu X C, Shi Y J, Yin J and Ouyang Y W 2015 Finite element analysis on mechanical performance of TEMCOR joints in aluminum alloy shell structures *Journal of Tianjin University (Science and Technology)* **48**(B07) 1-8.

# Preparation and Properties Research of Manganese Dioxide-polyaniline Electrode

**X L Zuo*, Q Yi, K C Gu, H Yan, J Wang and J H Shu**

Department of Petroleum, Army Logistics Academy, Chongqing 401331, China.
Email: zuoxl_cq@126.com

**Abstract**. Hydrogen energy is the cleanest energy at present because its combustion product and no carbon emission. Electrolysis of water to produce hydrogen is generally considered to be the most promising method to produce hydrogen in large quantities. But electrolysis of water to produce hydrogen is without the use of a catalyst. Manganese dioxide is an efficient and green catalyst. $MnO_2$ dioxide-polyaniline electrode was prepared by CV method and CV/CA method respectively. The properties of dioxide-polyaniline were studied with the linear sweep cyclic voltammetry test, electrochemical impedance test and tafel slope test, and the results were compared with those of $MnO_2$ electrode. It was found the $MnO_2$ dioxide-polyaniline electrode properties were outstanding. It was further showed that performance of manganese dioxide-polyaniline electrode was superior by preparation of CV/CA with the testes of XPS and SEM and stability test.
**Keywords.** Manganese dioxide electrode, manganese dioxide-polyaniline electrode, electrodeposition, electrochemical performance.

## 1. Introduction

It is urgent to develop reliable and efficient energy in the face of serious environmental problems and energy crisis. Hydrogen has the characteristics of high calorific value, easy storage and small weight. At present the main source of hydrogen is obtained by electrolysis of water. The smooth process of water electrolysis is inseparable from electrodes and catalysts. Manganese dioxide has many advantages, such as abundant reserves, low cost and no pollution to the environment compared with ruthenium oxide, and it has been widely concerned by researchers [1-3]. However, the main disadvantage of manganese dioxide is that its recycling ability is relatively poor and its activity is high only in alkaline environment. The cycling performance of $MnO_2$ and its application in neutral environment can be improved effectively by making the $MnO_2$ into nanometer structure and combining it with polyaniline or core-shell mechanism [4]. In this paper, preparation and properties of Manganese dioxide-polyaniline electrode were studied with CA and CV deposition methods, and then the linear sweep cyclic voltammetry (LSV), electrochemical impedance (EIS) and Tafel slope test were carried out.

## 2. Experimental

### *2.1. Experimental Reagents and Materials*

Aniline and manganese acetate are analytical pure. The concentration of sulfuric acid is 98wt%, and carbon cloth is battery grade. The deionized water used in the experiment was self-made in our laboratory. It was in table 1 for details.

**Table 1.** Experimental reagents.

| Reagent name | Purity |
|---|---|
| $C_6H_5NH_2$ | Analytical |
| $C_4H_6MnO_4·4H_2O$ | Analytical |
| $H_2SO_4$ | Analytical |

### *2.2. Experimental Equipment*
The experimental instruments and manufacturers were listed in table 2.

**Table 2.** Experimental instruments.

| Experimental equipment | Manufacturer |
|---|---|
| CHI760 electrochemical work station | Shanghai Chen-hua Instrument Co.,Ltd |
| AL204 electronic balance | Shenyang Long-teng Electronics Co.,Ltd |
| UC-6H ultrasonic cleaners | Shanghai Tai-tan Technology Co.,Ltd |
| DZF-6050 vacuum oven | Shanghai Jing-hong Laboratory Equiment Co.,Ltd |
| SEM | Japanese JEOL-JSM-7800F electron microscope |

### *2.3. Preparation of Carbon Cloth*
70 mL deionized water was poured into a 100 mL beaker, and 9.2 mL 65wt% $HNO_3$ and 2 mL 98wt% $H_2SO4$ were slowly added to mix the solution into 100 mL. The cut carbon cloth 1*1 $cm^2$ was immersed in the beaker for ultrasound for 2 h. Then 65 wt% $HNO_3$ was used for 6 h hydrothermal treatment at 80°C. After the hydrothermal reaction, it is washed with water, and then washed by ethanol, and dried at last. These measures were pre-oxidized the carbon cloth, and increased the number of oxygen-containing functional groups on the surface of the carbon cloth, improved the hydrophilicity of the carbon cloth with the purpose of adhering to the surface of the carbon cloth better of sediment.

### *2.4. Electrode Preparation*

*2.4.1. Preparation of $MnO_2$-polyaniline Electrode by CV.* 5.43 ml 98% concentrated sulfuric acid was taken to react with 5 mL aniline, and then diluted to 400 ml with deionized water. The mixed solution of manganese acetate and polyaniline was obtained with adding 0.368 g manganese acetate and further adding deionized water to 500 mL. The solution is evenly mixed by ultrasound, and 50 ml of the solution is placed in a beaker. Anodized carbon cloth was connected, and the carbon cloth was cut into a size of 1 cm*1 cm; the cathode is connected with platinum sheet electrode; it was connected with salt bridge in the middle. The electrochemical workstation adopts CV deposition, and the number of deposition cycles was 3. And the carbon cloth was removed and cleaned with deionized water, it was obtained the desired electrode.

*2.4.2. Preparation of MnO2-polyaniline Electrode by CV/CA.* Firstly, 0.368 g manganese acetate was weighed and dissolved in 500 ml deionized water to form a solution of 3 mM/L. Then, 50 mL of the solution was placed in a beaker for electrodeposition of manganese dioxide. Anodized carbon cloth was connected, and the carbon cloth was cut into a size of 1 cm*1 cm. The cathode is connected with platinum sheet electrode; it was connected with salt bridge in the middle. The electrochemical workstation adopts the method of CA deposition, and the deposition time is 50 s, 100 s, 200 s and 400 s respectively. Remove carbon cloth and clean with deionized water. Then 5.43 ml 98% concentrated sulfuric acid was taken to react with 5 mL aniline, and then diluted to 500 ml with deionized water. The solution is evenly mixed by ultrasound, 50 ml solution was placed in a beaker, and the treated carbon cloth was anodized. The carbon cloth was cut into a size of 1 cm*1 cm. The cathode is connected with a platinum electrode and a salt bridge is inserted in the middle. The electrochemical workstation adopts CV deposition, and the number of deposition cycles is 3.

### *2.5. Performance Characterization of Catalytic Materials*
Physical characterization is mainly used to determine the crystal shape, size, microstructure and other characteristics of materials. In this paper, scanning electron microscope and X-ray photoelectron spectroscopy are used for physical characterization. It was used for SEM scanning, and the scanning voltage was 12 kV.

## 3. Results and Discussion

### *3.1. Properties of MnO2-Polyaniline Electrode Prepared by CV*
The manganese dioxide-polyaniline electrode prepared by SEM scanning with a scanning voltage of 12 kV was shown in figure 1. Compared with the deposition of polyaniline alone, the deposition of manganese dioxide-polyaniline with the addition of manganese acetate in the solution is undoubtedly better. The objects on the carbon cloth also changed from the previously amorphous clumps and floccule objects to the qualitative sphericity, and the diameter of the spherical objects was mostly in the range of tens to hundreds of nanometers, which was basically consistent with the previous literature. In addition, as shown in the figure, more sediment indicates that its stability must be improved compared with pure polyaniline, which proves that the adsorption between manganese dioxide-polyaniline composite material and carbon cloth is greater than that between pure polyaniline and carbon cloth.

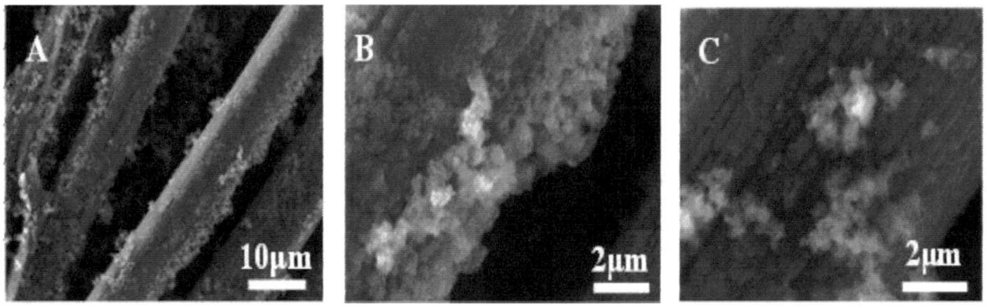


**Figure 1.** SEM image of manganese dioxide-polyaniline electrode.

As shown in figure 2, the linear sweep voltammetry (LSV), electrochemical impedance test (EIS) and tafel image slope test were carried out to analyze the electrochemical properties of $MnO_2$-Polyaniline electrode. Firstly, a linear sweep voltammetry test was shown in figure 2(A), the control group was the $MnO_2$ electrode with CA deposition of 200 s prepared previously. It can be seen that the catalytic effect of $MnO_2$-Polyaniline electrode was not good in the image of 1 V~2 V scanning interval, and the catalytic performance of $MnO_2$-Polyaniline electrode is worse than that of the pure $MnO_2$ electrode. And the potential was lower when it was reached 20mV overpotential in table 3.

Then, the electrochemical impedance spectroscopy (EIS) of $MnO_2$-Polyaniline electrode was tested. The test results were shown in figure 2(B) and figure 2(C). In figure 2(B), the slope of the manganese dioxide-polyaniline electrode prepared in this time was lower completely than that of the manganese dioxide electrode. In figure 2(C), it can be seen that the size of the semi-arc of $MnO_2$ electrode and $MnO_2$-Polyaniline electrode is similar, which proves that the size of the charge transfer resistance of the two electrode materials is similar. However, because the electrochemical performance of $MnO_2$ electrode material was superior to $MnO_2$-Polyaniline electrode material in the low frequency region, so the performance of $MnO_2$ electrode was still superior in the aspect of electrochemical impedance spectrum test [5-7]. Finally, tafel test was conducted on the prepared polyaniline electrode, and the test results were shown in figure 2(D). It can be seen that the tafel slope of $MnO_2$ electrode was 22.4 mV/Dec, which is slightly better than 25.2 mV/Dec of $MnO_2$-Polyaniline electrode. It can be seen from figure 2(D) that compared with the $MnO_2$ electrode, the effect of $MnO_2$ electrode is better, and the charge transfer speed of $MnO_2$ electrode was also faster. It was most of polyaniline sediments probably of the $MnO_2$-Polyaniline electrode, and it was a small amount of deposited $MnO_2$ products [8-9].

**Table 3.** Over potential of electrode material at critical potential.

| Electric materials | Potential (V) in 20mv overpotential | Overpotential (mV) in potential 2V |
|---|---|---|
| $MnO_2$ (200s) | 1.37 | 249.4 |
| $MnO_2$-Polyaniline (3circles) | 1.46 | 210.4 |

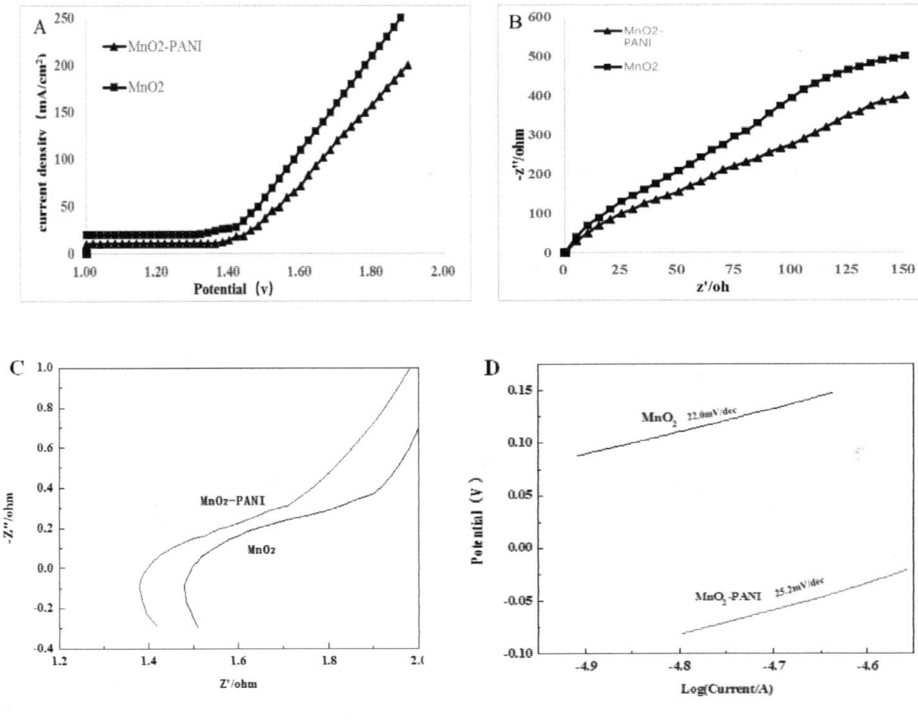


**Figure 2** .Electrochemical test of manganese dioxide electrode and manganese dioxide-polyaniline electrode; (A) LSV curve of the electrode; (B) the high frequency region of the electrode EIS test; (C) the low frequency region of the electrode EIS test; (D) The Tafel slope of the electrode.

### *3.2. Properties of MnO2-Polyaniline Electrode Prepared by CV/ CA*

The manganese dioxide-polyaniline electrode prepared by SEM was scanned with a scanning voltage of 12 kV. The scanning results are shown in figure 3. As can be seen from the figure, this deposition is undoubtedly very successful. On a carbon cloth, it can be seen that there are almost all spherical solid particles deposited densely, which are about several hundred nanometers in size. The spherical objects are tightly bonded to each other, ensuring that the spherical objects stick to the carbon cloth steadily. From this point of view, this is consistent with other test results [10-12].

The test results were shown in figure 4. Firstly, the prepared $MnO_2$-polyaniline electrode was tested by linear sweep voltammetry. The test results are shown in figure 4(A). It can be seen from the figure that there are some differences between the $MnO_2$-polyaniline electrode and the $MnO_2$-polyaniline electrode in the scanning range of 1 V~2 V. The current density even increased by 10 $mA/cm^2$ compared with the $MnO_2$-polyaniline electrode prepared in the previous electrode by CV. In addition, the current density of $MnO_2$-polyaniline electrode and $MnO_2$ electrode is basically equal in the potential range of 1-1.2 V. And from the image, the prepared $MnO_2$-polyaniline electrode was still very smooth compared with the manganese dioxide electrode. In general, the $MnO_2$-polyaniline electrode prepared was much more successful than that prepared in the previous chapter [13-15]. It may be that the deposited manganese dioxide can be well coated by the deposited polyaniline, and the potential of $MnO_2$-Polyaniline was close to the potential of $MnO_2$ when it was reached 20mV overpotential in table 4.

The EIS test results are shown in figure 4(B) and 4(C). As can be seen from figure 4(C) of the high-frequency region, the size of the semi-arc of the two materials is also very similar, which proves that the charge transfer resistance of the two electrode materials is also very similar, and further indicates that the conductivity of the two materials is very similar. It was illustrated further the preparation of manganese dioxide - manganese dioxide of the preparation of polyaniline electrode compared to the previous chapter, polyaniline electrode performance [16].

**Figure 3.** SEM image of manganese dioxide-polyaniline electrode.

Finally, tafel slope test was conducted on the prepared $MnO_2$-polyaniline electrode. The test results are shown in figure 4(D). It can be seen that the tafel slope of $MnO_2$-polyaniline electrode was 20.1 mV/ Dec, slightly better than 22.4 mV/ Dec of $MnO_2$-polyaniline electrode. It is known from literature that tafel slope represents the speed of charge transfer. The lower the tafel slope, the faster the charge transfer. This undoubtedly proves that the electrochemical performance of the $MnO_2$-polyaniline electrode prepared this time has been a qualitative leap compared with the $MnO_2$-polyaniline electrode prepared in the previous chapter [17-20].

**Table 4.** Over potential of electrode material at critical potential.

| Electric materials | Potential(V) in 20mv overpotential | Overpotential(mV) in potential 2V |
|---|---|---|
| $MnO_2$(200s) | 1.37 | 249.4 |
| $MnO_2$-Polyaniline(3circles) | 1.45 | 223.2 |

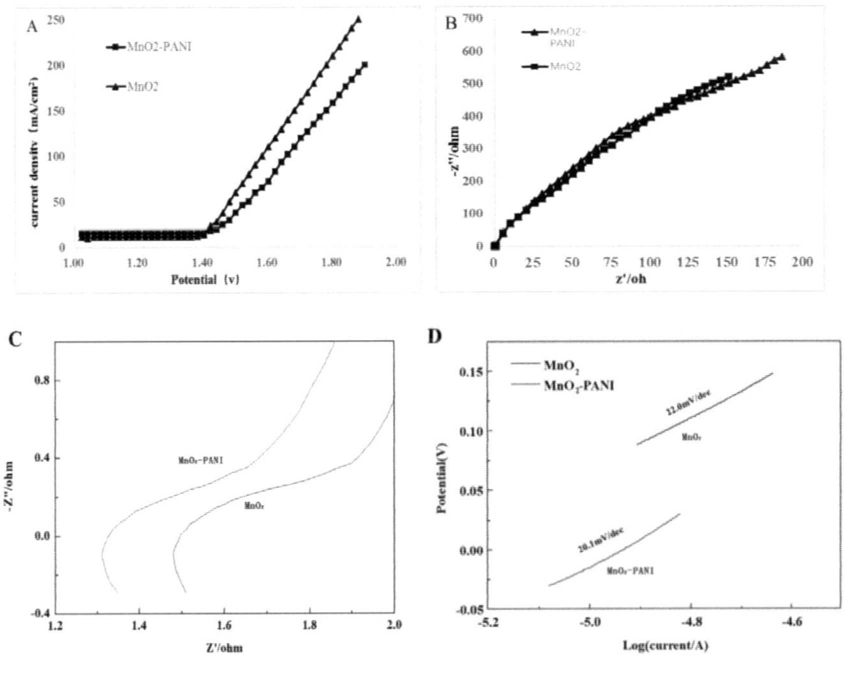


**Figure 4.** Electrochemical test of manganese dioxide electrode and manganese dioxide-polyaniline electrode; (A) LSV curve of the electrode; (B) the high frequency region of the electrode EIS test; (C) the low frequency region of the electrode EIS test; (D) The Tafel slope of the electrode.

### *3.3. XPS Analysis*

To further examine the composition of the $MnO_2$ electrode and $MnO_2$-Polyaniline electrode, the X-ray photoelectron spectroscopy (XPS) analyses were performed, Figure 5(A) and 5(B) showed the overall XPS spectra of $MnO_2$ electrode, Figure 5(C) and figure 5(D) were the XPS diagram of the manganese dioxide-polyaniline electrode. As can be seen from figure 5(A), manganese, oxygen and carbon have been detected. Figure 5(B) shows the typical spectral characteristics of hybrid Mn 2P with high resolution, with peaks of 654.1 and 642.4 eV, respectively, indicating that the spin splitting orbits of Mn 2P 1/2 and Mn 2P 3/2 are 11.7eV, indicating that the product contains manganese dioxide. Figure 5(C) contains oxygen, nitrogen and carbon, indicating that manganese dioxide can be well coated by polyaniline. Figure 5(D) was shown typical spectral characteristics of high yield of hybrid N 1S, in which three peaks are 400.6, 399.0 and 398.4eV, representing undoped imines (=N-), undoped amines (-NH-) and protic nitrogen (N+) respectively. These pyrolysis reactions indicated the coexistence of polyaniline and manganese dioxide in early hybridization [21-23].

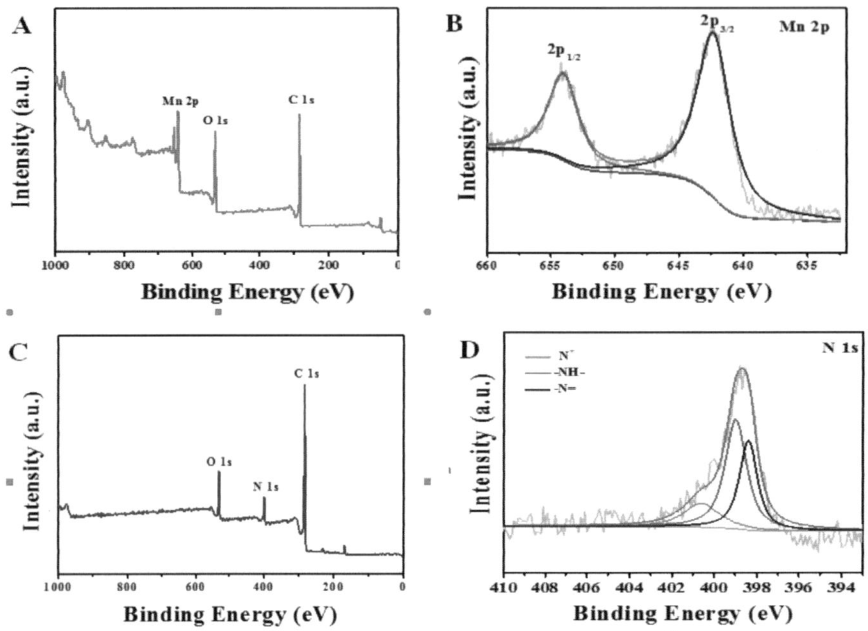


**Figure 5.** XPS diagram of manganese dioxide electrode (A, B); XPS image of manganese dioxide-polyaniline electrode (C, D).

*3.4. Stability of $MnO_2$-Polyaniline Electrode*

In order to verify the stability of the electrode after adding polyaniline, the stability of the prepared $MnO_2$ electrode and $MnO_2$-Polyaniline electrode were tested respectively. The test potential was 1.4V and the test time was 36000s. The test results are shown in figure 6. It can be seen from figure 6 that the current density of $MnO_2$-Polyaniline electrode is relatively stable in the first 25000 s, basically maintaining at 10 mA/cm$^2$. The current density begins to decline after 25,000 s. However, the current density of $MnO_2$ electrode is different from that of $MnO_2$ electrode. From the beginning, the current density of $MnO_2$ electrode decreases until it becomes stable after 27000 s.

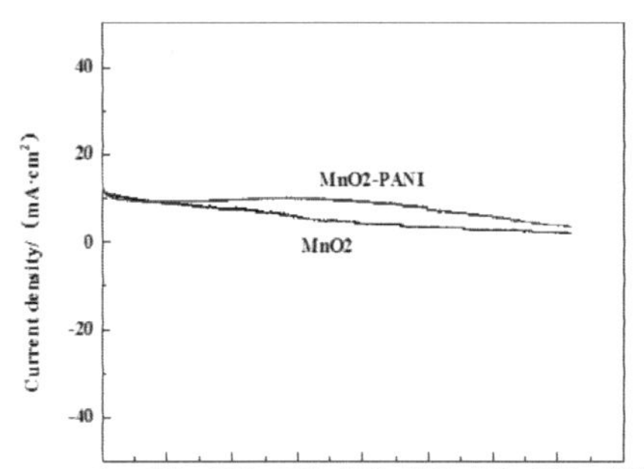


**Figure 6.** Stability test of manganese dioxide electrode and manganese dioxide-polyaniline electrode.

The solution of $MnO_2$-Polyaniline electrode reaction system was always clear and transparent, and the carbon cloth did not fall off. It is speculated that the manganese dioxide on the manganese dioxide electrode starts to fall off at about 10000 s and reacts with electrolytic liquid potassium hydroxide solution to generate manganese hydroxide, resulting in the solution turning yellow. And then all the potential is applied to the carbon cloth and eventually the carbon cloth falls off; The polyaniline on $MnO_2$-Polyaniline electrode protected the electrode well [24-25]. The experimental results strongly prove that the coating of polyaniline on $MnO_2$ electrode can improve the stability of the electrode.

## 4. Conclusion

(1) $MnO_2$-polyaniline electrode with better performance was prepared by one-step method and two-step method respectively. Linear sweep cyclic voltammetry, electrochemical impedance and Tafel slope tests were performed on the $MnO_2$-polyaniline electrode prepared by one-step method, and the results were compared with those of pure $MnO_2$ electrode. The $MnO_2$-polyaniline electrode prepared by the two-step method was scanned by SEM electron microscope. It was found that the product prepared was spherical, which was consistent with the morphology of literature.

(2) XPS characterization and stability test were carried out on the prepared $MnO_2$ electrode and $MnO_2$-polyaniline electrode, which proved that $MnO_2$-polyaniline electrode was excellent.

## Acknowledgements

This paper is supported by science and technology project of Chongqing Education Commission. Number: KJQN20191904.

## Reference

[1] Suen N T, Hung S F, Quan Q, Zhang N, Xu Y J and Chen H M 2017 Electrocatalysis for the oxygen evolution reaction: recent development and future perspectives *Chemical Society Reviews* **46** 337-365.

[2] Li P P, Zhao R B, Chen H Y, Wang H B, Wei P P, Huang H, Liu Q, Li T S, Shi X, Zhang X F, Liu M L and Sun X P 2019 Recent Advances in the Development of Water Oxidation Electrocatalysts at Mild pH *Small* **15**(13) 1805103.

[3] Kim J S, Kim B, Kim H and Kang K 2018 Recent Progress on Multimetal Oxide Catalysts for the Oxygen Evolution Reaction *Advanced Energy Materials* **8**(11) 1702774.

[4] Ghadge S D, Velikokhatnyi O I, Datta M, Shanthi P M, Tan S, Damodaran K and Kumta P N 2019 Experimental and Theoretical Validation of High Efficiency and Robust Electrocatalytic Response of One-Dimensional (1D) (Mn,Ir)O-2:10F Nanorods for the Oxygen Evolution Reaction in PEM-Based Water Electrolysis *Acs Catalysis* **9**(3) 2134-2157.

[5] Yu X B, Zhang S, Li C, Zhu C L, Chen Y J, Gao P, Qi L H and Zhang X T 2016 Hollow CoP nanopaticle/N-doped graphene hybrids as highly active and stable bifunctional catalysts for full water splitting *Nanoscale* **8**(21) 10902-10907.

[6] Liu Z H, Tan H, Liu D B, Liu X B, Xin J P, Xie J F, Zhao M W, Song L, Dai LM and Liu H 2019 Promotion of Overall Water Splitting Activity Over a Wide pH Range by Interfacial Electrical Effects of Metallic NiCo-nitrides Nanoparticle/$NiCo_2O_4$ Nanoflake/graphite Fibers *Advanced Science* **6**(5) 1801829.

[7] Yang Y, Su X S, Zhang L, Kerns P, Achola L, Hayes V, Quardokus R, Suib S L and He J 2019 Intercalating $MnO_2$ Nanosheets With Transition Metal Cations to Enhance Oxygen Evolution *Chemcatchem.* **11**(6) 1689-1700.

[8] Z M, Xu H, Chen Z G, She X J, Song Y H, Lian J B, Zhu X W, Yan P C, Lei Y C, Yuan S Q and Li H M 2019 Construction of $MnO_2$/Monolayer g-$C_3N_4$ with Mn vacancies for Z-scheme overall water splitting *Applied Catalysis B-Environmental* **241** 452-460.

[9] Najafpour M M, Sedigh D J , King'ondu C K K and Suib S L 2012 Nano-sized manganese oxide-bovine serum albumin was synthesized and characterized. It is promising and biomimetic catalyst for water oxidation *RSC Adv.* **2** (30)11253-11257.

[10] Najafpour M M and Pashaei B 2012 Nanoscale manganese oxide within Faujasite zeolite as an efficient and biomimetic water oxidizing catalyst *Dalton Trans.* **41**(34) 10156-10160.

[11] Najafpour M M, Ehrenberg T, Wiechen M and Kurz P 2010 Calcium Manganese(III) Oxides

($CaMn_2O_4$ center dot xH(2)O) as Biomimetic Oxygen-Evolving Catalysts *Angewandte Chemie-International Edition* **49**(12) 2233-2237.

[12] Zaharieva I, Chernev P, Risch M, Klingan K, Kohlhoff M, Fischer A and Dau H 2012 Electrosynthesis, functional, and structural characterization of a water-oxidizing manganese oxide *Energy Environ. Sci.* **5**(5) 7081-7089.

[13] El-Deab M S, Awad M I, Mohammad A M and Ohsaka T 2007 Enhanced water electrolysis: Electrocatalytic generation of oxygen gas at manganese oxide nanorods modified electrodes *Electrochem. Commun.* 9 2082-2087.

[14] Jiang N and Meng H M 2012 The durability of different elements doped manganese dioxide-coated anodes for oxygen evolution in seawater electrolysis *Surface & Coatings Technology* **206**(201) 4362-4367.

[15] Narita K, Kuwabara T, Sone K, Shimizu K I and Yagi M 2006 Characterization and activity analysis of catalytic water oxidation induced by hybridization of ($OH_2$)(terpy)Mn(mu-O)(2)Mn(terpy)($OH_2$) (3+) and clay compounds *Journal of Physical Chemistry B* **110**(46) 23107-23114.

[16] Nakagawa T, Beasley C A and Murray R W 2009 Efficient Electro-Oxidation of Water near Its Reversible Potential by a Mesoporous IrOx Nanoparticle Film *Journal of Physical Chemistry C* 113(30) 12958-12961.

[17] Nakamura R and Frei H 2006 Visible light-driven water oxidation by Ir oxide clusters coupled to single Cr centers in mesoporous silica *Journal of the American Chemical Society* **128**(33) 10668-10669.

[18] Takashima T, Hashimoto K and Nakamura R 2012 Mechanisms of pH-Dependent Activity for Water Oxidation to Molecular Oxygen by MnO2 Electrocatalyst *Journal of the American Chemical Society* **134**(3) 1519-1527.

[19] Takashima T, Hashimoto K and Nakamura R 2012 Inhibition of Charge Disproportionation of MnO2 Electrocatalysts for Efficient Water Oxidation under Neutral Conditions *Journal of the American Chemical Society* **134** 18153-18156.

[20] Chen S G, Wei Z D, Qi X Q, Dong L C, Guo Y G., Wan L J, Shao Z G and Li L 2012 Nanostructured Polyaniline-Decorated Pt/C@PANI Core-Shell Catalyst with Enhanced Durability and Activity, *Journal of the American Chemical Society* **134**(32) 13252-13255.

[21] Dominic J, David T, Vanaja A, Muralidharan G, Maheswari N and Kumar K K S 2018 Supercapacitor performance study of lithium chloride doped polyaniline *Applied Surface Science* **460** 40-47.

[22] Chen N N, Ni L, Zhou J H, Zhu G Y, Zhang Y, Chen S Y, Gao F J, Lu C L, Ji H M, Chen J, Wang X Z, Guo X F, Peng L, Ding WP and Hou W H 2018 Intercalation of alkylamines in layered $MoO_3$ and in situ carbonization for a high-performance asymmetric supercapacitor *Sustainable Energy & Fuels* **2**(12) 2788-2798.

[23] Wang J G, Kang F Y and Wei B B 2015 Engineering of $MnO_2$-based nanocomposites for high-performance supercapacitors *Progress in Materials Science* **74** 51-124.

[24] Shown I, Ganguly A, Chen L and Chen K H 2015 Conducting polymer-based flexible supercapacitor *Energy Science & Engineering* **3** (1)2-26.

[25] Shah A u H A, Khan M O, Bilal S, Rahman G and Van Hoang H 2018 Electrochemical co-deposition and characterization of polyaniline and manganese oxide nanofibrous composites for energy storage properties *Advances in Polymer Technology* **37** (6)2230-2237.

# Liquid-Infused Surfaces with Anti-Corrosion on Magnesium Alloy

**Qianqian Cai, Jinkai Xu*, Zhongxu Lian, Huadong Yu and Jian Li**

Ministry of Education Key Laboratory for Cross-Scale Micro and Nano Manufacturing, Changchun University of Science and Technology, Changchun 130022, China
Email: xujinkai2000@163.com

**Abstract.** The poor corrosion resistance of magnesium alloy has become the main reason restricting its application. In this paper, the layered double hydroxide (LDH) with good compactness was prepared on the surface of magnesium alloy by one-step hydrothermal method. The film thickness was 13 μm. The surface was injected with silicone oil to prepare the Liquid-infused surfaces (LIS). Electrochemical test studies show that the LIS has good corrosion resistance, and the lubricant and LDH on the surface can provide a double lasting physical barrier for magnesium alloy materials, effectively isolating chloride ions from corrosion to magnesium alloy substrates. In addition, the LIS is hydrophobic, which can reduce the contact area (CA) between the NaCl solution and the surface, and further improve the corrosion resistance of the magnesium alloy surface.
**Keywords.** Magnesium alloy, layered double hydroxide, Liquid-infused surface, corrosion resistance, physical barrier.

## 1. Introduction

Magnesium (Mg) alloy has a series of advantages, such as low density, high specific strength, strong ability to absorb plastic energy, excellent electrical shielding performance, good thermal conductivity and good biological compatibility [1-3]. However, compared with other engineering materials, its strength is low and plasticity is poor. In addition, Mg is the No. 12 element in the periodic table, and the standard equilibrium potential of Mg is $-2.4V_{NHE}$. Compared with other engineering materials, its standard potential is lower, so Mg and its alloys are more active than other engineering materials and are easily corroded [4, 5]. Therefore, the above shortcomings of Mg alloys limit its broader application prospects in various industries.

LDHs are also called hydrotalcite intercalation materials. Their main structure is octahedral brucite slab composed of divalent metal hydroxides [6, 7]. The trivalent metal elements located on the slab can be replaced by divalent metal ions with similar ion radius, so that the main slab is positively charged. Due to the balance between the interlayer anion and the positive charge of the plate, the LDH structure is neutral, and there will be some water molecules between the LDH plates [8, 9]. The LDH has good corrosion resistance due to its unique structural characteristics. Therefore, the preparation of LDH coating on the surface of Mg alloy can improve the corrosion resistance of Mg alloy [10, 11].

In this paper, LDHs were prepared on the surface of AZ91D Mg alloy by hydrothermal treatment. The LDHs were modified by injecting silicone oil into the surface to prepare a LIS with good corrosion resistance.

## 2. Experiment

### *2.1. Materials*
AZ91D Mg alloy materials were purchased from Dongguan Kuangyu Metal Co., Ltd., chemical reagents used in the test: ethanol, deionized water, and NaOH were purchased from Tianjin Dingshengxin Chemical Co., Ltd., and dimethyl silicone oil was purchased from Dow Corning.

### *2.2. Preparation of LISs*
After mechanical polishing, the Mg alloy sample with a diameter of Φ 14mm is ultrasonically cleaned with deionized water for 5 minutes, put it into the inner tank of the high-temperature kettle, add 30 ml of NaOH solution, heat the high-temperature kettle in a 180 °C drying oven for 180 minutes to prepare the LDH surface, wash and dry with deionized water, and inject silicone oil into the LDH surface with a vacuum device to obtain the oil injection surface.

### *2.3. Characterization*
Scanning electron microscope (SEM, JSM-7900F, Japan) was used to observe the microstructure of the Mg-Al LDH surface. The phase analysis of the sample surface was carried out by X-ray diffractometer (XRD, Empyrean, Netherlands). The contact angle measuring instrument (OCA20, Germany) is used to test the surface wettability of water droplets at room temperature.

### *2.4. Corrosion Test*
The electrochemical corrosion test of the treated sample was carried out by an electrochemical workstation (PARSTAT3000A-DX, USA). The test system adopted a three-electrode mode, the working electrode is the magnesium alloy sample treated by different methods, the reference electrode is saturated calomel, and the platinum electrode is used as the auxiliary electrode. The NaCl (0.9 wt.%) solution was prepared with deionized water as the test solution, and the open-circuit potential was measured and used as the initial potential for electrochemical impedance spectroscopy measurement with the scanning frequency from $10^5$ Hz to $10^{-2}$ Hz.

## 3. Results and Discussion

### *3.1. Surface Morphology, Chemical State and Wetting Behavior*
Figure 1 shows the surface topography and cross section energy spectrum of the Mg-Al LDH surface. It can be seen that after the hydrothermal treatment, the surface of the Mg alloy is randomly distributed with a six side shaped sheet structure with a side length of about 100-600 nm, which is stacked on the surface of the Mg alloy, and the thickness of the film is about 13 μm. EDS detection shows that the main elements of the coating are Mg, O and Al. Therefore, it can be concluded that except for $Mg(OH)_2$, the film may contain LDHs.

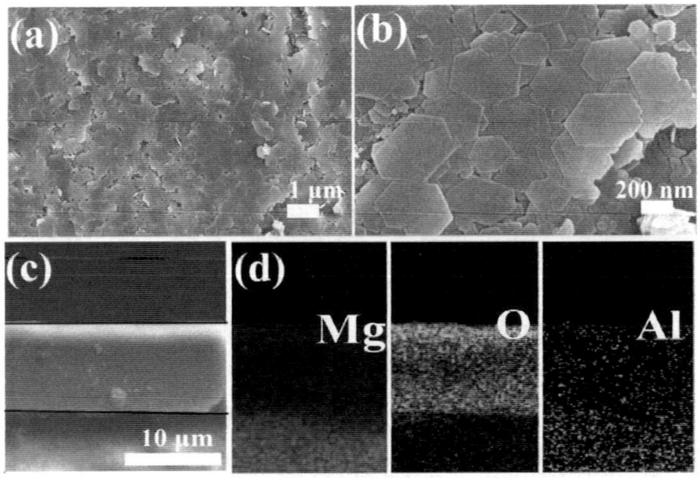


**Figure 1.** Morphology and cross-sectional energy spectrum of Mg-Al LDH surface.

Exploring the composition of the Mg-Al LDH surface, XRD analysis of the surface (figure 2). The diffraction peak of the surface mainly corresponds to Mg and $Mg(OH)_2$, but there is a strong diffraction peak at 11° corresponding to LDH . Note that the coating may contain Mg-Al LDHs. The reason for the LDH may be that the AZ91D Mg alloy contains 9% Al. Deduced that the main components of Mg-Al LDH film are $Mg(OH)_2$ and LDH.

Figure 3 shows the wettability of different sample surfaces. It can be seen that the Mg alloy surface is hydrophilic, and the static CA of water droplets on the surface is 57.3°. After hydrothermal treatment, the CA of Mg-Al LDH surface changes little (58.3°), but after silicone oil is injected into Mg-Al LDH surface, the CA increases, the LIS is hydrophobic, the static contact angle of water droplets is 105 °, and the sliding angle is greater than 10 °.

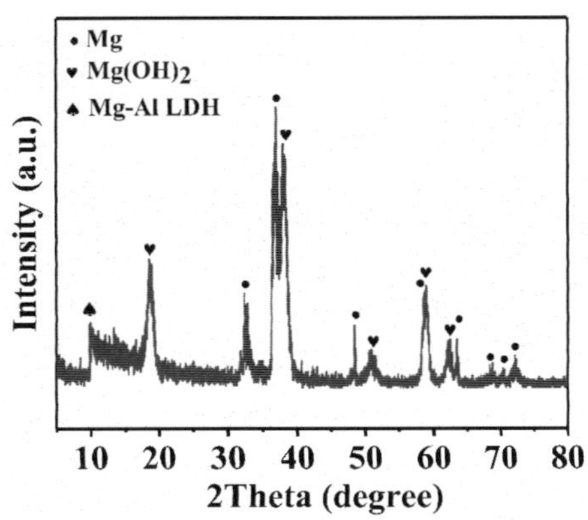


**Figure 2.** XRD pattern of Mg-Al LDH surface.

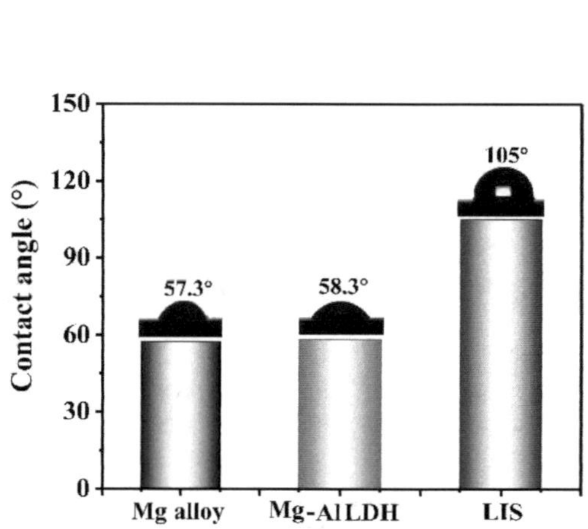


**Figure 3.** Wettability of different sample surfaces.

*3.2. Corrosion Resistance*

Potentiodynamic polarization (PDP) curve test is an effective method to evaluate the corrosion resistance of the sample surface. The polarization curve can directly obtain the corrosion potential ($E_{corr}$) and corrosion current density ($I_{corr}$) of the sample surface, and can directly reflect the corrosion resistance of the surface. Generally, lower $I_{corr}$ and higher $E_{corr}$ correspond to stronger corrosion resistance [12]. Figure 4 shows the PDP curves of AZ91D Mg alloy substrate surface, Mg-Al LDH surface and LIS. Tafel extrapolation method is used to obtain the $E_{corr}$ and $I_{corr}$ of the sample surface from figure 4. As shown in table 1, the Mg alloy substrate surface has the maximum $I_{corr}$ ($4.49 \times 10^{-4}$ A/cm$^2$) and the lowest $E_{corr}$ (-1.343 V). After hydrothermal treatment, the corrosion resistance of the Mg alloy surface is improved and the $I_{corr}$ is reduced by two orders of magnitude, which is due to the good corrosion resistance of LDH. The corrosion resistance of the LIS is the strongest. The $I_{corr}$ is reduced by 3 orders of magnitude compared with the surface of the Mg alloy substrate surface, and the $E_{corr}$ is increased from -1.343V to -0.361V. The lubricant on the LIS can effectively isolate the corrosion of the corrosion solution to the Mg alloy substrate. Especially in water solution, the lubricant and the solution are incompatible, which can provide lasting protection. The anti-corrosion efficiency ($\eta$) of the film surface is calculated as follows [13]:

$$\eta=\frac{I_0-I}{I_0}\times100\% \qquad (1)$$

Among them, $I_0$ is the $I_{corr}$ of the magnesium alloy surface, and I is the $I_{corr}$ of the sample surface with different treatment. Through equation (1), the anti-corrosion efficiency of Mg-Al LDH surface and LIS is 99.71% and 99.82% respectively.

**Table 1.** $E_{corr}$ and $I_{corr}$ of different sample surfaces in 0.9 wt.% NaCl solution.

| Sample | $E_{corr}$ (V) | $I_{corr}$ (A/cm$^2$) | $\eta$ |
|---|---|---|---|
| Base | -1.343 | $4.49\times10^{-4}$ | / |
| Mg-Al LDH | -0.419 | $1.31\times10^{-6}$ | 99.71% |
| LIS | -0.361 | $8.04 \times 10^{-7}$ | 99.82% |

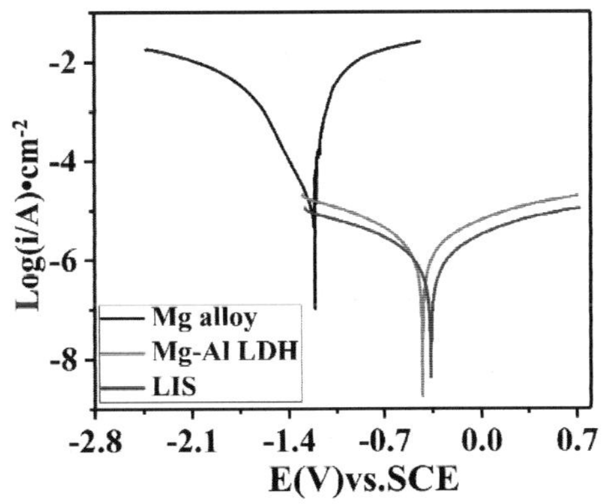


**Figure 4.** PDP curves of different sample surfaces.

Figure 5 is the Nyquist plots of Mg alloy substrate surface, Mg-Al LDH surface and LIS. The capacitive resistance arc diameter of LIS is the largest, Mg-Al LDH surface is the second, and the Mg alloy substrate surface is the smallest. The corrosion resistance sequence is the same as that obtained from the polarization curve. The capacitive resistance arc diameter is directly proportional to the corrosion resistance of the sample surface [14]. The larger the diameter of the capacitive arc, the stronger the corrosion resistance. It is further shown that the Mg-Al LDH surface and LIS prepared on the surface of Mg alloy can effectively improve the corrosion resistance of the substrate surface. The physical barrier provided by the film can effectively isolate the corrosion of chloride ion on Mg alloy, prolong the corrosion resistance of Mg alloy surface, and the LIS is hydrophobic, which can reduce the contact area between corrosion solution and surface.

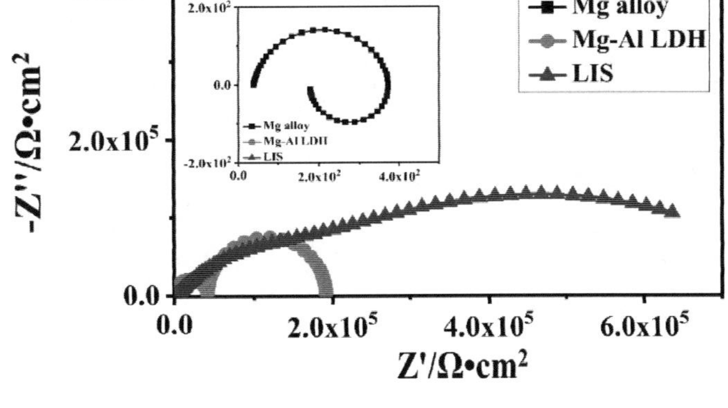


**Figure 5.** Nyquist plots of different sample surfaces.

## 4. Conclusions

In conclusion, a simple hydrothermal treatment method was used to prepare a LIS with water-repellency and corrosion resistance on the surface of magnesium alloys. The test results show that when the LIS is inclined by 10°, the water droplets on the surface can easily slide off. The LIS has better corrosion resistance. Compared with the Mg alloy substrate surface, the corrosion current density of the LIS is reduced by 3 orders of magnitude, the corrosion potential is increased by 0.98 V, and has the largest capacitive reactance arc diameter. The lubricant on the LIS can effectively isolate the corrosion of the corrosion solution on the Mg alloy substrate. Especially in the aqueous solution, the lubricant and the solution are incompatible with each other, therefore, the super-slippery surface is an effective method to prolong the service life of magnesium alloys.

## Acknowledgment

This work was supported by the National Natural Science Foundation of China (No. U19A20103), the China Postdoctoral Science Foundation (No. 2019M661184), the Jilin Province Scientific and Technological Development Program (No. YDZJ202101ZYTS025), and the Jilin Province Young Science and Technology Talent Lift Project (QT202030).

## References

[1] Xu J, Cai Q Lian Z, Yu Z, Ren W and Yu H 2021 *J. Bionic. Eng.* **18** 735-763
[2] Pan H, Ren Y, Fu H, Zhao H and Qin G 2016 *J. Alloy. Compd.* **663** 321-331
[3] Kiani M, Gandikota I, Rais-Rohani M and Motoyama K 2014 *J. Magnes. Alloy.* **2** 1-10
[4] Jin S, Zhang D, Lu X, Zhang Y, Tan L, Liu Y and Wang Q 2020 *J. Mater. Sci. Technol.* **47** 190-201
[5] Zomorodian A, Garcia M, Silva T, Fernandes J, Fernandes M and Montemor M 2013 *Acta Biomater.* **9** 8660-8670
[6] Jing C, Dong B and Zhang Y 2020 *Energy Environ. Mater.* **3** 346-379
[7] Jamil S, Alvi A, Khan S and Janjua M 2019 *Prog. Chem.* **31** 394-412
[8] Pang H, Wu Y, Wang X, Hu B and Wang X 2019 *Chem-Asian J.* **14** 2542-2552
[9] Wang J, Peng F, Wu X, Wang D, Zheng A, Cao L, Yu C, Liu X and Jiang X 2020 *Sci. China Mater.* **64** 460-473
[10] Peng F, Li H, Wang D, Tian P, Tian Y, Yuan G, Xu D and Liu X 2016 *ACS Appl. Mater. Inter.* **8** 35033-35044
[11] Peng F, Wang D, Zhang D, Yan B, Cao H, Qiao Y and Liu X 2018 *ACS Biomater. Sci. Eng.* **4** 4112-4121
[12] Li D, Wang H, Liu Y, Wei D and Zhao Z 2019 *Chem. Eng. J.* **367** 169-179
[13] Lian Z, Xu J, Yu Z, Yu P and Yu H 2019 *J. Alloy. Compd.* 793 326-335
[14] Zhang Z, An Y, Wang X, Cui L, Li S, Liu C, Zou Y, Zhang F and Zeng R 2022 *Bioact. Mater.* **18** 284-299

# Effect and Mechanism of $Ca^{2+}$ on the Corrosion of Hot-Rolled Strip in Turbid Ring Water Containing $Cl^-$ and $SO_4^{2-}$

**J W Xue[1], W Yu[1,2*] and Y L Chen[1**]**

[1] Insititute of Engnineering Technology, University of Science and Technology Beijing, Beijing 100083, China.
[2] National Research Center for Advanced Rolling and Artificial Manufacturing, University of Science and Technology Beijing, Beijing 100083, China.
Email: [*] yuwei@nercar.ustb.edu.cn, [**]yulaic@ustb.edu.cn.

**Abstract.** After the hot-rolled strip is cooled by laminar flow, $Cl^-$, $SO_4^{2-}$ and $Ca^{2+}$ in turbid ring water will remain on the surface of the steel plate, which makes the steel plate corrode during storage in a humid environment. In order to clarify the effect and mechanism of residual $Ca^{2+}$ on the corrosion of strip, simulations were carried out by setting different $Ca^{2+}$ concentrations in solutions containing $Cl^-$ and $SO_4^{2-}$ with different times of alternate immersion. Through the analyses of corrosion pit morphology, corrosion rate, electrochemical characteristics and corrosion products, it was found that the corrosion rate increased first and then decreased regardless of the $Ca^{2+}$ concentration. When the $Ca^{2+}$ concentration is 50~300 mg/L, the increase of $Ca^{2+}$ concentration will increase the adsorption rate of the matrix and increase the corrosion pits. In addition, the corrosion rate decreases, the depth of corrosion pit becomes shallow and the polarization resistance increases due to the inhibition of the calcium salt layer, and the increase of $Ca^{2+}$ concentration will enhance the inhibition of corrosion. When the $Ca^{2+}$ concentration is 1000 mg/L, the denseness of the calcium salt layer is easily destroyed and the corrosion is intensified due to the large amount of adsorption of ions by the matrix.
**Keywords.** Corrosion of steel, $Ca^{2+}$ concentration, calcium salts, alternate immersion test.

## 1. Introduction

Laminar flow cooling in the hot rolling process is a key factor in controlling rolling and controlling cooling [1], and water quality has an important impact on the surface quality of strip steel [2]. If the concentration of ions in the water is high, the ions remaining on the surface of the steel plate will make it corrode during storage, thus affecting its surface quality. $Cl^-$, $SO_4^{2-}$ and $Ca^{2+}$ are mainly present in the laminar cooling turbid ring water, and the concentration of these ions affects the pH, which has an effect on the corrosion rate of steel [3, 4]. In the studies of J K Boah [5], W Liu [6] and S Xiong [7], it was shown that $Cl^-$ and $SO_4^{2-}$ decreased the pH of the solution, which made the passivating film on the surface of the sample to be destroyed and increased the corrosion rate. For $Ca^{2+}$, on the one hand, the combination of cations and anions makes the pH in solution increase, on the other hand, the precipitation of calcium salts reduces the permeability of ions and oxygen during the corrosion process. K R Gancedo et al. [8] found that the combination of $Ca^{2+}$ and $SO_4^{2-}$ in solution caused a decrease in the concentration of acidic ions, which led to an increase in pH. This pattern was also found by K Y Kim et al. [9] in their study of the effect of Ca modification on the corrosion properties of weathering steel. Michael Grabowski et al. [10] suggested that the inhibitory effect of $Ca^{2+}$ on corrosion was mainly due to the increased thickness of corrosion products and the production of a more protective layer of calcium salts. In the studies of Kun-hu Wu [11] and Bei Wang [12], it was believed that $Ca^{2+}$ was the scaling ion, which generated calcium salts during corrosion, making the porosity of the corrosion film decrease and slowing down the corrosion rate.

The higher the concentration of ions in the turbidity ring water, the greater the residual ions on the surface of the strip. The ions remaining on the surface of the strip will make the hot-rolled strip susceptible to corrosion during storage in a humid environment. In this work, different amounts of $Ca^{2+}$ were added to the solution containing $Cl^-$ and $SO_4^{2-}$ to accelerate the corrosion of SPHC steel plates using the alternate immersion test, and with the change of $Ca^{2+}$ concentration, not only the corrosion rate but also the corrosion products as well as calcium salts would be affected. A comprehensive analysis was made to propose the corrosion mechanism of residual $Ca^{2+}$ in turbid ring water on hot-rolled strip in humid environment storage.

## 2. Experimental

The SPHC steel plate provided by a steel mill was used for the experiments, and its chemical composition is shown in table 1.

**Table 1.** Chemical composition of test steel (wt.%).

| Elements | C | Si | Mn | S | P | Als |
|---|---|---|---|---|---|---|
| Percentage | 0.05 | ≤0.03 | 0.2 | 0.02 | ≤0.02 | 0.015-0.06 |

The test solutions were prepared by adding different levels of $CaCl_2$ to the solutions containing 200 mg/L $Cl^-$, 130 mg/L $SO_4^{2-}$, so that the $Ca^{2+}$ concentrations were 50, 100, 300, and 1000 mg/L. The pH of the solutions was adjusted to 8. The size of the loss-in-weight sample was 60×40×3 mm, and the electrode exposure area of the electrochemical sample was 1 $cm^2$. The samples were polished to 800# sandpaper and cleaned by anhydrous ethanol and dried before the experiment. Alternate immersion test was used for the study, in which the wet and dry cycle was 60 min (10 min immersion time and 50 min drying time). When the corrosion time is 24 h, 72 h, 120 h, 168 h, 240 h, sampling was carried out.

FEI QUANTA FEG 450 environmental scanning electron microscope and LEXT QLS4100 laser confocal microscope were used to observe the microscopic morphology of corrosion pits on the surface of the samples after the removal of the rust layer. The corrosion rate was analyzed by weighing the samples before and after corrosion with an electronic analytical balance.

The RST5200F electrochemical workstation was used to test the polarization curves and electrochemical AC impedance spectra (EIS). A three-electrode system was used, with the auxiliary electrode being a platinum sheet electrode, the reference electrode being a saturated glycury electrode, and the working electrode being the rust layer sample obtained after the alternate immersion test. The test medium was a 3wt.% NaCl solution. Bruker D8 Advance X-ray diffractometer was used to analyze the corrosion products after different cycles of corrosion in different environments.

## 3. Results and Discussion

### *3.1. Microscopic Morphological Observation of Corrosion Pits*

Figure 1 shows the microscopic morphology of the corrosion pits of the descaled samples taken by scanning electron microscopy after the samples were alternately immersed in four different solutions for 168h. When the $Ca^{2+}$ concentration is the same, the initial stage of corrosion is the budding stage of pitting corrosion, the pits are mostly small and shallow. With the prolongation of corrosion time, corrosion pits not only deepen but also become larger in size, splicing occurs, and the surface of the sample is fully corroded. When the corrosion time is the same, the higher the concentration of $Ca^{2+}$, the more corrosion pits, the earlier the splicing occurs, the shorter the time from local corrosion to full-scale corrosion.

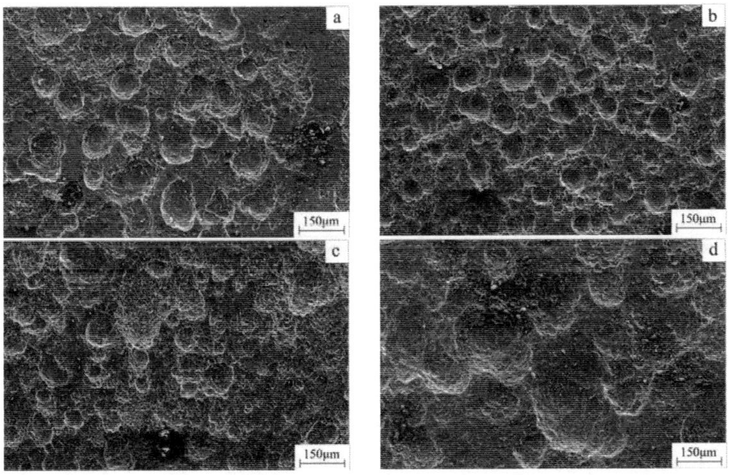


**Figure 1.** The microscopic morphology of the corrosion pits of the samples after corrosion with different $Ca^{2+}$ concentrations for 168 h. (a)-50 mg/L, (b)-100 mg/L, (c)-300 mg/L, (d)-1000 mg/L.

In order to analyze the morphology of corrosion pits more clearly, the depths of corrosion pits of the descaled samples after corrosion in different solutions for different times were measured and counted by laser confocal microscopy, as shown in figure 2. When the $Ca^{2+}$ concentration is the same, the depth of corrosion pits with the extension of corrosion time shows a trend of continuous deepening. By comparing the depth of corrosion pits at $Ca^{2+}$ concentrations of 50 mg/L and 100 mg/L, in the early stage of corrosion, the higher the concentration of ions, the greater the adsorption rate of the matrix, and the depth of corrosion pits deepens. In the middle and late stages of corrosion, the depth of corrosion pits becomes shallower with the increase of $Ca^{2+}$ concentration due to the inhibitory effect of $Ca^{2+}$ on corrosion. Through the analysis of the depth of corrosion pits at $Ca^{2+}$ concentrations of 50 to 300 mg/L, it is found that the depth of corrosion pits at $Ca^{2+}$ concentrations of 300 mg/L is shallower than the first two in the first and middle stages of corrosion, so it is believed that the increase in $Ca^{2+}$ concentration may enhance the inhibitory effect on corrosion. However, in the late stage of corrosion, it is considered that the corrosion pit deepening is the result of accelerated corrosion of the corrosion primary cell. The shallow depth of the corrosion pits at a $Ca^{2+}$ concentration of 1000 mg/L is due to the increased corrosion and the surface of the matrix being removed by corrosion.

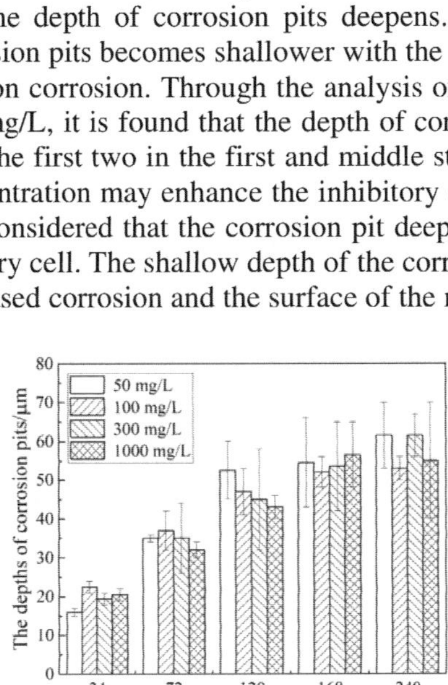


**Figure 2.** The depths of corrosion pits with different $Ca^{2+}$ concentrations for different times.

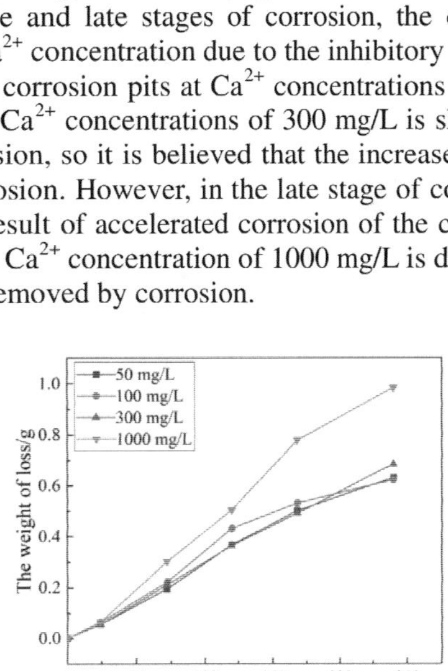


**Figure 3.** Corrosion rate of samples with different $Ca^{2+}$ concentrations.

### *3.2. Analysis of Corrosion Rate for Different $Ca^{2+}$ Concentrations*

Figure 3 shows the corrosion rates of the samples at different $Ca^{2+}$ concentrations. Regardless of the level of $Ca^{2+}$ concentration, with the extension of time, the corrosion rate shows a trend of increasing first and then decreasing. In the late stage of corrosion, the corrosion rate at $Ca^{2+}$ concentration of 100 mg/L is less than the corrosion rate at $Ca^{2+}$ concentration of 50 mg/L due to the inhibitory effect of $Ca^{2+}$ on corrosion. The increase in $Ca^{2+}$ concentration enhances this inhibitory effect, making the corrosion rate at a $Ca^{2+}$ concentration of 300 mg/L less than the corrosion rate at a $Ca^{2+}$ concentration of 100 mg/L in the middle of corrosion. For the corrosion rate at 1000 mg/L $Ca^{2+}$ concentration, it is shown that the continuous increase of $Ca^{2+}$ concentration increases the adsorption rate of ions by the matrix, which intensifies the corrosion.

### *3.3. Electrochemical Test Analysis of Samples with Rust*

The polarization curves of the samples were measured after corrosion in solutions with different $Ca^{2+}$ concentrations for different times, and the polarization resistance (Rp) was calculated using its parameters as well as equation (1).

$$Rp = \beta_a \beta_c \{2.303 i_{corr}(\beta_a + \beta_c)\}^{-1} \quad (1)$$

As shown in figure 4 is the trend of polarization resistance of the specimen with rust. When the $Ca^{2+}$ concentration is the same, with the corrosion time is extended, the polarization resistance are showing a trend of first decrease and then increase, indicating that the corrosion rate first increase and then decrease. This is consistent with the results of weight loss analysis.

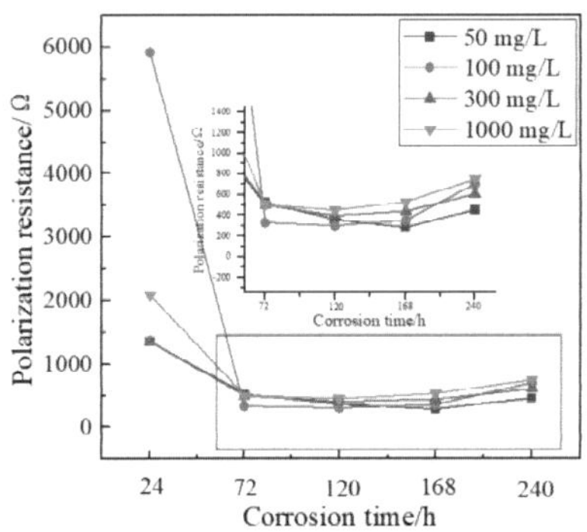


**Figure 4.** Polarization resistance with different $Ca^{2+}$ concentrations for different times.

When the corrosion time is 24 h, Rp100 > Rp1000 > Rp50 > Rp300. When the corrosion time is 72 h, Rp50 > Rp1000 > Rp300 > Rp100. When the corrosion time is 120 h, Rp1000 > Rp300 > Rp50 > Rp100. When the corrosion time is 168 h, Rp1000 > Rp300 > Rp100 > Rp50. When the corrosion time is 240 h, Rp1000 > Rp100 > Rp300 > Rp50. When the $Ca^{2+}$ concentration is 50~300 mg/L, the initial stage of corrosion is mainly the adsorption stage of ions, when the precipitation of calcium salts is less, the inhibition of corrosion is not obvious. When the corrosion time is 120 h, Rp300 > Rp50 > Rp100, and when the corrosion time is 168 h, Rp300 > Rp100 > Rp50, indicating that the increase in $Ca^{2+}$ concentration enhances the inhibition of corrosion, but when the corrosion time is 240 h, Rp100 > Rp300 > Rp50, the reason for this situation is likely to be the destruction of the calcium salt layer. When the $Ca^{2+}$ concentration is 1000 mg/L, in the early stage of corrosion, the polarization resistance is small due to the high ion concentration, which makes the corrosion increase, and in the late stage of corrosion, the corrosion rate is slowed down due to the precipitation of calcium salts.

### *3.4. Analysis of Corrosion Products*

Through the analysis of corrosion products, it was found that they included iron oxides and calcium salts. The corrosion products were analyzed semi-quantitatively, and the percentages of corrosion products for different corrosion times at different $Ca^{2+}$ concentrations are shown in table 2.

**Table 2.** Percentage content of phase with different $Ca^{2+}$ concentrations (%).

| $Ca^{2+}$ (mg/L) | Time(h) | $Fe_3O_4$ | α-FeOOH | $Ca(OH)_2$ | $CaSO_4$ | Fe |
|---|---|---|---|---|---|---|
| 50 | 72 | 41.92 | 6.25 | 7.92 | - | 43.91 |
| | 120 | 47.93 | 17.17 | 27.27 | - | 7.62 |
| | 168 | 39.02 | 26.49 | 34.14 | - | 0.36 |
| | 240 | 48.92 | 22.12 | 24.91 | - | 4.05 |
| 100 | 72 | 62.98 | 14.77 | 16.94 | | 5.31 |
| | 120 | 35.21 | 28.31 | 34.47 | - | 2.00 |
| | 168 | 26.63 | 31.28 | 39.30 | - | 2.79 |
| | 240 | 31.74 | 30.61 | 36.87 | - | 0.78 |
| 300 | 72 | 54.00 | 15.78 | 13.64 | - | 16.57 |
| | 120 | 42.59 | 28.86 | 22.46 | - | 6.09 |
| | 168 | 46.31 | 24.80 | 27.61 | - | 1.28 |
| | 240 | 33.33 | 19.65 | 31.52 | 14.86 | 0.65 |
| 1000 | 72 | 71.83 | 8.07 | 7.99 | 4.73 | 7.38 |
| | 120 | 58.80 | 10.26 | 18.17 | 6.80 | 5.96 |
| | 168 | 29.52 | 21.67 | 39.63 | 8.67 | 0.51 |
| | 240 | 27.13 | 26.62 | 33.62 | 11.97 | 0.66 |

When the $Ca^{2+}$ concentration is the same, with the prolongation of corrosion time, the percentage content of $Ca(OH)_2$ and $CaSO_4$ more and more, making the calcium salt layer thicker and dense, which has a certain hindrance and inhibition of corrosion. For 50 mg/L and 100 mg/L $Ca^{2+}$ concentration, the change in the content of $Fe_3O_4$ and α-FeOOH shows that in the early stage of corrosion, the content of 100 mg/L $Ca^{2+}$ concentration is higher than that of 50 mg/L $Ca^{2+}$ concentration, indicating a large corrosion rate, and in the late stage of corrosion, corrosion is inhibited and the corrosion rate at 100 mg/L $Ca^{2+}$ concentration is smaller than that at 50 mg/L $Ca^{2+}$ concentration. When $Ca^{2+}$ concentration is 300 mg/L, from the analysis of content changes can be seen that the corrosion rate is smaller than that of 100 mg/L$Ca^{2+}$ concentration. When the $Ca^{2+}$ concentration is 1000 mg/L, the adsorption of ions by the matrix increases, corrosion is accelerated, and also increases the generation rate of corrosion products. Although the precipitation of calcium salts is more, as the corrosion time is prolonged, too high ion concentration will have some damage to the calcium salt layer, which makes the corrosion to deeper occur.

When $Ca^{2+}$ is added to the solution containing $Cl^-$ and $SO_4^{2-}$, $Cl^-$ corrodes the substrate to produce $FeCl_2$, the hydrolysis of $FeCl_2$ promotes the ionization of water [13], $Ca^{2+}$ combines with $OH^-$ to produce calcium hydroxide precipitate, $Ca^{2+}$ combines with $SO_4^{2-}$ to produce calcium sulfate precipitate, these precipitates adhere to the surface of the substrate to form a calcium salt layer and thus inhibit corrosion [14]. Moreover, $Ca(OH)_2$ is often used as a softening agent to slow down corrosion [15]. In addition, the precipitation of the calcium salt layer affects the entry of external oxygen and thus reduces the corrosion rate [16, 17]. The addition of different levels of $Ca^{2+}$ in the solution will affect the reaction. The higher the ion concentration, the higher the adsorption rate of ions and the conductivity of the solution, which provides favorable conditions for pitting corrosion [18]. With the increase in $Ca^{2+}$ concentration, the density of corrosion pits on the surface of the substrate increases, and the splicing between corrosion pits is easy to occur, and the time required for local corrosion to become full-scale corrosion is shortened. The generation of iron oxides and calcium salts decreases the corrosion rate in the later stages of corrosion. Through the calculation of polarization resistance [19], the values of polarization resistance all show a trend of decreasing and then increasing, the corrosion rate increases and then decreases, and the increase of polarization resistance is the result of the precipitation of calcium salt layer. When the $Ca^{2+}$ concentration is 50~300 mg/L, the analyses of the corrosion pit depth, corrosion rate and polarization resistance reveal that the increase of $Ca^{2+}$ concentration enhances the inhibition of corrosion. However, when the $Ca^{2+}$ concentration is 1000 mg/L, corrosion intensifies, the rate of corrosion product generation is accelerated, the denseness of the calcium salt layer is destroyed, the porosity increases, and ions can easily penetrate into the interior of the matrix making the corrosion rate faster [20].

In summary, the corrosion mechanism of residual $Ca^{2+}$ on hot-rolled strip in a humid environment is proposed, as shown schematically in figure 5.

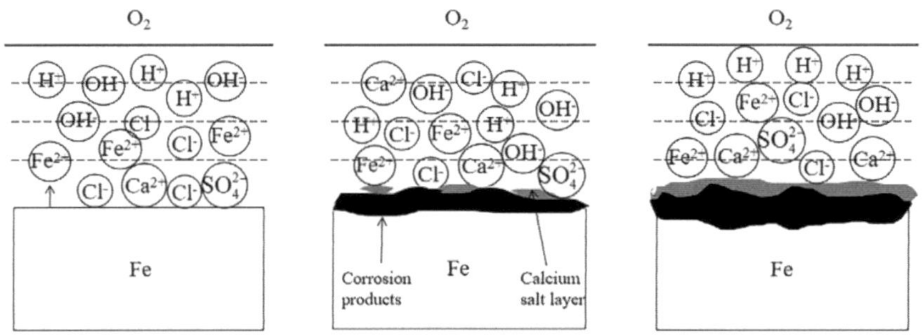


**Figure 5.** Corrosion mechanism of residual $Ca^{2+}$ on hot-rolled strip in a humid environment.

In the first stage, due to the small radius of $Cl^-$, it is preferentially adsorbed and the matrix is dissolved to produce $Fe^{2+}$, at which time the dissolution rate of $Fe^{2+}$ is greater than the adsorption rate of the matrix on $Ca^{2+}$ and $SO^{2-}_4$, and the corrosion of the substrate gradually increases, and the higher the concentration of ions, the greater the rate of this process. In the second stage, the corrosion proceeds with the generation of iron oxides and the precipitation of calcium salt layer. As the hydrolysis of $FeCl_2$ promotes the ionization of water, a large amount of $H^+$ and $OH^-$ is generated in the solution. On the one hand, the increase of $H^+$ concentration promotes the corrosion, on the other hand, when the concentration of $Ca^{2+}$ is low, $Ca^{2+}$ combines with $OH^-$ to produce $Ca(OH)_2$ with low solubility, and when the concentration of $Ca^{2+}$ increases to a certain value, $Ca^{2+}$ also combines with $SO^{2-}_4$ to produce $CaSO_4$, which will form a calcium salt layer. When the calcium salt layer does not fully cover the surface, the covered area and the uncovered area to form a corrosion cell, the uncovered area as the anode, deeper pitting corrosion occurs. In the third stage, the precipitation of calcium salts makes the calcium salt layer thicker and denser, which not only reduces the channels for ions to penetrate into the matrix, but also acts as an isolator for external oxygen. In this way, the corrosion rate decreases, but when the $Ca^{2+}$ concentration continues to rise, the denseness of the calcium salt layer decreases, weakening this inhibitory effect.

## 4. Conclusions

$Ca^{2+}$ has an inhibitory effect on corrosion, when the $Ca^{2+}$ concentration is the same, the corrosion rate shows a trend of first increasing and then decreasing, and the polarization resistance first decreases and then increases. When the $Ca^{2+}$ concentration is 50~300 mg/L, the higher the $Ca^{2+}$ concentration, the more the precipitation of calcium salts, the shorter the time required to form a full-coverage dense calcium salt layer, the earlier the inhibition of corrosion appears, and the more obvious the inhibition of corrosion.

When the $Ca^{2+}$ concentration is 1000 mg/L, in the early stage of corrosion, the large amount of adsorption of ions by the matrix causes the corrosion rate to accelerate. In the late stage of corrosion, the precipitation of calcium salts slows down the corrosion, but the continuous corrosion will destroy the denseness of the calcium salt layer, which in turn accelerates the corrosion.

## References

[1] Pian J X and Zhu Y L 2015 *Neurocomputing* **169** 457-465.
[2] Kolman D G and Taylor S R 1993 *Corrosion* **49(8)** 622-634.

[3] Bhatt H J and Phelps E H 1961 *Corrosion* **17(9)** 430-434.
[4] Klapper H S, Stevens J and Wiese G 2013 *Corrosion* **69(11)** 1095-1102.
[5] Boah J K, Somuah S K and Blanc P L 1990 *Corrosion* **46(2)** 153-158.
[6] Liu W, Cao F, Chen A, Chang L, Zhang J and Cao C 2012 *Corrosion* **68(4)** 1-14.
[7] Xiong S, Zhu Z, Zhang H and Jing L 2012 *Corrosion* **68(1)** 015001-1-015001-8.
[8] Gancedo K R, Alonso C, Andrade C and Gracia M 1989 *Corrosion* **45(12)** 976-977.
[9] Kim K Y, Chung Y H, Hwang Y H and Yoo J Y 2002 *Corrosion* **58(6)** 479-489.
[10] Grabowski M, Bluecher D, Korte M and Virtanen S 2015 *Corrosion* **71(6)** 703-725.
[11] Wu K H, Zhu L Q, Li W P and Liu H C 2010 *Corrosion Science* **52** 2244-2249.
[12] Wang B, Xu L N, Liu G Z and Lu M X 2018 *Corrosion Science* **136** 210-220.
[13] Turnbull A 1999 *Corrosion* **55(2)** 206-212.
[14] He X H, Ahn T and Gwo J P 2017 *Corrosion* **73(11)** 1381-1393.
[15] Lu Z, Dm A, Kz A, et al. 2020 *Water Research* 186.
[16] Sato F and Asakawa Y 1999 *Corrosion* **55(5)** 522-529.
[17] Novikova D, Kouřil M, MsallamováŠ, Stoulil J and Strnadová N 2016 *Koroze a Ochrana Materialu.* 68-73.
[18] Yabuki A and Murakami M 2007 *Corrosion* **63(3)** 249-257.
[19] Chen L and Su R 2020 *Construction and Building Materials* 267.
[20] Tavares L M, Costa E M D, Oliveira Andrade J J D, Hubler R and Huet B 2015 *Applied Surface Science* **359** 143.

# Optimization of Ultrasonic-Assisted Aqueous Two-Phase Extraction of Flavonoids from Hawthorn Leaves Using Response Surface Methodology

**Junsong Liang[1], Hao Zhang[1,2] and Chuanhe Zhu[1*]**

[1]Key Laboratory of Food Processing Technology and Quality Control in Shandong Province, College of Food Science and Engineering, Shandong Agricultural University, Taian 271018, China

[2]Department of Food Science and Formulation, Gembloux Agro-Bio Tech, University of Liège, Gembloux, 5030 Belgium

Email: chhzhu@sdau.edu.cn

**Abstract.** In this paper, using hawthorn leaves as raw materials, the effect of extracting total flavonoids from hawthorn leaves by aqueous two-phase system (ATPS) combined with ultrasonic assisted was studied. Through single factor test and response surface method Box-Behnken design (response surface method, BBD), according to the regression equation, the optimal process conditions were obtained as follows: the ratio of solid to liquid 1:37, ultrasonic time 40 min, ultrasonic power 360 w, ultrasonic temperature 65°C, and the yield of flavonoids was 2.869±0.0004%. This study shows that ultrasound combined with aqueous two-phase system (ATPS) is an effective method to extract total flavonoids from hawthorn leaves.

**Keywords.** Hawthorn leaves, flavonoids, aqueous two-phase system, ultrasound, response surface methodology.

## 1. Introduction

Hawthorn is generally grown in China as a medicinal and edible fruit [1]. Hawthorn leaf is a traditional medicinal material in China with biological and pharmacological properties, among which flavonoids play a key role in the pharmacological properties [2]. Some researchers have shown that hawthorn leaves are much higher in flavonoids than hawthorn berries and has special effects in regulating blood sugar and blood lipids, decreasing blood pressure, anti-atherosclerosis, and anti-oxidation [3, 4].

At present, traditional solvent extraction method is mainly used to extract flavonoids from plants, such as ethanol extraction. Although this method is simple, it has problems of long extraction time and low efficiency and is not conducive to the further purification of compounds [5]. Enzyme-assisted method has also been reported in recent years, but the conditions of this method are difficult to control [6]. Another method is microwave-assisted extraction, which can save the extraction time, but this method requires special equipment, thus raising costs in a disguised way [7]. Therefore, new extraction methods are urgently needed. As a new method of extracts, ATPS have been used to extract natural compounds [8]. In order to improve efficiency, ultrasonic assisted ATPS was used to extract flavonoids. According to previous literature, ultrasonic-assisted extraction has similar advantages to ATPS extraction, with short extraction time, less extraction solvent, and high extraction rate of flavonoids. Therefore, ultrasonic-assisted ATPS (UA-ATPS) extraction of flavonoids is feasible [9]. In this study, based on the single factor experiment, the UA-ATPS process conditions were optimized by response surface methodology.

## 2. Materials and Methods

### *2.1. Plant Material*
Hawthorn leaves were provided by Laiwu Wanbang Food Co., Ltd. After the hawthorn leaves are dried by hot air, they are ground into powder with an electric grinder, sieved through a 70-mesh sieve, and stored in a sealed bag.

### *2.2. Determination of Flavonoids*
The methods reported by Ma were used to determine flavonoids from hawthorn leave with some modifications [10].

### *2.3. UA-ATPS Extraction from Hawthorn Leaves*
1.0000g leaves powder was added into an Erlenmeyer flask and mixed with ATPS (PEG 11.5% and $(NH_4)_2SO_4$ 14%). UA-ATPS extraction was conducted in a numerical controlled ultrasonic cleaner (KQ-600DB, Kunshan). After extraction, centrifuge at 3800 rpm for 20 min. The formula for calculating the yield of flavonoids is:

$$\text{Yield of flavonoids (\%)} = \frac{100C_tV_t}{M}$$

In the formulae Ct and Vt, the measured apical phase concentration (mg/ml), volume (ml) and M are the weight (g) of the leaf powder.

### *2.4. Optimization of UA-ATPS Extraction*
Since the single factor test, the response surface test was designed by using the statistical analysis system. A Box-Behnken design (BBD) dominated by 4 factorial points, 3 layers, and 3 centers points was used to conduct 27 sets of tests to determine ideal conditions for UA-ATPS extraction. The results are shown in table 1.

**Table 1.** Factors and level table of RSM.

| Level | Factors | | | |
|---|---|---|---|---|
| | X1 The ratio of solid to liquid | X2 Ultrasonic time (min) | X3 Ultrasonic power (w) | X4 Ultrasonic temperature (°C) |
| -1 | 1: 30 | 30 | 300 | 50 |
| 0 | 1: 35 | 35 | 360 | 60 |
| 1 | 1: 40 | 40 | 420 | 70 |

### *2.5. Statistical Analysis*
All results were averaged from three repeated experiments. SPSS 23.0, Microsoft Excel 2020 and Design Expert8.0.6 were used for data collation, analysis and mapping.

## 3. Results and Discussion

### *3.1. Effect of the Ratio of Material to Solvent*
Figure 1 shows the effect of hawthorn leaf addition on flavonoid extraction rates. The results showed that with the increase of solid-liquid ratio, the flavonoid yield gradually increased from 1:15 to 1:35, and the flavonoid yield began to decrease after 1:35.

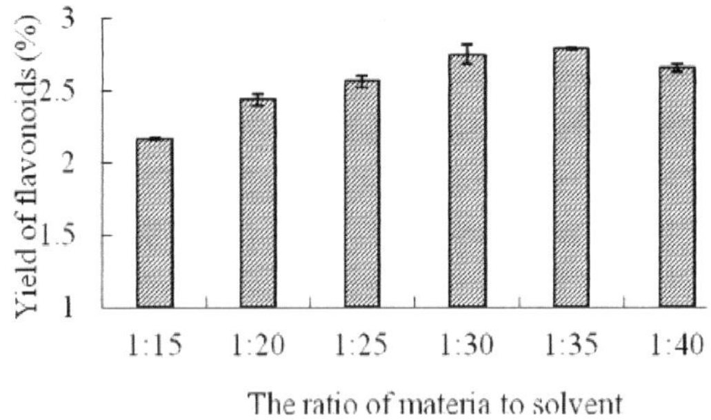


**Figure 1.** Effect of solid-liquid ratio on class yield.

### *3.2. Effect of Ultrasonic Time*

As shown in figure 2, in the range of 20-35 min, the flavonoids yield increased with time, but after more than 35 min, the yield of flavonoids decreased gradually. This is because the release of flavonoids in hawthorn leaves into the solvent is promoted at the beginning of ultrasonication, and the flavonoids are lost for too long, thereby reducing the yield of extraction [11].

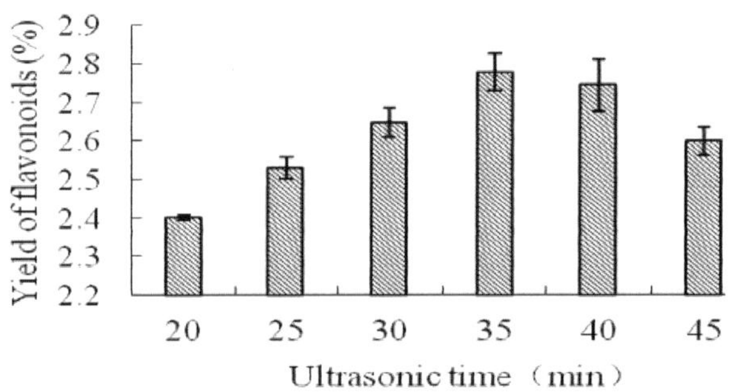


**Figure 2.** Effect of ultrasonic time on class yield.

### *3.3. Effect of Ultrasonic Power*

It can be seen from figure 3 that the effect of ultrasonic power on the extraction yield of flavonoids. Between 300 and 360 W, the extraction rate of flavonoids increased with the increase of ultrasonic power, and between 360 and 540 W, the extraction rate of flavonoids tended to decrease. More flavonoids are released. When the ultrasonic output increases, the amplitude will increase, resulting in the degradation of flavonoids [12].

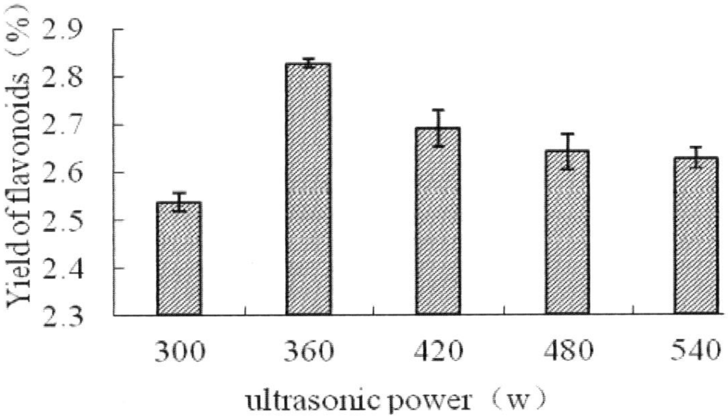


**Figure 3.** Effect of ultrasonic power on the class yield.

### *3.4. Effect of Ultrasonic Temperature*

The effect of ultrasonic temperature on the yield of flavonoids is shown in figure 4. Between 30 and 60°C, the production of flavonoids increased with increasing temperature. The yield of flavonoids decreased above 60°C (P>0.05). This phenomenon may be due to acoustic cavitation and diffusion through the cell wall. Both phenomena increase with increasing temperature. However, if the temperature is too high, the structure of flavonoids will be destroyed [13].

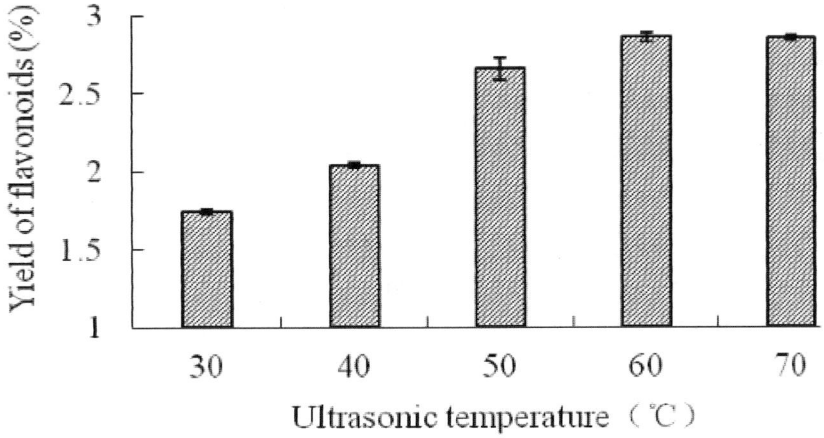


**Figure 4.** Effect of ultrasonic temperature on class yield.

### *3.5. Optimization of Extraction Conditions and Analyses of RSM*

Based on the above experimental results, a BBD experiment was designed and represented by X1, X2, X3, and X4 and total flavonoid yield (%) for the response value (Y), using Design Expert8.0.6 software to analyze the test results, get regression model parameters and variance analysis, the results were shown in table 2 and table 3.

**Table 2.** Results of RSM experiment.

| Run | Factor | | | | Yield of flavonoids (%) |
|---|---|---|---|---|---|
| | X1 | X2 | X3 | X4 | |
| 1 | 1:35 | 30 | 300 | 60 | 2.765 |
| 2 | 1:35 | 35 | 300 | 50 | 2.518 |
| 3 | 1:35 | 35 | 420 | 50 | 2.339 |
| 4 | 1:35 | 35 | 360 | 60 | 2.82 |
| 5 | 1:40 | 35 | 300 | 60 | 2.847 |
| 6 | 1:40 | 30 | 360 | 60 | 2.831 |
| 7 | 1:35 | 40 | 360 | 70 | 2.875 |
| 8 | 1:35 | 30 | 360 | 70 | 2.531 |
| 9 | 1:35 | 35 | 360 | 60 | 2.833 |
| 10 | 1:35 | 40 | 360 | 50 | 2.517 |
| 11 | 1:40 | 35 | 360 | 50 | 2.847 |
| 12 | 1:35 | 35 | 360 | 60 | 2.86 |
| 13 | 1:30 | 35 | 360 | 70 | 2.859 |
| 14 | 1:35 | 40 | 300 | 60 | 2.683 |
| 15 | 1:40 | 35 | 360 | 70 | 2.861 |
| 16 | 1:35 | 35 | 420 | 70 | 2.804 |
| 17 | 1:35 | 40 | 420 | 60 | 2.71 |
| 18 | 1:35 | 30 | 360 | 50 | 2.545 |
| 19 | 1:30 | 40 | 360 | 60 | 2.722 |
| 20 | 1:30 | 30 | 360 | 60 | 2.654 |
| 21 | 1:40 | 35 | 420 | 60 | 2.736 |
| 22 | 1:30 | 35 | 360 | 50 | 2.462 |
| 23 | 1:40 | 40 | 360 | 60 | 2.888 |
| 24 | 1:35 | 30 | 420 | 60 | 2.599 |
| 25 | 1:35 | 35 | 300 | 70 | 2.545 |
| 26 | 1:30 | 35 | 420 | 60 | 2.697 |
| 27 | 1:30 | 35 | 300 | 60 | 2.447 |

**Table 3.** ANOVA of RSM for the yield of flavonoids.

| Source | Sum of squares | df | Mean square | F value | P-value (Prob>F) |
|---|---|---|---|---|---|
| Model | 0.61 | 14 | 0.043 | 10.22 | 0.0001 |
| X1 | 0.11 | 1 | 0.11 | 26.82 | 0.0002 |
| X2 | 0.018 | 1 | 0.018 | 4.34 | 0.0594 |
| X3 | 0.00001 | 1 | 0.00001 | 0.13 | 0.7292 |
| X4 | 0.13 | 1 | 0.13 | 30.52 | 0.0001 |
| X1X2 | 0.00001 | 1 | 0.00001 | 0.00001 | 0.9341 |
| X1X3 | 0.033 | 1 | 0.033 | 7.67 | 0.017 |
| X1X4 | 0.037 | 1 | 0.037 | 8.64 | 0.0124 |
| X2X3 | 0.00001 | 1 | 0.00001 | 2.19 | 0.1644 |
| X2X4 | 0.035 | 1 | 0.035 | 8.15 | 0.0145 |
| X3X4 | 0.048 | 1 | 0.048 | 11.29 | 0.0057 |
| $X1^2$ | 0.00001 | 1 | 0.00001 | 0.1 | 0.752 |
| $X2^2$ | 0.017 | 1 | 0.017 | 4.12 | 0.0652 |
| $X3^2$ | 0.099 | 1 | 0.099 | 23.23 | 0.0004 |
| $X4^2$ | 0.096 | 1 | 0.096 | 22.68 | 0.0005 |
| Residual | 0.051 | 12 | 0.00001 | | |
| Lack of fit | 0.05 | 10 | 0.00001 | 12.04 | 0.0791 |

In this case, X1, X4, X1X3, X1X4, X2X4, X3X4, $X3^2$ and $X4^2$ were all significant for the yield of flavonoids. The final predictive equation was obtained by the means of neglecting the insignificant factors, using Yield of flavonoids as the response value, establish a quadratic response surface model, as shown below:

Y1=-4.30612+0.172303*X1+0.10816*X4-0.000105*X1*X3-0.001915*X1*X4+0.000146*X2*X4-0.000017*X3*X3+0.000266*X3*X4-0.001098*X4*X4

The response surface diagram was shown in figure 5.

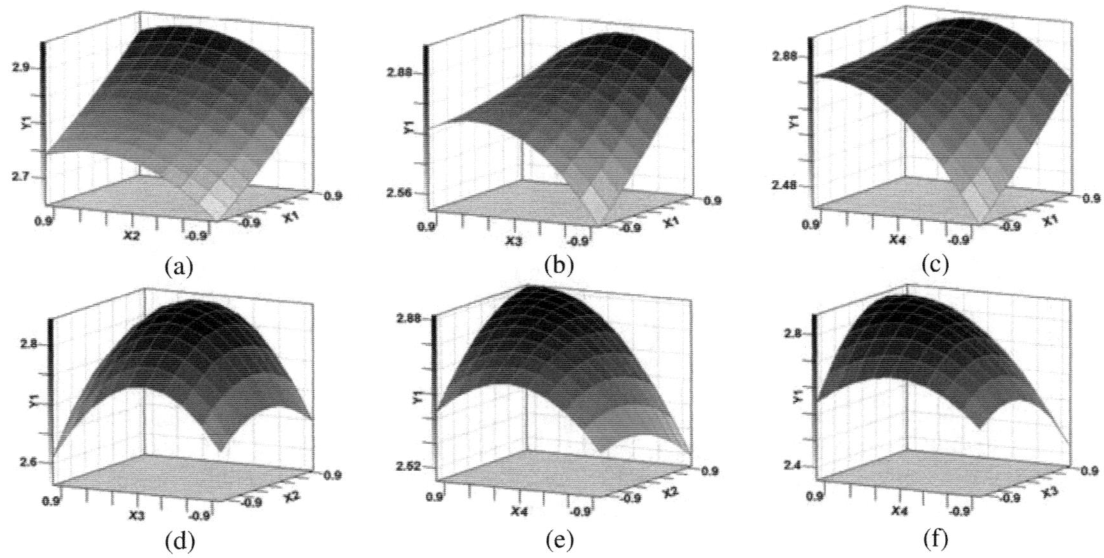

**Figure 5.** Response surface plots for the extract yield of flavonoids between different factors.

From figure 5(a), the yield of flavonoids increases slightly with the increase of solid-liquid ratio and sonication time. Figure 5(b) shows that the sonication time is fixed; the production of flavonoids increases with the solid-liquid ratio. Figure 5(c) shows the same trend as figure 5(b). Figure 5(d) shows that flavonoids first increased and then decreased with the increase of sonication time and sonication efficiency. As shown, figures 5(e) and 5(f) show that the production of flavonoids increased with increasing temperature between 50°C and 60°C.

According to the surface reaction curve method, the ideal process conditions for the extraction of conventional flavonoids from leaves are the solid-liquid ratio of 1:37.35, the ultrasonic time of 40.40 minutes, the ultrasonic power of 381.5 W, and the ultrasonic temperature of 67 degrees. The expected result is 2.921%. To facilitate the experiment, the following conditions were selected to confirm the surface pattern of the reaction curve: the ratio of solid to liquid was 1:37, the ultrasonic time was 40 minutes, the ultrasonic power was 360 W, and the ultrasonic temperature was 65 °C. The extraction efficiency of all flavonoids was 2869 ± 0.0004%, in line with.

## 4. Conclusions

This study shows that the use of ATP combined with ultrasound to extract total flavonoids from hawthorn leaves has superiority and defecation. The invention not only saves extraction time, but also improves extraction efficiency. The optimal extraction conditions were that the ratio of solid to liquid was 1:35, the ultrasonic time was 40 min, the ultrasonic power was 360 W, and the ultrasonic temperature was 65 °C. Under the optimum conditions, the yield of flavonoids was 2.869±0.0004 %. The UA-ATPS method shows the advantages of extracting flavonoids. The method provides a theoretical basis for the further purification of whole herb flavonoids, and can replace the traditional whole herb flavonoid extraction solvent method.

**References**

[1] Yin J J, Qu J G, Zhang W J, Lu D R, Gao Y C, Ying X X and Kang T G 2014 Tissue distribution comparison between healthy and fatty liver rats after oral administration of hawthorn leaf extract *Biomedical Chromatography* **28**(5) 637-47.

[2] Min Q, Bai Y, Zhang Y, Yu W, Zhang M, Liu D, Diao T and Lv W 2017 Hawthorn Leaf Flavonoids Protect against Diabetes-Induced Cardiomyopathy in Rats via PKC-α Signaling Pathway *Evid Based Complement Alternat Med* 2071952.

[3] Wang SY, Xu Q Y, Zhang W J, Liu X, Ying X X and Kang T G 2011 Study on effects of hawthorn leaves extract on fatty liver in rats *China Journal of Traditional Chinese Medicine and Pharmacy* **26** 950–955.

[4] Wang Y J, Han C H, Leng A J, Xue H F, Chen Y H, Yin J J, Lu D R and Ying X X 2012 Pharmacokinetics of vitexin in rats after intravenous and oral administration *African Journal of Pharmacy and Pharmacology* **6** 2368–2373.

[5] Mehmet B and Selin S 2013 Effects of geographical origin and extraction methods on total phenolic yield of olive tree (Olea europaea) leaves *Journal of the Taiwan Institute of Chemical Engineers* **44** 8-12.

[6] Chen S, Xing X H, Huang J J and Xu M S 2011 Enzyme-assisted extraction of flavonoids from Ginkgo biloba leaves: Improvement effect of flavonol transglycosylation catalyzed by Penicillium decumbens cellulose *Enzyme and Microbial Technology* **48** 100–105.

[7] Zhang H F, Zhang X, Yang X H, Qiu N X, Wang Y and Wang Z Z 2013 Microwave assisted extraction of flavonoids from cultivated Epimedium sagittatum: Extraction yield and mechanism, antioxidant activity and chemical composition *Industrial Crops and Products* **50** 857-865.

[8] Guo Y X, Han J, Zhang D Y, Wang L H and Zhou L L 2012 An ammonium sulfate/ethanol aqueous two-phase system combined with ultrasonication for the separation and purification of lithospermic acid B from Salvia miltiorrhiza Bunge *Ultrasonics Sonochemistry* **19** 719-724.

[9] Huang W, Xue A, Niu H, Jia Z and Wang J W 2009 Optimised ultrasonic-assisted extraction of flavonoids from Folium eucommiae and evaluation of antioxidant activity in multi-test systems in vitro *Food Chemistry* **114** 1147-1154.

[10] Ma X, Yang G and Li L 2012 Aqueous two-phase extraction of flavonoids from Artemisia argyi leaves using an ethanol-ammonium sulfate system *Advanced Materials Research* **527** 254-2258.

[11] Ying Z, Han X X and Li J R 2011 Ultrasound-assisted extraction of polysaccharides from mulberry leaves *Food Chemistry* **127** 1273-1279.

[12] Zhang D Y, Zu Y G, Fu Y J, Wang W, Zhang L, Luo M, Mu F S, Yao X H and Duan M H 2013 Aqueous two-phase extraction and enrichment of two main flavonoids from pigeon pea roots and the antioxidant activity *Separation and Purification Technology* **102** 26-33.

[13] Xiao W H, Han L J and Shi B 2008 Microwave-assisted extraction of flavonoids from Radix Astragali *Separation and Purification Technology* **62** 614-618.

# Study of $NH_3$-SCR Catalyst Combinations Based on Diesel Engine Exhaust Emission Mechanisms

**Xinyang Wang[1,a*], Yan Yan[2] and Cheng Zhang[3,b*]**

[1]Department of energy and power engineering, Beijing Institute of Technology, School of Mechanical Engineering, room 103, Engine test building, Haidian, Beijing, China.
[2]Department of Engineering Thermophysics, Beihang University, School of Energy and Power Engineering, room F626, National Laboratory, Changping, Beijing, China.
[3]Baotou Research Institute of Rare Earths, No.36 Huanghe Street, Rare Earth High-Tech Zone, Baotou, Inner Mongolia.
Email: [a] wangxinyang1998@outlook.com, [b] zhangchengbrire@163.com

**Abstract.** The overlap between the activity interval of a single catalyst and the exhaust gas temperature interval of several engines is small. In the industry, this problem is solved by increasing catalyst loadings. This has led to a certain amount of raw materials waste and increased catalyst production costs. In this study, Ce–W/Ti was introduced into Cu-SSZ-13 using solid phase grinding to change its activity range. Catalysts with different grinding ratios were subjected to small-sample performance experiments and characterization, and matching samples were selected by combining the results of the experiments with the emission characteristics of the MC07H diesel engine. World harmonized transient cycle (WHTC) and World harmonized steady-state cycle (WHSC) tests were conducted on the engine bench to investigate the causes of $NO_x$ concentration maxima during diesel engine operation using this catalyst combination strategy. The WHTC and WHSC emissions are 0.2464 and 0.1987 g/kWh. Based on the combined strategy, a performance match with diesel engines can be achieved while reducing the cost by $11/kg compared with the Cu-SSZ-13 catalyst, demonstrating the potential of the process for large-scale applications.
**Keywords.** Diesel engine, selective catalytic reduction, nitrogen emissions.

## 1. Introduction

Based on the Annual Report on Environmental Management of Mobile Sources in China (2021), $NO_x$ emissions from motor vehicles will reach 6.263 million tons in 2020. The $NO_x$ emissions from diesel vehicles account for more than 80% of the total $NO_x$ emissions from vehicles. Therefore, $NO_x$ treatment technology for diesel engines has received increased attention. The increasingly stringent $NO_x$ emission regulations from China's Stage IV Motor Vehicle Pollutant Emission Standards (China IV) to China's Stage VI Motor Vehicle Pollutant Emission Standards (China VI) led to the increasing cost of diesel exhaust treatment units. Based on a survey by He et al. [1], there is a phenomenon of some vehicles tampering with selective catalytic reduction (SCR). This falsification phenomenon is associated with the high cost of the post-treatment system. Therefore, society needs to reduce the cost of SCR catalysts while ensuring the denitrification performance.

Iwamoto et al [2] reported that Cu-ZSM-5 catalysts have an excellent De$NO_x$ performance. Subsequently, the research on copper molecular sieve catalysts has gradually increased. The chabazite

(CHA) molecular sieve has received much attention due to its superior resistance to hydrocarbon poisoning, hydrothermal stability, and anti-sulfur properties. The Cu-SSZ-13 catalyst is a commonly used CHA molecular sieve catalyst.

The element Ce ensures the dynamic oxygen equilibrium based on the storage and release of free oxygen during the conversion of $Ce^{4+}$ and $Ce^{3+}$, which ensures the stability of the SCR reaction. The presence of $Ce^{3+}$ generates more chemisorbed oxygen on the catalyst surface, increasing the adsorption and desorption capacity of $NH_3$ and enhancing the SCR activity [3]. The Ce–W/Ti catalyst is a commonly used Ce-based SCR catalyst. The subsequent introduction of W elements led to the appearance of more $Ce^{3+}$ and oxygen vacancies and an increase in the strength of the acid center, which improved the performance of the catalysts [4].

In this study, Ce–W/Ti and Cu-SSZ-13 catalysts were combined in different ratios by solid phase grinding. The suitable grinding ratio was determined according to the exhaust gas temperature range of the MC07H engine used in the bench test and combined with the performance and characterization experiments on small samples. The SCR unit with the combined catalyst was then installed in a diesel engine bench exhaust gas post-treatment system used for world harmonized transient cycle (WHTC) and world harmonized steady-state cycle (WHSC) tests. The potential of the large-scale application of this method was evaluated based on the performance and cost according to bench-scale emission data.

## 2. Materials and Methods

### *2.1. Catalyst Preparation*

The Ce–W/Ti catalyst powder used in this study was prepared in three steps: 1. Nano-$TiO_2$, tungstate, and $H_2O$ were mixed to form solution A; 2. Cerium gluconate, a copper-organic polyamine complex, and dispersant were mixed and added to solution A to obtain solution B; and 3. Solution B was aged at 150°C, filtered, washed, freeze-dried and then roasted in $N_2$ gas to obtain the powder.

The Cu-SSZ-13 catalyst powder was prepared in three steps: 1. $NaAlO_2$, NaOH, and $H_2O$ were mixed to prepare solution C; 2. $CuSO_4 \cdot 5H_2O$ and tetravinyl pentamine (TEPA) were added to solution C. The solution was stirred and silica sol was dropwise added to prepare solution D; and 3. Solution D was stirred and transferred to an autoclave for heating, filtered, collected, washed with deionized water, and dried to obtain the powder. Then the catalyst was prepared, as shown in figure 1. Measure the samples' specific surface area and particle size distribution.

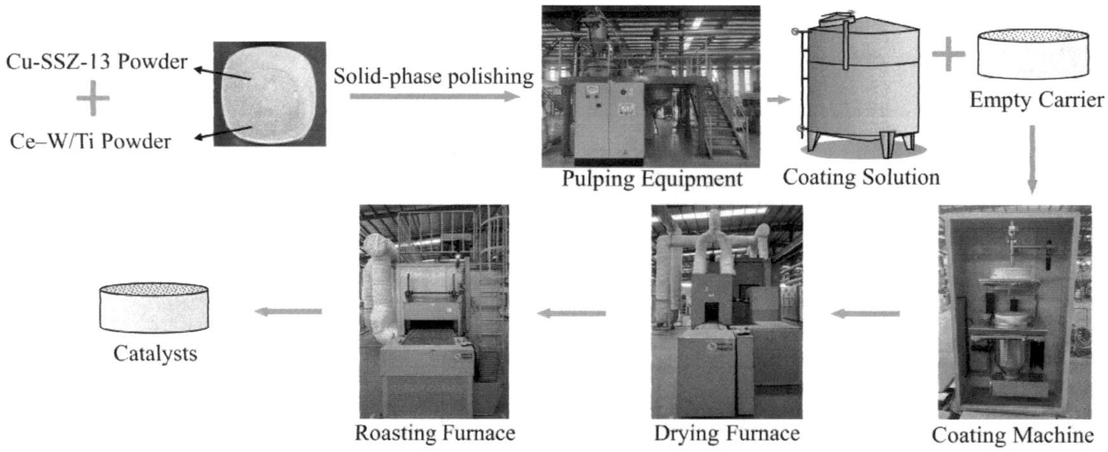


**Figure 1.** Catalyst production process.

### *2.2. Small-Sample Screening Tests for Catalysts*

The catalysts' $NO_x$ conversion and outlet ammonia concentration curves at different grinding ratios were measured using Fourier transform infrared spectroscopy to determine suitable grinding ratios in conjunction with the MC07H diesel engine. The parameter $M_1$ was defined as the mass of the Ce–W/Ti powder in the test catalyst and $M_2$ was defined as the mass of the Cu-SSZ-13 powder in the test catalyst. The grinding ratio N is defined as $M_2/M_1$. Five sets of experiments with N = 1, 2, 3, 4 and 5 were conducted. The $NO_x$ conversion rate of the experiments was calculated as follows:

$$\eta_{NO_x,Cur} = \frac{V_{NO_x,In} - V_{NO_x,Out}}{V_{NO_x,In}} \times 100\% \tag{1}$$

where $\eta_{NOx,Cur}$ is the $NO_x$ conversion efficiency, $V_{NOx,In}$ is the volume fraction of $NO_x$ at the inlet, and $V_{NOx,Out}$ is the volume fraction of $NO_x$ at the outlet. Only the $NO_x$ conversion efficiency of the standard reaction and $NH_3$ concentration at the outlet were tested during the small-sample performance test. The air speed was set to 60,000 $h^{-1}$. According to the standard reaction equation, the ratio of the NO and $NH_3$ concentrations introduced to the atmosphere tube furnace is 1:1. The NO and $NH_3$ concentrations were 670 and 380 mg/m$^3$, respectively. The $N_2$ was introduced as an equilibrium gas. Subsequently, 250 mg/m$^3$ CO, 94 mg/m$^3$ $C_3H_6$, 157 g/m$^3$ $CO_2$, 143 g/m$^3$ $O_2$, and 40 g/m$^3$ $H_2O$ were introduced to the tube atmosphere furnace. Because the MC07H heavy-duty diesel engine used for the bench test did not exceed an exhaust gas emission temperature of 450°C in previous experiments, the $NO_x$ conversion efficiency of the catalyst and the $NH_3$ concentration at the outlet were tested in the temperature range of 150–450°C.

### *2.3. Diesel Engine Bench Emission Testing*

The WHTC and WHSC cycle tests were conducted at the China National Heavy Duty Truck Engine Emission Laboratory. The engine model was an MC07H series heavy-duty diesel engine. The figure 2 shows that the SCR unit was installed in the exhaust gas post-treatment system of the diesel engine and a standard aqueous automotive urea solution with a mass fraction of 32.5% was utilized as the ammonia-supplying solution. The WHTC test consists of 1800 second-by-second operating conditions, among which approximately ~50%, 26%, and 24% are urban, suburban, and high speed. As shown in figure 3, the WHSC test consists of 13 steady-state conditions. The $DeNO_x$ pattern and performance of the catalyst in the heavy-duty diesel engine were analyzed based on the $NO_x$ emission data obtained in the WHTC and WHSC cycles, exhaust gas emission temperatures, and changes in the operating conditions.

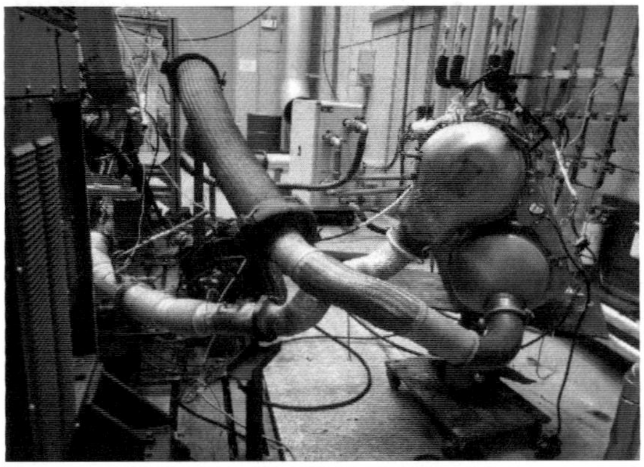

**Figure 2.** Diesel bench top experimental exhaust gas post-treatment system.

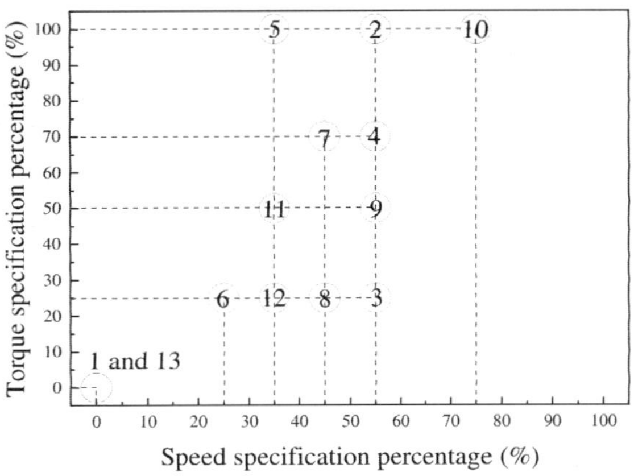


**Figure 3.** Distribution of the WHSC test points.

## 3. Results and Discussion

### *3.1. Results of the Catalyst Screening Experiments*

The specific surface areas of the powders with grinding ratios N ranging from 1 to 5 are shown in figure 4 and the particle size distribution of the slurry is displayed in figure 5. D($x$) represents the volume content of particles smaller than D($x$) as $x$% of all particles. Both the specific surface area and particle size distribution increase with increasing grinding ratio. The increasing trend of the specific surface area is notable, whereas, the increasing trend of D(10) and D(50) is not notable after the grinding ratio reaches 4 and the increasing trend of D(90) is not notable after the grinding ratio reaches 3.

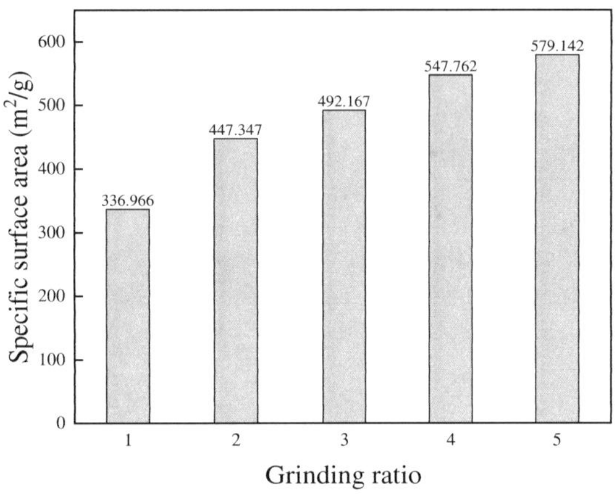


**Figure 4.** Specific surface areas of the powders at different grinding ratios.

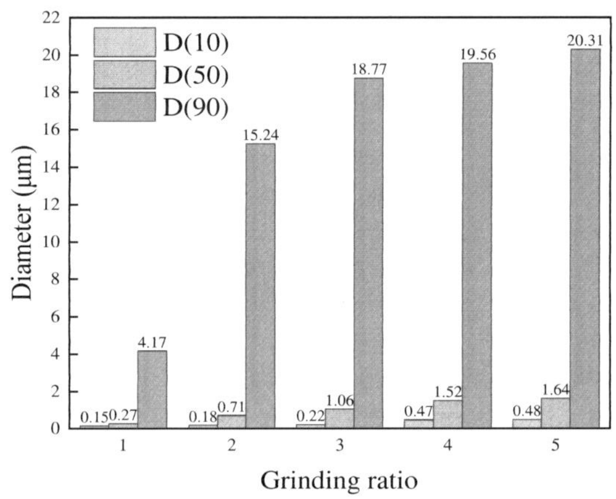


**Figure 5.** Particle size distribution of the slurry at different grinding ratios.

The $NO_x$ conversion efficiencies of the small samples of catalysts with powder grinding ratios N ranging from 1 to 5 are shown in figure 6. The catalyst with N = 3 performs best in the temperature interval of 150–200°C. Its $NO_x$ conversion efficiency reaches >90% at 200°C. As the mixing ratio increases, the SCR performance of the catalysts in the temperature range of 150–200°C first increases and then decreases. The best SCR performance in this temperature range corresponds to grinding ratio of ~3. In the temperature range of 400–450°C, the catalyst with N = 4 performs the best, without a significant decrease in the $NO_x$ conversion efficiency, whereas the catalyst with N = 3 exhibits the most significant decrease in the $NO_x$ conversion efficiency with increasing temperature, but it does not drop below 90% at 450°C. The best SCR performance in the 400–450°C temperature range corresponds to a grinding ratio of ~4. In summary, the catalyst with N = 3 has a better SCR performance in the temperature range of 150–450°C. The outlet $NH_3$ concentration curve of the catalyst with N = 3 is shown in figure 7.

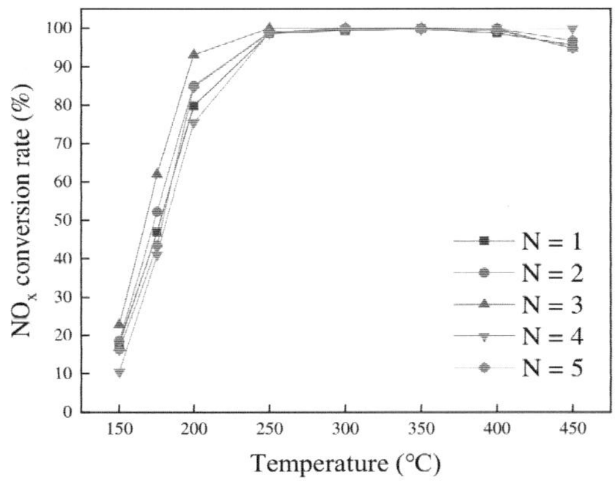


**Figure 6.** Comparison of the $NO_x$ conversions of samples with different powder mixing ratios.

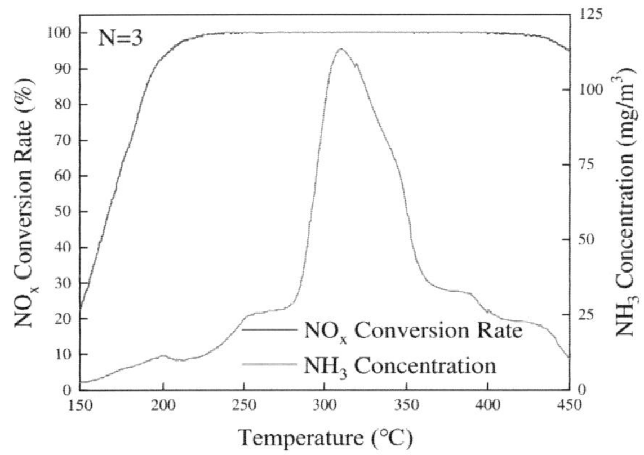


**Figure 7.** $NO_x$ conversion rate and $NH_3$ concentration.

The $NH_3$ adsorption performance of the sample with a ground ratio N = 3 was studied. When catalyzed by Cu-SSZ-13 only, $NH_3$ desorbs in the range of 238–319°C in 6MR and 463–512°C in 8MR [5], which correlates with the $NH_3$ adsorption in the Lewis and Brønsted acid centers, respectively [6]. When catalyzed by a Ce-based catalyst only, $NH_3$ is more easily desorbed in the Brønsted acid center than in the Lewis acid center at high temperatures [7]. As shown in figure 7. The $NO_x$ conversion in the range of 227–430°C reaches 100%, but ammonia leakage occurs, which is related to the desorption of $NH_3$ with increasing temperature. Because the ratio of $NH_3$ to NO of the substance is 1:1, NO does not completely react with the adsorbed $NH_3$ in the temperature range of 150–227°C. Therefore, when the temperature increases, the previously adsorbed but unreacted $NH_3$ is desorbed. In summary, the $NO_x$ conversion efficiency of the catalyst is high over a wide temperature range, and 227–430°C is the optimum temperature range for the combined catalyst with N = 3.

The catalyst with N = 3 was selected for subsequent bench experiments. The cost of the Ce–W/Ti and the Cu-SSZ-13 catalyst powders is ~$10.95/kg and ~$54.77/kg, respectively. Therefore, the cost of the combined catalyst powder with N = 3 is ~$43.82/kg.

### *3.2. Results of the Diesel Engine Cycle Test*

As shown in figure 8 and figure 9, the final calculated WHTC specific $NO_x$ concentration is 0.2464 g/kWh, which is lower than the emission limit of 0.4600 g/kWh in China VI. Under hot start conditions, the operating temperature can be maintained between 200 and 250°C during the first 400 s. Under cold start conditions, a temperature above 200°C can only be guaranteed after 400 s. The $NO_x$ conversion results in section 3.1 show that the $NO_x$ conversion of the catalyst below 200°C is low, which leads to the frequent $NO_x$ maxima in the first 700 s of the WHTC cycle test under cold start conditions. Note that the difference in the temperature profiles between the cold and hot start conditions after 700 s is relatively small, which indicates that the engine operating environment and condition associated with the two start conditions are basically the same after 700 s. However, the final $NO_x$ concentrations of the two conditions differ. Under cold start conditions, the $NO_x$ concentration is higher than under hot start conditions after 700 s. The average exhaust gas temperature in the first 700 s under the cold start conditions is 179.7°C, whereas it is 235.6°C under hot start conditions. The difference in the average exhaust gas temperature of 55.9°C leads to a difference in the convective heat transfer between the fluid and catalyst carrier pores under the two conditions, which affects the temperature field of the catalyst carrier and therefore, the reaction.

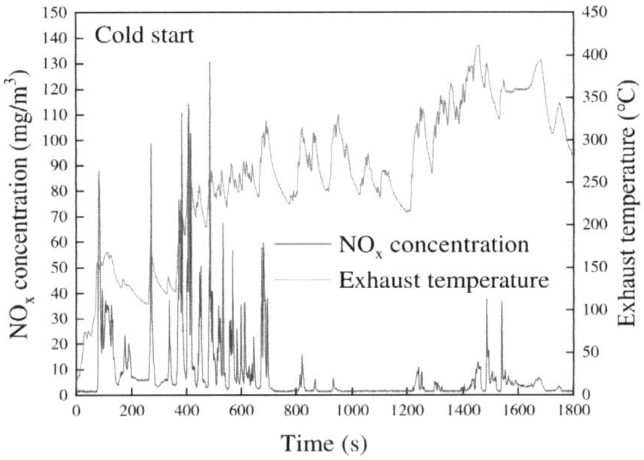


**Figure 8.** Final $NO_x$ concentration and temperature profile of the cold start WHTC test.

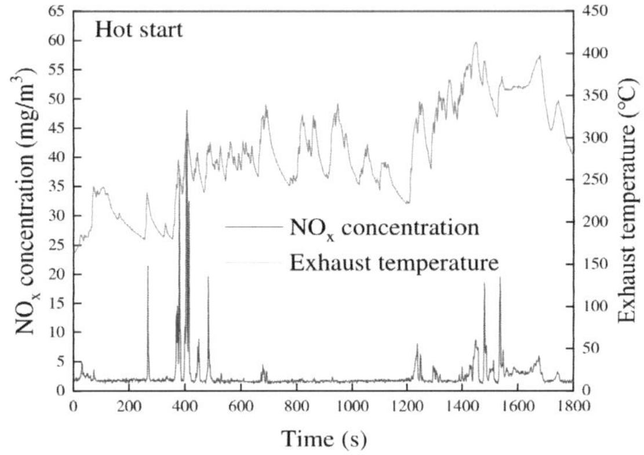


**Figure 9.** Final $NO_x$ concentration and temperature profile of the hot start WHTC test.

As shown in figure 10 and figure 11. The final calculated WHSC specific $NO_x$ concentration is 0.1987 g/kWh, which is lower than the emission limit of 0.4000 g/kWh in China VI.

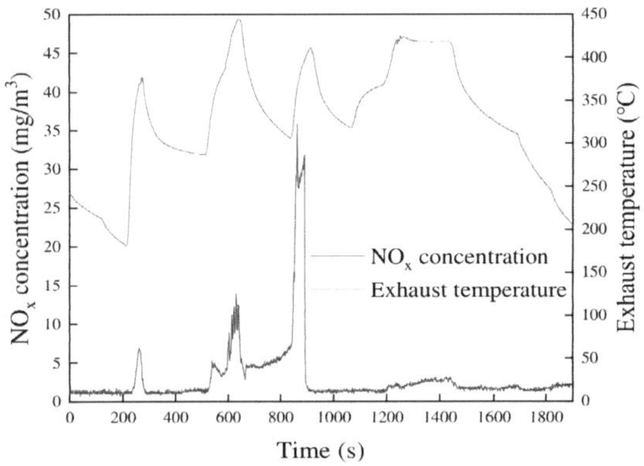


**Figure 10.** WHSC test-specific $NO_x$ concentration and temperature profile.

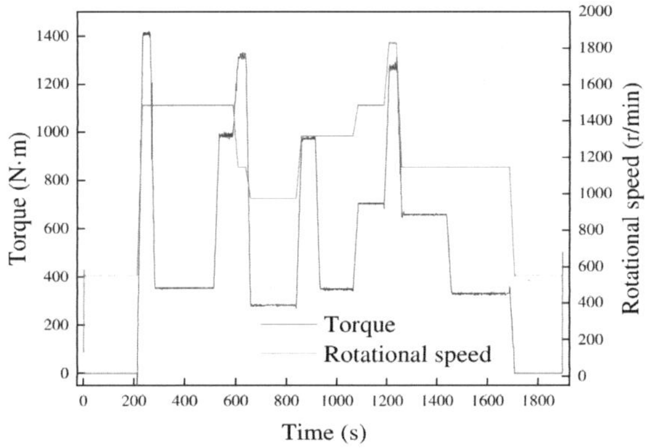


**Figure 11.** WHSC test-specific torque and speed curve.

In the WHTC and WHSC emission curves, it is found that the maximum $NO_x$ concentration is often accompanied by the maximum temperature.The explanation for this phenomenon is illustrated in figure 12. The $NO_x$ in an engine consists of (a) thermal $NO_x$ in the post-flame region, (b) fast $NO_x$ in the mid-flame region, and (c) fuel $NO_x$ based on the combination of impurities in the fuel with oxygen [8]. Due to the compression ignition characteristics of diesel engines,their $NO_x$ generation is dominated by thermal $NO_x$. Based on the Zeldovich chain reaction mechanism, the reactant concentration, temperature, and residence time of the reactants are the key factors affecting the generation of thermal $NO_x$ [9]. In the experiment, after the diesel engine received the torque-increase signal, the electronic control unit (ECU) sent an increase injection signal to the injector at 398 s. The increase in the in-cylinder injection leads to a rapid increase in the combustion temperature, at this time, the crank-connecting rod mechanism speed is unable to respond quickly. The work material remains in the cylinder for a longer period of time. Based on the Zeldovich chain reaction mechanism, the generation of $NO_x$ increases. The experiments in section 3.1 show that the rapid increase in the temperature of the catalyst from at 150–450°C leads to ammonia desorption. The rapid increase in the torque during the bench experiments leads to the rapid increase in the exhaust gas temperature, which in turn leads to the ammonia desorption of the catalyst in the post-treatment unit. This is not conducive to the denitrification reaction. The above-mentioned reasons lead to the occurrence of localized $NO_x$ concentration maxima.

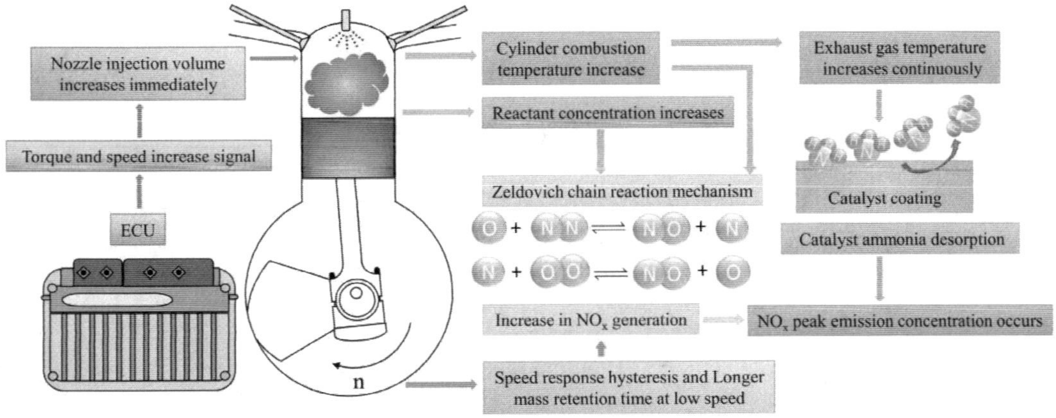


**Figure 12.** Bench top experiments on the $NO_x$ emission mechanisms.

## 4. Conclusion

It can be concluded that a combined catalyst can be prepared by changing the grinding ratio to vary the optimum temperature range and match diesel engines with different emission characteristics. This method meets the performance of heavy-duty vehicles on roads listed in China VI. It can reduce the overloading of a single catalyst powder, and can be used as an option for heavy-duty diesel engine post-treatment systems based on China VI. This method can potentially be used on a large scale in terms of the cost and performance.

## Acknowledgments

This work was supported by the Science and Technology Innovation Project of Inner Mongolia (No. XM2021CXZX02).

## References

[1] He L Q, Zhang S J, Hu J N, Li Z H, Zheng X, Cao Y H, Xu G Y, Yan M and Wu Y 2020 *Environ. Pollut.* **262** 114280.
[2] Iwamoto M, Furukawa H, Mine Y, Uemura F, Mikuriya S and Kagawa S 1986 *J. Chem. Soc. Chem. Commun.* **16** 1272–73.
[3] Zhao W Y, Li Z Q, Wang Y, Fan R R, Zhang C, Wang Y, Guo X, Wang R and Zhang S L 2018 *Catalysts* **8** 375.
[4] Fan R R, Li Z Q, Wang Y, Zhang C, Wang Y, Ding Z Y, Guo X and Wang R 2020 *RSC Adv.* **10** 5845–52.
[5] Leistner K, Xie K P, Kumar A, Kamasamudram K and Olsson L 2017 *Cat. Lett.* **147** 1882–90.
[6] Wang D, Gao F, Peden C H F, Li J H, Kamasamudram K and Epling W S 2014 *Chem. Cat. Chem.* **6** 1579–83.
[7] Zhao K, Han W L, Lu G X, Lu J Y, Tang Z C and Zhen X P 2016 *Appl. Surf. Sci.* **379** 316–322.
[8] Liu Y, Midkiff K C and Bell S R 2004 *Int. J. Engine Res.***5** 71–81.
[9] Palash S M, Kalam M A, Masjuki H H, Masum B M, Fattah I M R and Mofijur M 2013 *Renew. Sustain. Energy Rev.* **23** 473–490.

# Low Viscosity Polyurethane Modified Epoxy Resin Grouting for Steel Ring Reinforcement of Subway Tunnel

**Liang Huang[1], Ming Xiao[1], Weilong Li[1], Shuguang Bi[1,a*], Jiagong Tang[2], Xianmeng Luo[2] and Jianhua Ran[1,b*]**

[1]Hubei Key Laboratory of Biomass Fibers and Eco-Dyeing & Finishing, State Key Laboratory of New Textile Materials and Advanced Processing Technologies, College of Chemistry and Chemical Engineering, Wuhan Textile University, Wuhan 430200, China
[2]Wuhan Rongshengjimei Technology Co., Ltd., Wuhan 430050, China
Email: [a] sgbi@wtu.edu.cn, [b] jhran@wtu.edu.cn

**Abstract.** Steel ring reinforcement technology is an effective method to eliminate the damage of tunnel structure caused by ground load, surrounding foundation pit and underground water source. The grouting material used to fill the gap between the tunnel segment and the steel ring should have the characteristics of low viscosity, high strength, rapid curing at room temperature and anti-subway vibration. In this work, a special polyurethane modified epoxy resin grouting was fabricated to overcome the inherent brittleness of epoxy resin, and was combined with the reactive diluent and the amine curing agent to suit the above applications. The results showed that both tensile and impact strength of epoxy resin grouting modified by polyurethane were improved obviously compared with those without modification, presenting an obvious toughening effect. Although the viscosity increased slightly, it can still meet the perfusion requirements of steel ring reinforcement of subway tunnels.
**Keywords.** Epoxy resin grouting, polyurethane prepolymer, low viscosity, high strength, toughness.

## 1. Introduction

Epoxy resin adhesives are widely used in the fields of automotive [1-2], electronics [3], aircraft [4], and construction engineering [5-6] due to their advantages of high strength, low curing shrinkage, corrosion resistance, good thermal stability, and good electrical insulation performance. However, epoxy resin chains contain multiple benzene or heterocyclic rings, resulting in the low elasticity of molecular chains. Therefore, the cured epoxy resin adhesives have the disadvantages of high crosslinking structure, easy cracking, low shear strength and poor impact resistance, and need to be toughened [7]. Karthikeyan *et al.* [8] investigated the toughening effect of poly(etheretherketone)s terminated with hydroxyl (PEEKTOH) on the room temperature curing epoxy resin. The results showed that the shear strength and fracture toughness increased by 19% and 86% respectively when the additive amount was 5wt%. However, the increase of PEEKTOH content in the epoxy matrix led to the increase of viscosity, which reduced the wettability. Zeng *et al.* [9] used ethylene-glycidyl methacrylate-vinyl acetate solid rubber (EGVA) modified with carboxyl groups (EGVA-COOH) and amino groups (EGVA-$NH_2$) as reactive tougheners to increase the compatibility between epoxy resin matrix and rubbers. The results showed that the particle size of rubbers decreased by 71.4% and the impact toughness of epoxy resin increased by 130% with the addition of 5 phr EGVA-$NH_2$, while the tensile strength was not affected. However, the viscosity of epoxy precursors increased after adding the EGVA. In addition, there have been many reports on the interpenetrating polymer networks (IPNs) prepared by blending epoxy resin with polyurethane or hydroxyl-terminated polyurethane to improve the toughness performance of epoxy resin [10]. In this work, the room-temperature cured epoxy resin grouting with low viscosity, high strength and high toughness was prepared by using the active isocyanate-terminated polyurethane prepolymer as the toughening agent for repairing cracks in the subway tunnels.

## 2. Experimental

### *2.1. Materials*

Toluene diisocyanate (TDI) was purchased from Aladdin. Polyethylene glycol (PEG200), N-aminoethylpiperazine (AEP) and acetone were obtained from Sinopharm Chemical Reagent Co., Ltd. Benzyl glycidyl ether (BGE) was provided by Hubei Greenhome Materials Technology Co., Ltd. Bisphenol F epoxy resin (BPF-170) was purchased from China Petrochemical Co. Ltd. Their chemical structures and parameters were shown in table 1.

**Table 1.** Parameters of the epoxy resin, active diluent and curing agent.

| Materials | Molecular structure | Molecular weight | Viscosity (25 °C mPa·s) | Epoxy equivalent weight (g/eq) or Amine equivalent[a] (g/mol) |
|---|---|---|---|---|
| BPF-170 | 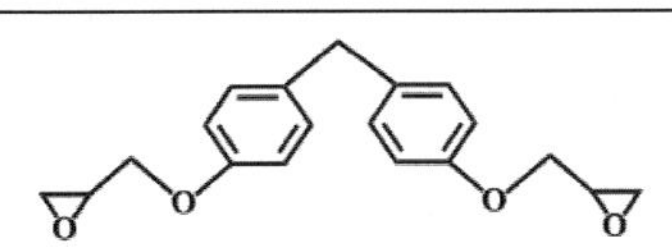 | 312 | 3050 ± 15 | 160-180 |
| BGE | 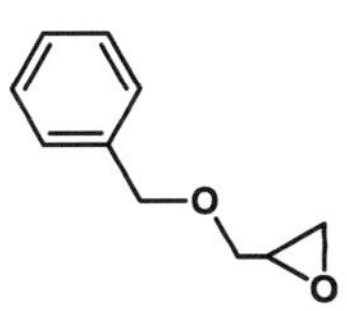 | 164 | 5-8 | ≥0.43 |
| AEP | 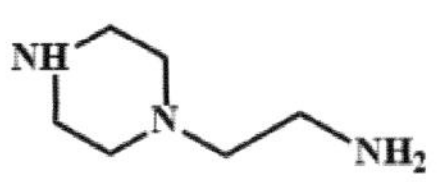 | 129 | 12.0 ± 2 | 43.0 |

[a] $\text{Amine equivalent} = \frac{\text{molecular weight of amine}}{\text{number of active hydrogens in an amine}}$

### *2.2. Preparation of Polyurethane Modified Epoxy Resin Grouting*

Firstly, isocyanate-terminated polyurethane prepolymer were synthesized at 70 °C for 3 h in the nitrogen atmosphere. The molar ratios of TDI and PEG200 were 1.5:1, 1.8:1 and 2:1, denoted as PUR-1, PUR-2 and PUR-3, respectively. Then, the epoxy resin of BPF-170 was added to make the account of polyurethane prepolymer for 10 wt%. Finally, 30 wt% of active diluent BGE and a certain amount of curing agent AEP were added, following by mixing evenly and pouring into the mechanical mold for vacuumization.

### *2.3. Characterization*

Fourier transform infrared spectroscopy (FTIR) was taken on the BRUKER Tensor 27 using KBr powder. Viscosity was measured using an NDJ-9S digital viscometer at 25 °C according to ASTM D-1084. Tensile and shear properties were tested using GOTECH AI-7000M universal testing machine according to ASTM D-638 and ISO 4587, respectively. Impact strength was tested by GOTECH impact testing machine according to ASTM D-256. The microstructure of the tensile section was observed by JSM IT500A scanning electron microscopy (SEM).

## 3. Results and Discussion

### *3.1. FTIR Spectra*

As shown in figure 1(a), the -OH vibration absorption peak of PEG200 and -NH- vibration absorption peak of PUR-3 appeared at the wavenumber of 3300 $cm^{-1}$. The -NCO vibration absorption peaks of TDI and PUR-3 appeared at 2270 $cm^{-1}$. The peak at 1720 $cm^{-1}$ was the characteristic peak of the -C=O group of PUR-3, indicating that the isocyanate-terminated polyurethane prepolymer was synthesized. As shown in figure 1(b), compared with PUR-3, the polyurethane modified epoxy resin (PUR-3/BPF-170) did not appear the vibration absorption peak of -NCO at 2270 $cm^{-1}$, and the vibration absorption peak of -NH- at 3300 $cm^{-1}$ and the vibration absorption peak of -C=O at 1720 $cm^{-1}$ were obviously weakened. This indicated that the polyurethane prepolymer was grafted onto the epoxy resin by end-isocyanate.

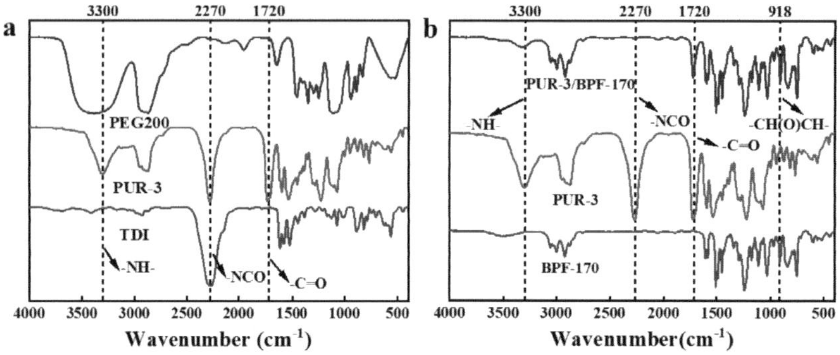


**Figure 1.** FTIR spectra of polyurethane prepolymer (a) and polyurethane modified epoxy resin (b).

### *3.2. Viscosity of Polyurethane Modified Epoxy Resin Grouting*

The viscosity of PUR-1/BPF-170/BGE/AEP, PUR-2/BPF-170/BGE/AEP, PUR-3/BPF-170/BGE/AEP and BPF-170/BGE/AEP were 96.1, 142.9, 173.3 and 88.4 mPa·s at 25 °C, respectively. Although polyurethane modification slightly increased the viscosity of epoxy resin grouting, the effect was small, less than 200 mPa·s, presenting good fluidity, which can be used in the pouring technology.

### *3.3. Mechanical Properties of Polyurethane Modified Epoxy Resin Grouting*

Compared with the unmodified epoxy resin grouting (BPF-170/BGE/AEP), the tensile strength of the three polyurethane modified epoxy resin grouting increased, up to 49.8 MPa, and the elongation at break increased 2-3 times, as shown in figure 2(a). Similarly, it can be seen from figure 2(b) and 2(c) that the shear elongation at break and impact strength also increased and the maximum shear strength and impact strength were 20.7 MPa and 3.18 $kJ/m^2$, respectively. These results showed that polyurethane modification increased the toughness of epoxy resin grouting.

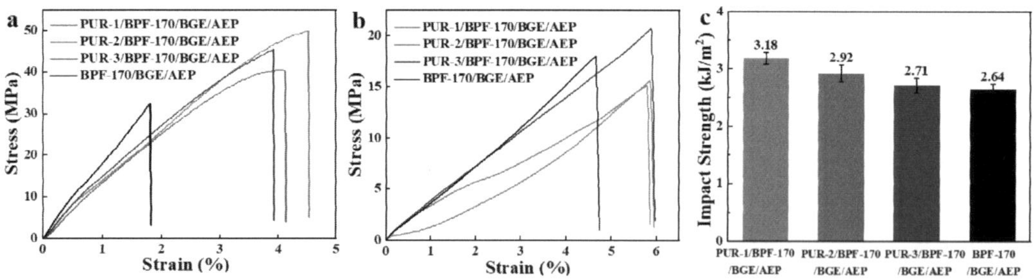


**Figure 2.** (a) Tensile stress-strain curves, (b) steel-steel shear stress-strain curves and (c) impact strength of polyurethane modified epoxy resin grouting.

### *3.4. Morphology of Polyurethane Modified Epoxy Resin Grouting*

As shown in figure 3, the tensile fracture section of the unmodified epoxy resin BPF-170/BGE/AEP was smooth and the crack direction was consistent, presenting the characteristics of brittle fracture. However, the tensile fracture section of polyurethane modified epoxy resins had more irregular and disordered cracks in the growth direction, due to the absorption and dispersion of the corresponding flexible chain segments in the polyurethane prepolymer, which further indicated that polyurethane had the toughening effect on epoxy resin.

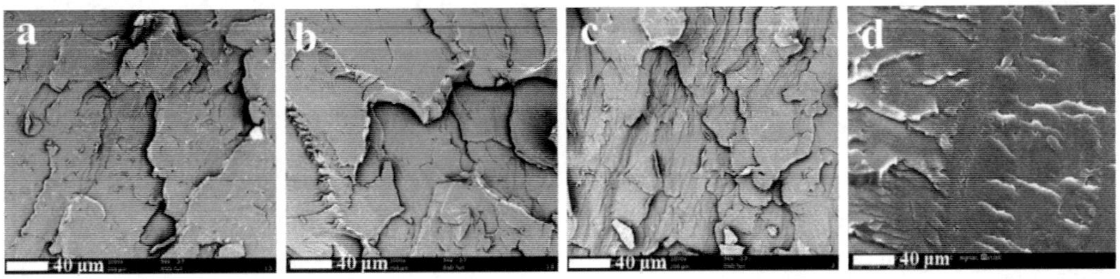


**Figure 3.** Microstructure of the tensile fracture section of (a) PUR-1/BPF-170/BGE/AEP, (b) PUR-2/BPF-170/BGE/AEP, (c) PUR-3/BPF-170/BGE/AEP and (d) BPF-170/BGE/AEP.

## 4. Conclusion

In this work, epoxy resin was modified with different proportions of isocyanate-terminated polyurethane prepolymer, and then prepared epoxy resin grouting with active diluent and amine curing agent at room temperature. After modification, the viscosity of the three kinds of epoxy resin grouting were all lower than 200 mPa·s, and the tensile strength and elongation at break were increased, presenting high strength and high toughness, which can be used as the grouting materials for repairing cracks in subway tunnels.

## Acknowledgments

The authors would like to express the appreciation for the funding supported by Key Project of "Science and Technology for Economy 2020" of National Key Research and Development Program, Science and Technology Research Project of Hubei Province Education Department (B2020073), Development of Science and Technology of Hubei Province (2021CFA034).

## References

[1] Maggiore S, Pedemonte M, Bazurro A, Stagnaro P, Utzeri R and Luciano G 2020 Characterization of the effect of an epoxy adhesive in hybrid FSW-bonding aluminium-steel joints for naval application *International Journal of Adhesion and Adhesives* **103** 102702.

[2] Ocaña R, Arenas J M, Alía C and Narbón J J 2015 Evaluation of degradation of structural adhesive jointsin functional automotive applications *Procedia Engineering* **132** 716-723.

[3] Ranji S and CheonLee M 2022 Study on synthesizing new urethane epoxy adhesives and their adhesive properties on different substrates *International Journal of Adhesion and Adhesives* **117 PartA** 103174.

[4] Encinas N, Oakley B R, Belcher M A, Blohowiak K Y, Dillingham R G, Abenojar J and Martínez M A 2014 Surface modification of aircraft used composites for adhesive bonding *International Journal of Adhesion and Adhesives* **50** 157-163.

[5] Balmori J A, Branco J M and Basterra L A 2021 Behaviour of the adhesive bond between low-grade wood and GFRP reinforcements using epoxy resin *Construction and Building Materials* **271** 121516.

[6] Modesti L A, Vargas A S D and Schneider E L 2020 Repairing concrete with epoxy adhesives *International Journal of Adhesion and Adhesives* **101** 102645.

[7] Cruz R, Correia L, Cabral-Fonseca S and Sena-Cruz J 2021 Effects of the preparation,curing and hygrothermal conditions on the viscoelastic response of a structural epoxy adhesive *International Journal of Adhesion and Adhesives* **110** 102961.
[8] Karthikeyan L, Robert T M, Mathew D, DesakumaranSuma D and Thomas D 2021 Novel epoxy resin adhesives toughened by functionalized poly(etheretherketone)s *International Journal of Adhesion and Adhesives* **106** 102816.
[9] Zeng Y, Wang Q, Xu P, Yang W, Hoch M and Ma P 2022 Toughening of epoxy resin with balanced performance by in situ constructing interfacial bonding and elastic segments into epoxynetwork *Materials Chemistry and Physics* **277** 125529.
[10] Qu C B, Wu T, Huang G W, Li N, Li M, Ma J L, Liu Y and Xiao H M 2021 Improving cryogenic mechanical properties of carbon fiber reinforced composites based on epoxy resin toughened by hydroxyl-terminated polyurethane *Composites Part B: Engineering* **210** 108569.

# Influence of Abrasive Hardness on Chemical Mechanical Polishing of 304 Stainless Steel

**Xiaojun Cao, Mingpu Xue, Zhichao Qi, Zhongtang Li and Jianxiu Su***

School of mechanical and electrical engineering, Henan Institute of Science and Technology, Xinxiang, 453003, The People's Republic of China
Email: dlutsu2004@126.com

**Abstract.** In order to improve the quality in chemical mechanical polishing (CMP) of stainless steel and the material removal rate (MRR), 6 kinds of abrasives were used to do the CMP test of 304 stainless steel. In this paper, the 6 kinds of abrasives have different hardness. The experimental results show that, in the CMP process, the MRR increases with the increase of the abrasive hardness, the surface roughness decreases firstly and then raises with the increase of the abrasive hardness. This surface roughness reduces the lowest value when the chromium oxide (CrO) abrasive is used. With low-hardness copper oxide abrasives and magnesium oxide abrasives, it can only remove partial scratches on the surface during the polishing process and obtained low MRR and poor surface quality. Silica abrasives were easy to condense in the polishing slurry and the Gelation hindered the cutting action of abrasive, So resulting in low MRR and poor surface quality. The chromium oxide and white corundum, their hardness are close to the stainless steel, can remove the surface oxide layer of the sample while not produce the damage of the workpiece surface, so the surface quality was very well. Considering that chromium oxide contains $Cr^{2+}$, it will pollute the environment. Therefore, in comprehensive consideration of the surface quality and MRR, take the white corundum as an abrasive in the CMP of 304 stainless steel.
**Keywords.** 304 stainless steel; abrasive hardness; chemical mechanical polishing; surface roughness; material removal rate

## 1. Introduction

Stainless steel has good corrosion resistance, high temperature resistance and abrasion resistance, and is widely used in aviation machinery, medical equipment, food engineering and other many fields. In order to further promote the surface quality and performance of stainless steel, it is generally necessary to perform ultra-precision processing. The traditional precision processing methods include the electrochemical polishing, the chemical polishing, the mechanical polishing, etc. For mechanical polishing, it is easy to damage the lattice structure of stainless steel, leaving damage to the surface, which is difficult to meet the requirements. The efficiency of chemical polishing is low. Many chemicals cause environmental pollution and create some substances that are harmful to the human body, and the surface quality after chemical processing is not perfect. The efficiency of electrochemical polishing and the surface quality of the workpiece after treatment have been improved, but the treatment process consumes a lot of energy, causes serious pollution and high cost [1-3]. As an ultra-precision machining method, the purpose of the chemical mechanical polishing (CMP) is to improve surface quality by reducing surface roughness. Therefore, applying CMP technology to ultra-precision processing of stainless steel can largely promote the overall performance of stainless steel [4-8].

At present, some scholars have studied the effect of the abrasive hardness on the material removal rate (MRR) and surface quality of chemical mechanical polishing. Mao Meijiao et al. [9] studied the influence of six different hardness abrasives (diamond, boron carbide, silicon carbide, alumina, zirconia, silicon oxide) on the CMP processing of cemented carbide blades, and the results showed that: boron carbide abrasive with grains large diameter dispersion produces more scratches on the surface of cemented carbide blades. Low-hardness silicon oxide, zirconia, and silicon carbide abrasives can only remove local scratches on the surface of cemented carbide blades. The hardness of Alumina which is close to the hardness of cemented carbide, can obtain better surface quality. The hardest diamond abrasive, can realize the higher MRR and surface quality than other abrasives. YUAN et al. [10] studied the influence of different abrasive hardness on the CMP of silicon carbide balls, and the results showed that: the cerium oxide abrasives and zirconia abrasives that are close to the hardness of the oxide layer have the best polishing effect, and the surface roughness of the polished silicon carbide balls is 10nm, followed by iron oxide abrasives and chromium oxide abrasives. Wang Jinpu [11] studied the influence of four abrasives (Al2O3, SiO2, CeO2, Fe2O3) on the chemical mechanical polishing of glass-ceramics under same conditions. The research results indicated that the surface quality of the glass-ceramics obtained by using the polishing liquid of zirconia abrasives. Preferably, the MRR reaches 100nm/min and the surface roughness Ra is 0.4nm. These studies show that the abrasive hardness is one of the key factors to influence the material removal rate and surface quality of workpiece in CMP. At present, the research on stainless steel CMP is basically to improve the surface quality by optimizing process parameters [12-13]. In addition, there is little study on the polishing removal mechanism of stainless steel CMP. This article uses abrasives of different hardness to perform CMP of stainless steel. By analyzing the surface roughness and material removal rate after polishing, it provides theoretical basis for the material removal mechanism of 304 stainless steel and improve the MRR and the surface quality.

## 2. Experiment

### *2.1. Experimental Conditions and Materials*

Using Bruker white light interferometer (Contour GT-K) with the vertical accuracy of 0.1nm, to detect the surface roughness Ra of 304 stainless steel, and using a Leica microscope (LEICA DM 2500) with an accuracy of 1μm to survey the contours of the stainless steel surface before and after polishing.

The mechanical properties of the 6 micron abrasives used in this paper are shown in table 1 [8], and in table 2 shows the mechanical properties of the 304 stainless steel used.

**Table 1.** Mechanical performance parameters of different abrasives [8].

| Abrasive type | Mohs hardness | Relative to 304 stainless steel hardness coefficient |
|---|---|---|
| CUO | 3.5 | 0.58 |
| MaO | 5.5 | 0.91 |
| $SiO_2$ | 7 | 1.1 |
| $Cr_2O_3$ | 8.5 | 1.3 |
| $Al_2O_3$ | 9 | 1.5 |
| Diamond | 10 | 1.7 |

**Table 2.** Mechanical properties of 304 stainless steel.

| Material | Mohs hardness | Elastic Modulus(E/GPa.) | Density $\rho/(g/cm^{-3})$ | Surface roughness Ra/μm |
|---|---|---|---|---|
| 304 stainless steel | 6 | 193 | 7.93 | 0.215 |

### *2.2. Polishing Experiment*

The experiment was carried out in a dust-free and ultra-clean laboratory with a constant temperature of 22 °C, using Shenyang Maike ZYP230 polishing machine (as shown in figure 1), the diameter of the polishing disc is 230mm, and the 304 stainless steel sample used in the experiment is a circle with a diameter of 50mm. During the polishing process, a gravity block is used to press the sample on the polishing pad so that the polishing pad is in full contact with the polishing surface of the sample. A precision electronic balance with an accuracy of 0.0001 g is used to measure the quality of the workpiece before and after polishing. The process parameters of the experiment are as follows. Take the rotating speed of the polishing machine as 60r/min, the polishing time as 15 min, the polishing pressure as 13.79 kPa, and the polishing liquid flow rate as 15 mL/min. The composition of polishing slurry is as shown in table 3.

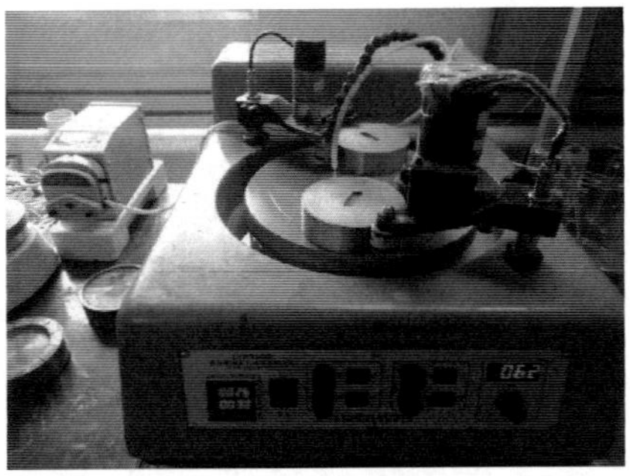

**Figure 1.** The ZYP230 polishing machine.

**Table 3.** Composition of polishing slurry (500 ml).

| Parameter | Value |
|---|---|
| Hydrogen peroxide quality $m_g$/g | 20 |
| Abrasive content $m_g$/g | 20 |
| NaPO $m_g$/g | 10 |
| C12H25SO4Na $m_g$/g | 1 |
| Deionized water | Margin |
| Oxalic acid | 4 |
| Abrasive particle diameter | 3.5μm |

Note. Copper oxide polishing liquid, magnesium oxide polishing liquid, silicon dioxide polishing liquid, chromium oxide polishing liquid, aluminum oxide polishing liquid, diamond polishing liquid based on $H_2O_2$.

Calculate the MRR of stainless steel by the following formula (1) [14]

$$MRR = \frac{\Delta m}{\rho \pi r^2 t} \tag{1}$$

which: the t is the polishing time (min); the Δm is the quality difference before and after polishing (g); the r is the radius of the material (mm); the ρ is the material density (g/cm3).

## 3. Results and Analysis

### *3.1. Influence of Abrasive Hardness on MRR in CMP of 304 Stainless Steel*

The effect of several abrasives with different hardness on the MRR of 304 stainless steel is shown in figure 2. From the figure 2, when the hardness of the abrasive increases, the material removal rate also increases. Because when the hardness of the abrasive increases, the mechanical action in the CMP process is enhanced, which increases the temperature of the polishing area, which in turn promotes the chemical reaction, which can increase the material removal rate; in addition, when the abrasive is changed by white corundum abrasive, the material removal rate increases, the trend has slowed down. This is because white corundum ($Al_2O_3$) is an amphoteric oxide, which can react with acids and alkalis. The oxalic acid used in the polishing liquid is an organic acid with acidic properties. It can react with white corundum abrasives, thus reducing the content of white corundum abrasives in the polishing liquid, thereby reducing the mechanical action. Therefore, when the abrasive is white corundum, the growth trend of the material removal rate is decreasing.

### *3.2. The Influence of Abrasives on Surface Quality of 304 Stainless Steel*

After each polishing experiment, the surface roughness of the workpiece was measured by the Bruker white light interferometer. Figure 3 shows the effect of several abrasives with different hardness on the surface roughness of 304 stainless steel. From the figure 3, with the abrasive hardness increases, the surface roughness of the workpiece decreases firstly, and then decreases to the lowest when the abrasive is chromium oxide, and then increases. After polishing with chromium oxide polishing liquid, the surface roughness of the 304 workpiece is the smallest, reaching Ra=17.468 nm, followed by white corundum abrasive, and the roughness of surface is Ra=32.966 nm, followed by diamond, silicon dioxide, magnesium oxide, copper oxide; because the hardness of chromium oxide is close to 304 stainless steel workpieces, on the surface of the workpiece, when the softened layer is removed, the matrix material will not be damaged, therefore, a good surface quality can be obtained; the hardness of silica is closer to 304 stainless steel, but silica is prone to gelation in the polishing liquid, which brings resistance to the abrasive and reduces the MRR, and the polished surface is damaged again. Silicon dioxide not only needs to remove the passive film on the stainless steel surface, but also cuts off the silicon-oxygen-silicon bond by mechanical action. Therefore, the surface quality of the workpiece after polishing with silicon dioxide abrasive is bad, and the material removal rate is also very low. The hardness of white corundum and diamond is much greater than that of 304 stainless steel workpiece. Therefore, after removing the softened layer on the workpiece surface, due to the high abrasive hardness, it will further cause damage to the polished surface and promote the surface roughness. The surface morphology before and after polishing is shown in figure 4.

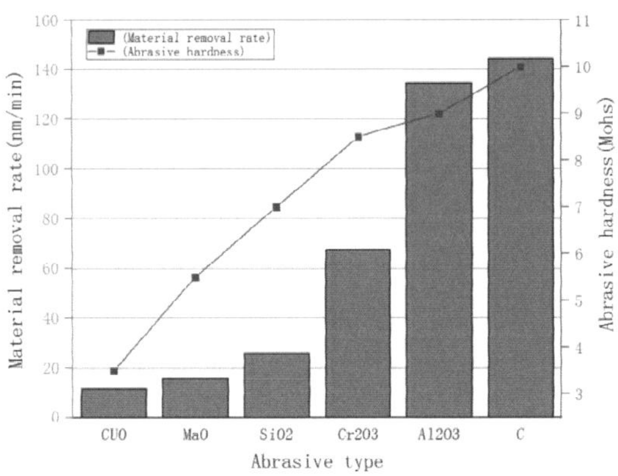


**Figure 2.** The effect of different abrasive hardness on the removal rate.

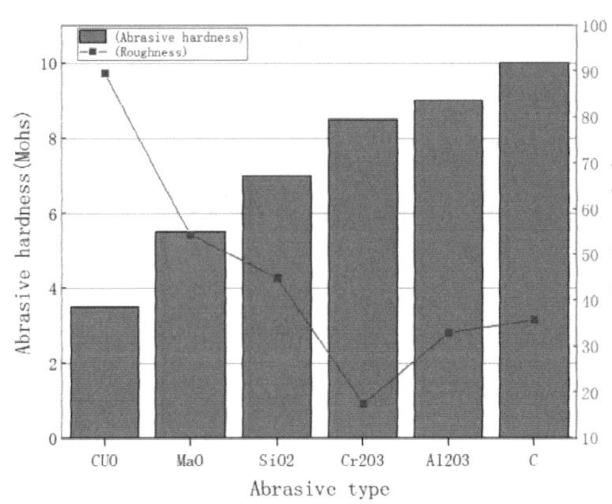


**Figure 3.** The influence of abrasive hardness on the surface roughness.

As shown in figure 4, the two-dimensional topography of the workpiece after polishing with different hardness abrasives. Figure 4(a) is the two-dimensional topography of the workpiece before polishing. From the figure 4(a), there are more large scratches on the sample surface. Figure 4(b) is the two-dimensional topography after polishing with copper oxide abrasive. Compared with the surface of the workpiece before polishing, it can be clearly seen that the scratches on the sample surface are evidently reduced, and the surface quality is effectively promoted; The hardness of magnesium oxide abrasive is less than 304 stainless steel, so the mechanical effect is small, the oxide layer on the sample surface cannot be completely removed and generate a residual oxide layer, finally resulting in poor surface quality, as shown in figure 4(c). The two-dimensional topography after polishing with silica abrasive is shown in the figure 4(d). By the figure 4(d), the surface quality is poor and the scratches are obvious, but compared to the use of copper oxide and magnesium oxide abrasives, the overall scratches are more uniform. This is because the hardness of silica abrasives is greater than 304 stainless steel. After removing the surface oxide layer, the silicon dioxide abrasive causes damage to the base material of 304 stainless steel, and because silicon dioxide is prone to gelation in the polishing liquid, it brings resistance to the abrasive and reduces the material removal rate; because the hardness of chromium oxide and white corundum is close to that of 304 stainless steel, it is only removed the oxide layer on the workpiece surface of the stainless steel, and the mechanical action of the abrasive will not damage the surface, so a better surface quality can be obtained, see the figure 4 (e) and the figure (f); the hardness of diamond is much greater than 304 stainless steel, so after polishing, it will penetrate the surface oxide layer, causing damage to the workpiece surface, and the surface quality is reduced. However, in the diamond polishing process, larger materials can be obtained. The removal rate can remove surface scratches as a whole [15].

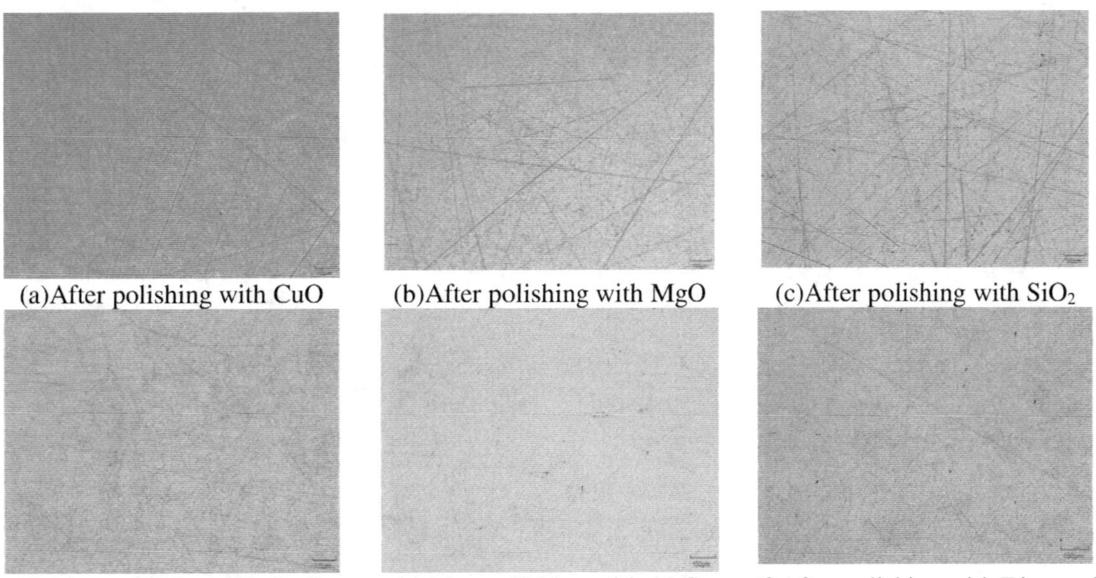

(a)After polishing with CuO (b)After polishing with MgO (c)After polishing with $SiO_2$

(d)After polishing with $Cr_2O_3$ (e)After polishing with $Al_2O_3$ (f)After polishing with Diamond

**Figure 4.** Two-dimensional surface map of the sample after polishing with different abrasives.

## 4. Conclusion

(1) When the hardness of the abrasive increases in the CMP process of 304 stainless steel, the MRR also increases. The surface quality decreases firstly and then rises with the increase of the abrasive hardness. When the chromium oxide abrasive is used, the surface roughness reaches the lowest, Ra=17.46 nm.

(2) Low-hardness copper oxide abrasives and magnesium oxide abrasives can only remove local scratches on the surface during the polishing process, with low material removal rate and poor surface quality; silica abrasives are easy to gel in the polishing liquid and hinders the cutting action of abrasive particles, resulting in low material removal rate and poor surface quality; the hardness of chromium oxide and white corundum is close to stainless steel, which can remove the surface oxide layer of the workpiece without damaging the surface of the workpiece and obtain good surface quality .

(3) Considering that chromium oxide contains $Cr^{2+}$will pollute the environment. Therefore, in comprehensive consideration, white corundum is used as abrasive in the polishing process of 304 stainless steel.

## References

[1] Cui H G, Wang Y C and Tang H 2018 Development of CMP polishing liquid for stainless steel *Machine Design and Manufacturing* **323**(1) 73-75.

[2] Zhang Q and Zhao Y W 2015 Effect of Process Parameters on Polishing Speed and Polishing Quality of SUS304 Stainless Steel *Journal of Jiangnan University (Natural Science Edition)* **14**(3) 321-325.

[3] Shen X A and Zhou Z Z 2009 Precision machining of free abrasive particles on stainless steel mirror surface *Aviation Precision Manufacturing Technology* **45**(4) 5-8.

[4] Yuan J L, Zhang F H, Dai Y F, et al. 2010 Research on the development of science and technology in the field of ultra-precision machining *Chinese Journal of Mechanical Engineering* **46**(15) 161-177.

[5] Li M, Yuan J L, Wu Z, et al. 2015 Research progress of ultra-precision machining methods for complex curved surface parts *Chinese Journal of Mechanical Engineering* **51**(5) 178-191.

[6] Abelev E, Smith A J, Hassel A W, et al. 2007 Potassium sorbate solutions as copper chemical mechanical planarization (CMP) based slurries *Electrochimica Acta* **52**(16) 5150-5158.

[7] Hu X K, Song Z T and Liu W L *A Precision Polishing Method for Stainless Steel Substrate* CN 101642893 A.

[8] Chen Y, Long R W, Chen Z G, et al. 2011 Synthesis and polishing characteristics of polystyrene/cerium oxide composite abrasives with different shell thicknesses *Chinese Journal of Mechanical Engineering* (14) 70-75.

[9] Mao M J, Liu S, Hu Z H, et al. 2019 The influence of different abrasives on the chemical mechanical polishing of cemented carbide blades *Lubrication and Sealing* **44**(10) 144-150.

[10] Yuan J L, Lü B H, Lin X, et al. 2002 Research on abrasives in the chemical–mechanical polishing process for silicon nitride balls *Journal of Materials Processing Technology* **129**(1-3) 171-175.

[11] Wang J P, Bai L S and Chu X F 2015 The Influence of Different Abrasives on the Chemical Mechanical Polishing of Glass Ceramics *Nanotechnology and Precision Engineering* **13**(1) 74-79.

[12] Zhang Q and Zhao Y W 2015 Effect of process parameters on polishing speed and polishing quality of SUS304 stainless steel *Journal of Jiangnan University* **14**(3) 321-325.

[13] Li Q, Chen S K, Peng Y N, et al. 2016 Research on 304 stainless steel chemical mechanical polishing process parameters *Diamond and Abrasive Engineering* **36**(5) 21-25.

[14] He X, Liu J, Gao S, et al. 2019 Preparation and polishing effect of mixed abrasive CMP polishing liquid *Henan Chemical Industry* **36**(5) 19-21.

[15] Lee H S, Kim D I, An J H, et al. 2010 Hybrid polishing mechanism of single crystal SiC using mixed abrasive slurry (MAS) *CIRP Annals - Manufacturing Technology* **59**(1) 333-336.

# Aerodynamic Characteristics of Supersonic Separator for Natural Gas Processing

**Chong Wang[1], Yizhen Li[1], Mengqi Zheng[2] and Ke Wang[2*]**

[1]Shandong Natural Gas Pipeline Co., Ltd, Shandong Jinan 250101, China
[2]Beijing Key Laboratory of Process Fluid Filtration and Separation, China University of Petroleum, Beijing 102249, China
Email: wang_ke@cup.edu.cn

**Abstract.** Natural gas is a key energy source to achieve the carbon peaking and carbon neutrality goals in China. Dehydration and de-hydrocarbon process is crucial for pipeline transportation of natural gas since water and hydrocarbons in the natural gas would condense to cause corrosion and freeze-blockage in the pipeline. The supersonic separator is innovative low-temperature separation equipment characterized by a simpler equipment structure and higher refrigeration efficiency than the traditional device. In this paper, the numerical analysis is employed to predict the performance of the supersonic separator. The results show that this separator can reach an advanced level compared with others. The separator cannot start-up as the inlet-outlet pressure ratio is lower than1.29, and the minimum pressure ratio needed for normal work is 1.39. Under normal operating conditions, total pressure loss is about 25.4%, the minimum temperature is about 213K, and the mean rotation speed is about 170m/s.
**Keywords.** Natural gas, Supersonic separator, Numerical simulation, dehydration.

## 1. Introduction

Natural gas has become the fastest-growing clean energy in the world due to its high quality, high efficiency, and green environmental protection. Especially in China, natural gas is a key energy source to achieve the carbon peaking and carbon neutrality goals. Generally, before entering the pipeline, natural gas needs to be dehydrated and de-hydrocarbon to prevent the condensation and precipitation of water and hydrocarbon components in the pipeline from corrosion and freezing blocking [1-3].

Traditional natural gas processing technology has disadvantages such as incomplete oil and gas separation, energy waste, low condensate recovery rate, large equipment space, high investment, heavy equipment maintenance workload, etc. In order to reduce operating costs and improve separation efficiency, it is urgent to develop new separation technology. Supersonic separation technology employs the theory of gas-liquid two-phase flow mechanics, thermodynamics and supersonic gas dynamics to achieve the purpose of dehumidification by consuming certain pressure energy [4-8]. Specifically, the LAVAL nozzle is first used to make high-pressure natural gas expand to a supersonic state. The temperature is lower than the condensation temperature of water vapour and heavy hydrocarbon components, so that water vapour and heavy hydrocarbon in natural gas condense in the expansion section of the nozzle, forming a large number of microdroplets and growing up. Then, a special device generates a high-speed rotating motion to achieve ultra-strong centrifugal separation. The high-density droplets move radial under the centrifugal position to leave the mainstream gas and gather towards the wall. Droplets gather on the wall surface to form a microlayer liquid film, which is carried by a small amount of airflow and leaves the separation device. The airflow in the centre is compressed from a supersonic to a subsonic state through the diffuser. Finally, all the kinetic energy of the airflow is converted into pressure energy and internal energy (higher temperature) and then exported. Compared with the traditional technology, it is a revolutionary major high and new technology. The technology has thermodynamic properties similar to that of a turbine expander. It is a compact pipe-like device that combines expansion, gas/liquid rotary centrifugal separation, and recompression. In contrast, the turbine expander converts pressure into axial power, while the hypersonic cyclone converts pressure into kinetic energy to achieve the condensation of the high dew point component [9-13]. Figure 1 is a schematic diagram of the principle and structure of the supersonic separator.

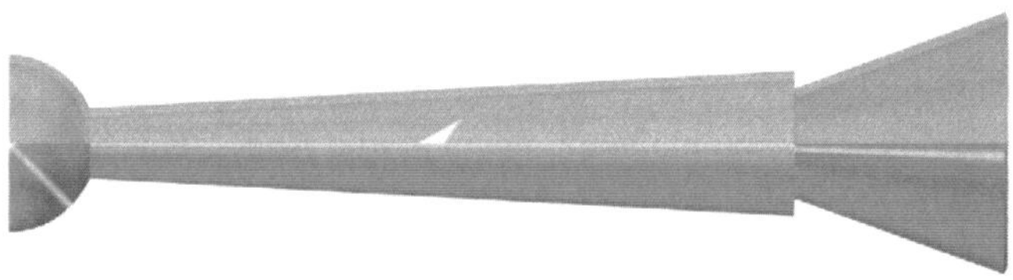

**Figure 1.** Schematic diagram of the structure of the supersonic separator.

TwisterBV company developed the first generation of Twister supersonic separators and proposed that the Laval nozzle could reduce the flow pressure to 30% of the initial pressure, forming a low-temperature and low-pressure environment. The back pressure portion can restore the outlet pressure to 60-80% of the initial pressure by separating the droplets from the delta wing by generating 30,000g of centrifugal acceleration. Based on the experimental verification, the dew point of water can be reduced to -9°C, and the dew point of hydrocarbon can be reduced to 10°C. Later on, TwisterBV developed the second-generation Twister separator with inlet cyclone. CFD simulation results show that the inlet swirl method can form a more stable eddy current field. According to experimental results, it is verified that the pressure loss is reduced from 33% to 25%, and the liquid separation efficiency reaches over 90% [14]. Russian scientists developed 3-S technology based on inlet swirl. Vortex generators are arranged in the inlet section, and the extended part can stabilize the flow, so that the vortex is in concentric state as far as possible and the total pressure loss caused by dissipation is reduced [15]. Empirical correlations have been constructed to predict separation efficiency. However, there is a dearth of data on the control of shock wave in the supersonic separator.

In this paper, the distribution of the flow field inside the supersonic separator is analyzed by numerical simulation method, and the distribution of pressure, temperature, and Mach number inside the supersonic separator is obtained. The factors that affect the total pressure loss and the impact of shock waves on the separation efficiency are also discussed.

## 2. Details of Numerical Simulation

### *2.1. Governing Equations*

In the present study, the flow in the pipe is simplified as a three-dimensional steady adiabatic compressible turbulent flow, as the governing equations of a fluid flow represent the mathematical statements of the conservation laws of physics. Therefore, for steady flow, the governing equations can be expressed as follow:

Continuity equation:

$$\frac{\partial u_x}{\partial x}+\frac{\partial u_y}{\partial y}+\frac{\partial u_z}{\partial z}=0 \tag{1}$$

Momentum equation:

$$u_x \frac{\partial u_i}{\partial x} + u_y \frac{\partial u_i}{\partial y} + u_z \frac{\partial u_i}{\partial z} = -\frac{1}{\rho}\frac{\partial p}{\partial x_i} + \upsilon\left(\frac{\partial^2 u_i}{\partial x^2} + \frac{\partial^2 u_i}{\partial y^2} + \frac{\partial^2 u_i}{\partial z^2}\right) \tag{2}$$

Among them, $u_i$ stands for $u_x$, $u_y$ and $u_z$, and $i$ stands for the selected coordinate.

Energy equation:

$$u_x \frac{\partial t}{\partial x} + u_y \frac{\partial t}{\partial y} + u_z \frac{\partial t}{\partial z} = \alpha\left(\frac{\partial^2 t}{\partial x^2} + \frac{\partial^2 t}{\partial y^2} + \frac{\partial^2 t}{\partial z^2}\right) \tag{3}$$

where $\rho$ is the density of the fluid, $v$ is the kinematic viscosity of the fluid, and $\alpha$ is the thermal diffusivity of the fluid.

Turbulent kinetic energy $k$ can be calculated by:

$$\frac{\partial}{\partial t}(\rho k) + \frac{\partial}{\partial \chi_i}(\rho k u_i) = \frac{\partial}{\partial \chi_j}\left(\alpha_k \mu_{eff} \frac{\partial k}{\partial \chi_j}\right) + G_k + G_b - \rho\varepsilon - Y_M + S_K \tag{4}$$

Turbulent kinetic energy dissipation $\varepsilon$ can be calculated by:

$$\frac{\partial}{\partial \tau}(\rho\varepsilon) + \frac{\partial}{\partial \chi_i}(\rho\varepsilon u_i) = \frac{\partial}{\partial \chi_j}\left(\alpha_k \mu_{eff} \frac{\partial \varepsilon}{\partial \chi_j}\right) + G_{1\varepsilon}\frac{\varepsilon}{k}(G_k + G_{3\varepsilon}G_b) - G_{2\varepsilon}\rho\frac{\varepsilon^2}{k} - R_\varepsilon + S_\varepsilon \tag{5}$$

where $G_k$ is the turbulent kinetic energy generated by the laminar velocity gradient, $G_b$ is the turbulent kinetic energy generated by buoyancy, $Y_M$ is the fluctuation generated by the transitional diffusion of the fluid, and $\alpha_k$ and $\alpha_\varepsilon$ are the turbulent Prandtl numbers of the $k$ equation and the $\varepsilon$ equation, respectively. Normally, $G_k$ can be defined as:

$$G_k = -\rho\overline{u'_i u'_i}\frac{\partial u_j}{\partial u_i} \tag{6}$$

Additional terms in the turbulent energy equation:

$$R_\varepsilon = \frac{C_\mu \rho \eta^3 (1 - \eta/\eta_0)}{1 + \beta\eta^3} \cdot \frac{\varepsilon^3}{k} \tag{7}$$

where $\eta = S_k/\varepsilon$, $\eta_0 = 4.38$, $\beta = 0.012$.

The turbulence model constant is given as:

$$G_{1\varepsilon}=1.42, \quad G_{2\varepsilon}=1.68, \quad C_\mu=0.09, \quad \sigma_k=1.0 \tag{8}$$

### *2.2. Computational Domain and Grid Construction*

Figure 2 shows the structure of the supersonic separator simulated in this paper, and its structure parameter is listed in table 1. The structure of the separator includes the tapered section, throat section, expansion section, rotation section, separation zone and pressure recovery section. The tetrahedral grids are deployed to the tapered section and rotation section and structured grids for the rest computational domain, as shown in figure 3. Referred to the grid-sensitive test illustrated in figure 4, a grid size of 0.87 million is sufficient to predict the single-phase flow in channels reasonably.

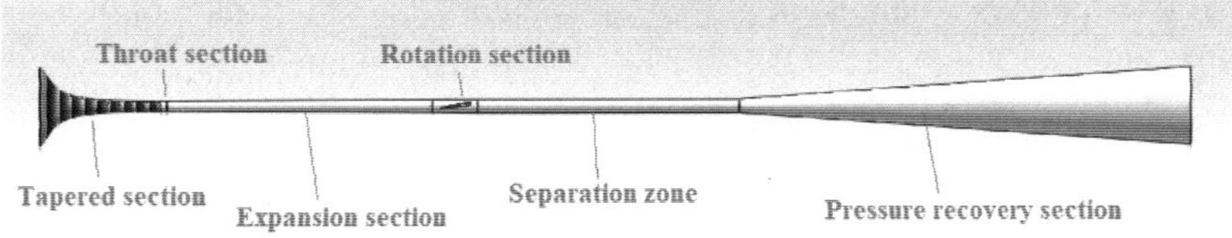


**Figure 2.** Structure of supersonic separator.

**Table 1.** The structure parameter of the supersonic separator.

| Parameter | | Value |
|---|---|---|
| Tapered section | Inlet diameter | 65 mm |
| | Curve type | Vitosinsky curve |
| | Length | 100 mm |
| Throat section | Diameter | 8.4 mm |
| | Length | 5 mm |
| Expansion section | Length | 300 mm |
| | Divergence angle | 0.22° |
| | Outlet diameter | 10.70 mm |
| Rotation section | Attack angle of delta wing | 8° |
| | Maximum thickness of wing root | 3 mm |
| Separation zone | Length | 138 mm |
| | Outlet diameter | 11 mm |
| Pressure recovery section | Outlet diameter | 65 mm |
| | Divergence angle | 4° |

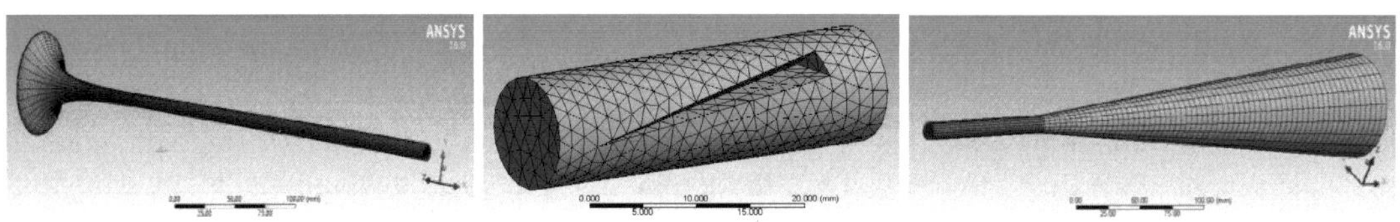

**Figure 3.** Mesh of supersonic separator.

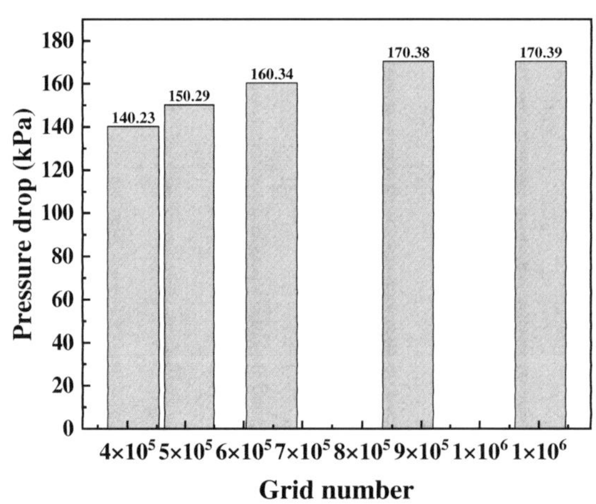

**Figure 4.** Grid independence verification.

*2.3. Boundary Conditions*

In the present study, ANSYS Fluent is employed to simulate the flow behaviors in the separator. The flow in the supersonic separator is very complicated, and the distribution of the flow field in the separator is very uneven due to transonic flow, shock wave and swirling flow. Taking methane as the working fluid, the simulation results indicate that the flow field obtained by using ideal gas property analysis is basically consistent with the flow structure obtained by using actual gas property analysis. As the convergence rate is very slow and the calculation time is several times longer than the ideal gas, methane ideal gas is selected as the working medium in the simulation, while the phase change and multiphase flow in the supersonic separator are ignored. Additionally, we believe that the existence time of fluid in the supersonic separation device is quite short, and the heat exchange between fluid and the environment is not considered. In this paper, the boundary conditions used in the calculation: the inlet of the simulated section is the pressure inlet condition, and the static pressure, total pressure and total temperature are 8 MPa, 8.000687 MPa and 287 K, respectively. The outlet is the pressure outlet condition, which is divided into gas phase outlet and liquid phase outlet. The static pressure and total temperature of both the gas phase outlet and liquid phase outlet are given at 3 MPa and 266 K. The inner wall of the separator and the surface of the delta wing are solid non-slip adiabatic walls, and the working medium is methane. Implicit and coupled solution methods are used in the calculation.

The separator is a typical turbulent flow. After the flow passes through the delta wing, the rotating flow is generated, which makes the streamline of the flow field bend, the distribution of pressure field and velocity field is not uniform, and there may be a backflow phenomenon. The standard $k$-$\varepsilon$ turbulence model is not suitable for swirling flow. The standard $k$-$\varepsilon$ model does not take into account turbulent vortex and cannot deal with flow line bending and swirling flow. The RNG $k$-$\varepsilon$ model improves the accuracy of the standard $k$-$\varepsilon$ model and provides an analytical formula for turbulent Prandtl numbers. Therefore, the RNG $k$-$\varepsilon$ model was used in this paper.

## 3. Results and Discussion

### *3.1. Flow Behaviors in a de Laval Nozzle*

Generally, a de Laval nozzle is employed to accelerate a hot, pressurized gas passing through it to a higher speed in the axial direction by converting the heat energy of the flow into kinetic energy. The flow behaviors in the de Laval nozzle are shown in figure 5. Accordingly, the flow in the throat reaches Mach=1 but is subsonic in the previous region. When the back pressure is low enough to have drawn a length of supersonic flow in the diverging section, a normal shock at the exit is terminated.

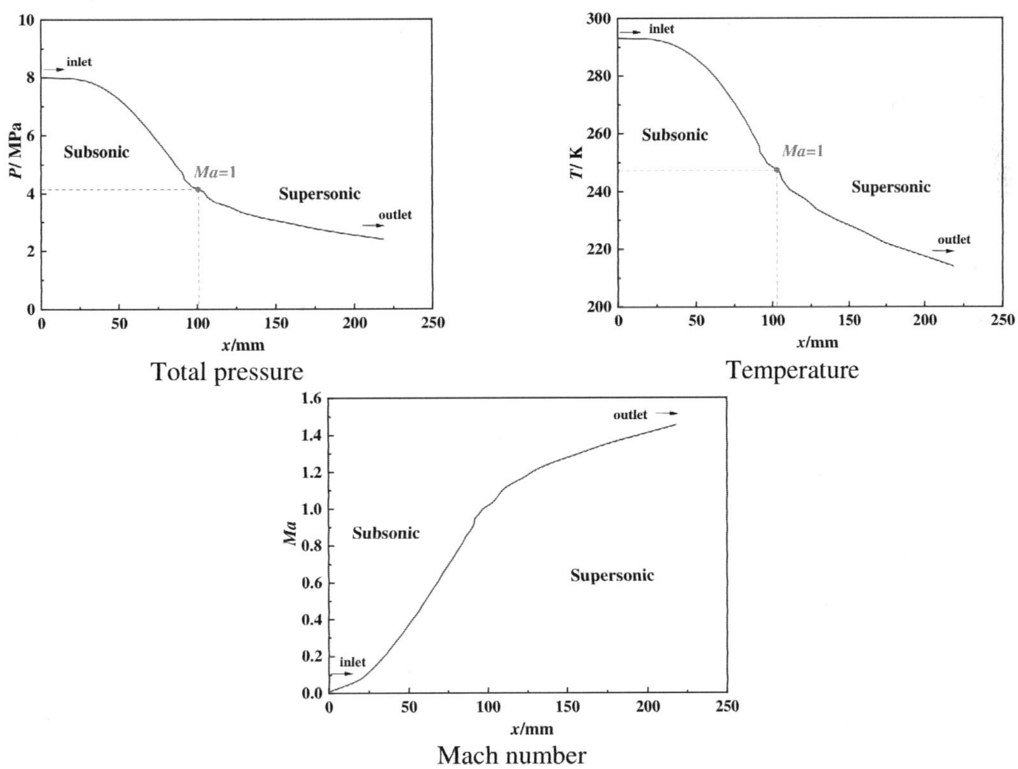


**Figure 5.** Variation of the flow field parameters.

Journal of Physics: Conference Series **2329** (2022) 012045 doi:10.1088/1742-6596/2329/1/012045

### *3.2. Flow Behaviors in the Separator*

Figure 6 shows the variation of the pressure and temperature in the separator. It can be inferred from the figure that the static pressure and temperature decrease to a minimum point and increase thereafter. Although the outlet temperature rises to be close to the inlet temperature, the outlet pressure is somewhat lower than the inlet pressure, indicating that the separator uses the loss of part of the pressure energy to achieve condensation and separation of water and heavy hydrocarbons. Figure 7 illustrates the variation of the static pressure and total pressure along the axis of the separator. The result indicates that the static pressure and total pressure of airflow in the nozzle section gradually decrease along the mainstream direction. In the swirl section, due to the great hindrance of the flow generated by the shock wave, the static pressure suddenly rises. The shock wave not only dissipates the kinetic energy but also increases the mechanical energy loss caused by the friction of the flow along the surface of the delta wing, resulting in a sudden drop in the total pressure. In the separation section, the static and total pressure change slightly. When the air enters the pressure recovery section, the static pressure increases, but the total pressure decreases by about 6% due to the shunt action of the discharge port and then rises in the backpressure to recover part of the total pressure.

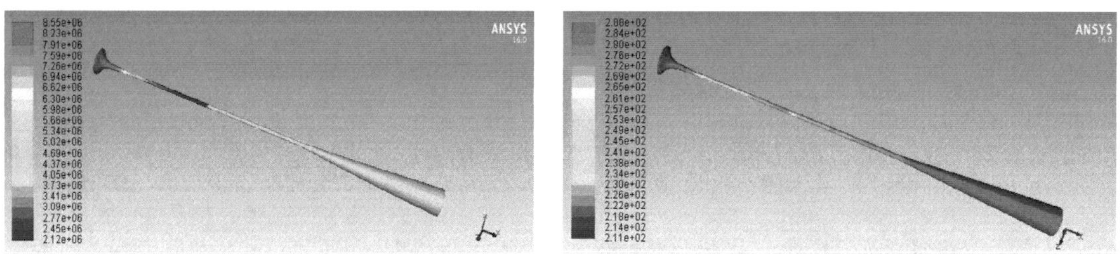


static pressure temperature

**Figure 6.** The distribution of the pressure and temperature in the separator.

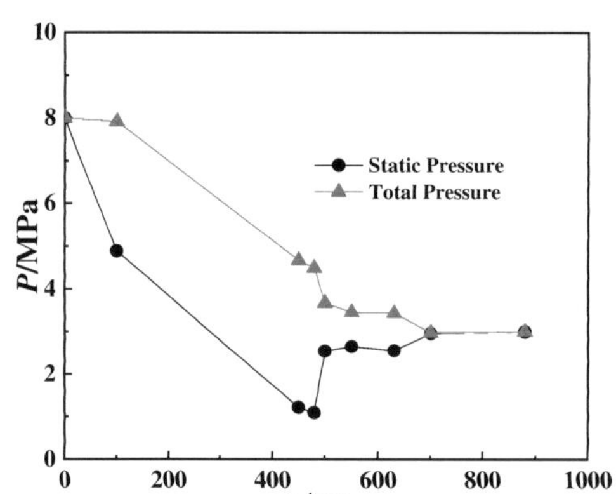


**Figure 7.** Variation of the static pressure and total pressure along the separator axis.

Table 2 lists the total pressure loss, pressure loss per unit, and ratio of total pressure loss in the nozzle section, swirl section, separation section and pressure recovery section in the separator. The results show that although the absolute value of total pressure loss and total pressure loss ratio is the largest in the nozzle section, the total pressure loss rate is small due to the long nozzle section (about 400mm). The loss of mechanical energy in the swirl segment is large, but the distance of the swirl segment is the smallest (about 50mm), so the total pressure loss rate in the swirl segment is the largest. In the separation and back pressure sections, the flow is subsonic, the swirl intensity decreases gradually, and the total pressure loss is small. Therefore, from the perspective of reducing the total pressure loss, the nozzle length and the configuration of the delta wing segment need to be further optimized to reduce the total pressure loss.

**Table 2.** Lost of total pressure in different section.

| | Nozzle section | Rotation section | Separation zone | Pressure recovery section |
|---|---|---|---|---|
| $\Delta P_{lost}^{'}$ (MPa) | 3.5 | 1 | 0.2 | 0.13 |
| $\Delta P_{lost}^{'}/L$ (MPa/m) | 0.78 | 2.22 | 0.14 | 0.05 |
| $\Delta P_{lost}^{'}/\Delta P_{lost}$ (%) | 70 | 20 | 4 | 2.6 |

Figure 8 depicts the distribution of the static pressure, temperature and Mach number on the internal $z$=0 plane of the supersonic separator. It can be seen from the figure that the pressure and temperature along the mainstream direction of airflow are gradually reduced in the nozzle. The static pressure decreases to 1 MPa at the trailing edge of the delta wing ($z$=495 mm), and the lowest temperature drops to 210 K. It is worth noting that the shock waves are generated due to the compressibility of the airflow. Accordingly, the shock wave leads to a sudden rise in the pressure and temperature, i.e., the pressure rises to 2.5 MPa while the temperature to 265 K. In the separation section, the delta wing with the angle of attack makes the air flow generate circumferential velocity, and the airflow deviates from the axial movement, forming a vortex. Driven by the axial velocity, the airflow moves spirally, and the distribution of pressure and temperature is asymmetric. After the airflow enters the backpressure, the pressure and temperature rise to 3MPa and 275 K, but the swirl strength of the airflow is not completely dissipated, so the pressure and temperature distribution in the backpressure section is not uniform. According to the distribution of the Mach number in the nozzle, the airflow expands and accelerates to supersonic speed. The maximum Mach number before the shock wave is about 1.6, indicating that the nozzle design is reasonable. Due to the hindrance of the shock wave to the airflow, the airflow in the separation section becomes a subsonic flow. After entering the back pressure chamber, the velocity further decreases, and the exit Mach number is about 0.1. Due to the effect of swirling flow, the airflow distribution is not uniform, and so is the Mach number distribution in the separation and pressure recovery sections.

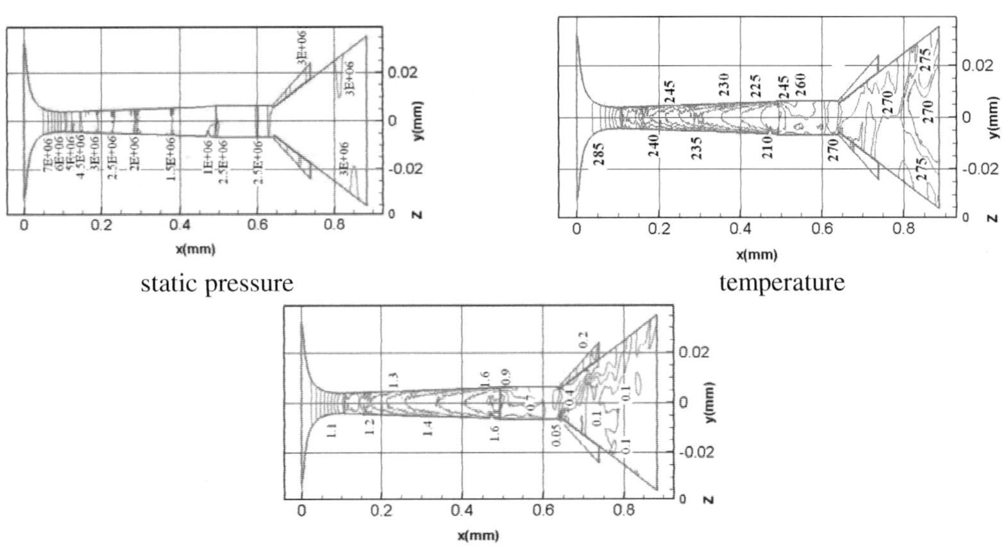


**Figure 8.** Variation of the flow parameters on the internal z=0 plane of the supersonic separator.

According to the simulation results, the shock wave appears at the trailing edge of the delta wing, which meets the requirements. Although the generation of shock waves reduces airflow speed and dissipates kinetic energy, it is beneficial to the whole separator if the shock wave is controlled behind the delta wing. This is because, in the separation section, the shock wave only reduces the axial velocity of the airflow but has little influence on the tangential velocity. Therefore, the swirl intensity decreases to a small degree, which increases the residence time of the airflow in the separation section and is conducive to improving the separation efficiency. The shock wave, without any doubt, is beneficial to improving the backpressure efficiency. However, disadvantages like adverse condensation and a large temperature rise intensifying the droplet re-evaporation are not conducive to dehumidification. As the shock waves will inevitably appear in the separator due to the structure of the delta wing, the flow passage and delta wing structure should be carefully designed to control the Mach number behind the delta wing is slightly greater than 1. Thus, the shock wave intensity can be reduced to improve the dehumidification efficiency.

For an airflow flying upwards at a certain angle of attack, under the action of pressure difference on the upper and lower surfaces of the delta wing, the airflow near the leading edge will stay close to the wall and flow from the pressure to the suction surface, resulting in separation of the boundary layer on the suction surface, thus forming vortexes that roll over from the pressure to the suction surface. Figure 9 shows the tangent velocity vector diagram of airflow on specific sections in the separation section. The length of the arrow indicates the magnitude of the absolute velocity, and the color indicates the magnitude of the tangent velocity. It can be clearly seen from the simulation results that the tangential vortex is strong when the air flows through the delta wing, and the tangential velocity is large far from the vortex center, while the velocity is small at the vortex center. It is worth noting that because there is a certain angle between the delta wing and the axis, the vortex generated is eccentric and has a higher intensity. As the vortex moves downstream, the intensity of the vortex decreases gradually, and the vortex center gets closer to the axis.

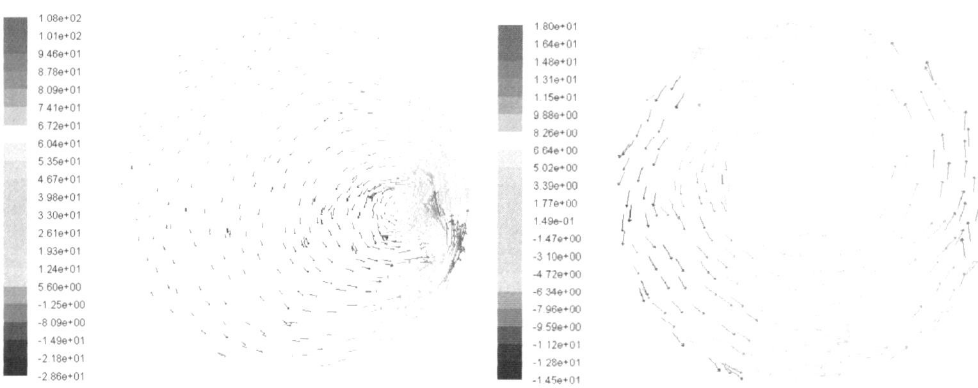


$x$=510 mm $x$=640 mm

**Figure 9.** Vector graphs of tangential velocities.

Figure 10 shows the variation of local maximum tangential velocity $V_{tan,max}$ and the ratio of local maximum centrifugal acceleration to gravitational acceleration $a_{max}/g$ from the leading edge of the delta wing to the outlet of the separator. As can be seen from the figure, the tangential velocity of airflow in the swirl segment gradually increases. In the delta wing region (495 mm away from the inlet of the separator), the local maximum tangential velocity $V_{tan,max}$ reaches the maximum value of 202 m/s, with the ratio $a_{max}/g$ of 620,000. Due to the generation of the shock wave, the local maximum tangential velocity $V_{tan,max}$ and $a_{max}/g$ decrease. After the airflow leaves the delta wing, the swirl intensity of the airflow decreases gradually due to swirl dissipation, and the tangential velocity of the airflow out of the separator decreases to 15 m/s.

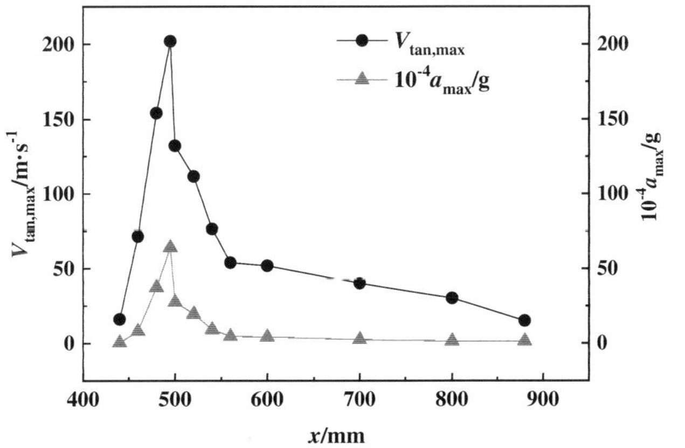


**Figure 10.** $V_{tan,max}$ and $a_m/g$ of bulk flow.

## 4. Conclusions

In this paper, the numerical simulation method is used to simulate the aerodynamic characteristics of the natural gas supersonic separator, and the distribution characteristics of pressure, temperature and Mach number in the separator are obtained. The results show that the maximum temperature drop in the separator reaches 73°C, creating conditions for the condensation of water and heavy hydrocarbon in natural gas components. The maximum Mach number is 1.6 near the trailing edge, and the maximum centrifugal acceleration is about 620,000 times of gravity acceleration, which provides conditions for the separation of condensed water droplets and heavy hydrocarbon droplets. In addition, the total pressure loss is the largest in the nozzle section, and the total pressure loss rate is the largest in the swirl section. Therefore, nozzle length and delta wing configuration require further optimization to reduce the total pressure loss. Inevitably, the shock wave appears at the rear of the delta wing, which is not conducive to dehumidification. In the present study, we propose a solution that the shock wave should be controlled in the separation section to reduce its strength by optimizing the nozzle size or reducing the process back pressure.

## Acknowledgments

The authors gratefully acknowledge research support from the Strategic Cooperation Technology Projects of CNPC and CUPB No. ZLZX2020-05, the Science Foundation of China University of Petroleum, Beijing No. ZX20200126.

## References

[1] Pau B, Kris A, Jamie S, et al. 2017 The Natural Gas Supply Chain: The Importance of Methane and Carbon Dioxide Emissions *ACS Sustainable Chem. Eng.* **5**(1) 3-20.
[2] Koku O, Perry S and Kim J K 2014 Techno-economic evaluation for the heat integration of vaporisation cold energy in natural gas processing *Appl. Energ.* **114** 250-261.
[3] Tian X and You F 2019 Carbon-neutral hybrid energy systems with deep water source cooling, biomass heating, and geothermal heat and power *Appl. Energ.* **250** 413-432.
[4] Liu H, Liu Z, Feng Y, et al. 2005 Characteristics of a Supersonic Swirling Dehydration System of Natural Gas *Chinese J. Chem. Eng.* **13**(1) 9-12.
[5] Alfyorov V, Bagirov L, Dmitriev L, et al. 2005 Supersonic nozzle efficiently separates natural gas components *Oil Gas J.* **103**(20) 53-58.
[6] Malyshkina M M 2008 The structure of gasdynamic flow in a supersonic separator of natural gas *High Temp.* **46**(1) 69-76.
[7] Karimi A and Abdi M A 2009 Selective dehydration of high-pressure natural gas using supersonic nozzles *Chem. Eng. Process.* **48**(1) 560-568.

[8] Shooshtari S and Shahsavand A 2017 Maximization of energy recovery inside supersonic separator in the presence of condensation and normal shock wave *Energ.* **120** 153-163.
[9] Zhang G, Wang F, Wang D, et al. 2019 Numerical study of the dehumidification structure optimization based on the modified model *Energ. Convers. Manage.* **181** 159-177.
[10] Cao X W and Bian J 2019 Supersonic separation technology for natural gas processing: A review *Chem. Eng. Process* **136** 138-151.
[11] Teixeira, Alexandre, Mendonca, et al. 2018 Recovery of thermodynamic hydrate inhibitors methanol, ethanol and MEG with supersonic separators in offshore natural gas processing *J. Nat. Gas Sci. Eng.* **52** 166-186.
[12] Jassim E, Abdi M A and Muzychka Y 2008 Computational Fluid Dynamics Study for Flow of Natural Gas through High-pressure Supersonic Nozzles: Part 1. Real Gas Effects and Shockwave *Petrol. Sci. Technol.* **26**(15) 1757-1772.
[13] Jassim E, Abdi M A and Muzychka Y 2008 Computational Fluid Dynamics Study for Flow of Natural Gas through High-pressure Supersonic Nozzles: Part 2. Nozzle Geometry and Vorticity *Petrol. Sci. Technol.* **26**(15) 1773-1785.
[14] Esmaeili A 2016 Supersonic Separation of Natural Gas Liquids by Twister Technology *Chem. Eng. Trans.* **56** 7-12.
[15] Wang X Q, Luo X M and Sun Y M 2007 Supersonic Separator: A New-type and High-effective Installations Used for Dehydration and De-hydrocarbon of Natural Gas *Nat. Gas Tech.* **5** 63-67 (in Chinese)

# Effect of Alkaline Earth Metals Doping on the Electronic Structure, Magnetic and Optical Properties of g-SiC Monolayer System by First Principles Calculation

**Deming Ma [1*], Hongbin Wei [1], Yang Shen [1], Zhen Cui [2], Enling Li [1] and Lin Zhang [1]**

[1] Department of Applied Physics, Xi'an University of Technology, No. 58 Yanxiang Road, Yanta District, Xi'an, 710054, China.
[2] School of Automation and Information Engineering, Xi'an University of Technology, No. 5 Jinhua South Road, Beilin District, Xi'an, 710048, China.
Email: xautmdm@163.com

**Abstract.** The band structure, electronic, magnetic, and optical properties of g-SiC monolayers doped with alkaline earth metals (Be, Mg, Ca, Sr, and Ba) are calculated by means of first principles based on density functional theory. Although the intrinsic g-SiC monolayer is nonmagnetic, it shows magnetic properties after doping with alkaline earth metals. The magnetic moments are, in the order of the above-named dopants, 1.583 $\mu_B$, 1.817 $\mu_B$, 2.000 $\mu_B$, 2.000 $\mu_B$, and 2.000 $\mu_B$. Charge transfer and the net spin charge occur mainly between the dopant atom and surrounding C atoms. The results show that the conductivity of g-SiC can be greatly improved by doping with alkaline earth metals, a technique that can be applied to the field of semiconductor spintronics. High absorption peaks in the ultraviolet indicate that the material has potential applications in UV optoelectronic devices.
**Key Words.** G-SiC; Alkaline earth metals; Dopant; First principles.

## 1. Introduction

Graphene was discovered in 2004 at the University of Manchester, UK, by Novoselov et al. [1]. Researchers followed up with numerous efforts to find other novel 2D nanomaterials, including conductors, semiconductors with varying bandgaps (e.g., $MoS_2$, GaN, and black phosphorus), and insulators (e.g., h-BN) [2-8]. These materials have broken the traditional perception of size for semiconductor electronic devices, and have been widely studied in optoelectronic, spintronics, and other devices [9, 10]. Increasingly, semiconductors with two-dimensional structures have been predicted and successfully prepared, and ultrathin SiC nanoflakes and nanosheets have been synthesized [11-13]. Theoretical studies based on first principles have shown that two-dimensional SiC has a graphene-like structure due to the $sp^2$ binding between Si and C, and shown that it can be stable [14].

Two-dimensional intrinsic SiC is a nonmagnetic semiconductor and has an indirect band gap. To expand its magnetic applications, methods such as doping, adsorption, construction of heterojunctions, and stressing have been used [15-21]. In 2010, E. Bekaroglu et al. revealed the dimensional effect of SiC. The phonon spectrum showed that two-dimensional SiC is stable in planar geometry, with a band gap of 2.53 eV [19]. In 2017, M. Luo et al. investigated the magnetic properties of SiC monolayers containing transition metals (Co, Cu, Mn, Fe, Ni) using density functional theory [18]. In 2019, Li et al. studied the co-doping of Mo and N on the magnetic local spin of two-dimensional SiC and found that Mo atoms can act as excellent local spin donor atoms in two-dimensional SiC [22]. In 2018, Deng et al. found that with the destruction of spatial symmetry in a matrix in two-dimensional materials, dopants will produce new quantum effects due to local magnetic anisotropy [23]. Baierle studied the adsorption of alkali metals (Li, K, and Na) and alkaline earth metals (Be, Mg, and Ca) on monolayer SiC using density functional theory, where the adsorbed atoms give rise to new and localized electronic levels [24]. However, there is little published research on nonmagnetic atoms doping 2D SiC materials, which can be treated theoretically and is of practical significance. In order to obtain local magnetic moments and magnetic ground states, dopants must be p-type and have large electronegativity differences from neighboring host atoms [25].

This work is a theoretical investigation of the doping of Group IIA elements (Be, Mg, Ca, Sr, and Ba) into an intrinsic g-SiC monolayer using density functional theory. Studying its electronic structure and its magnetic and optical properties and comparing its commonalities and characteristics will help to reveal possible applications in different fields, with potential theoretical and practical significance.

## 2. Calculation Method

The calculations are based on an iterative solution of the Kohn-Sham equations of density functional theory (DFT) in a plane wave basis set with the projector augmented wave (PAW) method [26]. The generalized gradient approximation (GGA) in the form of Perdew-Burke-Ernzerhof (PBE) [27] functions was used to describe the exchange-correlation interactions. The calculations have been done using VASP (Vienna Ab Initio Simulation Package) [28]. We set the plane wave cut-off energy to be 500 eV and a Γ-centered 3×3×1 Monkhorst-Pack k-point is used for geometry optimization. The convergence of energy is set at $10^{-5}$ eV, and the convergence of the force on each atom set to be less than 0.01 eV/Å. The vacuum-layer distance was set to 20 Å in the vertical direction.

## 3. Results and Discussion

### *3.1. The Structure and Properties of the Intrinsic g-SiC Monolayer*

The optimized lattice parameter and bond length of the SiC monolayer are taken to be 3.094 Å and 1.785 Å, respectively [14], consistent with experimental data [11].The optimized model of the intrinsic g-SiC monolayer is shown in figure 1(a). The band structure and the density of states of the structure are shown in figure 1(b). The calculation shows that the valence band maximum (VBM) is 2.36 eV at the K point, and the conduction band maximum (CBM) is -0.22 eV at the Γ point. Therefore, it finds that the intrinsic monolayer is an indirect band-gap semiconductor with a band gap $E_{gap}$ of 2.58 eV, which is similar to previous results [14, 22]. PBE functional and GGA approach usually underestimate the band gaps of semiconductors [29]. The intrinsic g-SiC monolayer is nonmagnetic.

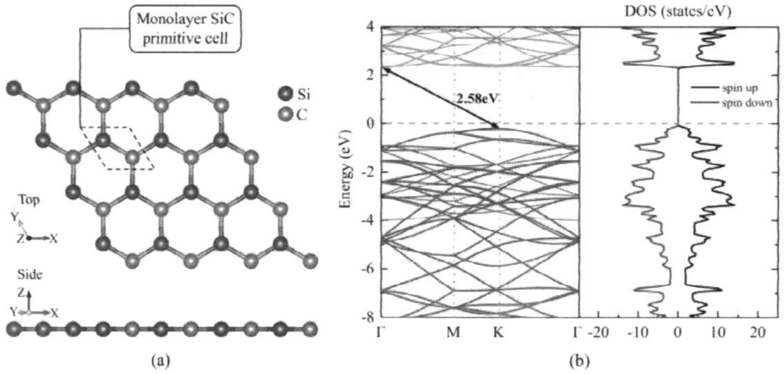


**Figure 1.** (a) Calculated crystalline structures and (b) band structure and density of states of an intrinsic g-SiC monolayer.

### *3.2. Structure of g-SiC Monolayer Doped with Alkaline Earth Metals*

Figure 2 shows the model of an alkaline earth metal-doped g-SiC monolayer. An alkaline earth metal atom X (Be, Mg, Ca, Sr, or Ba) replaces a Si atom or a C atom, the replacement being denoted by $X_{Si}$ or $X_C$, and the effect of position on the crystal structure is observed by different substitution doping of metal atoms.

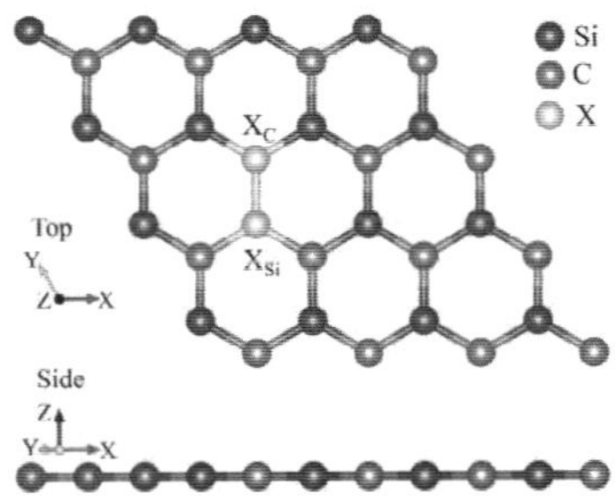


**Figure 2.** Alkaline earth metal atoms substituted for a Si or C atom of a g-SiC monolayer.

The binding energies and other parameters are shown in table 1. The binding energy of a g-SiC monolayer doped with different metals is calculated as:

$$E_b = E_T - E_0 + n(\mu_i - \mu_X) \quad (1)$$

In equation (1), $E_T$ is the total energy of the doped g-SiC monolayer and $E_0$ is the total energy of the intrinsic g-SiC monolayer; $n$ (= 1) is the number of substituted alkaline earth metal atoms; $\mu_i$ is the chemical potential of a single Si or C atom; and $\mu_X$ is the chemical potential of an alkaline earth metal atom. The analysis shows that the doping metal atoms replace Si atoms more easily than C atoms, because the system is more stable after replacing Si atoms.

**Table 1.** Binding energies ($E_b$), total magnetic moments ($\mu_{total}(\mu_B)$), and bond lengths in the alkaline earth metal atoms X (Be, Mg, Ca, Sr, and Ba) doping a g-SiC monolayer.

| Structure | Be | Mg | Ca | Sr | Ba |
|---|---|---|---|---|---|
| **$Si_{16}C_{15}X_1$** | | | | | |
| $E_b$ (eV) | -1.12 | -0.93 | -0.02 | 1.64 | 3.03 |
| $\mu_{total}(\mu_B)$ | -0.002 | 0.001 | 1.587 | 3.999 | 5.984 |
| X-Si (Å) | 1.95 | 2.17 | 2.33 | 2.41 | 2.46 |
| **$Si_{15}X_1C_{16}$** | | | | | |
| $E_b$ (eV) | -5.02 | -4.51 | -2.94 | -1.81 | -0.23 |
| $\mu_{total}(\mu_B)$ | 1.583 | 1.817 | 2.000 | 2.000 | 2.000 |
| X-C (Å) | 1.73 | 1.98 | 2.13 | 2.22 | 2.28 |

### *3.3. Electronic Structure*

The band structure resulting from substitution of a Si atom in a g-SiC monolayer with an alkaline earth metal atom is shown in figure 3; black and blue lines correspond to the spin-up and spin-down electron motion, respectively, and the red part shows the energy bands of the dopant atoms in the g-SiC monolayer. When Be and Mg atoms are doped in g-SiC [figure 3(b) and (c)], the spin-up and spin-down energy levels of the two systems all pass through the Fermi level. Compared with the intrinsic energy band in figure 3(a), the valence band and the conduction band both move up. The doping of Ca, Sr, and Ba atoms are shown in figure 3(d), (e), and (f). The band structure near the Fermi energy level is narrower, demonstrating that the electrons in this band structure are very localized and have a relatively large effective mass. Among all five doping structures, the g-SiC monolayer with a Ba atom occupying a Si site has the strongest hybridization.

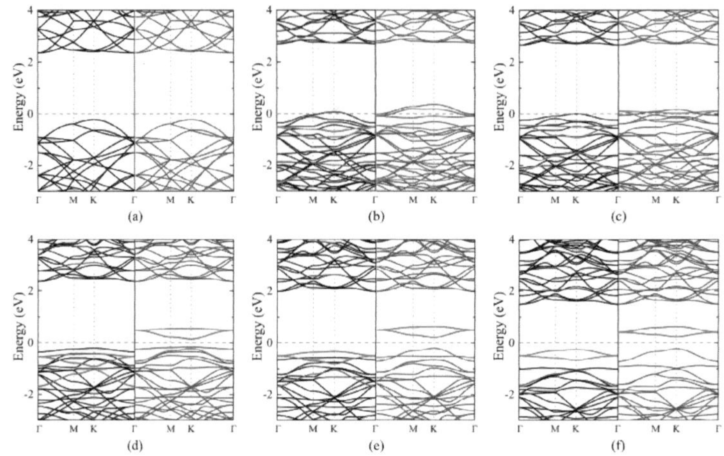


**Figure 3.** Band structures of (a) intrinsic, (b) Be-doped, (c) Mg-doped, (d) Ca-doped, (e) Sr-doped, and (f) Ba-doped g-SiC monolayers.

Figure 4 reveals the intrinsic g-SiC monolayer electronic properties and the effects on magnetism of different nonmagnetic metals by showing the total density of states (TDOS) and projected density of states (PDOS) for the above six cases. In comparison with the intrinsic g-SiC properties shown in figure 4(a), the spin-up and spin-down peaks shown in figure 4(b)-(f) doped with alkaline earth metal elements are no longer symmetric. These asymmetric impurity states are mainly derived from the hybridization of dopants (Be, Mg, Ca, Sr, and Ba) and their neighboring three C atoms. The contributing states of dopant atoms [figure 4(d)-(f)] are significantly increased compared with those shown in figure 4(b) and (c). It is noteworthy that the impurity levels increase the number of conduction electrons and conduction holes in the system, so that this material has potential application in electronic and optoelectronic devices.

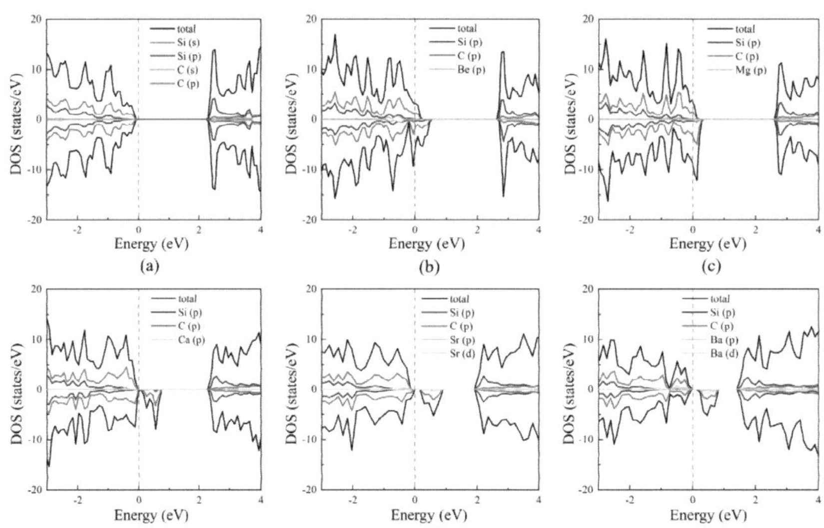


**Figure 4.** Density of states of (a) intrinsic, (b) Be-doped, (c) Mg-doped, (d) Ca-doped, (e) Sr-doped, and (f) Ba-doped g-SiC monolayers.

Next, the transfer charge of the system was analyzed by the specific spatial distribution of charge accumulation or depletion in the differential charge density:

$$\Delta\rho = \rho_{\text{total}} - \rho_0 - \rho_{\text{i}} \tag{2}$$

In equation (2), $\rho_{\text{total}}$, $\rho_0$ and $\rho_{\text{i}}$ are, respectively, the total charge density of an alkaline earth metal atom-doped g-SiC monolayer; the charge density of a g-SiC monolayer of one vacancy without dopants; and the charge density with one dopant atom. The yellow color indicates charge accumulation; the cyan color indicates depletion. It can be seen in figure 5 that after doping with alkaline earth metals, charge transfer occurs mainly around the doped atoms and three C atoms, showing a strong local state. In addition, the quantitative charge transfer of five doping systems has been calculated by Bader charge. Since charge moves from the dopant atom to the C atoms and gathers between the C and the dopant atom. The charge depletes mainly on the dopant atom, which acts as a donor. The accumulations of three host C atoms in the five studied systems acting as acceptors are, respectively, 6.967 |e|, 7.028 |e|, 6.635 |e|, 6.700 |e|, and 6.605|e|.

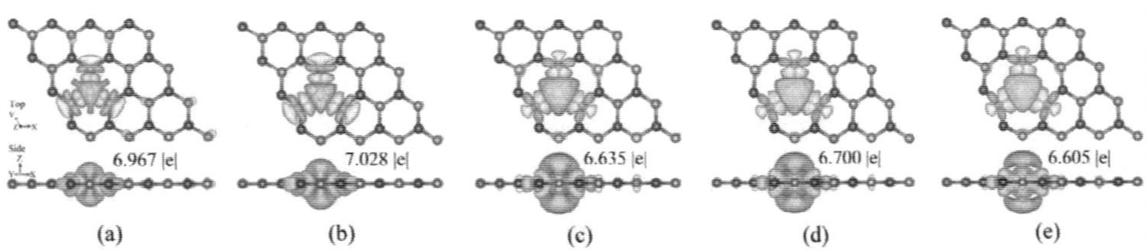


**Figure 5.** Differential charge densities of (a) Be-doped, (b) Mg-doped, (c) Ca-doped, (d) Sr-doped, and (e) Ba-doped g-SiC monolayers.

### *3.4. Magnetic Properties*

To further illuminate the spin-polarized electron density distributions for the alkaline earth metal-doped g-SiC system, we explored its net spin charge density distribution, as shown in figure 6. The net charge density is described by the following formula:

$$\Delta\rho = \rho_{\text{up}} - \rho_{\text{down}} \tag{3}$$

In equation (3), $\rho_{\text{up}}$ and $\rho_{\text{down}}$ represent the spin-up (red part) and spin-down (green part) charge densities, respectively. A major part of what is shown in red in figure 6 is distributed at three host C atoms adjacent to the dopant and is mainly derived from the contributions of the C atoms, showing strong local states. The total magnetic moments corresponding to the doped Be, Mg, Ca, Sr, and Ba atoms are, respectively, 1.583 $\mu_B$, 1.817 $\mu_B$, 2.000 $\mu_B$, 2.000 $\mu_B$, and 2.000 $\mu_B$.

For intrinsic g-SiC, both Si and C atoms belong to the IVA group. There are four electrons outside the electron core, but because of the high electronegativity of the C atom, charge is transferred from the Si atom to the C atom. The alkaline earth metal atoms contain two valence electrons, which are in 2s orbitals, and Si and C atoms both have four valence electrons. Therefore, when the Si atom is replaced by an alkaline earth metal atom, the metal atom only provides two electrons in the 2s orbitals, which makes the three surrounding C atoms unable to acquire the original coordination electrons.

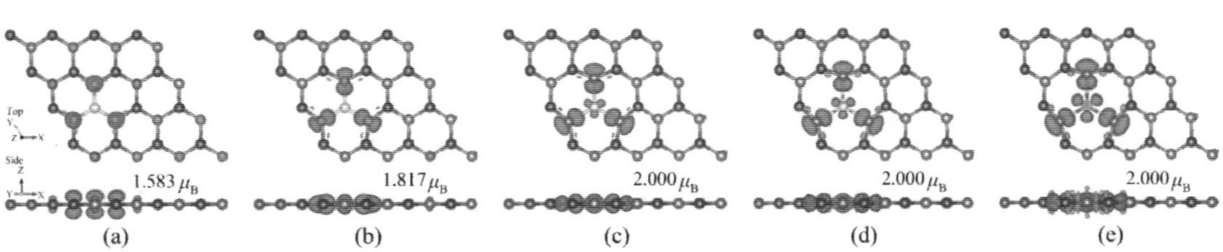


**Figure 6.** Net spin charge density distributions of (a) Be-doped, (b) Mg-doped, (c) Ca-doped, (d) Sr-doped, and (e) Ba-doped g-SiC monolayers.

### 3.5. Optical Properties

The optical properties of materials are represented by the transverse dielectric function $\varepsilon(\omega)$ [30, 31]. equation (4) is an expression for the dielectric function, where the real part is $\varepsilon_1(\omega)$, the imaginary part is $\varepsilon_2(\omega)$, and $\omega$ is the photon frequency of the incident light. In additional, the absorption coefficient $\alpha(\omega)$ [equation (5)] can be obtained from the $\varepsilon_1(\omega)$ and $\varepsilon_2(\omega)$ [32]. The optical absorption spectrum of alkaline earth metals doped g-SiC monolayers are shown in figure 7.

$$\varepsilon(\omega) = \varepsilon_1(\omega) + i\varepsilon_2(\omega) \tag{4}$$

$$\alpha(\omega) = \sqrt{2}\omega\left[\frac{\sqrt{\varepsilon_1^2(\omega)+\varepsilon_2^2(\omega)}-\varepsilon_1(\omega)}{2}\right]^{\frac{1}{2}} \tag{5}$$

Both the absorption spectra of the intrinsic g-SiC monolayer and the doped systems have strong absorption in the ultraviolet region. Comparing the doped (coloured lines) with the intrinsic spectra (black lines), the doping of Be and Mg atoms results in strong absorption peaks in the UV region at 145.18 nm and 144.35 nm, with intensities of $8.44\times10^5$ nm$^{-1}$ and $8.42\times10^5$ nm$^{-1}$, respectively, which are better than those of the intrinsic g-SiC monolayer. The positions of the absorption peaks of all five doped systems are blue shifted. The doping leads to many electron-hole interactions in systems, resulting in large exciton energy. The studies will provide theoretical help for potential applications of two-dimensional SiC in spintronics and UV optoelectronic devices such as UV exciton lasers.

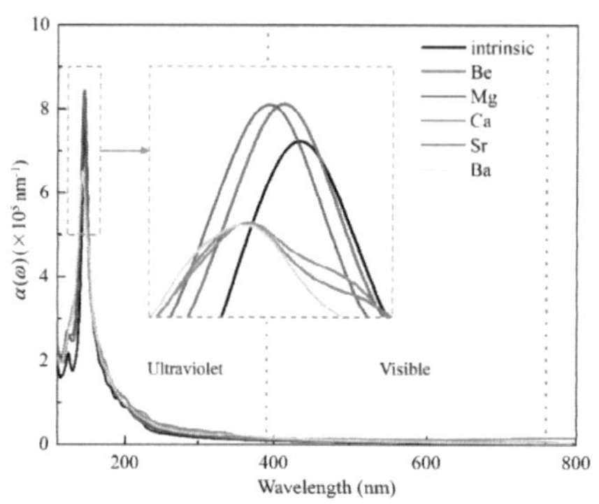


**Figure 7.** Optical absorption spectra of alkaline-earth metals doped g-SiC monolayers.

## 4. Conclusion

Using first principles calculations, we calculated the properties of g-SiC monolayers—intrinsic and doped with alkaline earth metals (Be, Mg, Ca, Sr, and Ba). Our results can be summarized as follows:

(1) We have considered different doping sites as well as different dopant atoms. The dopants introduce impurity energy levels near the Fermi energy level, which increases the probability of electron leap.

(2) The larger the atomic radius, the more pronounced the distortion of the structure near the dopant site, the weaker the binding ability of the nucleus to the electrons outside the nucleus, and the stronger the interaction between the outermost electron of the dopant atoms and the three neighboring C atoms.

(3) The doping of alkaline earth metals all shows magnetic properties, which are derived from the doped atoms and their surrounding C atoms. The doping can effectively modulate the spin properties of g-SiC monolayer.

(4) Compared with the intrinsic g-SiC monolayer, the optical absorption spectrum of the doped system shows a significant blue shift, and the doping of Be and Mg atoms results in strong

absorption peaks in the UV region, which are higher than the peak absorption spectra of the intrinsic g-SiC monolayer.

**Acknowledgments**
Scientific Research Program Funded by the National Natural Science Foundation of China (Grant No. 11904285), the Natural Science Basic Research Program of Shaanxi Province (Grant No. 2021JM-336), and Natural Science Foundation of Shaanxi Provincial Education Department (Grant No.21JK0789).

**References**

[1] Novoselov K S, Geim A K, Morozov S V, Jiang D E, Zhang Y, Dubonos S V, Grigorieva I V and Firsov A A 2004 *Science* **306**(5696) 666-9
[2] Li X, Tao L, Chen Z, Fang H, Li X, Wang X, Xu J and Zhu H 2017 *Appl. Phys. Rev*. **4**(2) 021306
[3] Dean C R, Young A F, Meric I, Lee C, Wang L, Sorgenfrei S, Watanabe K, Taniguchi T, Kim P, Shepard K L and Hone J 2010 *Nat. Nanotechnol.* **5**(10) 722-6
[4] Splendiani A, Sun L, Zhang Y, Li T, Kim J, Chim C Y, Galli G and Wang F 2010 *Nano Lett.* **10**(4) 1271-5
[5] Zhao L, Chang H, Zhao W, Luan Z, Tian X, Tan C and Huang Y 2019 *Superlattice. Microst.* **130** 93-102
[6] Menezes M G and Capaz R B 2018 *Comp. Mater. Sci.* **143** 411-7
[7] Bai K, Cui Z, Li E, Ding Y, Zheng J, Liu C and Zheng Y 2020 *Vacuum* **175** 109293
[8] Ma X, Lv Y, Xu J, Liu Y, Zhang R and Zhu Y 2012 *J. Phys. Chem. C* **116**(44) 23485-93
[9] Xu H, Akbari M K and Zhuiykov S 2021 *Nanoscale Res. Lett.* **16**(1) 94
[10] Chaves A, Azadani J G, Alsalman H, Da Costa D R, Frisenda R, Chaves A J, Song S H, Kim Y D, Koester S J, He D, Zhou J, Castellanos-Gomez A, Peeters F M, Liu Z, Hinkle C L, Oh S-H, Ye P D, Lee Y H, Avouris Ph, Wang X and Low T 2020 *Npj 2D Mater. Appl.* **4(1)** 29
[11] Lin S S 2012 *J. Phys. Chem. C* **116**(6) 3951-5
[12] Chabi S, Chang H, Xia Y and Zhu Y 2016 *Nanotechnology* **27**(7) 075602
[13] Susi T, Skákalová V, Mittelberger A, Kotrusz P, Hulman M, Pennycook T J, Mangler C, Kotakoski J and Meyer J C 2017 *Sci. Rep.* **7**(1) 4399
[14] Bekaroglu E, Topsakal M, Cahangirov S and Ciraci S 2010 *Phys. Rev. B* **81**(7) 075433
[15] Luo M, Shen Y H and Yin T L 2017 *J. Supercond. Nov. Magn.* **30**(5) 1269-73
[16] Zhao Q, Xiong Z, Qin Z, Chen L, Wu N and Li X 2016 *J. Phys. Chem. Solids* **91** 1-6
[17] Drissi L B and El Yahyaoui F 2017 *B. Mater. Sci.* **40**(6) 1081-6
[18] Luo M, Shen Y H and Yin T L 2017 *Physica E Low Dimens. Syst. Nanostruct.* **85** 280-4
[19] Hussain T, Hankel M and Searles D J 2017 *J. Phys. Chem. C* **121**(27) 14393-400
[20] Arellano L G, de Santiago F, Miranda Á, Salazar F, Trejo A, Pérez L A and Cruz-Irisson M 2021 *Int. J. Hydrog. Energy* **46**(38) 20266-79
[21] Luo M, Xu Y E and Song Y X 2018 *J. Supercond. Nov. Magn.* **31**(2) 597-601
[22] Li W, Yan Z, Ban L, Zhai X, Chen L, Liu W, Wang Y and Zhou G 2020 *Appl. Surf. Sci.* **509** 145193
[23] Deng X, Yang H, Qi S, Xu X and Qiao Z 2018 *Front. Phys* **13**(5) 137308
[24] Baierle R J, Rupp C J and Anversa J 2018 *Appl. Surf. Sci.* **435** 338-45
[25] Liu Z and Ni J 2010 *Sci. China Phys. Mech.* **53**(1) 1-10
[26] Kresse G and Joubert D 1999 *Phys. Rev. B* **59**(3) 1758
[27] Perdew J P, Burke K and Ernzerhof M 1996 *Phys. Rev. Lett.* **77**(18) 3865-8
[28] Kresse G and Furthmüller J 1996 *Comp. Mater. Sci.* **6**(1) 15-50
[29] Tran F and Blaha P 2009 *Phys. Rev. Lett.* **102**(22) 226401
[30] Ehrenreich H and Cohen M H 1959 *Phys. Rev.* **115**(4) 786-90
[31] Toll J S 1956 *Phys. Rev.* **104**(6) 1760-70
[32] Fox M 2002 *Optical Properties of Solids* vol 3, ed G F Bertsch (New York: Oxford University Press) p 305

# Preparation and Characterization of Fluorine-free Superhydrophobic and UV-resistant Cotton Fabric

**Runlian Liu [1], Pingping He [1], Zhongwei Xiao[1], Shiwei Li [1,2, a*] and Shuguang Bi[1, b*]**

[1] Hubei Key Laboratory of Biomass Fibers and Eco-dyeing & Finishing, Wuhan research center of eco-dyeing &finishing and functional textile, College of Chemistry and Chemical Engineering, Wuhan Textile University, Wuhan 430200, China.
[2] Jiangsu Engineering Research Center of Textile Dyeing and Printing for Energy Conservation, Discharge Reduction and Cleaner Production (ERC), Soochow University, Suzhou 215123, China.
Email: [a] lishiwei1988@126.com, [b] SGBi@wtu.edu.cn

**Abstract.** In this work, the superhydrophobic and UV-resistant cotton fabrics were fabricated via combining the nano titanium dioxide ($TiO_2$) and silicone oil. The nano-$TiO_2$ makes cotton fabrics possess outstanding ultraviolet (UV) resistance property, while increases the micro and nano size rough structure on the surface of cotton fabric. Then the cotton fabric was hydrophobically treated with silicone oil to make it superhydrophobic. The chemical constituents, structures, superhydrophobic and UV-resistant performance of cotton fabrics were analyzed. The cotton fabric treated with titanium dioxide and silicone oil not only showed excellent superhydrophobicity, but also displayed good UV-resistant ability. Water contact angle and UPF value can reach 161.32° and 55.17 respectively. Such multifunctional cotton fabric has broad commercial application prospects.
**Keywords.** $TiO_2$, superhydrophobic, UV-resistant, cotton fabric.

## 1. Introduction

Textiles are omnipresent materials that protect our bodies from the environment. Cotton is widely used in all kinds of textiles. It is because cotton has many advantages, such as softness, breathability, biodegradability and extreme moisture absorbance. However, the main disadvantages of cotton fabrics are that they are easily stained vulnerability to bacterial and microbial attack, and poor shielding to ultraviolet radiation. In order to overcome these shortcomings of cotton and expand its application field, it is necessary to endow cotton fabrics with different functions through finishing [1].

Inspired by the lotus leaves phenomenon in nature, researchers found that superhydrophobicity is due to the combination of the low surface energy materials and the micro/nano rough surface [2]. Therefore, the superhydrophobic surface can be obtained by constructing a structure similar to lotus leaves on the surface of the material [3, 4]. As markets in outdoor textiles have been expanded, multifunctional textiles have got more great interests continuously, in which superhydrophobic textiles with UV-resistant property that can broaden practical areas of the finial product [5].

$TiO_2$ nanoparticles have many advantages, which are widely used in textile field [6-8]. In all kinds of substances, fluorine compounds are extensively used for superhydrophobic properties due to their ultra-low surface energy provided by fluorine groups. However, fluorinated chemicals have certain disadvantages. Consequently, fluorine-free alternatives are usually used to prepare superhydrophobic self-cleaning cotton textiles [9].

In this work, the cotton fabric with superhydrophobic and UV-resistant properties were prepared by combining the nano-$TiO_2$ and subsequently hydrophobized with linear bulk silicone oil via a conventional pad-dry-cure method. Firstly, the 3-aminopropyltriethoxy silane-coupling agent (KH-550) was applied to modify nano-$TiO_2$ to increase its dispersibility. Then the modified nano-$TiO_2$ is treated to the cotton fabric surface. The modified nano-$TiO_2$ not only endowed the fabric with UV-resistant property, but also constructed micro/nano rough structure on cotton fabric surface. Finally, linear bulk silicone oil was applied to the cotton fabric surface. After hydrophobic treatment of linear bulk silicone oil, superhydrophobic cotton fabric was obtained. The chemical constituents, structures, superhydrophobic and UV-resistant properties of treated fabrics were studied.

## 2. Experimental

### *2.1. Materials and Reagents*

Nano-$TiO_2$ was provided by Aladdin Industrial Corporation. Linear silicone oil 0156 was purchased from Dow Corning Corporation. (3-Aminopropyl) triethoxysilane (KH550) was purchased from Aladdin Industrial Corporation. Other chemicals were used without further purification. Cotton fabrics were purchased from local markets.

### *2.2. Synthesis of Modified Nano-$TiO_2$*

A certain amount of KH550 was dropped into the mixture of distilled water and ethanol, and the solution was poured into a round-bottom flask for hydrolysis for 1h. Then, dry nano-$TiO_2$ particles and anhydrous ethanol were added into the round-bottom flask for ultrasonic dispersion for 1h. Adjust the pH value of the solution to 4 with acetic acid. Finally, the round-bottom flask was placed in 75°C water bath for 4h to obtain the modified nano-$TiO_2$.

### *2.3. Preparation of Superhydrophobic Cotton Fabric*

The cotton fabric was immersed in the modified nano-$TiO_2$ dispersion, two dipping and two rolling, drying at 80°C for 5min, baking at 120°C for 3min, and the modified nano-$TiO_2$ treated cotton fabric was obtained. The silicone oil was emulsified on the high shear dispersion emulsifier, and the modified nano-$TiO_2$ treated cotton was immersed in the silicone oil for 1h, then obtained the superhydrophobic cotton fabric after dried 1h at 120°C.

### *2.4. Characterization*

The chemical constituents of cotton fabric were determined by Tensor 27 FTIR spectroscopy. The crystal structures of cotton fabric can be obtained by Rigaku Ultima IV X-ray diffractometry. The water contact angle (WCA) was carried out by utilizing a professional DSA20 contact angle measurement instrument with 5μL deionized water droplet. UV-resistant performance of cotton fabric was tested via UV2000F ultraviolet transmittance spectrometer.

## 3. Results and Discussion

### *3.1. FTIR Analysis*

FTIR analysis was used to verify the deposition effect of modified nano-$TiO_2$ and silicone oil over cotton fabric. Figure 1 showed the FTIR spectra of pristine cotton, $TiO_2$@cotton and $TiO_2$-silicone oil@cotton. The FTIR spectrum of pristine cotton showed some absorption peaks at 3425 $cm^{-1}$, 2900 $cm^{-1}$, 2845 $cm^{-1}$, 1433 $cm^{-1}$, 1312 $cm^{-1}$,1166 $cm^{-1}$, 1112 $cm^{-1}$, 1063 $cm^{-1}$, and 1020 $cm^{-1}$, which were all the typical vibrational modes of cellulose [10, 11]. There are new peaks at 498~669 $cm^{-1}$ in the FTIR spectra of $TiO_2$@cotton and $TiO_2$-silicone oil@cotton $TiO_2$-silicone oil @cotton, which corresponds to Ti-O and Ti-O-Ti characteristic peaks, confirmed modified nano $TiO_2$ was deposited on the cotton fabric surface. Furthermore, the peaks at 498 $cm^{-1}$ which corresponds to Si-O bending extension vibration peak in the FTIR spectra of $TiO_2$-silicone oil @cotton, confirmed silicone oil was successfully attached onto the $TiO_2$@cotton.

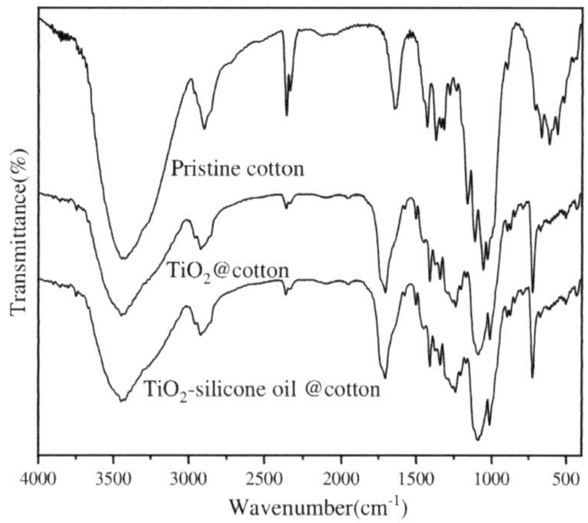


**Figure 1.** FTIR of pristine cotton, $TiO_2$@cotton and $TiO_2$-silicone oil @cotton.

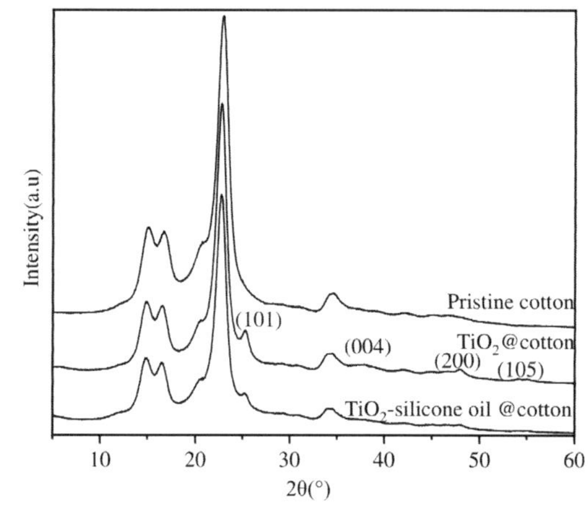


**Figure 2.** The XRD of pristine cotton, $TiO_2$@cotton and $TiO_2$-silicone oil @cotton.

*3.2. XRD Analysis*

The crystalline phases of pristine cotton, $TiO_2$@cotton and $TiO_2$-silicone oil @cotton have been analyzed by XRD presented in figure 2. Figure 2 shows that the pristine cotton, $TiO_2$@cotton and $TiO_2$-silicone oil @cotton has typical cellulose I crystalline form, such as 2θ diffraction peaks at 15.00°, 16.68°, 22.86°, and 34.55°. It reveals that the superhydrophobic functional finishing does not change the main crystalline form of cotton fabric. Characteristic peaks of $TiO_2$@cotton and $TiO_2$-silicone oil@cotton appeared at 2θ=25.31°, 37.99°, 48.07° and 54.72° correspond to the (101), (004), (200) and (105) crystal faces of anatase, respectively [12, 13]. It has been demonstrated that modified nano-$TiO_2$ was successfully deposited onto the cotton fabric surface.

*3.3. Surface Wettability*

Figure 3 shows the WCA of pristine cotton and the treated cotton fabric. It was observed that the pristine cotton, $TiO_2$@cotton and Silicone oil@cotton fabrics were highly hydrophilic and completely absorbed the water droplets. The WCA results showed that the wetting behavior of cotton could not be significantly changed by $TiO_2$ or silicone oil alone, and the WCA of fabric was 0°. However, the wettability of cotton changed from completely hydrophilic to hydrophobic by a combination of $TiO_2$ and silicone oil. The WCA of $TiO_2$-silicone oil@cotton was 161.32°, which show that it has excellent superhydrophobic properties. The surface wettability results showed that the superhydrophobic cotton fabric could not be obtained by constructing micro/nano rough structure or low surface energy material treatment on the cotton fabric surface alone. The superhydrophobic properties of cotton fabric must be obtained by constructing micro/nano rough structure and treating with low surface energy material simultaneously [14].

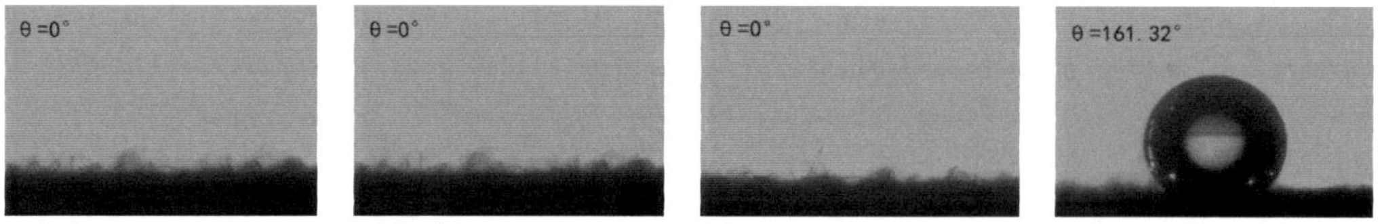


Pristine cotton TiO₂@cotton Silicone oil@cotton TiO₂-silicone oil@cotton

**Figure 3.** The water contact angles of pristine cotton and the treated cotton.

### *3.4. UV-Resistant Property*

The superhydrophobic and UV-resistant cotton fabric was prepared by coating the nano-$TiO_2$ followed by silicone oil modification. The UV transmittances curves of the different cotton and $TiO_2$-silicone oil@cotton were shown in figure 4 and figure 5, respectively. The UPF values of cotton fabric are shown in table 1. Apparently, the UV transmittances of treated cotton fabric is lower than the pristine cotton. The UV transmittances of $TiO_2$@cotton and $TiO_2$-silicone oil@cotton is lower than the silicone oil@cotton. And the UV transmittances of $TiO_2$-silicone oil@cotton is lowest, which possess the best UV-resistant property. As shown in figure 5, as the increase of $TiO_2$ amount, the UV transmittances of $TiO_2$ silicone oil@cotton notably decreased. It indicated that UV-resistant property of the cotton was improved when $TiO_2$ amount increased. It can be seen from table 1, the UPF value of pristine fabric were 4.67, presenting a poor UV-resistant property. The UPF value of silicone oil@cotton, $TiO_2$@cotton and $TiO_2$-silicone oil@cotton were 5.57, 20.64 and 42.84, respectively. The results indicated the $TiO_2$-silicone oil@cotton fabric had excellent UV-resistant property. While the $TiO_2$ amount was 0.5g, the UPF of the treated fabric reached 20.63. With the increase of $TiO_2$ amount, the UV-resistant property of the fabric was improved. When the $TiO_2$ amount was 3g, the UPF of the treated fabric reached 55.17. The test results exhibit that the cotton fabric has excellent UV-resistant property. The pristine cotton fabric shows a very poor UV-resistant property owing to the inability of cellulose in UV range absorption. The effectiveness of UV protection is attributed to the scattering properties and high UV absorption of nano-$TiO_2$ located majorly on the coating surface [15-17]. To sum up, the cooperation of $TiO_2$ surface rough micro/nano structure and low surface energy layer with silicone oil imparted good UV-resistant property to the $TiO_2$-silicone oil@cotton fabric.

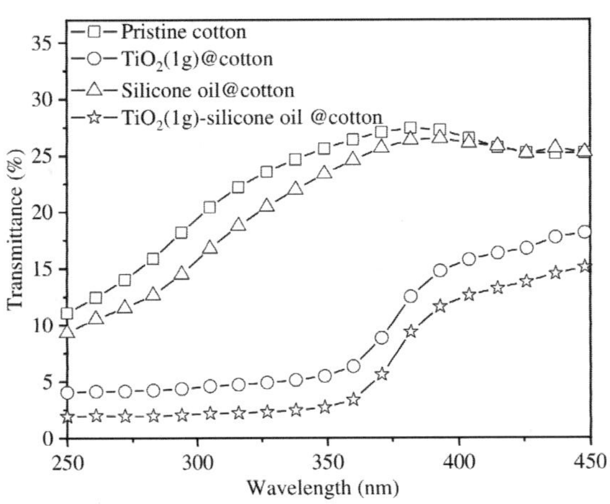


**Figure 4.** UV transmittance curves of the different cotton.

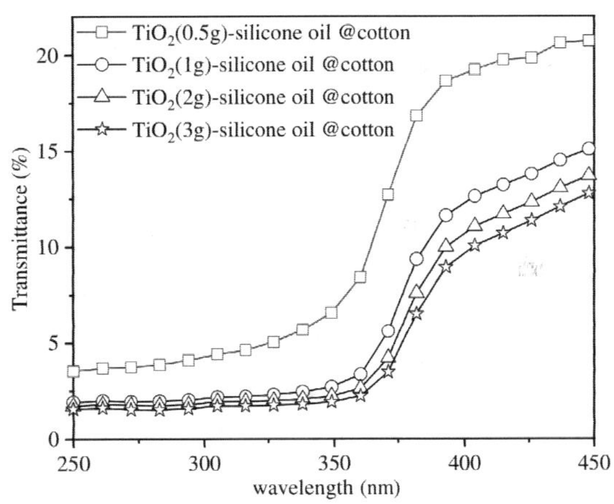


**Figure 5.** UV transmittance curves of $TiO_2$-silicone oil@cotton.

**Table 1**. UPF values of pristine and treated cotton.

| Samples | UPF value |
|---|---|
| Pristine cotton | 4.67 |
| $TiO_2$(1g)@cotton | 20.64 |
| Silicone oil@cotton | 5.57 |
| $TiO_2$(0.5g)-silicone oil@cotton | 20.63 |
| $TiO_2$(1g)-silicone oil@cotton | 42.84 |
| $TiO_2$(2g)-silicone oil@cotton | 48.65 |
| $TiO_2$(3g)-silicone oil@cotton | 55.17 |

## 4. Conclusions

In this study, the superhydrophobic and UV-resistant cotton fabrics were fabricated by combining the nano $TiO_2$ and the silicone oil. The WCA of $TiO_2$-silicone oil@cotton surfaces could reach 161.32°, which is due to the synergistic effect of the rough surface structure caused by the modified nano-$TiO_2$ and the low surface energy adhesive layer with silicone oil. Nano-$TiO_2$ particles on the cotton fabric surface can not only cause a roughness surface structure for improving the hydrophobicity, but also lead to excellent UV-shielding property. This approach reveals a true potential for fluorine-free organic/inorganic composite materials to generate superhydrophobic surfaces. Such multifunctional cotton fabric has broad commercial application prospects.

## Acknowledgments

This work was financially supported by the Opening Project of Jiangsu Engineering Research Center of Textile Dyeing and Printing for Energy Conservation, Discharge Reduction and Cleaner Production (ERC) (SDGC2104); Scientific Research Program Guiding Project of Hubei Provincial Department of Education(B2021088); Hubei Key Laboratory of Biomass Fibers and Eco-Dyeing & Finishing (STRZ202123), Wuhan research center of eco-dyeing &finishing and functional textile (EDFT2021003), Wuhan Textile University.

## References

[1] Suryaprabha T and Sethuraman M G 2017 *Cellulose* **24** 395-407
[2] Dinh T H, Ngo C V and Chun D M 2018 *Nanomaterials* **8** 766
[3] Richard E, Lakshmi R V, Aruna S T and Basu B J 2013 *Applied surface science* **277** 302-9
[4] Liu H, Yang L, Zhan Y, Lan J, Shang J, Zhou M and Lin S 2021 *Cellulose* **28** 1715-29
[5] Xu L, Shen Y, Ding Y and Wang L 2018 *Journal of Nanoscience and Nanotechnology* **18** 6879-86
[6] Li S, Huang J, Ge M, Cao C, Deng S, Zhang S and Lai Y 2015 *Advanced Materials Interfaces* **2** 1500220
[7] Stan M S, Nica I C, Popa M, Chifiriuc M C, Iordache O, Dumitrescu I, Diamandescu L and Dinischiotu A 2019 *Journal of Industrial Textiles* **49** 277-93
[8] Zhang X, Guo Y, Zhang Z and Zhang P 2013 *Applied Surface Science* **284** 319-23
[9] Jiang C, Liu W, Yang M, Liu C, He S, Xie Y and Wang Z 2018 *Colloids and Surfaces A: Physicochemical and Engineering Aspects* **559** 235-42
[10] Suryaprabha T and Sethuraman M G 2018 *Cellulose* **25** 3151-61
[11] Zhang M and Wang C 2013 *Carbohydrate Polymers* **96** 396-402
[12] He T, Zhao H, Liu Y, Zhao C, Wang L, Wang H, Zhao Y and Wang H 2020 *Colloids and Surfaces A: Physicochemical and Engineering Aspects* **585** 124080.
[13] Zhang Y, Li S, Huang F, Wang F, Duan W, Li J, Shen Y and Xie A 2012 *Russian Journal of Physical Chemistry A* **86** 413-17
[14] Ghasemi N, Seyfi J and Asadollahzadeh M J 2018 *Cellulose* **25** 4211-22
[15] Xiong M, Ren Z and Liu W 2020 *Journal of Dispersion Science and Technology* **41** 1703-10
[16] Zhang M, Li J, Zang D, Lu Y, Gao Z, Shi J and Wang C 2016 *Carbohydrate Polymers* **137** 264-70
[17] Raeisi M, Kazerouni Y, Mohammadi A, Hashemi M, Hejazi I, Seyfi J, Khonakdar H A and Davachi S M 2021 *International Journal of Biological Macromolecules* **171** 158-65

# AUTHOR INDEX

**Institute of Physics**
Dirac House, Temple Back
Bristol BS1 6BE UK

ISSN: 1742-6588
ISBN 978-1-7138-6048-8

CURRAN ASSOCIATES INC.
proceedings.com